Karsten König (Ed.)

Multiphoton Microscopy and Fluorescence Lifetime Imaging

Also of Interest

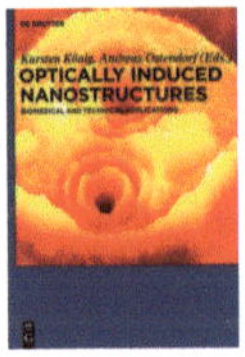

Optically Induced Nanostructures.
Biomedical and Technical Applications
Karsten König, Andreas Ostendorf, 2015
ISBN 978-3-11-033718-1, e-ISBN (PDF) 978-3-11-035432-4,
e-ISBN (EPUB) 978-3-11-038350-8

Biomedical Imaging.
Principles of Radiography, Tomography and Medical Physics
Tim Salditt, Timo Aspelmeier, Sebastian Aeffner, 2017
ISBN 978-3-11-042668-7, e-ISBN (PDF) 978-3-11-042669-4,
e-ISBN (EPUB) 978-3-11-042351-8

Medical Physics.
Volume 1: Physical Aspects of Organs and Imaging
Hartmut Zabel, 2017
ISBN 978-3-11-037281-6, e-ISBN (PDF) 978-3-11-037283-0,
e-ISBN (EPUB) 978-3-11-037285-4

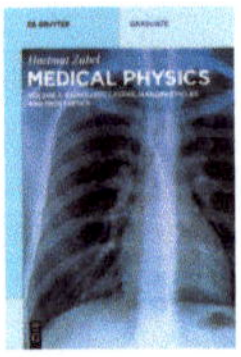

Medical Physics.
Volume 2: Radiology, Lasers, Nanoparticles and Prosthetics
Hartmut Zabel, 2017
ISBN 978-3-11-055310-9, e-ISBN (PDF) 978-3-11-055311-6,
e-ISBN (EPUB) 978-3-11-055317-8
Volume 1 & Volume 2: also available as a set.
Set-ISBN 978-3-11-055957-6

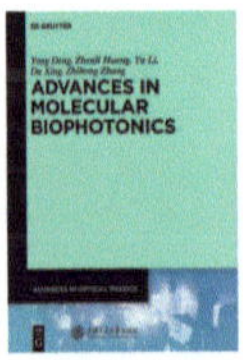

Advances in Molecular Biophotonics
Yong Deng, Zhenli Huang, Yu Li, Da Xing, Zhihong Zhang, 2017
ISBN 978-3-11-030438-1, e-ISBN (PDF) 978-3-11-030459-6,
e-ISBN (EPUB) 978-3-11-038803-9

Multiphoton Microscopy and Fluorescence Lifetime Imaging

Applications in Biology and Medicine

Edited by
Karsten König

DE GRUYTER

UNIVERSITÄT DES SAARLANDES

JenLab

Editor
Prof. Dr. Karsten König
Saarland University
Department of Biophotonics and Laser Technology
66123 Saarbruecken
Germany
k.koenig@blt.uni-saarland.de

ISBN 978-3-11-043898-7
e-ISBN (PDF) 978-3-11-042998-5
e-ISBN (EPUB) 978-3-11-043007-3

Library of Congress Cataloging-in-Publication Data
A CIP catalog record for this book has been applied for at the Library of Congress.

Bibliographic information published by the Deutsche Nationalbibliothek
The Deutsche Nationalbibliothek lists this publication in the Deutsche Nationalbibliografie; detailed bibliographic data are available on the Internet at http://dnb.dnb.de.

Cover image: Karsten König
Typesetting: PTP-Berlin, Protago-TEX-Production GmbH, Berlin
Printing and binding: CPI books GmbH, Leck
♾ Printed on acid-free paper
Printed in Germany

www.degruyter.com

Dedicated to Professor Wolfgang Kaiser
Pioneer of Laser Technology and Nonlinear Optics

Preface

The observation of two-photon excited fluorescence by Kaiser and Garrett and of second-harmonic generation (SHG) by Franken et al. in 1961 marked the birth of Nonlinear Optics (NLO). They used a ruby laser emitting in the red spectral region at 694.3 nm. This novel light source was invented just some months before.

The NLO theory of two-photon absorption was provided by the PhD student and later Nobel Prize winner named Maria Göppert in Göttingen in 1929. However, at that "pre-laser" time, there was no intense light source to prove her hypothesis of two-quantum transitions.

In 1961, three decades later, Wolfgang Kaiser called up Maria Goeppert-Mayer, who had married the chemist Joseph Edward Mayer and worked as professor in San Diego. He told that her theory of two-photon absorption was finally confirmed. In order to honor her outstanding PhD thesis, the unit of two-photon absorption cross sections was termed "GM".

When combined with an optical microscope, the laser beam can be confined to an intense sub-wavelength light spot. According to Abbe's diffraction law, the spot diameter can be as small as 200 nanometers when using visible light.

In 1962, intense ruby laser spots were used to generate a nonlinear induced micro-emission from materials. Laser-induced breakdown spectroscopy (LIBS) was born. Later on, Berns et al. used the laser microscope to perform intracellular laser surgery and Ashkin employed the laser microscope non-destructively to create optical traps and laser tweezers.

The pixel-by-pixel exposure across a sample with a focused laser spot by stage scanning or beam scanning in order to induce signals such as fluorescence led to the invention of laser scanning microscopy. Confocal laser scanning microscopes (CLSM) provide optical sections for three-dimensional (3D) imaging. Typically, non-pulsed (cw) visible laser beams are used.

The integration of ultrafast picosecond and femtosecond lasers into scanning microscopes was a milestone in optical microscopy. It allowed fluorescence lifetime imaging (FLIM) by time-correlated single-photon counting (TCSPC) and efficient non-linear excitation within the focal spot, in particular two-photon microscopy.

The first integration of an ultrafast laser with a confocal laser scanning microscope happened to my knowledge in the late 1980s in Jena in East Germany. The microscope manufacturer VEB Carl Zeiss Jena developed together with our Department of Physics at the Friedrich Schiller University a prototype of a picosecond laser scanning microscope for 4D microscopy with high spatial and temporal resolution based on a mode-locked argon ion laser and a TCSPC module. Using that unique prototype, we performed the first FLIM laser scanning microscopy in Life Sciences in 1988 on living cancer cells.

https://doi.org/10.1515/9783110429985-001

FLIM microscopy opened the possibility to introduce a further image contrast modality, to probe the microenvironment of the fluorophore, and to detect protein-protein interactions with high spatial resolution.

However, working in the NLO group under the guidance of the ultrashort laser pulse expert Bernd Wilhelmi, we missed the chance to detect two-photon excited fluorescence with this very first ultrafast laser scanning microscope.

One year later in 1989, Denk, Strickler, and Webb from Cornell University used a sub-picosecond dye laser microscope to perform two-photon imaging and two-photon photochemistry. Cornell University filed a patent on November 14, 1989 on two-photon laser microscopy. Licensed to the UK company Bio-Rad, this patent started a major legal fight between the major microscope producers. In 2004, ZEISS acquired the Cell Science Business of Bio-Rad to get access to the worldwide exclusive rights of multiphoton microscopy.

Femtosecond near-infrared (NIR) laser two-photon microscopes became the major tool for live cell imaging. Not only intracellular exogenous fluorophores and fluorescent proteins in transfected cells have been visualized over longer periods of time noninvasively. Also a variety of endogenous fluorophores has been imaged. Piston and König et al. introduced two-photon autofluorescence FLIM on live cells.

Two-photon optical sections were performed on tissue biopsies, and later mounted in the skull through a glass window within the brain of live transgenic mice. Two-photon fluorescence microscopes thus became now unique 3D imaging systems to study intratissue neurons in situ.

In 1998, Masters et al. used a home-built two-photon microscope to image the skin of the author's arm noninvasively and label-free. They were able to monitor the complete skin architecture of the epidermis and upper dermis.

Finally in 2003, König and coworkers introduced the first certified in vivo multiphoton tomograph for clinical imaging based on two-photon excited autofluorescence, SHG, and FLIM. Meanwhile the skin of thousands of volunteers and patients has been examined with these multiphoton tomographs to detect skin cancer and skin aging signs, to trace cosmetic and pharmaceutical compounds, and even to measure skin alterations of astronauts after long-term space trips.

Instead of using a single femtosecond laser beam, Stefan Hell added a second optically shaped laser beam to come up with the Nobel Prize winning STED microscopy. STED is the abbreviation for *stimulated emission depletion microscopy*, which opened the door for super-resolution microscopy far beyond Abbe's diffraction limit, and is also called optical nanoscopy.

Two ultrafast laser beams were also employed to realize 3D microscopy with chemical fingerprints based on *Coherent Anti-Stokes Raman Spectroscopy* (CARS). Meanwhile, CARS tomographs are in clinical use in hospitals in Germany, USA, and China.

Besides imaging, nonlinear microscopes became novel micromachining tools in material production and refractive eye surgery. Within the last ten years, femtosecond laser nanoscopes have "turned" micromachining into nanomachining. When using one-beam NIR nanoscopes, only the central part of the intense laser spot is employed to create sub-100 nm structures based on two-photon lithography, multiphoton ionization, and plasma formation. In bulk 3D nanoprocessing became feasible. When using two-beam nanoscopes, the 3D nanostructures with a feature size down to 9 nm have been produced with nonlinear lithography where one beam is used for photoinduction and the other one for photoinhibition.

Undoubtedly, nonlinear microscopy/nanoscopy and multiphoton tomography have revolutionized the imaging of live cells and tissues. It all started with the pioneering work of Kaiser and Garrett who demonstrated the two-photon fluorescence for the first time. It is a great pleasure to dedicate this book to Professor Wolfgang Kaiser on his 92nd birthday, which is on July 17, 2017.

Berlin, July 2017
Karsten König
Department of Biophotonics and Laser Technology, Saarland University, Germany

Wolfgang Kaiser with Karsten König at the *10th International Workshop and Conference on Advanced Multiphoton and Fluorescence Lifetime Imaging Techniques FLIM2015* on June 17, 2015.

Foreword

The publication of this book is timely, because it is likely that multiphoton imaging is about to make a serious impact in clinical diagnosis, after more than two decades of successful applications in research in the life sciences. And who better to edit the book than Karsten König, who has been active in the field almost since its experimental demonstration?

Two-photon absorption was first proposed by Maria Göppert in 1929. It is interesting to recall that Paul Dirac spent three months in Göttingen in early 1927, visiting Max Born, Göppert's supervisor. In this same period he submitted two papers, 'The quantum theory of dispersion', and 'The quantum theory of emission and absorption of radiation', which introduce the quantum theory of creation and annihilation of photons. The simultaneous absorption of multiple photons was predicted by his theory, but he rejected this result as 'These terms correspond to processes in which two light-quanta are emitted or absorbed simultaneously, and cannot rise in a light-quantum theory in which there are no forces between the light quanta. The effects of these terms will be found to be negligible, so that the disagreement with the light-quantum theory is not serious.' Later, Dirac made reference to multiphoton absorption in his book *The Principles of Quantum Mechanics*.

Various different nonlinear processes were observed experimentally soon after the invention of the laser. In a paper submitted in 1976, we (Sheppard and Kompfner: Resonant optical scanning microscope. Appl Opt. 1978;17:2879–2882) proposed that the high field strength in a tightly focused laser beam could be used to excite nonlinear effects and produce images by scanning of the laser spot. Harmonic generation, two-photon fluorescence, and coherent Raman scattering were processes that were specifically mentioned. Second-harmonic images were presented at a conference in 1977 (Sheppard et al. The second harmonic generation (SHG) microscope. IEEE J Quant Electron. 1977;QE13:100D, post-deadline). Although the advantage of using short pulses was mentioned in the paper, the harmonic images were, in fact, produced using a continuous wave laser. We had also experimented on using picosecond pulses from a mode-locked argon ion laser. We thought that we had observed a second-harmonic signal, but further investigation showed it was fluorescence in the infrared. The patent granted to Denk, Strickler, and Webb (US5034613) was for a two-photon fluorescence microscope, specifically one using pulses with a pulse length shorter than one picosecond. Interestingly, a patent was also granted to Hänninen and Hell (WO1995030166) for a microscope system using pulses longer than one picosecond. These patents have now expired, and several companies manufacture multiphoton microscopes.

https://doi.org/10.1515/9783110429985-002

The scope of the present book includes multiphoton manipulation and material processing, and also fluorescence lifetime imaging. These associated methods are also very useful in biological and medical applications. The book satisfies a need for an up-to-date treatment of all these related techniques.

Wollongong, NSW, Australia, August 2017
Colin Sheppard

Contents

Part II: **Modern nonlinear microscopy of live cells**

Part III: **Nonlinear tissue imaging**

List of contributing authors

Thérèse Baldeweck
L'Oréal Research and Innovation
1 avenue Eugène Schueller
93601 Aulnay-sous-Bois
France
tbaldeweck@rd.loreal.com

Mihaela Balu
Laser Microbeam and Medical Program
Beckman Laser Institute
University of California
Irvine, CA, 92612
USA
mbalu@uci.edu

Ana Batista
JenLab GmbH
Schillerstr. 1
07745 Jena
Germany
batista@jenlab.de

Wolfgang Becker
Becker & Hickl GmbH
Nahmitzer Damm 30
12277 Berlin
Germany
becker@becker-hickl.com

Vsevolod V. Belousov
Shemyakin-Ovchinnikov
Institute of Bioorganic Chemistry RAS
16/10 Miklukho-Maklaya St.
Moscow 117997
Russia
vsevolod.belousov@gmail.com

Hans Georg Breunig
JenLab GmbH
Schillerstrasse 1
07745 Jena
Germany
breunig@jenlab.de

Jasmin Breymayer
Ulm University
Core Facility Confocal and Multiphoton
Microscopy
Albert-Einstein-Allee 11
89081 Ulm
Germany
jasmin.breymayer@uni-ulm.de

Sébastien Brizion
L'Oréal Research and Innovation
1 avenue Eugène Schueller
93601 Aulnay-sous-Bois
France
sbrizion@rd.loreal.com

Marco Castello
Molecular Microscopy and Spectroscopy,
Nanophysics, Istituto Italiano di Tecnologia
via Morego 30
16163, Genoa
Italy
Marco.Castello@iit.it

Dhyan Chandra
Department of Pharmacology and Therapeutics
BLSC-Centre for Genetics and Pharmacology
Roswell Park Cancer Institute
Buffalo, NY 14263
USA
dhyan.chandra@roswellpark.org

Heejin Choi
Department of Mechanical and Biological
Engineering
Massachusetts Institute of Technology
Cambridge, MA 02139
USA
choihj@mit.edu

Etienne Decencière
MINES ParisTech, PSL Research University
CMM - Centre for Mathematical Morphology
Fontainebleau
France
Etienne.Decenciere@ensmp.fr

Alberto Diaspro
Nanoscopy, Nanophysics, Istituto Italiano di Tecnologia
via Morego 30
16163, Genoa
Italy
alberto.diaspro@iit.it

Department of Physics
University of Genoa
via Dodecaneso 33
16146 Genoa
Italy

Nikon Imaging Center
Istituto Italiano die Tecnologia
via Morego 30
16163 Genoa
Italy

Christoph Donitzky
JenLab GmbH
Schillerstr. 1
07745 Jena
Germany
donitzky@jenlab.de

Varvara V. Dudenkova
Institute of Biomedical Technologies
Nizhny Novgorod State Medical Academy
Minin and Pozharsky Square 10/1
Nizhny Novgorod, 603005
Russia
orannge@mail.ru

Institute of Biology and Biomedicine
Nizhny Novgorod State University
Gagarin Avenue 23
Nizhny Novgorod, 603950
Russia

Irina N. Druzhkova
Institute of Biomedical Technologies
Nizhny Novgorod State Medical Academy
Minin and Pozharsky Square 10/1
Nizhny Novgorod, 603005
Russia
danirin@yandex.ru

Ronald M. Harris
Department of Dermatology
University of California
Irvine, CA, 92697
USA
rmharris@uci.edu

Robert M. Hoffman
AntiCancer Inc.
7917 Ostrow St
San Diego, CA 92111
USA
all@anticancer.com

University of California San Diego
Department of Surgery
California
USA

Amy Holmes
School of Pharmacy and Medical Sciences
The University of South Australia
5001 Adelaide
Australia
Amy.Holmes@unisa.edu.au

Volker Huck
University Medical Center Hamburg-Eppendorf
Center for Internal Medicine
SWG Multiphoton Tomography
Martinistr. 52
20246 Hamburg
Germany
v.huck@uke.de

Heidelberg University
Medical Faculty Mannheim
Experimental Dermatology
Theodor-Kutzer-Ufer 1–3
68167 Mannheim

Suchitra Joshi
Department of Neurology
University of Virginia
Charlottesville, VA 22908
USA
sj9z@virginia.edu

Wolfgang Kaiser
Technische Universität München
Fakultät für Physik
James-Franck-Str. 1
85748 Garching
Germany
wkaiser@tum.de

Sviatlana Kalinina
Ulm University
Core Facility Confocal and Multiphoton Microscopy
Albert-Einstein-Allee 11
89081 Ulm
Germany
Sviatlana.Kalinina@uni-ulm.de

Sven R. Kantelhardt
Dept. of Neurosurgery
Johannes-Gutenberg-University Mainz
55101 Mainz
Germany
sven.kantelhardt@unimedizin-mainz.de

Jaideep Kapur
Department of Neurology
University of Virginia
Charlottesville, VA 22908
USA
jk8t@virginia.edu

Kristen M. Kelly
Department of Dermatology
University of California
Irvine, CA, 92697
USA
kmkelly@uci.edu

Karsten König
Saarland University
Department of Biophotonics and Laser Technology
Campus A5.1
66123 Saarbruecken
Germany
k.koenig@blt.uni-saarland.de

JenLab GmbH
Schillerstr. 1
07745 Jena
Germany
koenig@jenlab.de

Serge Koudoro
MINES ParisTech, PSL Research University
CMM - Centre for Mathematical Morphology
Fontainebleau
France
serge.koudoro@mines-paristech.fr

Arnd Krueger
MKS Instruments, Inc.
Newport Spectra-Physics GmbH
Guerickeweg 7
64291 Darmstadt
Germany
arnd.krueger@newport.com

Luca Lanzano
Nanoscopy, Nanophysics, Istituto Italiano di Tecnologia
via Morego 30
16163, Genoa
Italy
Luca.Lanzano@iit.it

Xiaowen Liang
Therapeutics Research Centre
School of Medicine
The University of Queensland
Princess Alexandra Hospital
QLD 4102 Woolloongabba
Australia
x.liang@uq.edu.au

Xin Liu
Therapeutics Research Centre
School of Medicine
The University of Queensland
Princess Alexandra Hospital
QLD 4102 Woolloongabba
Australia
xin.liu@uq.edu.au

Maria M. Lukina
Institute of Biomedical Technologies
Nizhny Novgorod State Medical Academy
Minin and Pozharsky Square 10/1
Nizhny Novgorod, 603005
Russia
kuznetsova.m.m@yandex.ru

Anna Letizia Allegra Mascaro
European Laboratory for Nonlinear Spectroscopy (LENS)
Via Nello Carrara, 1
50019 Sesto Fiorentino (FI)
Italy
allegra@lens.unifi.it

Aleksandra V. Meleshina
Institute of Biomedical Technologies
Nizhny Novgorod State Medical Academy
Minin and Pozharsky Square 10/1
Nizhny Novgorod, 603005
Russia
almele@ya.ru

Institute of Biology and Biomedicine
Nizhny Novgorod State University
Gagarin Avenue 23
Nizhny Novgorod, 603950
Russia

Meghan J. O. Melia
The W.M. Keck Center for Cellular Imaging
Departments of Biology and Biomedical Engineering
Physical and Life Sciences Building
University of Virginia
Charlottesville,VA 22904
USA
mjo3vp@virginia.edu

Christian Mess
University Medical Center Hamburg-Eppendorf
Center for Internal Medicine
SWG Multiphoton Tomography
Martinistr. 52
20246 Hamburg
Germany
c.mess@uke.de

University of Münster
Department of Dermatology
Von-Esmarch-Str. 58
48148 Münster
Germany

Nataliya M. Mishina
Shemyakin-Ovchinnikov
Institute of Bioorganic Chemistry RAS
16/10 Miklukho-Maklaya
St., Moscow 117997
Russia
natsafr@yandex.ru

Francesco S. Pavone
European Laboratory for Non Linear Spectroscopy (LENS)
Via Nello Carrara 1
50019 Sesto Fiorentino
Italy
francesco.pavone@unifi.it

Ammasi Periasamy
The W.M. Keck Center for Cellular Imaging
Departments of Biology and Biomedical Engineering
Physical and Life Sciences Building
University of Virginia
Charlottesville,VA 22904
USA
ap3t@virginia.edu

Shagufta Rehman Alam
The W.M. Keck Center for Cellular Imaging
Departments of Biology and Biomedical Engineering
Physical and Life Sciences Building
University of Virginia
Charlottesville, VA 22904
USA
sr9zz@virginia.edu

Michael S. Roberts
School of Pharmacy and Medical Sciences
The University of South Australia
5001 Adelaide
Australia
m.roberts@uq.edu.au

Therapeutics Research Centre
School of Medicine
The University of Queensland
Princess Alexandra Hospital
Woolloongabba
Australia

Ana-Maria Pena
L'Oréal Research and Innovation
1 avenue Eugène Schueller
93601 Aulnay-sous-Bois
France
ampena@rd.loreal.com

Christopher Rowlands
Department of Bioengineering
Imperial College London
South Kensington
SW7 2AZ
United Kingdom
c.rowlands@imperial.ac.uk

Angelika Rück
Ulm University
Core Facility Confocal and Multiphoton Microscopy
Albert-Einstein-Allee 11
89081 Ulm
Germany
angelika.rueck@uni-ulm.de

Leonardo Sacconi
European Laboratory for Nonlinear Spectroscopy (LENS)
Via Nello Carrara, 1
50019 Sesto Fiorentino (FI)
Italy
sacconi@lens.unifi.it

Vladislav I. Shcheslavskiy
Becker & Hickl GmbH
Nahmitzer Damm 30
12277 Berlin
Germany
vis@becker-hickl.de

Lorenzo Scipioni
Nanoscopy, Nanophysics, Istituto Italiano di Tecnologia
via Morego 30
16163, Genoa
Italy
Lorenzo.Scipioni@iit.it

Department of Computer Science, Bioengineering, Robotics and Systems Engineering
University of Genoa
Italy

Tatiana F. Sergeeva
Institute of Biomedical Technologies
Nizhny Novgorod State Medical Academy
Minin and Pozharsky Square 10/1
Nizhny Novgorod, 603005
Russia
tatiana.f.sergeeva@gmail.com

Colin Sheppard
Honorary Professoral Fellow
School of Chemistry
Faculty of Science, Medicine and Health
University of Wollongong
Northfields Avenue
NSW2522
Australia

Marina V. Shirmanova
Institute of Biomedical Technologies
Nizhny Novgorod State Medical Academy
Minin and Pozharsky Square 10/1
Nizhny Novgorod, 603005
Russia
shirmanovam@mail.ru

Institute of Biology and Biomedicine
Nizhny Novgorod State University
Gagarin Avenue 23
Nizhny Novgorod, 603950
Russia

Ludovico Silvestri
European Laboratory for Nonlinear Spectroscopy (LENS)
Via Nello Carrara, 1
50019 Sesto Fiorentino (FI)
Italy
silvestri@lens.unifi.it

Peter T. C. So
Department of Mechanical and Biological Engineering
Massachusetts Institute of Technology
Cambridge, MA 02139
USA
ptso@mit.edu

Hauke Studier
School of Pharmacy and Medical Sciences
The University of South Australia
5001 Adelaide
Australia

Becker & Hickl GmbH
Nahmitzer Damm 30
12277 Berlin
Germany
studier@becker-hickl.de

Zdenek Svindrych
The W.M. Keck Center for Cellular Imaging
Departments of Biology and Biomedical Engineering
Physical and Life Sciences Building
University of Virginia
Charlottesville,VA 22904
USA
zs4d@virginia.edu

Emmanuelle Tancrède-Bohin
L'Oréal Research and Innovation
1 avenue Eugène Schueller
93601 Aulnay-sous-Bois
France
etancrede@rd.loreal.com

Camilla Thorling
Therapeutics Research Centre
School of Medicine
The University of Queensland
Princess Alexandra Hospital
QLD 4102 Woolloongabba
Australia
camilla.thorling@gmail.com

Bruce J. Tromberg
Laser Microbeam and Medical Program
Beckman Laser Institute
University of California
Irvine, CA, 92612
USA
bjtrombe@uci.edu

Aisada Uchugonova
Saarland University
Department of Biophotonics & Laser Technology
Campus A5.1
66123 Saarbruecken
Germany
a.uchugonova@blt.uni-saarland.de

Giuseppe Vicidomini
Molecular Microscopy and Spectroscopy, Nanophysics, Istituto Italiano di Tecnologia
via Morego 30
16163, Genoa
Italy
giuseppe.vicidomini@iit.it

Steeve Victorin
L'Oréal Research and Innovation
1 avenue Eugène Schueller
93601 Aulnay-sous-Bois
France
svictorin@rd.loreal.com

Horst Wallrabe
The W.M. Keck Center for Cellular Imaging
Departments of Biology and Biomedical Engineering
Physical and Life Sciences Building
University of Virginia
Charlottesville, VA 22904
USA
hw5m@virginia.edu

Haolu Wang
Therapeutics Research Centre
School of Medicine
The University of Queensland
Princess Alexandra Hospital
QLD 4102 Woolloongabba
Australia
h.wang21@uq.edu.au

Martin Weinigel
JenLab GmbH
Schillerstrasse 1
07745 Jena
Germany
weinigel@jenlab.de

Elijah Yew
Department of Mechanical and Biological Engineering
Massachusetts Institute of Technology
Cambridge, MA 02139
USA
elijah.yew@gmail.com

Christopher B. Zachary
Department of Dermatology
University of California
Irvine, CA, 92697
USA
czachary@uci.edu

Elena V. Zagaynova
Institute of Biomedical Technologies
Nizhny Novgorod State Medical Academy
Minin and Pozharsky Square 10/1
Nizhny Novgorod, 603005
Russia
ezagaynova@gmail.com

Institute of Biology and Biomedicine
Nizhny Novgorod State University
Gagarin Avenue 23
Nizhny Novgorod, 603950
Russia

Part I: **Basics**

Karsten König

1 Brief history of fluorescence lifetime imaging

Abstract: This review gives an overview of the history of fluorescence lifetime imaging (FLIM) in life sciences. FLIM microscopy based on an ultrafast laser scanning microscope and time-correlated single photon counting (TCSPC) was introduced in Jena/Germany in 1988/89. FLIM images of porphyrin-labeled live cells and live mice were taken with an unique ZEISS confocal picosecond laser microscope. Five years later, the first *in vivo* FLIM on human volunteers started with time-gated cameras to detect dental caries based on one-photon wide-field pulsed laser excitation of autofluorescent bacteria. Another five years later, two-photon FLIM of autofluorescent skin was performed on a volunteer with a lab microscope in the frequency domain. The first clinical non-invasive optical, two-photon 3D FLIM biopsies were obtained fifteen years ago in patients with dermatological disorders using a certified clinical multiphoton tomograph based on a tunable femtosecond titanium:sapphire laser and TCSPC.

A current major FLIM application in cell biology is the study of protein-protein interactions in transfected cells by FLIM-FRET microscopy. Clinical FLIM applications are still on a research level and include preliminary studies on (i) one-photon FLIM autofluorescence microscopy of patients with ocular diseases using picosecond laser diodes, (ii) time-gated imaging in brain surgery using a nanosecond nitrogen laser, and (iii) two-photon clinical FLIM tomography of patients with skin cancer and brain tumors with near-infrared femtosecond lasers and TCSPC.

1.1 Introduction

FLIM is an imaging methodology with high specificity and high sensitivity to the nano- and microenvironment. It is relatively insensitive to concentration and signal intensity artefacts. The acronym FLIM stands for *Fluorescence Lifetime Imaging* (sometimes also for fluorescence lifetime microscopy) and means that the fluorescence lifetime τ is depicted with spatial resolution. FLIM images are typically generated with microscopes, endoscopes, and tomographs but can be also acquired on a macroscopic level. FLIM images can be taken in the frequency domain (measurement of phase shift and demodulation with cw or pulsed laser) as well as in the time domain using time-gated cameras, streak cameras, and time-correlated single photon counting (TCSPC) units [1]. They are often false-color coded where the colors reflect certain τ *values*.

FLIM applications in cell biology are based on endogenous (intrinsic) and exogenous fluorophore detection in a specific nano- and microenvironment. FLIM can be used to probe intermolecular interactions within a 10 nm distance, such as bindings, by Förster resonance energy transfer (FRET) and anisotropy measurements. Animal

https://doi.org/10.1515/9783110429985-003

studies, artworks, microfluidics, and forensic science such as fingerprint detection are other applications of FLIM [2–8].

However, the most challenging and exciting FLIM application is *in vivo* clinical imaging of patients. Clinical FLIM with several hundreds of patients is currently based on TCSPC using

(i) picosecond laser diodes in the visible spectral range (one-photon excitation) and
(ii) near-infrared (NIR) femtosecond Ti:sapphire lasers (two-photon excitation).

1.2 Time-resolved spectroscopy and first time-resolved fluorescence microscopes

For decades, time-resolved fluorescence spectroscopy has been employed to investigate photophysical, photochemical, and photobiological processes of fluorescent molecules in solution, such as the primary steps of vision and photosynthesis (e.g. [9]). In order to gain time-resolved fluorescence information on living cells, nanosecond and sub-nanosecond fluorometers were combined with microscopes. Typically, a fixed laser beam was focused to a micrometer-sized laser spot [10–17]. Applications of these time-resolved spectrometers with some spatial information included the analysis of chlorophyll fluorescence from single chloroplasts, the coenzyme fluorescence in bacteria, and the porphyrin fluorescence in cells after administration of fluorescent photosensitizers.

1.3 First FLIM laser scanning microscope

In 1988, a real breakthrough was obtained by the combination of an ultrafast pulsed laser in the picosecond range with a laser scanning microscope. Now the object was illuminated point-by-point by the focused pulsed laser beam and the corresponding time-resolved fluorescence response was detected at each point by time-correlated single photon counting (TCSPC). The fluorescence intensity I_F and the fluorescence decay time τ were used as parameters to generate FLIM images. This first prototype of the FLIM laser scanning microscope became operational in the Department of Physics at the Friedrich Schiller University Jena in 1988 [18], shortly before the unification of Germany. Two years earlier, the patent on time-resolved pulsed laser scanning microscopy had been filed by Gröbler from the company VEB Carl Zeiss Jena [19].

This unique FLIM system was based on a confocal laser scanning microscope (scanning stage with 0.5 µm steps) equipped with a 120 MHz mode-locked argon ion laser (100 ps). The fluorescence was detected with photomultipliers (FEU-77) with short picosecond rise time from the Soviet Union in combination with the time-resolved single photon counting unit SPC100 (ZOS Berlin, Fig. 1.1).

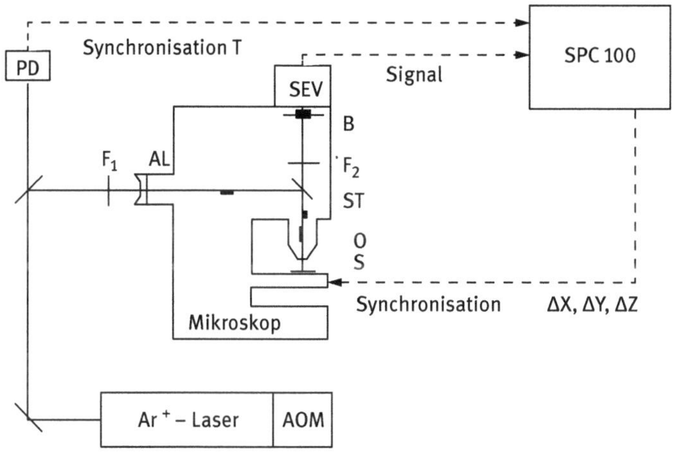

Fig. 1.1: First FLIM laser scanning microscope. The apparatus is based on a ZEISS confocal stage scan fluorescence microscope equipped with a mode-locked 120 MHz argon ion laser and the TCSPC module SPC100 [20–22]. Single photon events are accumulated in a 256-channel memory, maximum 65 535 counts per channel, channel width 48 ps or 20 ps, with a maximum count rate of 2×10^5/s. Temporal resolution: 300 ps, 50× NA 0.9 objective. AOM: acousto-optic mode-locker, PD: photodiode, F_1: attenuator, F_2: blocking filter, ML: matching lens, DM: dichroic mirror, O: objective, S: scanning stage with 0.5 μm steps, SPC: single photon counting.

König et al. used this unique novel TCSPC-FLIM tool to perform the first laser scanning FLIM in life science [20–22]. In 1988, our group generated FLIM images with submicron resolution (50×, NA 0.95) from living cancer cells. In particular, red-emitting intracellular fluorescent porphyrin photosensitizers were imaged as line scans and 2D plots with sub-nanosecond temporal resolution. Bi-exponential as well as global fitting was performed to obtain data on porphyrin monomers, dimers, and higher aggregates as well as to study photodynamic reactions (Fig. 1.2 and 1.3).

The mode-locked 120 MHz argon ion laser was also used to pump a tunable dye laser. Unfortunately, no two-photon excited fluorescence images (that would require the use of short pass filters) were taken at that time in Jena with this remarkable tunable picosecond laser scanning microscope working even in the red spectral range.

In 1989, Wang et al. reported on fluorescence lifetime distribution measurements by phase-resolved detection with an image dissector tube [23] and one year later on time-resolved fluorescence microscopy by multichannel photon counting [24].

Denk, Strickler, and Webb realized the two-photon microscope based on a sub-picosecond dye laser in 1989/1990 [25, 26]. With the availability of the more stable and user-friendly femtosecond titanium:sapphire laser, two-photon laser scanning microscopes with their inherent optical sectioning rapidly became a major tool for cell biologists. Since these microscopes already have the expensive laser source, it is relatively simple to add on a TCSPC module in order to perform FLIM.

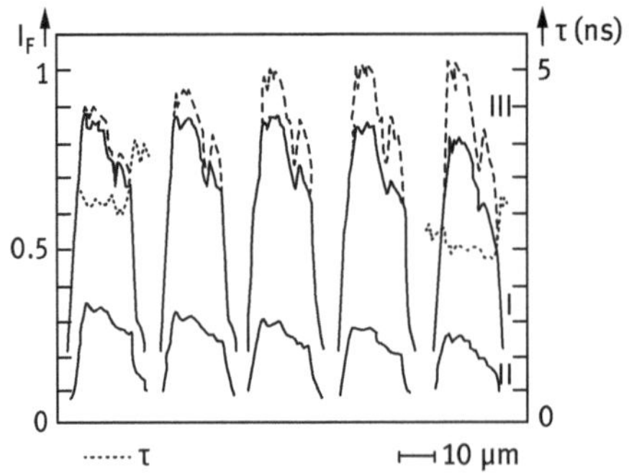

Fig. 1.2: Subsequent line scans (5 times the same cell, step width: 0.5 μm, laser spot size: 0.6 μm) with the FLIM microscope were taken to study laser-induced bleaching and photoproduct formation in a live cancer cell labeled with the photosensitizer HpD (mean power: 7 μW, interval I: 0.7–3.6 ns, interval II: 3.6–6.5 ns, interval III: whole excitation pulse period). The intensity increase in interval III and the decrease of the mean fluorescence lifetime reflect the formation of a short-lived porphyrin photoproduct during five laser scans [20–22].

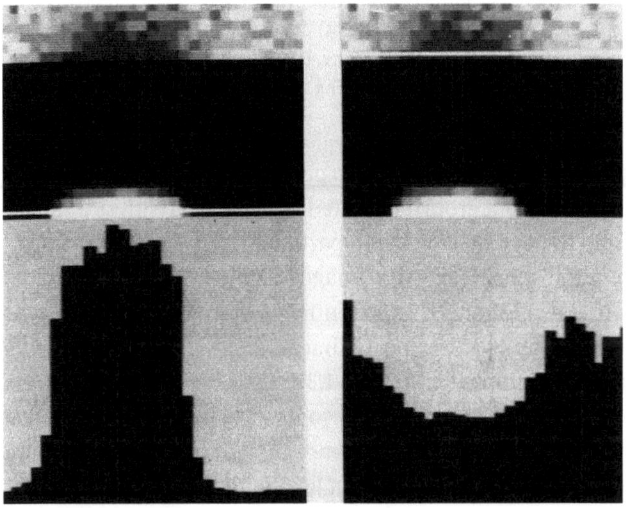

Fig. 1.3: First FLIM images using a confocal picosecond laser scanning microscope. The figure shows a 2D scan of the fluorescence of the intracellular photosensitizer HpD in a part of an *Ehrlich Ascites Carcinoma* (EAC) cell. Upper part: $\tau(x, y)$ values (1 μm steps), middle: fluorescence intensity values, lower part: histograms along the lines [20–22].

1.4 Two-photon FLIM microscopy

A drawback of one-photon FLIM with visible light sources is its inability to image the most important intracellular fluorescent biomolecule, the coenzyme NAD(P)H in the mitochondria with its absorption maximum around 340 nm. Furthermore, problems occur due to out-of-focus fluorescence excitation and the limited light penetration depth in most tissues. These drawbacks can be overcome when using multiphoton imaging with NIR lasers based on two-photon excited fluorescence, second harmonic generation (SHG), and coherent anti-Stokes Raman spectroscopy (CARS).

Some years after the invention of the two-photon microscope, Piston et al. [27], So et al. [28], and König et al. [29] performed two-photon FLIM microscopy on living cells. Interestingly, a variety of first two-photon FLIM microscopes used a frequency domain approach whereas the majority of today's two-photon FLIM devices are employing fast TCSPC. Soon after their introduction, the first two-photon FLIM images from animal tissues were taken and finally from the fingers of a microscope user. Masters et al. used a lab two-photon microscope operating in the frequency domain in 1997/1998 to obtain the first two-photon FLIM data from *in vivo* human skin [30–32].

1.5 First wide-field FLIM in humans

Some years earlier in the mid-1990s, the first wide-field macroscopic FLIM images from humans were obtained by König and Schneckenburger. *In vivo* autofluorescence images of the tooth region of two volunteers with carious lesions and dental plaque were taken. In particular, the distribution of the red-emitting, porphyrin-producing bacteria *Actinomyces odontolyticus* with fluorescence lifetimes of 10 ns and longer has been studied in non-healthy teeth of volunteers with wide-field FLIM [33, 34]. A time-gated camera and wide-field illumination with a frequency-doubled two nanosecond Nd:YAG laser was employed. Time-gated "snapshots" were acquired at various time delays after the excitation. When using "snapshots" several nanoseconds after the excitation pulse, dental plaque could be clearly detected due to the suppression of the short-lived autofluorescence of normal tooth material and the scattered laser light (Fig. 1.4).

Today, clinical wide-field macroscopic FLIM is employed to detect tumor borders during brain surgery. A nanosecond nitrogen laser at 337 nm is used to excite the fluorescence of the tissue and to allow "snapshots" for the identification of the tumor [35].

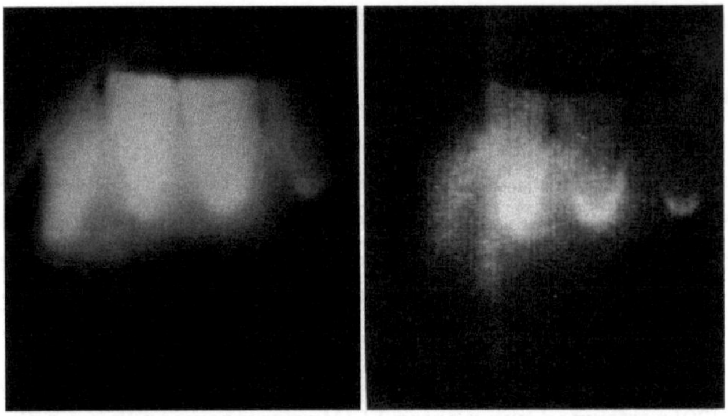

Fig. 1.4: *In vivo* time-gated autofluorescence images from anterior teeth of a volunteer with dental plaque and caries. Bacteria (*Actinomyces odontolyticus*) produce coproporphyrin and PP IX with long > 10 ns fluorescence lifetimes. Therefore, time windows with a long >10 ns delay compared to the excitation light exhibited the porphyrin autofluorescence in caries and dental plaque. The short-lived tooth autofluorescence is suppressed. Left: time gate: 0–5 ns, right: 30–55 ns [34].

1.6 Clinical FLIM tomography

A patent on a device and a method for clinical time-resolved two-photon imaging and treatment of skin disorders was filed by König in 2000 [36]. The first prototype of a medical two-photon tomograph "DermaInspect" with picosecond temporal resolution was realized based on a tunable Ti:sapphire laser, x/y-galvoscanners, a piezodriven NA 1.3 objective, and a TCSPC module in 2002 [37, 38]. Soon after it became CE-marked for clinical use and commercialized by JenLab GmbH, Jena, Germany. The first studies were performed on patients with skin cancer at the University Hospital Jena.

Today, the portable certified multiphoton tomographs "MPTflex" and "MPTflex-CARS" with their optomechanical arms are in clinical use in Australia, Japan, Russia, US, and Europe. These medical tomographs simultaneously depict FLIM images, SHG images, two-photon fluorescence intensity images, and CARS images, respectively (Fig. 1.5 and 1.6). Clinical FLIM applications include early detection of skin cancer, tumor border recognition during brain surgery, and detection of intratissue inflammation sites (e.g. [39–56]).

▸ **Fig. 1.6:** Clinical two-photon FLIM images from human skin are based on two-photon excitation of endogenous fluorophores using TCSPC. The arrival times of some fluorescence photons per pixel are depicted as fluorescence decay curve (lower part). Bi-exponential fitting provides amplitudes and lifetimes of two components as well as the mean fluorescence lifetime per pixel.

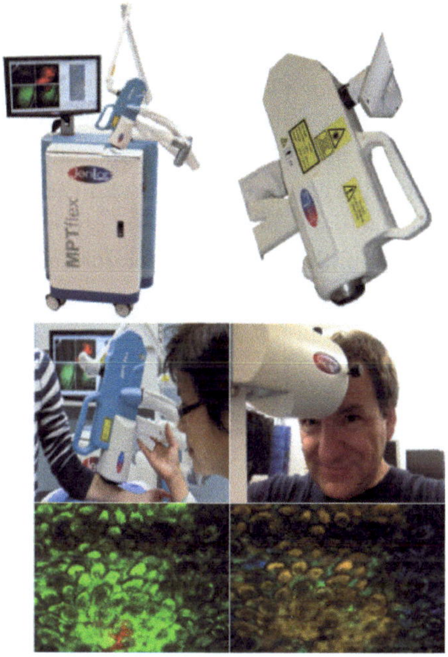

Fig. 1.5: The Prism Award winning certified clinical multiphoton FLIM tomograph *MPTflex*™ with its flexible optomechanical arm with active beam stabilizer, the compact scan/detection head, the tunable femtosecond laser, and the TCSPC module. The head consists of a 3D scanning system, the high NA focusing optics, and two single photon counting sensors. Wide-field (up to 5 × 5 mm^2) images can be taken by mosaic scanning. Horizontal, vertical and diagonal FLIM sections are possible. The lower image left shows an emission intensity image based on autofluorescence AF (green) and SHG (red). The right image shows the time-resolved AF image with pseudocolors representing the fluorescence lifetimes. FLIM of tissue AF provides a significantly better contrast than intensity images.

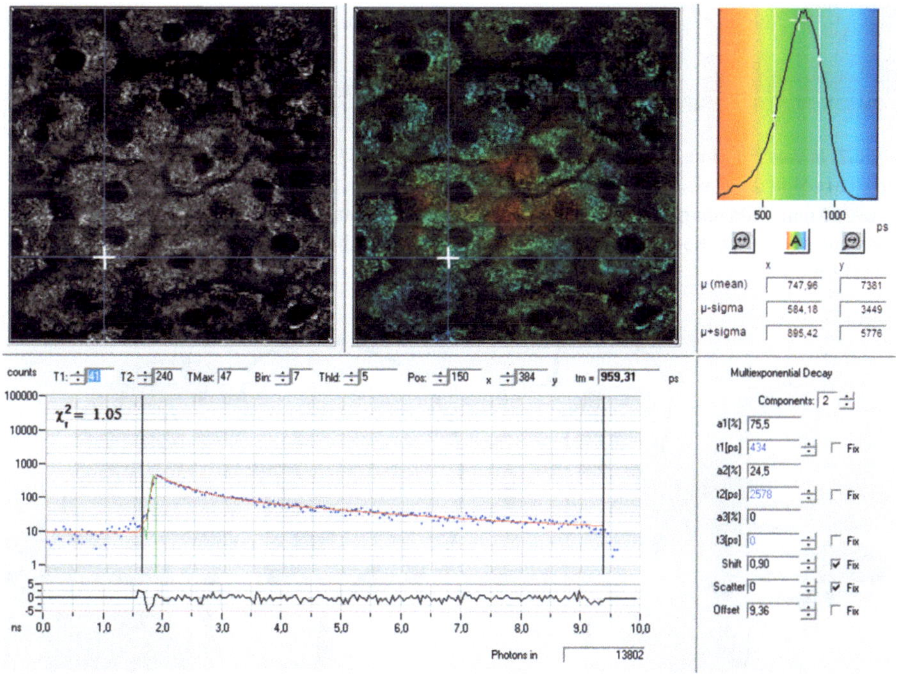

1.7 One-photon FLIM in ophthalmology

The major clinical one-photon FLIM application today is in the field of ophthalmology. Picosecond laser diodes in the visible range are employed to excite the ocular autofluorescence. FLIM images are taken by the use of TCSPC units (Fig. 1.7 and 1.8). The company *Heidelberg Engineering* has produced the first prototypes for clinical use. They are being tested, e.g., for the detection of macular degeneration and other diseases. This pioneering FLIM work for the field of ophthalmology was performed by Schweitzer and coworkers in Jena, Germany [57–59].

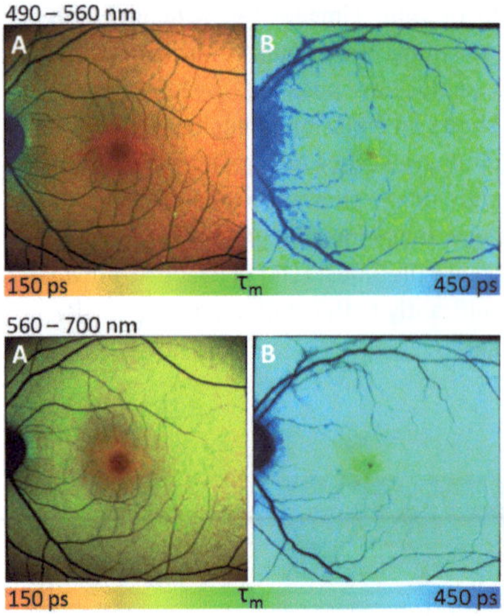

τ_1 → retinal pigmented epithelium
τ_2 → neural retina
τ_3 → influence of lens fluorescence

		a_1 [%]	τ_1 [ps]	a_2 [%]	τ_2 [ps]	a_3 [%]	τ_3 [ps]
490 - 560 nm	*A*	85.3	87.7	11.5	388.4	2.7	2691.7
	B	84.5	137.8	11.5	586.8	4	3767.9
560 - 700 nm	*A*	72.3	83.7	24.8	320.6	2.9	1723.1
	B	74.8	128.5	22.3	475.1	2.9	2388.2

Fig. 1.7: *In vivo* FLIM images from the eyes of two volunteers. Clinical one-photon FLIM is based on picosecond laser excitation (473 nm, 80 MHz) and TCSPC in two channels (498–560 nm and 560–720 nm). FLIM images of the editor (B) and a younger female student (A) were taken under the same conditions. Note the longer lifetimes in the case of the older volunteer. Image acquisition: 1.5 min.

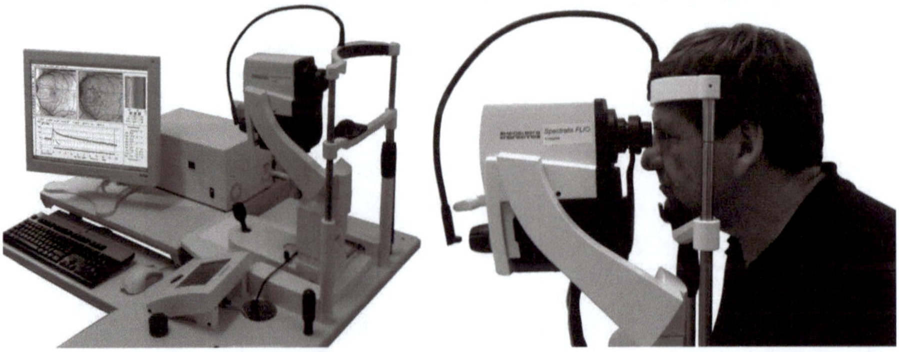

Fig. 1.8: The fluorescence lifetime ophthalmoscope FLIO.

1.8 Endoscopic FLIM

In the late 1990s, the first FLIM endoscopes, such as the system by Mizeret et al. that worked in the frequency domain [60], were also developed. Later on in 2004, video-rate FLIM with a potentially portable flexible one-photon FLIM endoscope was reported based on gated optical image intensifier technology [61].

Already in 2002, a compact two-photon fluorescence microscope based on a single-mode fiber coupler was reported and one year later, two-photon low-weight endoscopic systems were employed to study brain tissue in live mice [62].

Finally in 2007, the first clinical time-resolved two-photon fluorescence endoscopy was performed using special high NA GRIN microendoscopes on patients with chronic and acute wounds ([63], Fig. 1.9).

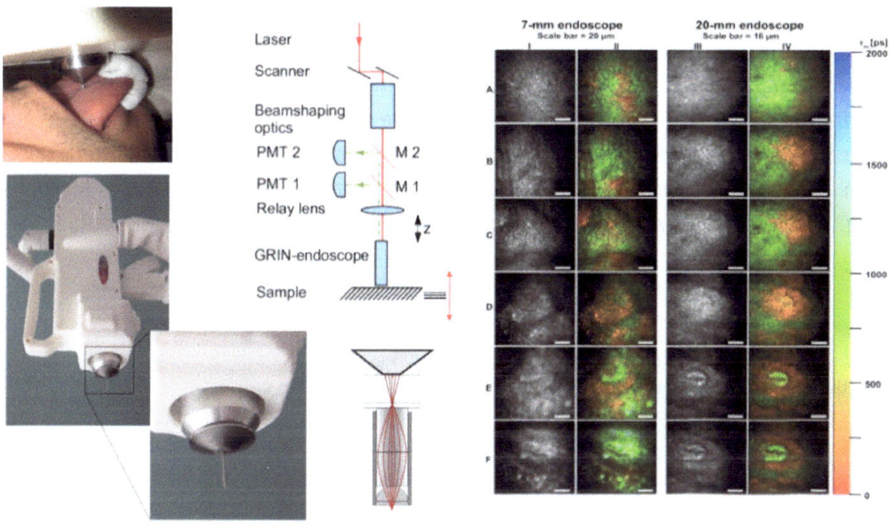

Fig. 1.9: Clinical two-photon GRIN microendoscope with high NA (0.8). The endoscope is connected to a multiphoton tomograph. The images show two-photon sections of the lip of a volunteer.

1.9 FLIM-FRET

One current major application of FLIM microscopes in cell biology is FLIM-FRET. Protein-protein interactions and the micro- and nano-environment of a molecule are studied by the detection of the shortening of the fluorescence lifetime of a donor molecule during Förster resonance energy transfer (FRET). During that transfer, the energy of the light-excited fluorescent donor is transferred non-radiatively to a nearby molecular chromophore in the ground state. A major condition is the spectral overlap of the donor's emission with the acceptor's absorption. The probability of fluorescence is thereby diminished and as a result, the fluorescence lifetime is shortened. Förster showed in the 1940s that the rate of energy transfer is proportional to R^{-6}, where R is the distance between the chromophore centers [64]. Practically, it means that the distance should be less than 10 nanometers and binding should occur.

FRET can be measured with a normal steady-state fluorescence microscope. However, it is difficult to quantify the FRET measurements because standards must be used for calibration of the fluorescence intensities and problems are faced due to concentration gradients and dynamics of the proteins within the cell as well as due to photobleaching phenomena. FLIM can overcome these difficulties in making reliable FRET studies independent of fluorescence intensities by measuring the modifications of the fluorescence decay of the donor [5, 6, 65].

Most FLIM-FRET studies are performed on green fluorescent proteins (GFP) and other genetic constructs (Fig. 1.10). In some cases, FRET between natural fluorescent chromophores such as NADH and tryptophan has also been studied. Often FLIM-FRET is performed with one-photon donor excitation such as UV and blue light. Two-photon excitation may cause a minor problem because of the broad two-photon excitation spectra and the subsequent excitation of the donor as well as the non-desired excitation of the acceptor. However, this problem can be solved by deconvolution methods (e.g., global fitting). More information on FLIM-FRET can be found in [66–68].

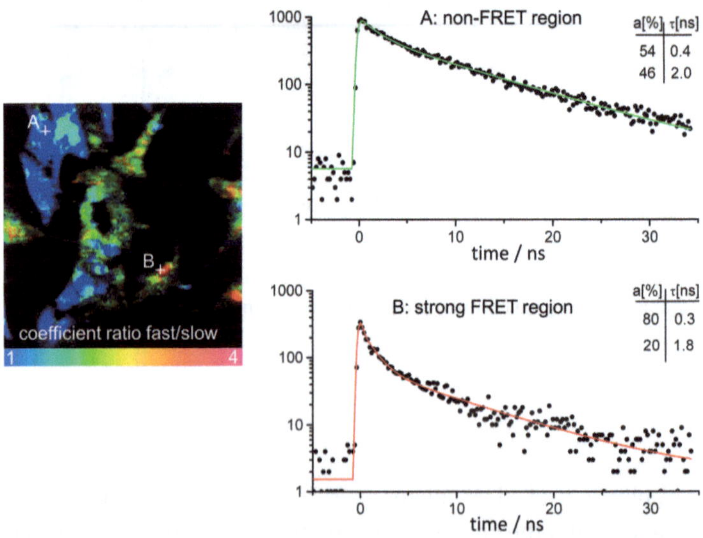

Fig. 1.10: FLIM-FRET measurement from 2001. Our group observed FRET in Vero cells containing green fluorescent protein (GFP, acceptor) and yellow fluorescent protein (YFP, donor). The shortening of the YFP lifetime and the increase of the ratio fast to slow component a_1/a_2 indicated FRET. ZEISS LSM-410, Multichannel plate detector, SPC-730 TCSPC module [5].

1.10 Conclusion

Since its introduction in life science in the late 1980s, FLIM scanning microscopy has become an important tool for cell biologists. In particular, FLIM microscopy gained a significant boost with the introduction of two-photon femtosecond laser scanning microscopes in the 1990s. Nowadays, TCSPC-FLIM add-on modules can easily be implemented in commercial ultrashort laser scanning microscopes.

Ten years after the invention of the two-photon microscope, the first commercial certified clinical FLIM tomographs ("DermaInspect") for medical diagnostics became available [36–38].

Current FLIM microscopes, FLIM tomographs, and FLIM endoscopes still have the drawback of being expensive and relatively bulky systems. However, with the availability of ultracompact low cost femtosecond laser sources within the next years, FLIM devices will likely become a common tool in research and application labs. Still, it remains a major goal to identify promising applications in the medical field [69].

References

[1] O'Connor DV, Phillips D, editors. Time-correlated single photon counting. Academic Press; 1984.
[2] Schneckenburger H, König K. Fluorescence decay kinetics and imaging of NAD(P)H and flavins as metabolic indicators. Opt Eng. 1992;31:1447–1451.
[3] So P, König K, Berland K, Dong CY, French T, Buehler C, Ragan T, Gratton E. New time-resolved techniques in two-photon microscopy. Cell Mol Biol. 1998;44:771–794.
[4] Bastiaens PIH, Squire A. Fluorescence lifetime imaging microscopy: spatial resolution of biochemical processes in the cell. Trends Cell Biol. 1999;9:48–52.
[5] Becker W, Benndorf K, Bergmann A, Biskup C, König K, Tirlapur U, Zimmer T. FRET measurements by TCSPC laser scanning microscopy. Proc SPIE 4431; 2001.
[6] Becker W, Bergmann A, König K, Tirlapur U. Picosecond Fluorescence Lifetime Microscopy by TCSPC Imaging. Proc SPIE 4262; 2001.
[7] Cunall RB, Dale RE, editors. Time-resolved spectroscopy in biochemistry and biology. New York, London: Plenum Press; 1983.
[8] Suhling K, et al. Fluorescence lifetime imaging (FLIM). Basic concepts and some recent developments. Medical Photonics. 2015;27:3–40.
[9] Venetta BD. Microscope phase fluorometer for determining the fluorescence lifetimes of fluorochromes. Rev Sci Instr. 1959;6:450–457.
[10] Docchio F, Ramponi R, Sacchi CA, Bottiroli G, Freita. An automatic pulsed laser microfluorometer with high spatial and temporal resolution. J Microsc. 1984;134:151–160.
[11] Bottiroli G, Cionini PG, Dicchio F, Sacchi CA. "In situ" evaluation of the functional state of chromatin by means of quinacrine mustard staining and time-resolved fluorescence microscopy. Histochem J. 1984;16:223–233.
[12] Schneckenburger H. Time-resolved microfluorescence in biomedical diagnosis. Opt Engineering. 1985;24:1042–1044.
[13] Rodgers MAJ, Firey PA. Instrumentation for fluorescence microscopy with picosecond time resolution. Photochem Photobiol. 1985;42:613–616.

[14] Minami T, Kawahigashi M, Sakai Y, Shimamoto K, Hirayama S. Fluorescence lifetime measurements under a microscope by the time-correlated single photon counting technique. J Luminescence. 1986;35:247–253.
[15] Pamponi R, Rodgers MAJ. An instrument for simultaneous acquisition of fluorescence spectra and fluorescence lifetimes from single cells. Photochem Photobiol. 1987;45:161–165.
[16] Schneckenburger H, Frenz M, Tsuchiya Y, Denzer U, Schleinkofer L. Picosecond fluorescence microscopy for measuring chlorophyll and porphyrin components in conifers and cultured cells. Lasers in the Life Sciences. 1987;1:229–307.
[17] Kinoshita S, Seki T, Fu Liu T, Kushida T. Fluorescence of hematoporphyrin in living cells and in solution. J Photochem Photobiol B. 1988;2:195–208.
[18] Gärtner W, Gröbler B, Schubert D, Wabnitz H, Wilhelmi B. Fluorescence scanning microscopy combined with subnanosecond time resolution. Experimental Techniques in Physics. 1988;36:443–451.
[19] Gröbler B. Apparatus for the graphic representation and analysis of fluorescence signals. Vorrichtung zur Analyse und bildlichen Darstellung des bei einer punktweisen Anregung eines Präparates durch Laserlicht entstehenden zeitlichen Intensitätsverlaufes der Fluoreszenzstrahlung. German patent DE 3614359C2, United States patent US4786170A. Filed April 28, 1986 (July 26, 1985)
[20] König K. Beiträge zur Photochemotherapie und optischen Diagnostik von Tumoren [PhD thesis]. Translation: Contributions to photochemotherapy and optical diagnostics of tumors. Jena, Germany: University of Jena Archive, Germany; 1989.
[21] Bugiel I, König K, Wabnitz H. Investigation of cells by fluorescence laser scanning microscopy with subnanosecond time resolution. Lasers Life Sciences. 1989;3:1–7.
[22] König K, Wabnitz H. Fluoreszenzuntersuchungen mit hoher zeitlicher, spektraler und räumlicher Auflösung. Labortechnik. 1990;23:26–31.
[23] Wang XF, Uchida T, Minami S. A fluorescence lifetime distribution measurement system based on phase-resolved detection using an image dissector tube. Appl Spectrosc. 1989;43:840–845.
[24] Wang XF, Kitajima S, Uchida T, Coleman DM, Minami S. Time-resolved fluorescence microscopy using multichannel photon counting. Appl Spectrosc. 1990;44:25–29.
[25] Denk W, Strickler JP, Webb WW. Two-photon laser microscopy. United States patent Patent US 5,034,613. Filed Nov 14, 1989.
[26] Denk W, Strickler JH, Webb WW. Two-photon laser scanning fluorescence microscopy. Science. 1990;248:73–76.
[27] Piston DW, Sandison DR, Webb WW. Time-resolved fluorescence imaging and background rejection by two-photon excitation in laser scanning microscopy. Proc SPIE 1640; 1992.
[28] So PTC, French T, Yu WM, Berland K, Dong CY, Gratton E. Time-resolved fluorescence microscopy using two-photon excitation. Bioimaging. 1995;3:49–63.
[29] König K, So PTC, Mantulin WW, Tromberg BJ, Gratton E. Two-photon excited lifetime imaging of autofluorescence in cells during UVA and NIR photostress. J Microsc. 1996;183:197–204.
[30] Masters BR, So PTC, Gratton E. Multiphoton excitation fluorescence microscopy and spectroscopy of in vivo human skin. Biophys J. 1997;72:2405.
[31] Masters BR, So PTC, Gratton E. Multiphoton excitation microscopy of in vivo human skin. Functional and morphological optical biopsy based on three-dimensional imaging, lifetime measurements and fluorescence spectroscopy. Ann NY Acad Sci. 1998;838:58–67.
[32] Masters BR, So PTC, Gratton E. Optical biopsy of in vivo human skin: multi-photon excitation microscopy. Laser Med Sci. 1998;13:196–203.

[33] König K, Schneckenburger H. Laser-induced dental caries and plaque diagnosis on patients by sensitive autofluorescence spectroscopy and time-gated video imaging. Preliminary studies. Proc SPIE 2128: Lasers in Surgery. 1994;403–408.
[34] König K, Schneckenburger H, Hibst R. Time-gated in vivo imaging of dental caries. Cell Mol Biol. 1999;45:233–239.
[35] Sun Y, Hatami N, Yee M, Phipps J, Elson DS, Gorin F, Schrot RJ, Marcu L. Fluorescence lifetime imaging microscopy for brain tumor image-guided therapy. J Biomed Opt. 2010;15:056022.
[36] König K. Verfahren und Anordnung zur nicht-invasiven dreidimenisonalen optischen Untersuchung und Therapie der Haut. German patent DE 10065146A1. Filed December 22, 2000.
[37] König K, Wollina U, Riemann I, Peuckert C, Halbhuber KJ, Fünfstück V, Fischer TW, Elsner P. Optical tomography of human skin with subcellular spatial and picosecond time resolution. Proc SPIE 4620; 2002.
[38] König K, Riemann I. High-resolution multiphoton tomography of human skin with subcellular spatial resolution and picosecond time resolution. J Biomed Opt. 2003;8:432–439.
[39] König K, Riemann I, Ehrlich G, Ulrich V, Fischer P. Multiphoton FLIM and spectral imaging of cells and tissues. Proc SPIE 5323; 2004.
[40] König K. Clinical multiphoton tomography. J Biophotonics. 2008;1:13–23.
[41] Dimitrow E, Riemann I, Ehlers A, Koehler JM, Norgauer J, Elsner P, König K, Kaatz, M. Spectral fluorescence lifetime detection and selective melanin imaging by multiphoton laser tomography for melanoma diagnosis. Exp Dermatol. 2009;18:509–515.
[42] Huck V, Gorzelanny C, Thomas K, Niemeyer V, Luger T A, König K, Schneider SW. Intravital multiphoton tomography as a novel tool for non-invasive in vivo analysis of human skin affected with atopic dermatitis. Proc SPIE 7548; 2010.
[43] Talbot CB, Patalay R, Munro I, Breunig HG, König K, Alexandrov Y, Warren S, Chu A, Stamp GW, Neil MAA, French PMW, Dunsby C. A multispectral FLIM tomograph for in-vivo imaging of skin cancer. Proc SPIE 7903; 2011.
[44] Patalay R, Talbot C, Munro I, Breunig HG, König K, Alexandrov Y, Warren S, Neil MAA, French PMW, Chu A, Stamp GW, Dunsby C. Fluorescence lifetime imaging of skin cancer. Proc SPIE 7883; 2011.
[45] Benati E, Bellini V, Borsari S, Dunsby C, Ferrari C, French P, Guanti M, Guardoli D, König K, Pellacani G, Ponti G, Schianchi S, Talbot C, Seidenari S. Quantitative evaluation of healthy epidermis by means of multiphoton microscopy and fluorescence lifetime imaging microscopy. Skin Res Technol. 2011.
[46] Seidenari S, Arginelli F, Bassoli S, Cautela J, French PMW, Guanti M, Guardoli D, König K, Talbot C, Dunsby C. Multiphoton laser microscopy and fluorescence lifetime imaging for the evaluation of the skin. Dermatol Res Pract. 2012;810749.
[47] Manfredini M, Arginelli F, Dunsby C, French PMW, Talbot C, König K, Ponti G, Pellacani G, Seidenari S. High-resolution imaging of basal cell carcinoma: a comparison between multiphoton microscopy with fluorescence lifetime imaging and reflectance confocal microscopy. Skin Res Techn. 2013;19:e433–e443.
[48] Arginelli F, Manfredini M, Bassoli S, Dunsby C, French PMW, König K, Magnoni C, Ponti G, Talbot C, Seidenari S. High resolution diagnosis of common nevi by Multiphoton Laser Tomography and Fluorescence Lifetime Imaging. Skin Res Tech. 2013;19:194–204.
[49] Seidenari S, Arginelli F, Dunsby C, French PMW, König K, Magnoni C, Manfredini M, Talbot C, Ponti G. Multiphoton laser tomography and fluorescence lifetime imaging of basal cell carcinoma: morphological features for non-invasive diagnostics. Exp Dermatol. 2012;21:831–836.
[50] Koehler MJ, Preller A, Elsner P, König K, Hipler UC, Kaatz M. Non-invasive evaluation of dermal elastosis by in vivo multiphoton tomography with autofluorescence lifetime measurements. Exp Dermatol. 2012;21:48–51.

[51] Patalay R, et al. Quantification of cellular autofluorescence of human skin using multiphoton tomography and fluorescence lifetime imaging in two spectral detection channels. Biomed Opt Exp. 2011;2:3295–3308.
[52] Patalay R, Talbot C, Alexandrov Y, Lenz MO, Kumar S, Warren S, Munro I, Neil MAA, König K, French PMW, Chu A, Stamp GW, Dunsby C. Multispectral fluorescence lifetime imaging using multiphoton tomography for the evaluation of basal cell carcinomas. Plos One. 2012
[53] König K. Hybrid multiphoton multimodal tomography of in vivo human skin. IntraVital. 2012;1:11–26.
[54] Roberts MS, Dancik Y, Prow TW, Thorling CA, Lin LL, Grice JE, Robertson TA, König K, Becker W. Non-invasive imaging of skin physiology and percutaneous penetration using fluorescence spectral and lifetime imaging with multiphoton and confocal microscopy. Eur J Pharm Biopharm. 2011;77:469–488.
[55] Huck V, Gorzelanny C, Thomas K, Getova V, Niemeyer V, Zens K, Unnerstall TR, Feger JS, Fallah MA, Metze D, Ständer S, Luger TA, König K, Mess C, Schneider SW. From morphology to biochemical state-intravital multiphoton fluorescence lifetime imaging of inflamed human skin. Sci Rep. 2016;6:22789.
[56] Kantelhardt S, Kalasauskas D, König K, Kim E, Weinigel M, Uchugonova A, Giese A. In vivo multiphoton tomography and fluorescence lifetime imaging of human brain tumor tissue. J Neurooncol. 2016;127:473–482.
[57] Schweitzer D, Kolb A, Hammer M, Thamm E. Tau-mapping of the autofluorescence of the human ocular fundus. Proc SPIE 4164; 2000.
[58] Schweitzer D, Kolb A, Hammer M. Autofluorescence lifetime measurement in images of the human ocular fundus. Proc SPIE 4432; 2001.
[59] Schweitzer D, Hammer M, Schweitzer F, Anders R, Doebbecke T, Schenke S, Gaillard ER. In vivo measurement of time resolved auto-fluorescence at the human ocular fundus. J Biomed Opt. 2004;9(6):1214–1222.
[60] Mizeret J, Stepinac T, Hansroul M, Studzinski A, van den Bergh H, Wagnières G. Instrumentation for real-time fluorescence lifetime imaging in endoscopy. Rev Sci Instrum. 1999;70:4689.
[61] Bird D, Gu M. Compact two-photon fluorescence microscope based on a single-mode fiber coupler. Optics Letters. 2002;27:1031–1033.
[62] Jung JC, Schnitzer MJ. Multiphoton endoscopy. Optics Letters. 2003;28:902–903.
[63] König K, Ehlers A, Riemann I, Schenkl S, Bückle R, Kaatz M. Clinical two-photon microendoscopy. Microscopy Research and Technique. 2007;70(5):398–402.
[64] Förster T. Energiewanderung und Fluoreszenz. Energy migration and fluorescence. Naturwissenschaften. 1946;6:166–175.
[65] Wouters FS, Bastiaens PIH. Fluorescence lifetime imaging of receptor tyrosine kinase activity in cells. Current Biol. 1999;9:1127–1130.
[66] Becker W. Advanced time-correlated single-photon counting techniques. Berlin, Heidelberg, New York: Springer; 2005.
[67] Periasamy A, Clegg RM, editors. FLIM in Biology and Medicine. Taylor & Francis (CRC Group); 2009
[68] Becker W. Advanced time-correlated single photon counting applications. Springer; 2015.
[69] König K, Uchugonova A. Multiphoton fluorescence lifetime imaging at the dawn of clinical application. In: Periasamy A, Clegg RM, editors. FLIM in biology and medicine. Taylor & Francis (CRC Group); 2009. p. 165–187.

Wolfgang Kaiser

2 The long journey to the laser and its use for nonlinear optics

Abstract: Einstein introduced the basic principle of the laser in 1917. However, the possibility of light amplification was not recognized for decades. Eventually, Maiman from the Hughes Aircraft Co. realized the first laser based on ruby as active medium. This unexpected result excited our group at the Bell Labs and we realized the first intense and highly directional (< 1 degree) laser beam. On October 5, 1960 we demonstrated the red laser beam fired from the radar tower in New Jersey and its detection 25 miles away. The new light source initiated the field of Nonlinear Optics (NLO). We realized the first two-photon fluorescence in 1961.

The title of this article suggests a substantial history of the laser. For you the laser is a well-known and well-established optical light source. But please note, that the laser started its existence just 55 years ago.

In 1900, Max Planck published a theory about the experimentally known black body radiation [1]. In his derivation he introduced the light quantum $E = hf$ for the transition between two energy states. This revolutionary idea was used by the – then young – Einstein in 1917 in a completely different derivation of the same black body radiation [2]. In this paper the basic principle of the laser was introduced: the stimulated emission of light. An excited quantum state can be de-excited by a passing light quantum of the proper frequency, i.e., two photons leave the system or – in other words – the stimulating photon is amplified. During the following decades the process of stimulated light emission received surprisingly little attention in the scientific community. In a few papers the dispersion of light in excited gases was investigated and named negative dispersion. The possibility of light amplification was not recognized for 34 years.

Finally, in 1951, Charles Townes, a professor at Columbia University in New York and consultant at the Bell Laboratories had the ingenious idea to build a microwave amplifier (with ammonia gas at 1.25 cm) using the stimulated excitation process. It is interesting to read in the biography of Townes how he found the maser (microwave amplifier by stimulated emission of radiation). He was on his way to a meeting concerning new microwave amplifiers. He knew that no proposals existed. Under a certain pressure the idea occurred to him to study the stimulated light emission process quantitatively for the application in a new device. In fact, in 1953 the first maser was in operation. The desirable extension to higher frequencies – preferentially to the visible – was first discussed theoretically by Schawlow and Townes in 1958 [3]. Tab. 2.1 provides an overview of the historical development of the maser and laser.

https://doi.org/10.1515/9783110429985-004

Tab. 2.1: Overview of the historical development of the maser and laser.

1900	Planck	Quantum of radiation $E = hf$
1917	Einstein	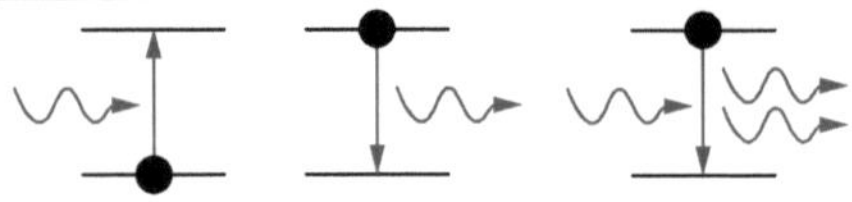
1923	Tolman	
1927	Ladenburg	
1941	Fabrikant	absorption emission stimulated emission
1951	Townes	
1952–1955	Townes	MASER
(1955)	Basov	Microwave Amplification by Stimulated
	Prokhorov	Emission of Radiation
1958	Schawlow	LASER
	Townes	Theory: *Physical Review* 112, 1940 (1958)

T. H. Maiman [4], a physicist at the Hughes Aircraft Co., realized the first laser based on ruby as active medium. He talked on the new light source during a press conference in New York on July 7, 1960. The newspaper *New York Times* reported on it one day later. The first scientific paper “Stimulated optical radiation in ruby” was published in *Nature* in August 6, 1960. Maiman pumped a ruby crystal with a powerful flash lamp and observed narrowing of the emitting beam to 55 degrees. The beam was delivered through a hole in the silver coating.

This unexpected result excited our group at the Bell Laboratories to repeat the ruby experiment. After three weeks we saw the expected extremely intense, highly directional (< 1 degree) beam of coherent highly narrowed radiation and sent a manuscript on our findings to the journal *Physics Review Letters* on August 26, 1960 ([5], Fig. 2.1).

There was no doubt: This observation was a true laser emission as expected from theory. We used a semitransparent silver coating instead of a hole for out-coupling. On October 5, 1960 we gave a press conference (Fig. 2.2) and demonstrated a red beam of the new light source, fired from the radar tower in New Jersey and detected 25 miles away at Crawfort Hill using a photomultiplier (Fig. 2.3 and 2.4).

The same year, the He:Ne gas laser became operational at Bell [6] and the 4-level-laser was introduced [7, 8].

The very strong light emission by the new light source laser initiated the field of Nonlinear Optics (NLO). In 1961, second harmonic radiation was found in non-centrosymmetric crystals (e.g. quartz) by Franken et al. at the University of Michigan [9] and by Giordmaine at Bell [10]. First two-photon fluorescence was seen in $Eu^{2+}CaF_2$ by Kaiser and Garrett at Bell [11].

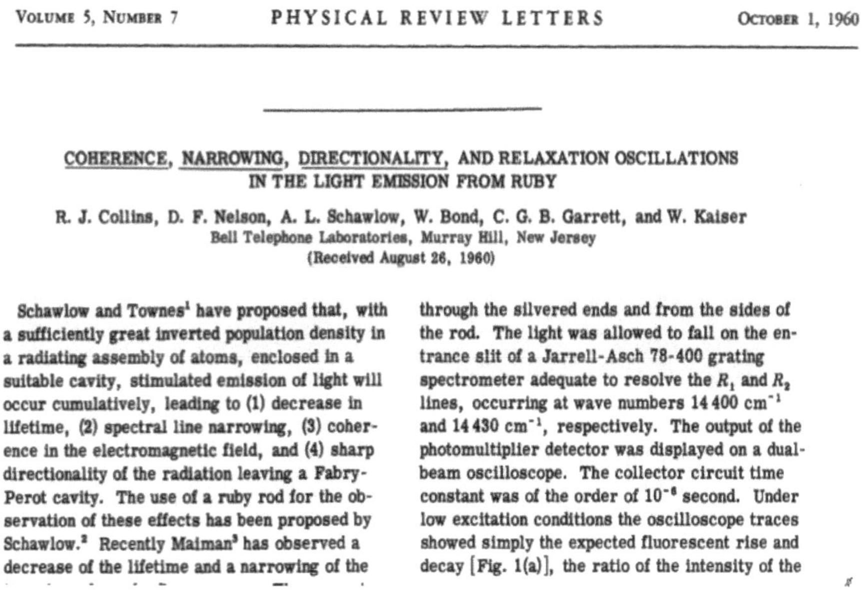

VOLUME 5, NUMBER 7 PHYSICAL REVIEW LETTERS OCTOBER 1, 1960

COHERENCE, NARROWING, DIRECTIONALITY, AND RELAXATION OSCILLATIONS IN THE LIGHT EMISSION FROM RUBY

R. J. Collins, D. F. Nelson, A. L. Schawlow, W. Bond, C. G. B. Garrett, and W. Kaiser
Bell Telephone Laboratories, Murray Hill, New Jersey
(Received August 26, 1960)

Schawlow and Townes[1] have proposed that, with a sufficiently great inverted population density in a radiating assembly of atoms, enclosed in a suitable cavity, stimulated emission of light will occur cumulatively, leading to (1) decrease in lifetime, (2) spectral line narrowing, (3) coherence in the electromagnetic field, and (4) sharp directionality of the radiation leaving a Fabry-Perot cavity. The use of a ruby rod for the observation of these effects has been proposed by Schawlow.[2] Recently Maiman[3] has observed a decrease of the lifetime and a narrowing of the

through the silvered ends and from the sides of the rod. The light was allowed to fall on the entrance slit of a Jarrell-Asch 78-400 grating spectrometer adequate to resolve the R_1 and R_2 lines, occurring at wave numbers 14 400 cm^{-1} and 14 430 cm^{-1}, respectively. The output of the photomultiplier detector was displayed on a dual-beam oscilloscope. The collector circuit time constant was of the order of 10^{-6} second. Under low excitation conditions the oscilloscope traces showed simply the expected fluorescent rise and decay [Fig. 1(a)], the ratio of the intensity of the

Fig. 2.1: Publication in *Physics Review Letters* on the first stable laser by Collins, Nelson, Schawlow, Bond, Garrett, and Kaiser on October 1, 1960 (manuscript received on August 26, 1960) [12].

Fig. 2.2: Press conference on October 8, 1960. Reused with permission of Nokia Corporation [13].

Fig. 2.3: First presentation of a stable laser beam to the public. Williard Boyle and Robert Collins "fired" the ruby laser from the radar tower at Murray Hill, close to the Bell Labs, to Crawfort Hill in New Jersey, 25 miles away, where it was detected by Walter Bond and Donald Nelson. Reused with permission of Nokia Corporation [13].

While we studied the two-photon transition, we learned that already in 1927 a theoretical paper was written on the same topic by Mrs. Göppert in Göttingen/Germany. She calculated correctly that the probability to see two-quantum transitions is too small for the light sources existing at her time. With our powerful laser light we saw the two-quantum transition quite readily in good agreement with her calculations. In 1961 I called her up and told that we were able to prove her theory of two-photon transitions using the laser. However, at that time her research focused on the nuclear shell.

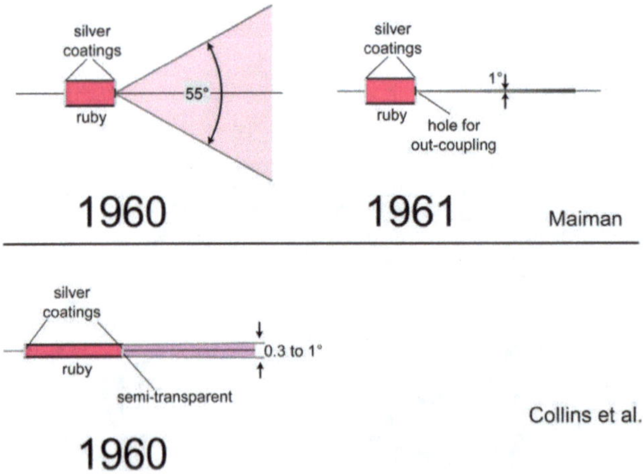

Fig. 2.4: The ruby laser beam from Bell Labs was highly directional (< 1°) and therefore more intense than the first laser beam by Maiman at Hughes Aircraft Co.

During the past 55 years the number of new lasers has increased to an impressive number and the number of general laser applications has grown tremendously. Besides high speed data communication, industrial fabrication and environmental monitoring, only two further topics should be cited here. First, femtosecond laser spectroscopy is applied to study the fast carrier dynamics in semiconductors as well as primary reactions in photosynthesis (Fig. 2.5). With respect to regenerative energy sources it is worth mentioning that the yearly turnover of photosynthesis is 2×10^{18} kJ on earth which corresponds to the six-fold primary energy needed currently by humans. Second, new medical laser applications have significant value for our health. Especially diagnoses of eyes and skin – discussed further in this book – are of great importance for all of us. In 1963 Dr. Goeppert-Mayer received the Nobel Prize for theoretical physics [13, 14]. Further information and the development of the laser can be obtained from references [13] and [14].

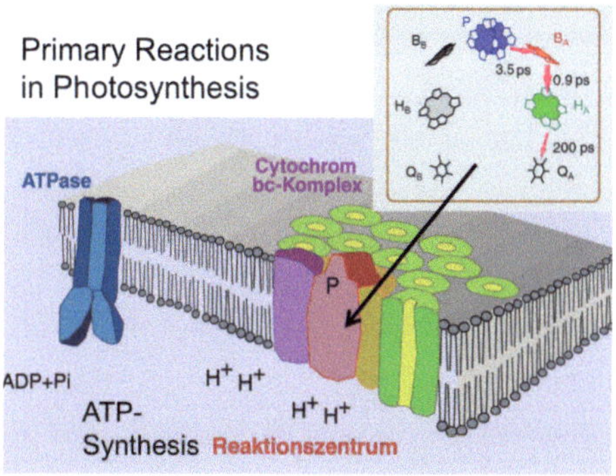

Fig. 2.5: Femtosecond spectroscopy is employed to study rapid primary processes during photosynthesis.

References

[1] Planck M. On the law of distribution of energy in the normal spectrum. Annalen der Physik. 1901;4:553–563.

[2] Einstein A. Zur Quantentheorie der Strahlung. Phys Zeit. 1917;18:121–128.

[3] Schawlow AL, Townes CH. Infrared and optical masers. Phys Rev. 1958;112:1940–1949.

[4] Maiman TH. Stimulated optical radiation in ruby. Nature. 1960;187:493–494.

[5] Collins RJ, Nelson DF, Schawlow AL, Bond W, Garrett CGB, Kaiser W. Coherence, narrowing, directionality, and relaxation oscillations in the light emission from ruby. Phys Rev Lett. 1960;5:303–305.

[6] Javan A, Bennet WR, Herriot DR. Population inversion and continuous optical maser oscillation in a gas discharge containing a He-Ne mixture. Phys Rev Lett. 1961;6:106–110.

[7] Sorokin PP, Stevenson MJ. Solid-state optical maser using divalent samarium in calcium fluoride. IBM Journal. 1961;5:56–58.

[8] Kaiser W, Garrett CGB, Wood DL. Fluorescence and optical maser effects in CaF2:Sm++. Phys Rev. 1961;123:766–776.

[9] Franken PA, Hill AE, Peters CW, Weinreich G. Generation of optical harmonics. Phys Rev Lett. 1961;7:118–119.
[10] Giordmaine JA. Mixing of light beams in crystals. Phys Rev Lett. 1962;8:19–20.
[11] Kaiser W, Garrett CGB. Two-photon excitation in $CaF_2:Eu^{2+}$. Phys Rev Lett. 1961;7:229–231.
[12] Collins RJ, Nelson DF, Schawlow AL, Bond W, Garrett CGB, Kaiser W. Coherence, Narrowing, Directionality, and Reflaxation oscillations in the light emission from ruby. Phys. Rev. Lett. 1960;5.
[13] Nelson DF, Collins RJ, Kaiser W. Bell Labs and the ruby laser. Physics Today. 2010;63:40–45.
[14] Zinth W, Laubereau A, Kaiser W. The long journey to the laser and its rapid development after 1960. Eur Phys J. H 2011;36:153–181.

Wolfgang Becker

3 Advanced TCSPC-FLIM techniques

Abstract: Fluorescence lifetime imaging (FLIM) is aiming at the measurement of molecular parameters in the local environment of the fluorophores. It exploits the fact that the fluorescence decay function of a fluorophore depends on its molecular environment but not on the fluorophore concentration. A fluorescence lifetime imaging (FLIM) technique for application in biology has to combine high photon efficiency, high lifetime accuracy, resolution of multi-exponential decay profiles, simultaneous recording in several wavelength intervals, high spatial resolution, and suppression of out-of-focus fluorescence and scattered light. We will show that the combination of multidimensional time-correlated single-photon counting (TCSPC) with confocal or multiphoton laser scanning meets these requirements almost ideally. FLIM by multidimensional TCSPC is based on scanning the sample by a high repetition rate pulsed laser beam and the detection of single photons of the fluorescence signal returning from the sample. In the simplest case, each photon is characterized by its time in the laser pulse period and the coordinates of the laser spot in the scanning area in the moment of its detection. The recording process builds up a photon distribution over these parameters. The result is an array of pixels, each containing a full fluorescence decay curve in a large number of time channels. TCSPC-FLIM has got a new push from the introduction of 64-bit data acquisition software. Due to the large memory space available in the 64-bit environment FLIM can now be recorded at unprecedented pixel numbers. This allows a large number of cells to be imaged simultaneously under identical experimental conditions. The technique can further be extended by including additional parameters in the photon distributions. Such parameters can be the wavelength of the photons, additional spatial coordinates, the time after a stimulation of the sample, or the time within the period of an additional modulation of the laser. These advanced techniques can be used to record lateral mosaics of FLIM images, FLIM Z-stacks, multiwavelength FLIM images, images of physiological effects occurring in the sample, and to simultaneously record fluorescence and phosphorescence lifetime images.

3.1 The fluorescence decay function as an indicator of molecular parameters

Fluorescence lifetime imaging (FLIM) is aiming at the measurement of molecular parameters by recording changes in the fluorescence lifetime of endogenous or exogenous fluorophores. It uses the fact that the fluorescence decay function of a fluorophore depends on its molecular environment but, within reasonable limits, not on its concentration. The decay function is also independent of the absorption of the

https://doi.org/10.1515/9783110429985-005

fluorescence within the sample and its spectral variation, and of the spectral behavior of the detection system. Molecular effects can thus be investigated more quantitatively than by using techniques based on fluorescence intensities or fluorescence spectra [1–4]. There are various mechanisms which cause environment-dependent changes in the fluorescence lifetime.

3.1.1 Ion concentrations

An excited molecule can dissipate the absorbed energy by interaction with other molecules. The effect is called fluorescence quenching. The fluorescence lifetime then becomes shorter than the natural fluorescence lifetime. One quenching mechanism is 'collisional' quenching, i.e., the collision of the excited molecule with another molecule, atom, or ion. Typical quenchers are oxygen, halogens, metal ions, and a variety of organic molecules. For fluorescence lifetime microscopy it is important that the rate constant of fluorescence quenching depends linearly on the concentration of the quencher. The concentration of the quencher can therefore be directly obtained from the decrease in the fluorescence lifetime. The classic example is quinine, whose lifetime follows the reciprocal Cl^- concentration over two orders of magnitude. A practical example is shown in Fig. 3.1.

A slightly different mechanism is 'static quenching'. In that case, there is a fraction of the fluorophore that is bound to the ion, and another fraction that is not. Both fractions have different fluorescence lifetimes. The ratio of bound and unbound fluorophore depends on the ion concentration. The apparent lifetime of the mixture

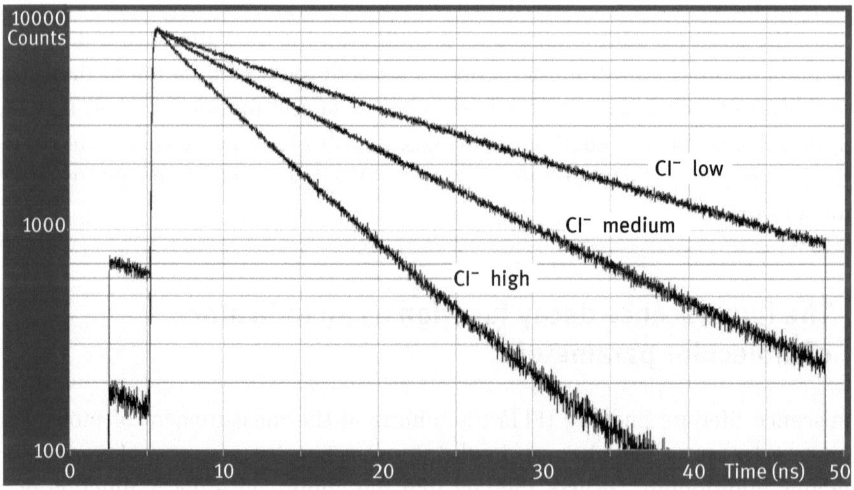

Fig. 3.1: Left: Fluorescence quenching. The fluorescence lifetime of quinine sulfate decreases with increasing Cl^- concentration.

thus depends on the concentration of the ion. A comprehensive overview about ion quenching can be found in [3]. Of particular interest to cell biology are Ca^{++} and Cl^{-}, both of which are important to the function of the neuronal system. Applications of fluorescence lifetime recording and FLIM to the measurement of Ca^{++} and Cl^{-} concentrations can be found in [5–7] and [8–11], respectively.

3.1.2 pH sensors

Many fluorescent molecules have a protonated and a deprotonated form. The equilibrium between both depends on the pH. If the protonated and deprotonated form have different lifetimes the apparent lifetime is an indicator of the pH [3, 12–14].

3.1.3 Binding to proteins, protein configuration

Many fluorescent molecules, including endogenous fluorophores, form complexes with biomolecules, in particular proteins. The fluorescence spectra of these different conformations can be virtually identical, but the fluorescence lifetimes are often different. The exact mechanism of the lifetime changes is not always clear. The most likely mechanisms are fluorescence quenching and a change in the configuration of the fluorophore which, in turn, changes the rate of the internal nonradiative decay. In practice, for almost all dyes the fluorescence lifetime depends more or less on the binding to proteins, DNA or lipids [3, 15, 16]. When a fluorophore is bound to a protein also the protein configuration may have an influence on the fluorescence lifetime [17]. A glucose sensor based on this mechanism has been described in [18]. An effect of target protein configuration has even been found for EGFP which had long believed to be insensitive to its environment [19]. Lifetime changes of diphenylhexatriene (DPH) depending on the cholesterol concentration were used in [20]. Environment-dependent changes of fluorescence parameters have even been observed for quantum dots and nanoparticles [21–24]. There is an enormous amount of fluorophores used in microscopy today [25], and only a few of them have been tested for target-specific lifetime changes. The most promising ones are those of medium and low quantum efficiency. These are most likely to be influenced by the local molecular environment [26]. FLIM users are encouraged to keep their eye on lifetime changes that may be suitable as probe functions, see also Section 3.3.4.

3.1.4 Förster Resonance Energy Transfer: FRET

Resonance energy transfer was theoretically explained by Theodor Förster in 1946 [27, 28]. Förster resonance energy transfer is an interaction of two molecules in which the emission band of one molecule overlaps the absorption band of the other. In this case the energy from the first molecule, the donor, can transfer into the second one, the acceptor. FRET can result in an extremely efficient quenching of the donor fluorescence and, consequently, in a considerable decrease of the donor lifetime. The energy transfer rate from the donor to the acceptor decreases with the sixth power of the distance. Therefore it is noticeable only at distances shorter than 10 nm [3]. FRET is used as a tool to investigate protein-protein interaction. Different proteins are labeled with the donor and the acceptor, and FRET is used as an indicator of the binding between these proteins. Donor and acceptor may also be attached to the same protein. FRET intensity is then a function of the protein conformation. A number of sensors for molecular parameters are based on the FRET effect, please see [25] for an overview. FRET experiments are the most frequent FLIM application. The number of FLIM-based FRET papers has literally exploded in the last decade. Please see [29–32] for more references.

3.1.5 Endogenous fluorophores, effect of metabolic activity

Most endogenous fluorophores are mixtures of slightly different compounds, with almost identical spectra but different fluorescence lifetimes [33]. The composition of the mixture may vary, and thus induce a change in the amplitudes of the decay components and in the apparent lifetime. Moreover, the lifetimes depend on the binding to proteins, on the oxygen concentration [34], or other metabolic parameters. The best reporters of metabolic activity are NAD(P)H and FAD. NAD(P)H and FAD exist in a bound and in an unbound form. The bound/unbound ratio depends on the type of metabolism and on metabolic activity. The fluorescence lifetimes of the forms are different, and the different fractions can be distinguished by FLIM [35, 36]. Both the fluorescence lifetimes of the decay components and the relative amplitudes have been found sensitive to the metabolic state of cells and tissue [37–46]. The lifetimes and amplitudes change also during cell maturation, apoptosis, and necrosis [47–49]. Recording and interpretation of the signals from endogenous fluorophores is the basis of clinical FLIM, please see [4, 50–53].

3.1.6 Oxygen

Oxygen is an efficient fluorescence quencher for a large number of fluorophores [3], especially those of longer fluorescence lifetime. Strong effects are observed for pyrene, anthracene, and coronene. Unfortunately these compounds absorb in the UV. They are therefore of limited usefulness in microscopy. Strong oxygen quenching is observed for the phosphorescence of organic ruthenium, europium, platinum and palladium compounds [3, 54–58]. Phosphorescence lifetimes are long – from hundreds of nanoseconds to milliseconds. Phosphorescence decay curves are therefore difficult to record in combination with the high laser pulse-repetition rates and fast scan rates used in modern laser scanning microscopes. Phosphorescence lifetime imaging (PLIM) is, however, possible by a technique based on laser modulation and dual-timebase recording [29, 59], see Section 3.4.6.

3.1.7 Redox potential

There are fluorophores that change their fluorescence behavior depending on the redox potential of the local environment. Sensors based on GFP variants are described in [61]. An improved redox sensor based on a modified GFP (roGFPiE) was developed by Avezov et al. Test results obtained from various cells by TCSPC-FLIM are described in [60].

3.1.8 Electron transfer

At distances smaller than about 1 nm between an electron donor and an electron acceptor photo-induced electron transfer can occur [3]. The excited molecule can either be the electron donor or the electron acceptor. The direction of the electron transfer is determined by the oxidation potential of the donor and the reduction potential of the acceptor. Electron transfer delivers two radicals, both in the ground (S0) state. Although electron transfer may be the basic mechanism of a number of quenching effects it is not explicitly used as a probing function. In biological imaging, photoinduced electron transfer and the radicals produced by it, may be a source of photobleaching and photodamage.

3.1.9 Other parameters influencing fluorescence lifetime

There are other molecular parameters that have an influence on the fluorescence decay functions. There is an effect of local viscosity, local refractive index, proximity to metal surfaces or nanoparticles, solvent polarity, and aggregation of the fluorophore. Please see [2, 3, 25, 29] for details.

3.1.10 Requirements for a FLIM technique in biology

The measurement of molecular parameters requires that the measurement does not induce changes in the molecular environment of the fluorophores. The photon efficiency, i.e., the number of photons required to obtain a given accuracy of the fluorescence lifetime is therefore extremely important. Moreover, there may be several fluorophores of different fluorescence lifetime, or several fractions of the same fluorophores in different states of interaction with the environment. The fluorescence decay functions then become multi-exponential. The biological information is often in the composition of the decay curve rather than in its average lifetime. The FLIM technique should therefore be able to resolve multi-exponential decay functions into their components. This requires high time resolution, and the recording of the decay data in a sufficiently large number of time channels. The decay data must be obtained from a defined plane in the sample, without contamination from out-of-focus fluorescence or lateral and longitudinal scattering. Therefore, the optical system must have optical sectioning capability, i.e., must be able to record fluorescence from a defined focal plane and a defined lateral position inside the sample.

3.2 TCSPC-FLIM with laser scanning systems

3.2.1 The advantage of scanning

In the 1950s it was discovered that scanning improves image contrast. Consider a focused beam of light scanning over the sample, a single-point detector measuring the light intensity returned from the excited spot, and some kind of image reconstruction device, such as a CRT screen, that builds up an image from the intensity values at each point of the scan. A device like that records scattered light *only from the excited pixel* but not from the other pixels within the sample. The contrast is thus better than for wide-field illumination and detection by an image sensor: In a wide-field system, the image sensor in every pixel detects scattered light from *all other pixels* of the illuminated area. The difference between scanning and wide-field imaging can be dramatic, see Fig. 3.2.

Scanning alone, however, does not solve another problem of microscopy of thick samples: The image contains light not only from the focal plane but from the en-

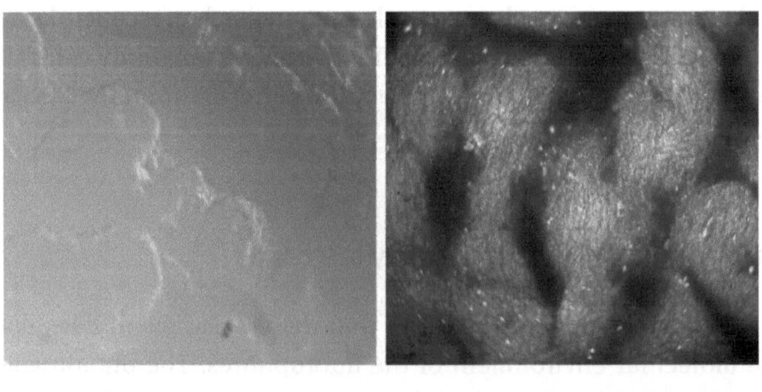

Fig. 3.2: Images obtained from a pig skin in the same microscope. Left: wide-field image. Right: scan image. Focal plane about 20 μm below the tissue surface, image size 200 μm × 200 μm.

tire depth of the sample. The images therefore quickly lose contrast with increasing thickness of the sample. The problem was solved with the invention of the confocal microscope by Marvin Minsky [62, 63]. Minsky combined scanning with detection through a confocal pinhole. Light from outside the focal plane is not focused into the pinhole and thus substantially suppressed. Although Minski used confocal imaging for transmission microscopy the principle can be applied to fluorescence microscopy as well.

The next step came with the invention and introduction of multiphoton microscopy [64, 65]. Different to confocal detection, which avoids out-of-focus *detection*, multiphoton excitation avoids out-of-focus *excitation*. A confocal pinhole is therefore not needed, and photons from deep layers of the sample are easier to detect. Other advantages are that the NIR excitation light penetrates into deep tissue layers, and that possible photoeffects in the sample are restricted to the focal plane. Multiphoton excitation is therefore the approach to deep tissue imaging.

3.2.2 FLIM by multidimensional TCSPC

It is a lucky coincidence, or perhaps a result of the intuition of the designers, that the most accurate and sensitive technique of fluorescence decay measurement, multidimensional time-correlated single-photon counting (TCSPC), is almost ideally compatible with laser scanning techniques. TCSPC goes back to developments in the 1950s [66]. In its original form, the technique was based on the detection of single photons, the measurement of the times of the photons after an excitation pulse, and the build-up of the distribution of the photons over these times [67]. Multidimensional TCSPC was developed in the 1990s. Unlike the classic technique, it builds up photon distributions not only over the times of the photons, but also over additional parameters, such as the coordinates of the excitation beam in a scan area, the wavelength of the photons, or the time after a stimulation of the sample [29, 68, 69]. The principle of FLIM by multidimensional TCSPC is shown in Fig. 3.3.

A laser scanning microscope (or another scanning device) scans the sample with a focused beam of a high repetition rate pulsed laser. Depending on the laser used, the fluorescence in the sample can either be excited by one-photon or by multiphoton excitation. The TCSPC detector is attached either to a confocal or a non-descanned port of the scanning microscope, please see [29] or [70] for details. For every detected photon the detector sends an electrical pulse into the TCSPC module. From this pulse, the TCSPC module determines the time, t, of the photon within the laser pulse sequence (i.e., in the fluorescence decay). Moreover, the TCSPC module receives scan clock signals (pixel, line, and frame clock) from the scan controller of the microscope. A system of counters (the scanning interface) counts the pixels within each line and the lines within each frame. The counter outputs deliver the position of the laser beam, x and y, in the scan area.

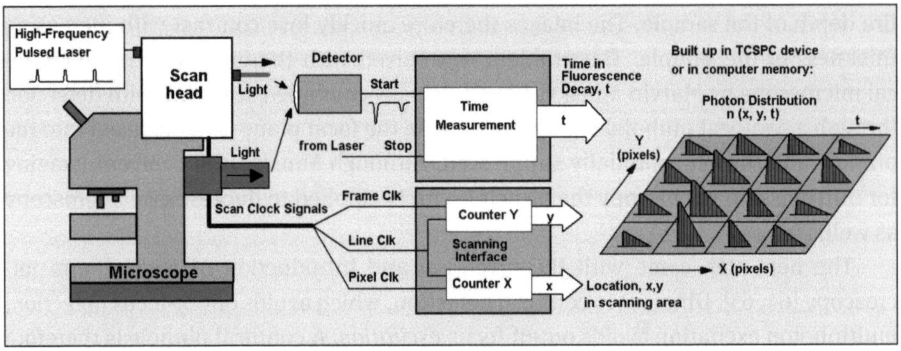

Fig. 3.3: Multidimensional TCSPC architecture for FLIM, from [29].

The information, t, x, y, addresses a location in a memory in which the detection events are accumulated. Thus, in the memory the distribution of the photon density over x, y, and t builds up. The result is a data array representing the pixels of the scan, with every pixel containing a large number of time channels with photon numbers for consecutive times after the excitation pulse. In other words, the result is an image that contains a fluorescence decay curve in each pixel [68, 71–73].

The procedure described above does not require that the laser beam stays in the same position until enough photons have been acquired in the decay curve of the corresponding pixel. It is only necessary that the *total pixel time*, over a large number of subsequent frames, is large enough to record a reasonable number of photons per pixel. In other words, the signal-to-noise ratio depends on the total acquisition time, not on the scan rate. Thus, TCSPC-FLIM works at even the highest scan rates available in laser scanning microscopes. At pixel rates used in practice, the recording process is more or less random: A photon is just stored in a memory location according to its time in the fluorescence decay and the position of the laser spot at the moment of its detection.

The process of TCSPC-FLIM delivers a near-ideal detection efficiency, and an excellent time resolution. There is no time-gating or gain modulation of the detector, and thus no suppression of photons. Every photon that causes an output pulse of the detector contributes to the result. The time resolution is given by the transit time jitter in the detector, not by the width of the single-electron response (SER) of the detector. The jitter is normally an order of magnitude smaller than the width of the SER. The time resolution, or the width of the temporal instrument response function, IRF, of a TCSPC-FLIM system is thus significantly better than for a system with analog recording of the signals. The combination of high detection efficiency, short IRF, and large number of time channels results in a near-ideal photon efficiency, i.e., in a near-ideal lifetime accuracy for a given number of recorded photons [29, 70, 74].

As an argument against TCSPC-FLIM it is often stated that it is 'slow' in terms of acquisition time. The technical issue behind this opinion is the 'pile-up' effect. A TCSPC device cannot detect several photons per excitation period, so that the number of photons (and thus the lifetime accuracy) obtained within a given period of time is limited. The pile-up effect was indeed a limitation in early TCSPC systems which worked with low repetition rate excitation sources. Moreover, there is a mistake in the technical literature: It is usually stated that the count rate must be lower than 0.1 % of the excitation rate. Correct is 0.1 × excitation rate, which is a factor of 100 higher than commonly believed. With modern high repetition rate excitation sources the pile-up effect is no longer a real limitation. A consideration of pile-up error, photon rate, lifetime accuracy, and acquisition time can be found in [29] and [68].

FLIM results are normally displayed as pseudo-color images. The brightness represents the number of photons per pixel. The color can be assigned to any parameter of the decay profile: the lifetime of a single-exponential approximation of the decay, the average lifetime of a multi-exponential decay, the lifetime or amplitude of a decay component, or the ratio of such parameters [29]. An example is shown in Fig. 3.4. The color shows the amplitude-weighted mean lifetime of a double-exponential decay. Decay curves in two selected pixels are shown on the right. Similar curves are contained in any pixel of the image. FLIM images from three typical applications are shown in Fig. 3.5 and 3.6.

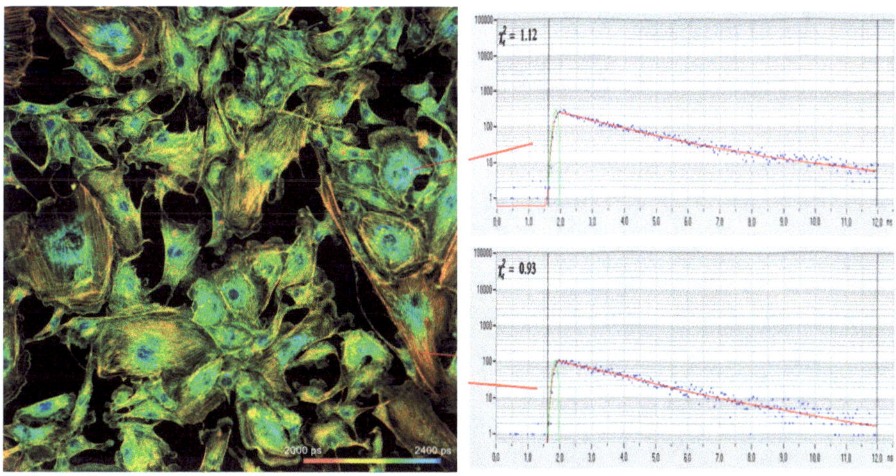

Fig. 3.4: Lifetime image of a BPAE cell, stained with Alexa 488 and MitoTracker Red. FLIM data 2048 × 2048 pixels, 256 time channels per pixel. Fluorescence decay shown for two selected pixels. Becker & Hickl DCS120 confocal scanner with Simple-Tau 152 TCSPC-FLIM system.

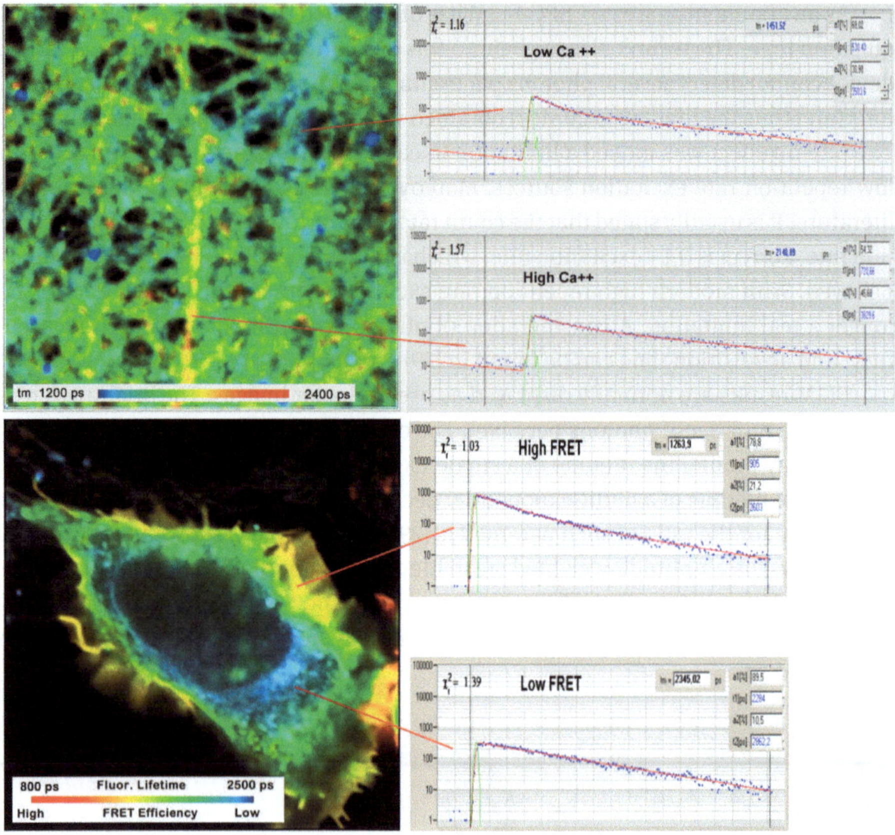

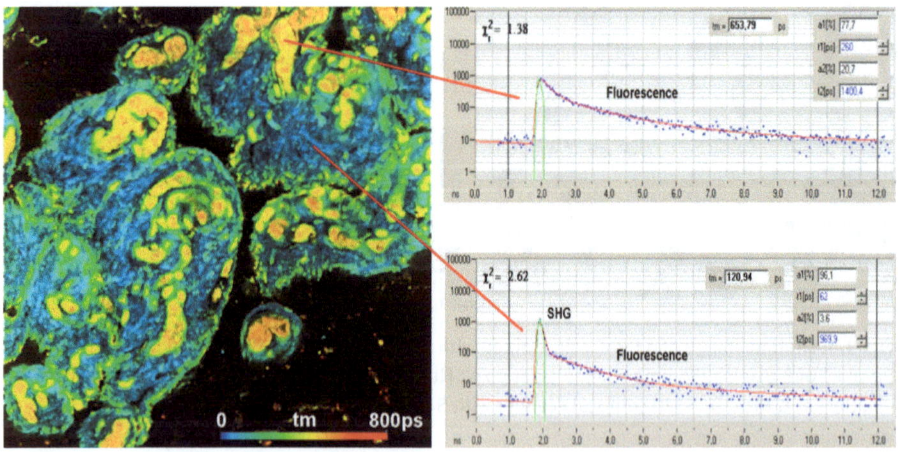

Fig. 3.6: Autofluorescence of pig skin. Two-photon excitation at 800 nm. Amplitude-weighted lifetime of double-exponential decay. Emission mainly comes from SHG of collagen and from fluorescence of NADH and FAD. Leica SP 5 MP, bh SPC-830 TCSPC-FLIM module.

◂ **Fig. 3.5:** Top: Ca^{2+} concentration measurement. Cultured neurons, incubated with Oregon Green BAPTA. Two-photon excitation, Zeiss LSM 7 MP, bh Simple-Tau 150 FLIM system. Bottom: FRET measurement. Cell contains a GFP fusion protein and a Cy3-labeled antibody. Lifetime image of donor (GFP). Excitation by ps diode laser, 445 nm, Zeiss LSM 710 with bh Simple-Tau 152 FLIM system.

3.3 Combination with different optical scanning techniques

3.3.1 One-photon excitation confocal FLIM

One-photon FLIM uses picosecond diode lasers or a super-continuum laser with a tunable filter as excitation sources. Because the one-photon process excites fluorescence through the whole depth of the sample the fluorescence light must be fed back into the scanner, descanned, and projected into a confocal pinhole, please see [29] and [70]. The optical principle of confocal detection is not suitable for detecting photons that are scattered on the way out of the sample. Confocal FLIM therefore works well for single cells or thin samples but not for deep layers of thick tissue. An example of confocal FLIM is shown in Fig. 3.7.

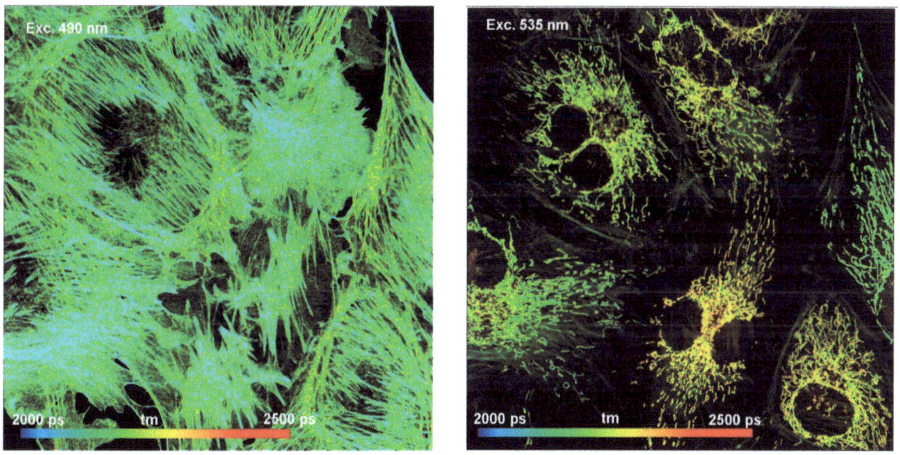

Fig. 3.7: Confocal FLIM with LSM Zeiss 710, tunable 'Intune' laser, and bh SPC150 TCSPC-FLIM module. BPAE cells stained with Alexa 488 phalloidin and MitoTracker Red. Excitation at 490 nm, detection at 500 to 700 nm (left) and excitation at 535 nm, detection at 550 to 700 nm (right).

3.3.2 Multiphoton FLIM

Multiphoton FLIM uses a near-infrared femtosecond laser, normally a Ti:sapphire laser, for excitation. Fluorescence is excited by a multiphoton absorption process. Simultaneous absorption of two (or more) photons was theoretically predicted by Maria Göppert-Mayer [75], suggested for laser scanning microscopy by Wilson and Sheppard in 1984 [65] and practically introduced by W. Denk, J.H. Strickler and W. Webb [64] in 1990. Multiphoton excitation is nonlinear, and excites fluorescence only in the focus of the microscope lens. A pinhole to suppress out-of-focus signals is therefore not required. Instead, the fluorescence light is diverted from the excitation beam directly behind the microscope lens and projected on a large area detector. The principle is called 'non-descanned detection', or 'NDD' [29, 70]. The NDD beam path transmits photons to the detector also if they are scattered on the way out of the sample. Multiphoton FLIM with NDD therefore works not only for single cells but also for thick samples. Multiphoton NDD FLIM is also able to excite and detect second-harmonic generation (SHG) signals from collagen structures. An example is shown in Fig. 3.8.

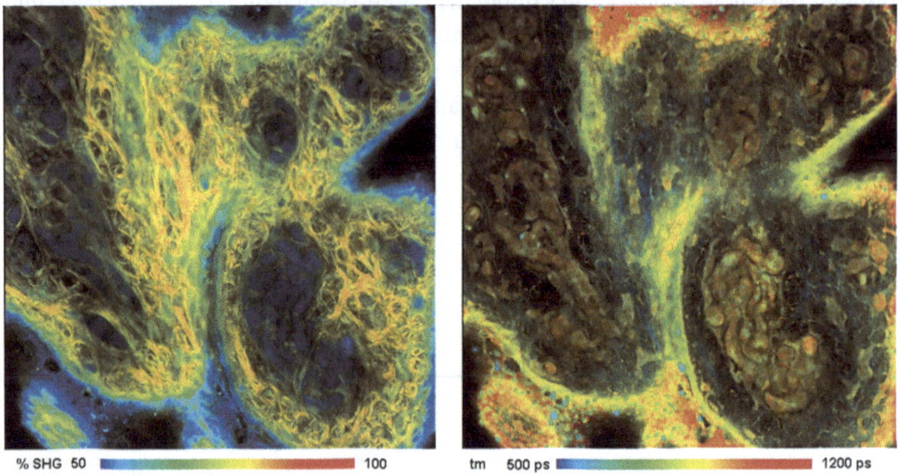

Fig. 3.8: Two-photon NDD FLIM of pig skin. Left: wavelength channel < 480 nm, color shows percentage of SHG in the recorded signal. Right: wavelength channel > 480 nm, color shows amplitude-weighted mean lifetime. Microscope Zeiss LSM 710 NLO, with bh hybrid detectors and Simple-Tau 152 dual-channel FLIM system. 512 × 512 pixels, 256 time channels.

3.3.3 FLIM with excitation wavelength multiplexing

A scanning system with several ps diode lasers or with a super-continuum laser and an acousto-optical filter (AOTF) is able to multiplex the excitation wavelength synchronously with the pixels, lines, or frames of the scan. The TCSPC-FLIM process is able to route photons excited by different lasers or laser wavelengths into different photon distributions. The result is a set of FLIM images for different excitation wavelengths, or, if the FLIM system has several detection channels, images for different excitation/emission combinations. Technical details are described in [29]. An example is shown in Fig. 3.9.

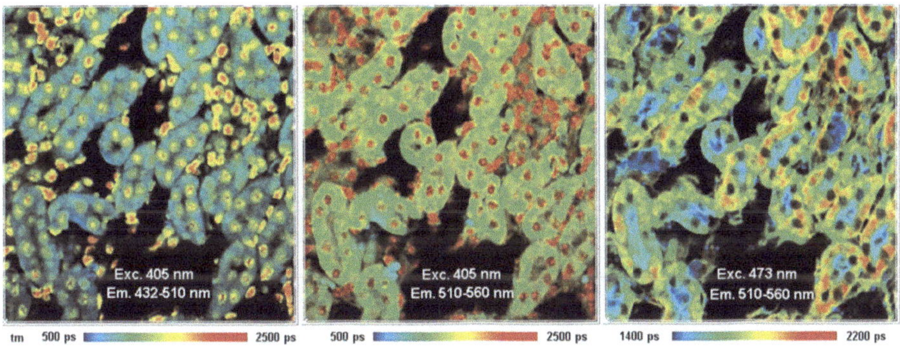

Fig. 3.9: FLIM with excitation wavelength multiplexing, 405 nm and 473 nm. Detection wavelength 435 nm to 510 nm and 510 nm to 550 nm. Mouse kidney section, stained with Alexa 488 WGA, Alexa 568 phalloidin, and DAPI. BH DCS-120 scanner with SPC-152 dual-channel TCSPC-FLIM system.

3.3.4 Near-infrared FLIM

Near-infrared dyes are used as contrast agents and as fluorophores in combination with diffuse optical imaging techniques. For these applications it is important to have information about binding of the dyes to proteins, DNA, collagen, and other cell constituents available. It is also important to know whether the dyes change their fluorescence lifetimes on binding, and whether these lifetime changes depend on the binding targets. Other NIR applications aim at avoiding autofluorescence. NIR detection is not immediately possible with most confocal microscopes. The microscopes do not have the right excitation lasers, and neither the dichroic mirrors in the detection beam path nor the detectors are suitable for NIR wavelengths [26]. One way to obtain NIR-FLIM is to use the titanium-sapphire laser of a multiphoton microscope for one-photon excitation. The fluorescence is detected by an NIR-sensitive FLIM detector through the confocal beam path. Another way is to use a scanner that is designed for confocal NIR-FLIM, such as the bh DCS-120 system. NIR-FLIM with both principles is shown in Fig. 3.10.

NIR-FLIM has been demonstrated for up to 1064 nm excitation and 1100 to 1700 nm detection wavelength [29]. The detector was an Id Quantique id 220 InGaAs SPAD, the FLIM system a bh SPC-150 module. NIR-FLIM can also be obtained by two-photon excitation. The excitation source is normally an OPO that is pumped by a titanium-sapphire laser. This makes the system expensive. Results are shown in [26, 29] and Fig. 3.18.

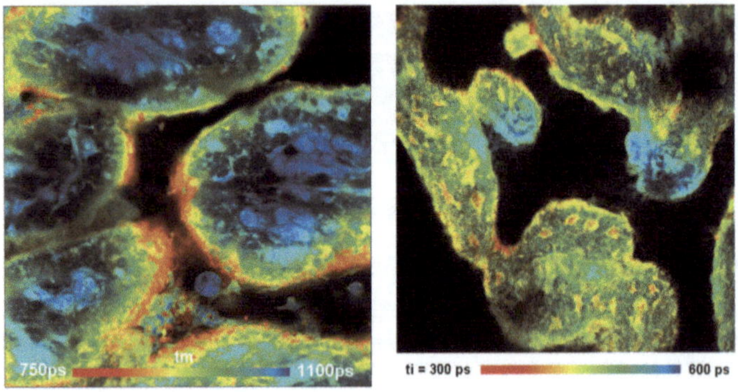

Fig. 3.10: NIR-FLIM. Left: LSM 710 NLO multiphoton microscope, Ti:Sa laser used as one-photon excitation source. Right: DCS-120 system, excitation by 685 nm diode laser. HPM-100-50 hybrid detector with GaAs cathode, bh Simple-Tau 150 TCSPC-FLIM system. Images from [29].

3.3.5 STED-FLIM

Stimulated Emission Depletion (STED) microscopy exploits the nonlinearity of stimulated emission to obtain optical super-resolution [76, 77]. The fluorescence excited by the excitation beam in a scanning microscope is depleted by emission stimulated by a second (STED) laser beam. The wavefront of the STED beam is manipulated to obtain a diffraction pattern that either has a doughnut shape laying in the x-y-plane or a dumbbell shape oriented along the z-axis. Fluorescence remains undepleted in the center part of the doughnut or the dumbbell. Because stimulated emission is highly nonlinear the undepleted volume can be made considerably smaller than the point spread function of the excitation beam. By scanning both beams together images with optical super-resolution are obtained. Because STED is a scanning technique it can be combined with TCSPC-FLIM [78, 79, 97]. An example is shown in Fig. 3.11.

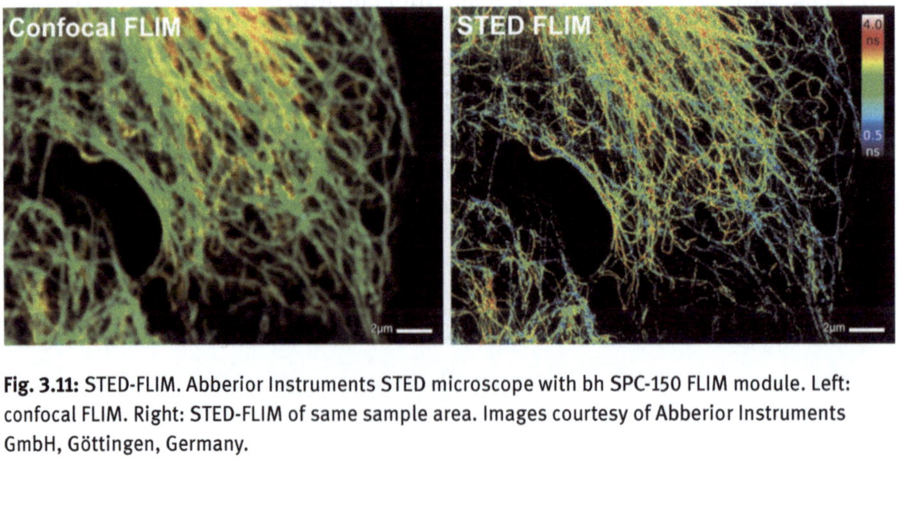

Fig. 3.11: STED-FLIM. Abberior Instruments STED microscope with bh SPC-150 FLIM module. Left: confocal FLIM. Right: STED-FLIM of same sample area. Images courtesy of Abberior Instruments GmbH, Göttingen, Germany.

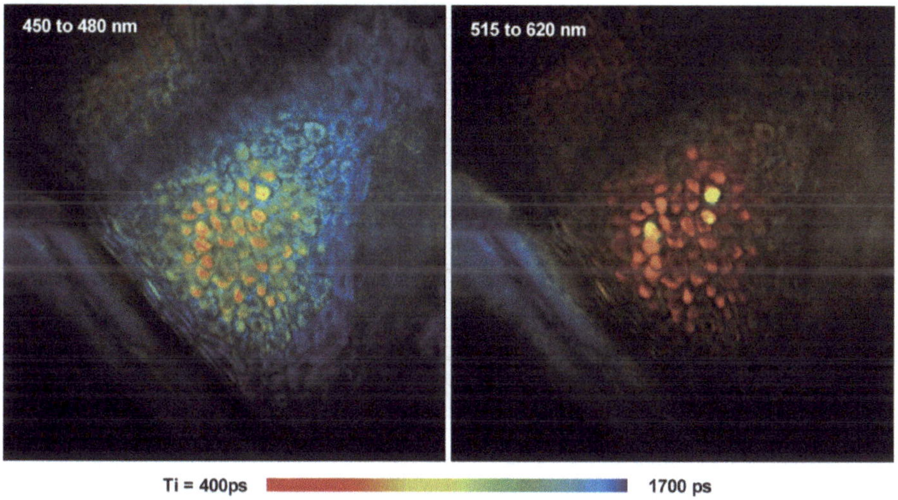

Fig. 3.12: FLIM of the stratum granulosum at the forearm of a human volunteer. Data courtesy of Washington Sanchez and Michael Roberts, University of Queensland, Brisbane, Australia.

3.3.6 Multiphoton tomography of human skin

Commercial instruments for clinical skin tomography have been designed by JenLab, Jena, Germany. Excitation is performed by a Ti:sapphire laser, the fluorescence is detected via an NDD beam path. The principle goes back to the work of K. König, P. So, E. Gratton, B. Tromberg and B. Masters in the 1990s [80, 81]. FLIM is detected in up to four spectral channels by separate detectors [82], or in 16 channels by a bh MW FLIM detector [98]. A typical result is shown in Fig. 3.12.

An impressive number of papers is related to FLIM data obtained by these instruments, please see [29]. A summary of clinical multiphoton tomography of skin is given in [51, 53, 83] and [82], The combination with microendoscopes is described in [52].

3.3.7 Ophthalmic FLIM

Ophthalmic imagers use a scanning technique with the pupil of the eye placed in the exit pupil of the scanning system. This reduces image distortion and blurring due to the poor optical quality of the lens of the eye. Confocal detection is used to suppress reflection, scattering, and fluorescence signals from the lens of the eye. Ophthalmic FLIM dates back to the work of D. Schweitzer and M. Hammer in the mid-1990s. Recent ophthalmic FLIM systems use excitation by ps diode lasers, resonance scanning, confocal detection in two spectral channels, and eye-motion compensation. The optical principle is described in [29] and [84]. Typical images are shown in Fig. 3.13. More results can be found in [33, 34, 85–89].

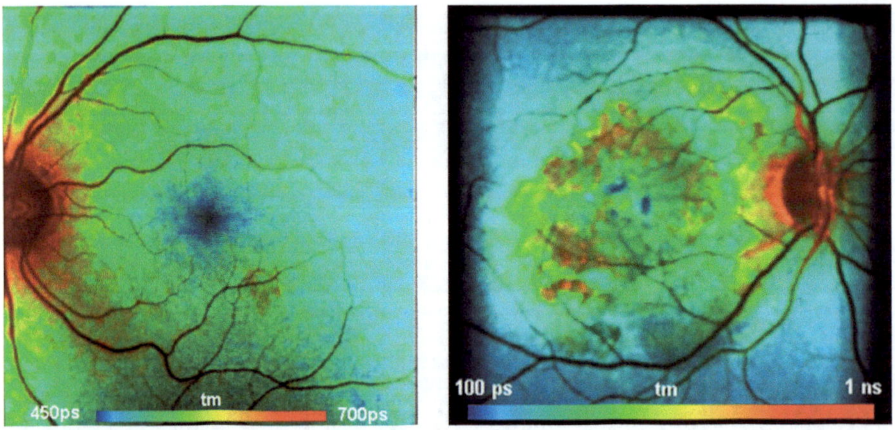

Fig. 3.13: Ophthalmic FLIM, fundus of the human eye. Left: healthy volunteer. Right: AMD patient. Heidelberg Engineering FLIO ophthalmic FLIM system, data courtesy of Dietrich Schweitzer, University Jena, Germany. Images from [29].

3.3.8 TCSPC-FLIM with other scanning techniques

Laser scanning microscopes use galvanometer mirrors for beam deflection. Galvanometer mirrors are surprisingly fast. Scan rates of more than 1000 lines per second can be reached. Faster scan rates can be obtained by polygon scanners and by resonance scanners. The disadvantage of polygon and resonance scanners is that they have a fixed scan field or a fixed scan rate, respectively. Nevertheless, they can be used with TCSPC-FLIM. Please see [29, 70].

A galvanometer scanner, polygon scanner, or resonance scanner can scan a macroscopic object directly in the focal plane of the scan lens. The scan area is on the order of 1 cm × 1 cm. It is also possible to scan through an endoscope. The optical principle and typical results are described in [29] and [70]. With modified scanner optics, samples as large as 10 cm × 10 cm can be scanned, please see [90].

Some TCSPC-FLIM systems are moving the sample instead of the excitation and detection beam. Such 'stage scanning' systems have the advantage that the beam is always on the optical axis. Compared with a beam scanning system the requirements to the optics are therefore modest. However, moving the sample induces motion in the sample. Stage scanning systems are therefore problematic for live cell imaging. They are also not suitable for clinical applications. The biggest disadvantage of stage scanning is the long scan time. The maximum speed is on the order of a few lines per second. A fast preview (as in beam scanning FLIM systems) is not possible. Focusing into the right plane and selecting the desired scan region is almost impossible. Stage scanning is therefore mainly used in applications where beam scanning is not applicable. An example is near-field optical microscopy (NSOM). For combination of TCSPC-FLIM with NSOM please see [29].

3.4 Recent advances in TCSPC-FLIM

Within the past 15 years, TCSPC-FLIM made significant progress. The recording efficiency was increased by new detectors [99], the wavelength range was extended [26], new excitation sources became available [29], and TCSPC-FLIM was introduced into clinical applications [33, 51, 53]. The most significant progress came, however, from computer development. The increase in available memory and the increase in processing speed not only paved the way to larger pixel numbers but also to entirely new multidimensional FLIM techniques. TCSPC-FLIM is, in principle, able to record a photon distribution not only over x, y and t but, additionally, over other parameters of the photons. The memory space required for these new applications is provided by 64-bit operating systems and 64-bit software. Some of these multidimensional FLIM techniques are described in the sections below.

3.4.1 Megapixel FLIM

FLIM data are extremely voluminous: A single FLIM data set can be considered a stack of (at least) 256 images for different times after the excitation pulse. Early TCSPC-FLIM systems built up the photon distribution in their own memory [71, 72]. The maximum image size under these circumstances was about 256 × 256 pixels, with 265 time channels per pixel.

The way to record larger images is software accumulation. The data of the individual photons and the scan clock pulses are transferred from the TCSPC electronics to the computer, and the photon distribution is built up by software. The required hardware architecture and suitable acquisition modes were available even in early TCSPC modules. However, the bus transfer speed and the processing capabilities of the PCs of the 1990s was insufficient for this kind of operation.

The situation changed with the introduction of fast multicore PCs. The tasks of data transfer and data processing could now be shared between the CPU cores. Typical images sizes with 32-bit operating systems were 512 × 512 pixels, with 256 time channels per pixel. With the introduction of Windows 7, 64 bit, and corresponding 64-bit data acquisition software, the maximum image size increased to 2048 × 2048 pixels × 256 time channels and more [29, 100]. Pixel numbers this large are enough to record images of the maximum field of view even of the best microscope lenses without compromising spatial or temporal resolution. An example is shown in Fig. 3.14.

The capability of recording FLIM data from a large field of view is a clear advantage in tissue imaging applications. Megapixel FLIM is, however, also useful in cell imaging. FLIM of a large number of cells can be recorded simultaneously, and under exactly identical experimental conditions. The data of the individual cells within one FLIM data set are therefore better comparable than images of individual cells taken in separate scans.

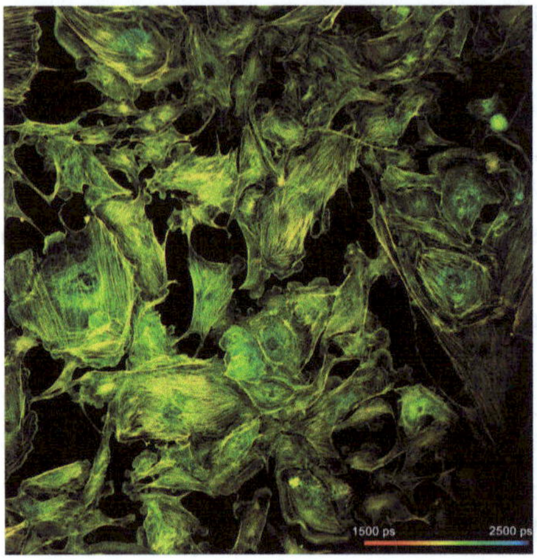

Fig. 3.14: BPAE sample (Invitrogen) scanned with 2048 × 2048 pixels and 256 time channels. Excitation at 473 nm, detection from 485 to 560 nm. Becker & Hickl DCS-120 system, image from [105].

3.4.2 Multiwavelength FLIM

Multiwavelength FLIM uses a detector that determines the wavelength of the individual photons, and, in addition to the single-photon pulse, generates a wavelength-channel number for the individual photons. Typically, the wavelength is encoded in a 4-bit data word, corresponding to 16 wavelength channels. For each photon, the wavelength channel word is transferred into the TCSPC device together with the photon time. The recording process uses the wavelength information as an additional coordinate of the photon distribution. Please see [29, 68, 69] for details.

Compared with single-wavelength FLIM, multiwavelength FLIM requires 16 times more data space. Although multiwavelength FLIM is around since 2002 [101] it reached its real performance only with the introduction of 64-bit software.

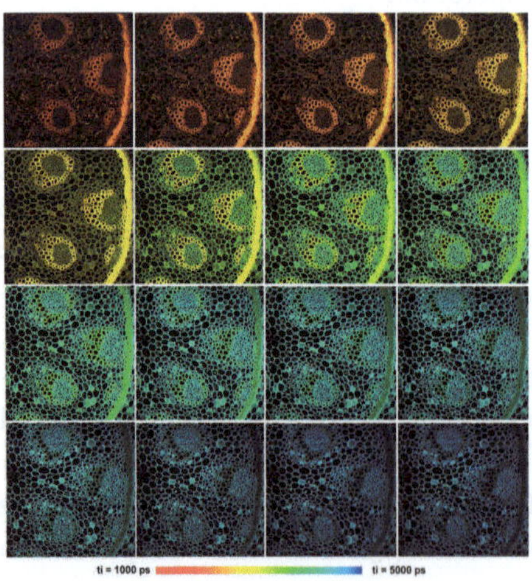

Fig. 3.15: Multiwavelength FLIM with a bh MW-FLIM GaAsP 16-channel detector. 16 images with 512 × 512 pixels and 256 time channels were recorded simultaneously. Wavelength from upper left to lower right, 490 nm to 690 nm, 12.5 nm per image. DCS-120 confocal scanner, Zeiss Axio Observer microscope, ×20 NA = 0.5 air lens.

An example of multiwavelength FLIM obtained by 64-bit data acquisition software is shown in Fig. 3.15. The data were recorded by a bh DCS-120 confocal scanning system with a bh MW-FLIM GaAsP 16-channel detector. The data were recorded in 16 wavelength channels at a resolution of 512 × 512 pixels and 256 time channels per pixel. The wavelength channel width is 12.5 nm, the center wavelengths of the channels range from 490 nm to 690 nm. The file size of the entire multiwavelength data is 2 GB.

Fig. 3.16, left demonstrates the true spatial resolution of the data. The image from one wavelength channel, 565 nm, of the data shown in Fig. 3.15 is displayed at larger scale and with individually adjusted lifetime range. The spatial resolution is comparable with what previously was reached for FLIM at a single wavelength. Nevertheless, the decay data are recorded at a temporal resolution of 256 time channels. The decay curve for an arbitrary selected pixel of the image is shown in Fig. 3.16, right.

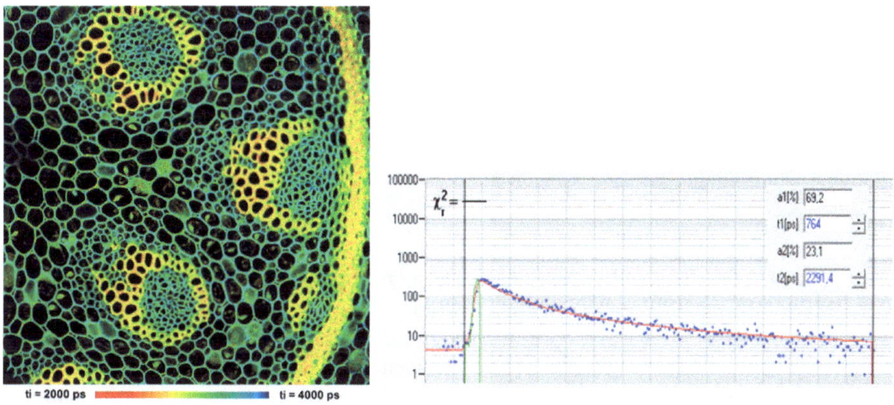

Fig. 3.16: Single image from the data array shown in Fig. 3.15, displayed at larger scale and with individually adjusted lifetime range. Wavelength channel 565 nm. The image has 512 × 512 pixels and 256 time channels.

3.4.3 *X*-*Y* mosaic FLIM

To record images larger than the maximum field of view of the microscope lens laser scanning microscopes use a 'tile imaging' function. The microscope scans an image at one position of the sample, then offsets the sample by the size of the scan area, and scans a new image. The process is repeated, and the images of the individual frames are stitched together. This way, a sample area larger than the field of view of the microscope lens is imaged.

The large memory size available in the 64-bit environment allows tile imaging to be combined with FLIM. In the memory for the FLIM data, a two-dimensional mosaic of FLIM data blocks is defined. The number and arrangement of the data blocks is the same as for the tiles in the imaging procedure of the microscope. Moreover, for each element of the FLIM mosaic a number of frames is defined which is identical with the number of frames per tile defined in the microscope. When the microscope starts the scan the FLIM system starts to record an image in the first mosaic element. It continues to do so until the defined number of frames has been completed, and then switches to the next mosaic element. This way, the FLIM system switches through all mosaic elements, filling all of them with data of the corresponding tiles [29]. The procedure ends when the microscope has completed the last frame of the last tile.

An example of *X*-*Y* mosaic FLIM is shown in Fig. 3.17. The image was recorded by a Zeiss LSM 710 Intune (tunable excitation) system in combination with a bh Simple-Tau 150 FLIM system. The mosaic has 4×4 elements, each element has 512×512 pixels with 256 time channels. The complete mosaic has thus 2048×2048 pixels, each pixel holding 256 time channels. The entire mosaic covers an area of 2.5 mm × 2.5 mm.

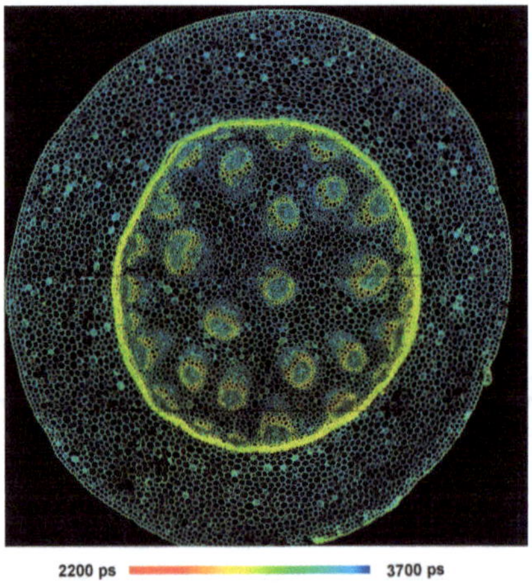

Fig. 3.17: Mosaic FLIM of a Convallaria sample. The mosaic has 4 × 4 elements, each element has 512 × 512 pixels, each pixel has 256 time channels. The total image size is 2048 × 2048 pixels, 256 time channels. From [29], Zeiss LSM 710 Intune system with bh Simple-Tau 150 FLIM system.

3.4.4 *Z*-stack FLIM by mosaic recording

Mosaic FLIM recording can be used to record a *Z*-stack of FLIM images. Instead of shifting the sample in *X* and *Y* the focal plane is shifted in *Z*. The number of mosaic elements is made identical with the number of *Z*-planes the microscope scans. The number of frames per mosaic element is selected according to the number of frames per *Z*-plane [29]. Compared with a *Z*-stack FLIM procedure by a record-and-save procedure mosaic FLIM has the advantage that the synchronization between the microscope and the FLIM system is easier, and no time has to be reserved for the saving actions. Moreover, the data of all *Z*-planes are recorded in one and the same photon distribution. The data analysis can therefore be run for the data of all *Z*-planes together. This makes it easier to maintain exactly the same fit conditions and model parameters for all *Z*-planes, and to extract global parameters from the entire array. An example of a *Z*-stack recording by mosaic FLIM is shown in Fig. 3.18.

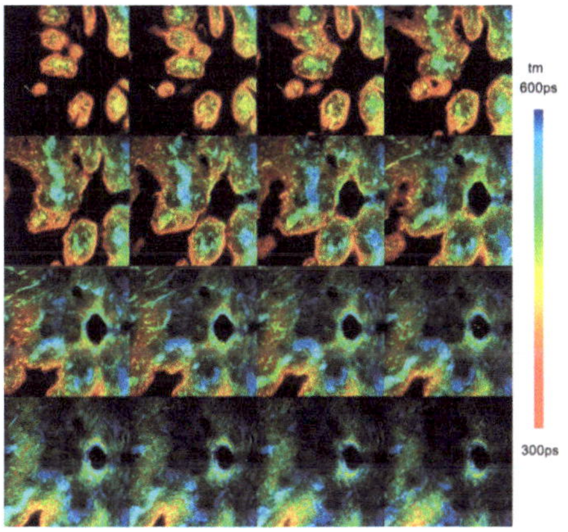

Fig. 3.18: *Z*-stack recorded by mosaic FLIM, SPCM 64-bit software. Pig skin stained with Indocyanin Green. Zeiss LSM 7 MP OPO system, 16 planes from 0 to 60 µm from top of sample, each plane 512 × 512 pixels, 256 time channels. Amplitude-weighted lifetime of double-exponential decay. Depth increases from upper left to lower right. Image from [102].

3.4.5 Temporal mosaic FLIM

Mosaic FLIM provides a new, extremely efficient way to record fast FLIM time series. The sample is scanned repetitively, and the photons of subsequent frames (or several accumulated frames) are recorded in consecutive elements of the FLIM mosaic. The result is a time series, in which each time step is the frame time or a multiple of it.

An example is shown in Fig. 3.19. The sample was a fresh moss leaf, the microscope a bh DCS-120 confocal FLIM system. The time runs from lower left to upper right. The excitation light causes a nonphotochemical transient in the chlorophyll fluorescence [91, 92], resulting in a decrease in the fluorescence lifetime with the time of exposure.

Mosaic time series recording has several advantages over the conventional record-and-save procedure. First, the transition from one mosaic element to the next occurs instantaneously. No time has to be reserved for saving the data of the individual steps of the series. Moreover, the result is not a sequence of files but a single, large photon

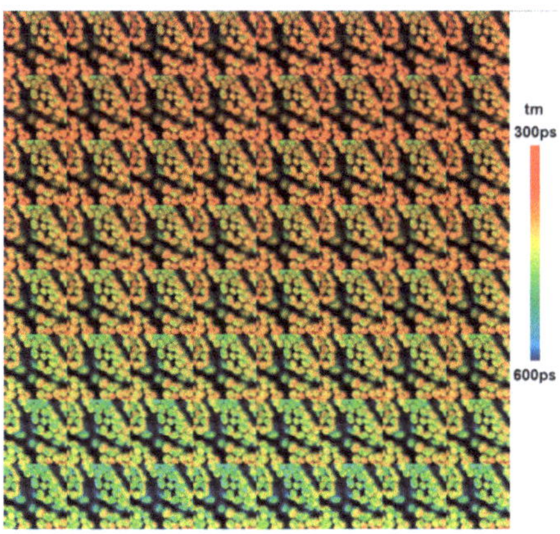

Fig. 3.19: Time series recorded by mosaic imaging. Nonphotochemical transients in chloroplasts of a moss leaf. 64 mosaic elements for consecutive times after turn-on of excitation light. Acquisition time per element 1 second, total time of sequence 64 seconds, image size of each element 128 × 128 pixels, 256 time channels. Time runs from lower left to upper right. bh DCS-120 confocal FLIM system with SPC-150 TCSPC modules, SPCImage data analysis software.

distribution. The data can thus be analyzed by a single data analysis run. This makes it easier to use fit procedures with global parameters.

The most important advantage is, however, that mosaic time series data can be accumulated. A sample would be repeatedly stimulated by an external event, and the start of the mosaic recording be triggered by the stimulation. With every new stimulation the recording procedure runs through all elements of the mosaic, and accumulates the photons. Accumulation allows data to be recorded without the need for trading photon number and lifetime accuracy against speed. The signal-to-noise ratio just depends on the total acquisition time of the experiment. Consequently, the time per step of the series (i.e., per mosaic element) is only limited by the minimum frame time of the scanner. For pixel numbers on the order of 128 × 128 pixels and small scan areas galvanometer scanners reach frame times on the order of 50 milliseconds. Resonance scanner are even faster. This brings the time series resolution into the range where even lifetime changes induced by Ca^{2+} transients in neurons can be recorded.

Fig. 3.20 shows a mosaic FLIM recording of the Ca^{2+} transient in cultured neurons after stimulation with an electrical signal. Oregon Green was used as a calcium sensor. A Zeiss LSM 7 MP multiphoton microscope was used to scan the sample. The scan format was 64 × 64 pixels. With 64 × 64 pixels and a zoom factor of 5, the LSM 7 MP reaches a frame time of 38 ms. The FLIM data were recorded by a bh SPC-150 module. A FLIM mosaic size of 8 × 8 elements was defined. The stimulation period was 3 seconds. 150 milliseconds before every stimulation a recording through the entire 64-element mosaic was started. With the frame time of 38 ms, the acquisition thus runs through the entire mosaic in 2.43 seconds. 100 such acquisition runs were accumulated. The result shows clearly the increase in the fluorescence lifetime of the Ca^{2+} sensor in the mosaic elements 4 to 6, and a return to the resting state over the next 10 to 15 mosaic elements (380 to 570 ms).

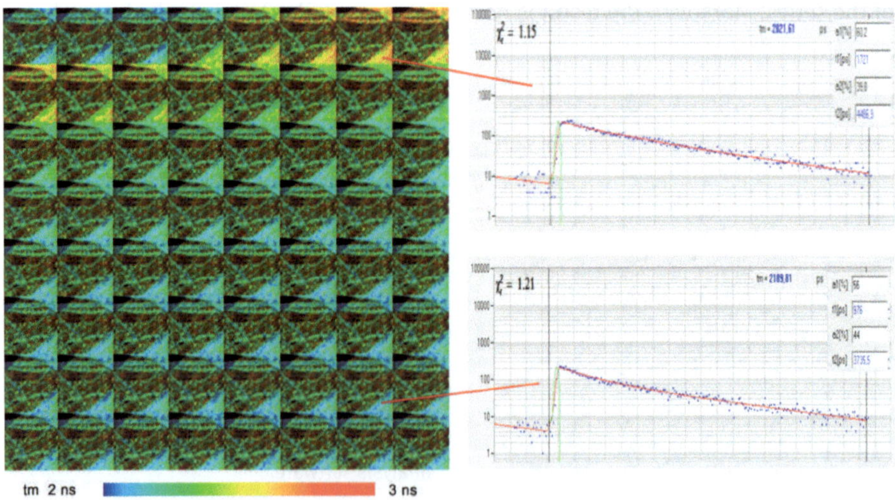

3.4.6 Simultaneous fluorescence and phosphorescence lifetime imaging

Phosphorescence is a slow process, with lifetimes on the order of microseconds or even milliseconds. A commonly used approach to phosphorescence recording is to reduce the excitation pulse rate and extend the timescale of decay recording to microseconds or milliseconds. Except for fluorophores of high intersystem crossing rate and for high fluorophore concentration, the results obtained this way are disappointing.

There are two reasons why simply decreasing the repetition rate does not work well. Both are related to the low S1–T1 transition rate. The excitation energy injected into the molecules is preferentially dissipated by the S1-S0 transition, and not deposited in the T1 state. The T1 population therefore remains low, and so does the phosphorescence intensity. Simply increasing the peak power of the excitation pulse meets technical constraints, or causes nonlinear effects in the sample. The second problem is that high excitation peak power also causes high peak intensity of fluorescence. It can become so strong that it causes temporary overload in the detector. These problems can be solved by exciting phosphorescence by several laser pulses instead of a single one. The principle is shown in Fig. 3.21.

A high-frequency pulsed laser is modulated on/off at a period a few times longer than the phosphorescence lifetime. The laser pulses within the 'on' phase of the modulation period excite fluorescence and build up phosphorescence. The 'off' phase contains only phosphorescence. The advantage of this excitation principle is not only that it can be used for multiphoton excitation but also that it can be used *to record fluores-*

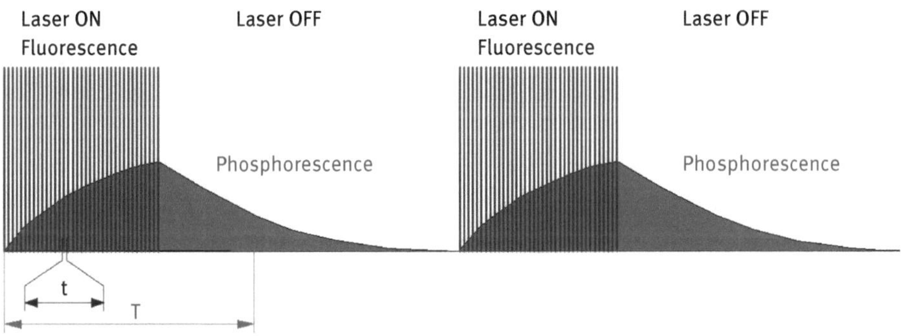

Fig. 3.21: Excitation principle for combined FLIM/PLIM.

◂ **Fig. 3.20:** Temporal mosaic FLIM of the Ca^{2+} transient in cultured neurons after stimulation with an electrical signal. The time per mosaic element is 38 milliseconds, the entire mosaic covers 2.43 seconds. Experiment time runs from upper left to lower right. Photons were accumulated over 100 stimulation periods. Recorded by Zeiss LSM 7 MP and bh SPC-150 TCSPC module. Data courtesy of Inna Slutsky and Samuel Frere, Tel Aviv University, Sackler Faculty of Medicine.

cence and phosphorescence decay data simultaneously. This is achieved by assigning two times to every photon, one of which is the time in the laser pulse period, t, the other the time in the laser modulation period, T. Two separate photon distributions are built up, one over t, the other over T. The distribution over t is the fluorescence decay, the distribution over T the phosphorescence decay [29, 59]. Please see [29] for further technical details.

An example of a FLIM/PLIM measurement is shown in Fig. 3.22. The figure shows yeast cells stained with tris (2,2′-bipyridyl) dichlororuthenium (II) hexahydrate. The FLIM image is shown in Fig. 3.22, left. It is dominated by autofluorescence of the cells. PLIM is shown in Fig. 3.22, right. The phosphorescence comes from the ruthenium dye. Decay curves in selected pixels are shown at the bottom.

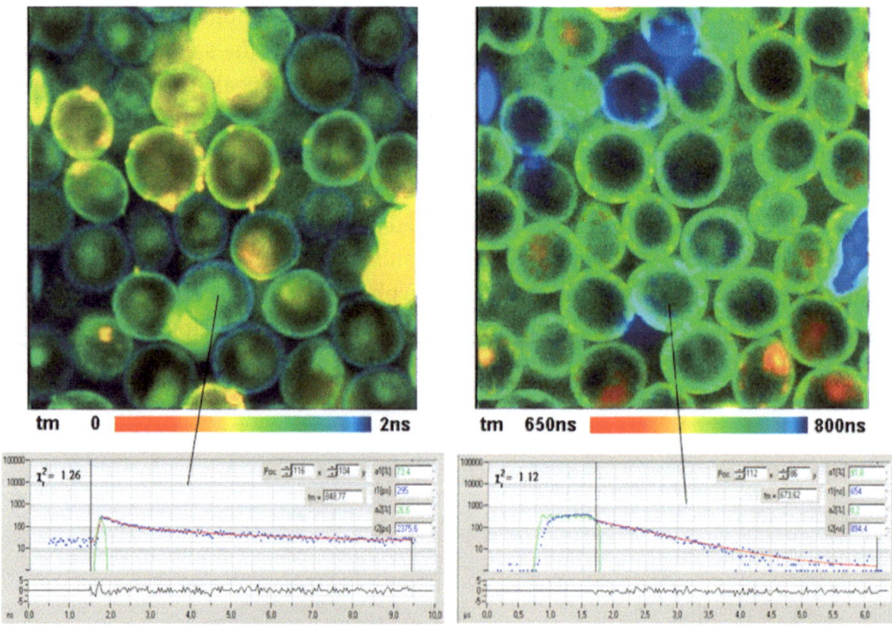

Fig. 3.22: Combined FLIM/PLIM recording. Yeast cells stained with a ruthenium dye. FLIM (left) shows autofluorescence, PLIM (right) phosphorescence of the ruthenium dye.

PLIM applications aim at suppression of autofluorescence [103, 104], detection of nanoparticles in biological tissue [82], and, importantly, oxygen-concentration sensing [95]. Phosphorescence from almost any phosphorescing compound is strongly quenched by oxygen. The phosphorescence lifetime can therefore be used to determine oxygen concentrations [55, 93–96]. In these applications, the FLIM/PLIM technique described above is able to record metabolic information via FLIM while monitoring the oxygen concentration via PLIM.

3.5 Summary

TCSPC-FLIM combines single-photon sensitivity, near-ideal photon efficiency, high time resolution, and recording of the full decay profiles in the pixels with high-contrast imaging, depth resolution, and suppression of laterally and longitudinally scattered light. Due to this unique combination of features the technique has found wide application in imaging of molecular environment parameters, protein interaction experiments by FRET techniques, and autofluorescence imaging. The first clinical instruments based on TCSPC-FLIM are on the market.

In the past 10 years, 64-bit computer technology and increased detector efficiency have resulted in substantial improvements in the performance of FLIM. Images of the entire field of view of the microscope lens can now be recorded without compromising spatial or temporal resolution. Multiwavelength FLIM can be recorded without restriction in the number of pixels or time channels. The large memory space available in the 64-bit environment has led to entirely new multidimensional recording procedures like lateral mosaic FLIM, fast Z-stack FLIM, and fast time series FLIM. By keeping the data of a large number of individual FLIM recordings in the memory, time series FLIM can be combined with periodic stimulation of the sample and data accumulation. Extremely fast time series can thus be recorded at high signal-to-noise ratio. A combined FLIM/PLIM technique based on modulation of a pulsed laser and dual-time-base recording opens new ways of metabolic imaging. Metabolic information can now be recorded together with oxygen concentration, and the effect of variation in oxygen supply can be directly studied.

References

[1] Becker W. Fluorescence lifetime imaging—techniques and applications. J Microsc. 2012;247(2).

[2] Berezin MY, Achilefu S. Fluorescence lifetime maesurement and biological imaging. Chem Rev. 2010;110(5):2641–2684.

[3] Lakowicz JR. Principles of fluorescence spectroscopy. 3rd edition. Springer; 2006.

[4] Roberts MS, Dancik Y, Prow TW, Thorling CA, Li L, Grice JE, Robertson TA, König K, Becker W. Non-invasive imaging of skin physiology and percutaneous penetration using fluorescence spectral and lifetime imaging with multiphoton and confocal microscopy. European Journal of Pharmaceutics and Biopharmaceutics. 2011;77:469–488.

[5] Becker W, Shcheslavkiy V, Frere S, Slutsky I. Spatially resolved recording of transient fluorescence-lifetime effects by line-scanning TCSPC. Microsc Res Techn. 2014;77:216–224.

[6] Kuchibhotla KV, Lattarulo CR, Hyman B, Bacskai BJ. Synchronous hyperactivity and intercellular calcium waves in astrocytes in Alzheimer mice. Science. 2009;323:1211–1215.

[7] Wilms CD, Schmidt H, Eilers J. Quantitative two-photon Ca^{2+} imaging via fluorescence lifetime analysis. Cell Calcium. 2006;40:73–79.

[8] Funk K, Woitecki A, Franjic-Würtz C, Gensch T, Möhrlein F, Frings S. Modulation of chloride homeostasis by inflammatory mediators in dorsal ganglion neurons. Molecular Pain. 2008;4:32.

[9] Gensch T, Untiet V, Franzen A, Kovermann P, Fahlke C. Determination of intracellular chloride concentrations by fluorescence lifetime imaging. In: Becker W, editor. Advanced time-correlated single photon counting applications. Berlin, Heidelberg, New York: Springer; 2015.
[10] Gilbert D, Franjic-Würtz C, Funk K, Gensch T, Frings S, Möhrlen F. Differential maturation of chloride homeostasis in primary afferent neurons of the somatosensory system. Int J Devl Neuroscience. 2007;25:479–489.
[11] Kaneko H, Putzier I, Frings S, Kaupp UB, Gensch T. Chloride accumulation in mammalian olfactory sensory neurons. J Neurosci. 2004;24:7931–7938.
[12] Hanson KM, Behne MJ, Barry NP, Mauro TM, Gratton E. Two-photon fluorescence imaging of the skin stratum corneum pH gradient. Biophys J. 2002;83:1682–1690.
[13] Sanders R, Draaijer A, Gerritsen HC, Houpt PM, Levine YK. Quantitative pH Imaging in cells using confocal fluorescence lifetime imaging microscopy. Analytical Biochemistry. 1995;227:302–308.
[14] Anderssen RM, Carlsson K, Liljeborg A, Brismar H. Characterization of probe binding and comparison of its influence on fluorescence lifetime of two ph-sensitivy benzo[c]xantene dyes using intensity-modulated multiple-wavelength scanning technique. Analytical Biochemistry. 2000;283:104–110.
[15] Knemeyer JP, Marmé N, Sauer M. Probes for detection of specific DNA sequences at the single-molecule level. Anal Chem. 2002;72:3717–3724.
[16] Van Zandvoort MAMJ, de Grauw CJ, Gerritsen HC, Broers JLV, Egbrink MGA, Ramaekers FCS, Slaaf DW. Discrimination of DNA and RNA in cells by a vital fluorescent probe: Lifetime imaging of SYTO13 in healthy and apoptotic cells. Cytometry. 2002;47:226–232.
[17] Benesch J, Hungerford G, Suhling K, Tregidgo C, Mano JF, Reis RL. Fluorescence probe techniques to monitor protein adsorption-induced conformation changes on biodegradable polymers. Journal of Colloid and Interface Science. 2007;312:193–200.
[18] Saxl T, Khan F, Matthews DR, Zhi ZL, Rolinski O, Ameer-Beg S, Pickup J. Fluorescence lifetime spectroscopy and imaging of nano-engineered glucose sensor microcapsules based on glucose/galactose-binding protein. Biosensors and Bioelectronics. 2009;24:3229–3234.
[19] Ghukasyan V, Hsu CC, Liu CR, Kao FJ, Cheng TH. Fluorescence lifetime dynamics of enhanced green fluorescent protein aggregates with expanded polyglutamine. J Biomed Opt. 2010;15(1) 016008-1–016008-11.
[20] Haluska CK, Schröder AP, Didier P, Heissler D, Duportail G, Mely Y, Marques CM. Combining fluorescence lifetime and polarization microscopy to discriminate phase separated domains in giant unilamellar vesicles. Biophys J. 2008;95:5737–5747.
[21] Carlini L, Nadeau JL. Uptake and processing of semiconductor quantum dots in living cells studied by fluorescence lifetime imaging microscopy (FLIM). Chem Commun (Camb). 2013;49(17):1714–1716 .
[22] Cooper DR, Kudinov K, Tyagi P, Hill CK, Bradforth SE, Nadeau JL. Photoluminescence of cerium fluoride and cerium-doped lanthanum fluoride nanoparticles and investigation of energy transfer to photosensitizer molecules. Phys Chem Chem Phys. 2014;16(24):12441–12453.
[23] Georgin M, Carlini L, Cooper D, Bradforth SE, Nadeau JL. Differential effects of β-mercaptoethanol on CdSe/ZnS and InP/ZnS quantum dots. Phys Chem Chem Phys. 2013;15(25):10418–10428.
[24] Zhang X, Shastry S, Bradforth SE, Nadeau JL. Nuclear uptake of ultrasmall gold-doxorubicin conjugates imaged by fluorescence lifetime imaging microscopy (FLIM) and electron microscopy. Nanoscale. 2014;7(1):240–251.
[25] Suhling K, Hirvonen LM, Levitt JA, Chung PH, Tregido C, le Marois A, Rusakov D, Zheng K. Fluorescence Lifetime Imaging (FLIM): Basic Concepts and Recent Applications. In: Becker W,

editor. Advanced time-correlated single photon counting applications. Berlin, Heidelberg, New York: Springer; 2015.

[26] Becker W, Shcheslavskiy V. Fluorescence lifetime imaging with near-infrared dyes. Photonics and Lasers in Medicine. 2015;4:73–83

[27] Förster T. Zwischenmolekulare Energiewanderung und Fluoreszenz. Ann Phys (Serie 6). 1948;2:55–75.

[28] Förster T. Energy migration and fluorescence. Translated by Klaus Suhling. J Biomed Opt. 2012;17: 011002-1–011002-10.

[29] Becker W. The bh TCSPC Handbook. 6th edition. Becker & Hickl GmbH; 2015.

[30] Periasamy A, Clegg RM, editors. FLIM microscopy in biology and medicine. CRC Press; 2009.

[31] Periasamy A, Mazumder N, Sun Y, Christopher KG, Day RN. FRET microscopy: basics, issues and advantages of FLIM-FRET imaging. In: Becker W, editor. Advanced time-correlated single photon counting applications. Berlin, Heidelberg, New York: Springer; 2015.

[32] Vogel SS, Thaler C, Koushik SV. Fanciful FRET. Sci STKE. 2006:re2.

[33] Schweitzer D, Schenke S, Hammer M, Schweitzer F, Jentsch S, Birckner E, Becker W. Towards metabolic mapping of the human retina. Micr Res Tech. 2007;70:403–409.

[34] Schweitzer D, Hammer M, Schweitzer F, Anders R, Doebbecke T, Schenke S, Gaillard ER. In vivo measurement of time-resolved autofluorescence at the human fundus. J Biomed Opt. 2004;9:1214–1222.

[35] Lakowicz JR, Szmacinski H, Nowaczyk K, Johnson ML. Fluorescence lifetime imaging of free and protein-bound NADH. PNAS. 1992;89:1271–1275.

[36] Paul RJ, Schneckenburger H. Oxygen concentration and the oxidation-reduction state of yeast: Determination of free/bound NADH and flavins by time-resolved spectroscopy. Naturwissenschaften. 1996;83:32–35.

[37] Bird DK, Yan L, Vrotsos KM, Eliceiri KE, Vaughan EM. Metabolic mapping of MCF10A human breast cells via multiphoton fluorescence lifetime imaging of coenzyme NADH. Cancer Res. 2005;65:8766–8773.

[38] Blacker TS, Mann ZF, Gale JE, Ziegler M, Bain AJ, Szabadkai G, Duchen MR. Separating NADH and NADPH fluorescence in live cells and tissues using FLIM. Nature Communications. 2014;5:3936-1–3936-6.

[39] Chorvat D, Chorvatova A. Multi-wavelength fluorescence lifetime spectroscopy: a new approach to the study of endogenous fluorescence in living cells and tissues. Laser Phys Lett. 2009;6:175–193.

[40] Chorvat D Jr, Mateasik A, Cheng Y, Poirier N, Miro J, Dahdah NS, A. Chorvatova. Rejection of transplanted hearts in patients evaluated by the component analysis of multi-wavelength NAD(P)H fluorescence lifetime spectroscopy. J Biophotonics. 2010;3:646–652.

[41] Chorvatova A, Elzwiei F, Mateasik A, Chorvat D. Effect of ouabain on metabolic oxidative state in living cardiomyocytes eavaluated by time-resolved spectroscopy of endogenous NAAD(P)H fluorescence. J Biomed Opt. 2012;17(10):101505-1–101505-7.

[42] Chorvatova A, Aneba S, Mateasik A, Chorvat D Jr, Comte B. Time-resolved fluorescence spectroscopy investigation of the effect of 4-hydroxynonenal on endogenous NAD(P)H in living cardiac myocytes. J Biomed Opt. 2013;18(6):067009-1–067009-11

[43] Ghukassian V, Kao FJ. Monitoring cellular metabolism with fluorescence lifetime of reduced nicotinamide adenine dinucleotide. J Phys Chem C. 2009;113:11532–11540.

[44] Skala MC, Riching KM, Bird DK, Dendron-Fitzpatrick A, Eickhoff J, Eliceiri KW, Keely PJ, Ramanujam N. In vivo multiphoton fluorescence lifetime imaging of protein-bound and free nicotinamide adenine dinucleotide in normal and precancerous epithelia. J Biomed Opt. 2007;12: 02401-1–02401-10.

[45] Skala MC, Riching KM, Gendron-Fitzpatrick A, Eickhoff J, Eliceiri KW, White JG, Ramanujam N. In vivo multiphoton microscopy of NADH and FAD redox states, fluorescence lifetimes, and cellular morphology in precancerous epithelia. PNAS. 2007;104:19494–19499.

[46] Walsh AJ, Shah AT, Sharick JT, Skala MC. Fluorescence Lifetime measurements of NADH in live cells and tissue. In: Becker W, editor. Advanced time-correlated single photon counting applications. Berlin, Heidelberg, New York: Springer; 2015.

[47] König K, Uchugonova A, Gorjup E. Multiphoton Fluorescence Lifetime Imaging of 3D-Stem Cell Spheroids During Differentiation. Microsc Res Techn. 2011;74:9–17.

[48] Sanchez WY, Prow TW, Sanchez WH, Grice JE, Roberts MS. Analysis of the metaboloic deterioration of ex-vivo skin, from ischemic necrosis, through the imaging of intracellular NAD(P)H by multiphoton tomography and fluorescence lifetime imaging microscopy (MPT-FLIM). J Biomed Opt. 2010;15:046008.

[49] Wang HW, Ghukassyan V, Chen CT, Wei YH, Guo HW, Yu JS, Kao FJ. Differentiation of appoptosis from necrosis by dynamic changes of reduced nicotinamide adenine dinucleotide fluorescence lifetime in live cells. J Biomed Opt. 2008;13(5):054011-1–054011-9.

[50] Dimitrow E, Riemann I, Ehlers A, Koehler MJ, Norgauer J, Elsner P, König K, Kaatz M. Spectral fluorescence lifetime detection and selective melanin imaging by multiphoton laser tomography for melanoma diagnosis. Experimental Dermatology. 2009;18:509–515.

[51] Koenig K. Clinical multiphoton tomography. J Biophoton. 2008;1:13–23.

[52] König K, Weinigel M, Hoppert D, Bückle R, Schubert H, Köhler MJ, Kaatz M, Elsner P. Multiphoton tissue imaging using high-NA microendoscopes and flexible scan heads for clinical studies and small animal research. J Biophoton. 2008;1:506–513.

[53] Koenig K, Uchugonova A. Multiphoton fluorescence lifetime imaging at the dawn of clinical application. In: Periasamy A, Clegg RM, editors. FLIM microscopy in biology and medicine. CRC Press; 2009.

[54] Gerritsen HC, Sanders R, Draaijer A, Levine YK. Fluorescence lifetime imaging of oxygen in cells. J Fluoresc. 1997;7:11–16.

[55] Kurokawa H, Ito H, Inoue M, Tabata K, Sato Y, Yamagata K, Kizaka-Kondoh S, Kadonosono T, Yano S, Inoue M, Kamachi T. High resolution imaging of intracellular oxygen concentration by phosphorescence lifetime. Scientific Reports. 2015;5:1–13.

[56] Lebedev AY, Cheprakov AV, Sakadzic S, Boas DA, Wilson DF, Vinogradov SA. Dendritic phosphorescent probes for oxygen imaging in biological systems. Applied Materials & Interfaces. 2009;1:1292–1304.

[57] Papkovsky, Zhdanov AV, Fercher A, Dmitriev RI, Hynes J. Phosphorescent oxygen-sensitive probes. Springer; 2012.

[58] Papkovsky DB, Dmitriev RI. Biological detection by optical oxygen sensing, Chem Soc Rev. 2013;42:8700–8732.

[59] Becker W, Su B, Bergmann A, Weisshart K, Holub O. Simultaneous Fluorescence and Phosphorescence Lifetime Imaging. Proc SPIE 7903; 2011. p. 790320.

[60] Avezov E, Cross BCS, Kaminski Schierle GS, Winters M, Harding HP, Pinho Melo E, Kaminski CF, Ron D. Lifetime imaging of a fluorescent protein sensor reveals surprising stability of ER thiol redox. J Cell Biol. 2013;201:337–349.

[61] Hanson GT, Aggeler R, Oglesbee D, Cannon M, Capaldi RA, Tsien RY, Remington SJ. Investigating mitochondrial redox potential with redox-sensitivegreen fluorescent protein indicators. J Biol Chem. 2004;279:13044–13053.

[62] Minsky M, inventor. United States patent US 3013467. 1957.

[63] Minsky M. Memoir on inventing the confocal microscope. Scanning. 1988;10:128–138.

[64] Denk W, Strickler JH, Webb WWW. Two-photon laser scanning fluorescence microscopy. Science. 1990;248:73–76.

[65] Wilson T, Sheppard C. Theory and practice of scanning optical microscopy. London: Academic Press; 1984.

[66] Bollinger LM, Thomas GE. Measurement of the time tependence of scintillation intensity by a delayed coincidence method. Rev Sci Instrum. 1961;32:1044–1050.

[67] O'Connor DV, Phillips D. Time-correlated single photon counting. London: Academic Press; 1984.

[68] Becker W. Advanced time-correlated single-photon counting techniques. Berlin, Heidelberg, New York: Springer; 2005.

[69] Becker W. Introduction to multi-dimensional TCSPC. In: Becker W, editor. Advanced time-correlated single photon counting applications. Berlin, Heidelberg, New York: Springer; 2015.

[70] Becker W, Shcheslavskiy V, Studier H. TCSPC FLIM with different optical scanning techniques. In Becker W, editor. Advanced time-correlated single photon counting applications. Berlin, Heidelberg, New York: Springer; 2015.

[71] Becker W, Benndorf K, Bergmann A, Biskup C, König K, Tirlapur U, Zimmer T. FRET measurements by TCSPC laser scanning microscopy. Proc SPIE 4431; 2001. p. 94–98.

[72] Becker W, Bergmann A, König K, Tirlapur U. Picosecond fluorescence lifetime microscopy by TCSPC imaging. Proc SPIE 4262; 2001. p. 414–419.

[73] Becker W, Bergmann A, Hink MA, König K, Benndorf K, Biskup C. Fluorescence lifetime imaging by time-correlated single photon counting. Micr Res Techn. 2004;63:58–66.

[74] Gerritsen HC, Asselbergs MAH, Agronskaia AV, van Sark WGJHM. Fluorescence lifetime imaging in scanning microscopes: acquisition speed, photon economy and lifetime resolution. J Microsc. 2002;206:218–224.

[75] Göppert-Mayer M. Über Elementarakte mit zwei Quantensprüngen. Ann Phys. 1931;9:273–294.

[76] Hell SW, Wichmann J. Breaking the diffraction resolution limit by stimulated emission: stimulated-emission-depletion fluorescence microscopy. Opt Lett. 1994;19:780–782.

[77] Klar TA, Hell SW. Subdiffraction resolution in far-field fluorescence microscopy. Opt Lett. 1999;24:954–956.

[78] Bückers J, Wildanger D, Vicidomini G, Kastrup L, Hell SW. Simultaneous multi-lifetime multi-color STED imaging for colocalization anlysis. Opt Expr. 2011;19:3130–3143.

[79] Lesoine MD, Bose S, Petrich JW, Smith EA. Supercontinuum stimulated emission depletion fluorescence lifetime imaging. J Phy Chem B. 2012;116:7821–7826.

[80] König K, So PTC, Mantulin WW, Tromberg BJ, Gratton E. Two-Photon excited lifetime imaging of autofluorescence in cells during UVA and NIR photostress. J Microsc. 1996;183:197–204.

[81] König K. Multiphoton microscopy in life sciences. J Microsc. 2000;200:83–104.

[82] Sanchez WY, Pastore M, Haridass I, König K, Becker W, Roberts MS. Fluorescence Lifetime Imaging of the Skin. In: Becker W, editor. Advanced time-correlated single photon counting applications. Berlin, Heidelberg, New York: Springer; 2015.

[83] Roberts MS, Dancik Y, Prow TW, Thorling CA, Li L, Grice JE, Robertson TA, König K, Becker W. Non-invasive imaging of skin physiology and percutaneous penetration using fluorescence spectral and lifetime imaging with multiphoton and confocal microscopy. European Journal of Pharmaceutics and Biopharmaceutics. 2011;77:469–488.

[84] Schweitzer D, Hammer M. Fluorescence Lifetime Imaging in Ophthalmology. In: W. Becker, editor. Advanced time-correlated single photon counting applications. Berlin, Heidelberg, New York: Springer; 2015.

[85] Dysli C, Quellec G, Abegg M, Menke MN, Wolf-Schnurrbusch U, Kowal J, Blatz J, La Schiazza O, Leichtle AB, Wolf S, Zinkernagel MS. Quantitative analysis of fluorescence lifetime measurements of the macula using the fluorescence lifetime imaging ophthalmoscope in healthy subjects. IOVS. 2014;55:2107–2113.

[86] Jentsch S, Schweitzer D, Schmidtke KU, Peters S, Dawczynski J, Bär KJ, Hammer M. Retinal fluorescence lifetime imaging ophthalmoscopy measures depend on the severity of Alzheimer's disease. Acta Ophthalmol. 2014;93(4):e241-7
[87] Schweitzer D. Quantifying fundus autofluorescence. In: Lois N, Forrester JV, editors. Fundus autofluorescence. Wolters Kluwer, Lippincott Willams & Wilkins; 2009.
[88] Schweitzer D. Metabolic Mapping. In: Holz FG, Spaide RF, editors. Medical retina, Essential in Opthalmology. Springer; 2010.
[89] Schweitzer D, Quck S, Klemm M, Hammer M, Jentsch S, Dawczynski J. Zeitaufgelöste Autofluoreszenz bei retinalen Gefäßverschlüssen. Der Ophthalmologe. 2010;12:1145–1152.
[90] Wabnitz H, Mazurenka M, Di Sieno L, Boso G, Becker W, Fuchs K, Contini D, Dalla Mora A, Tosi A, Macdonald R, Pifferi A. Time-domain diffuse optical imaging of tissue by non-contact scanning. In: Becker W, editors. Advanced time-correlated single photon counting applications. Berlin, Heidelberg, New York: Springer; 2015.
[91] Govindjee. Sixty-three years since Kautsky: chlorophyll α fluorescence. Aust J Plant Physiol. 1995;22:131–160.
[92] Kautsky H, Hirsch A. Neue Versuche zur Kohlensäureassimilation. Naturwissenschaften. 1931;19:964.
[93] Dmitriev RI, Zhdanov AV, Nolan YM, Papkovsky DB. Imaging of neurosphere oxygenation with phosphorescent probes. Biomaterials. 2013;34:9307–9317.
[94] Dmitriev RI, Kondrashina AV, Koren K, Klimant I, Zhdanov AV, Pakan JMP, McDermott KW, Papkovsky DB. Small molecule phosphorescent probes for O2 imaging in 3D tissue models. Biomater. Sci. 2014;2:853–866.
[95] Jenkins J, Dmitriev RI, Papkovsky DB. Imaging Cell and Tissue O_2 by TCSPC-PLIM. In: Becker W, editor. Advanced time-correlated single photon counting applications. Berlin, Heidelberg, New York: Springer; 2015.
[96] Sakadžic S, Roussakis E, Yaseen MA, Mandeville ET, Srinivasan VJ, Arai K, Ruvinskaya S, Devor A, Lo EH, Vinogradov SA, Boas DA. Two-photon high-resolution measurement of partial pressure of oxygen in cerebral vasculature and tissue. Nature Methods. 2010;7(9):755–759.
[97] Auksorius E, Boruah BR, Dunsby C, Lanigan PMP, Kennedy G, Neil MAA, French PMW. Stimulated emission depletion microscopy with a supercontinuum source and fluorescence lifetime imaging. Opt Lett. 2008;33:113–115.
[98] Dimitrow E, Riemann I, Ehlers A, Koehler MJ, Norgauer J, Elsner P, König K, Kaatz M. Spectral fluorescence lifetime detection and selective melanin imaging by multiphoton laser tomography for melanoma diagnosis. Experimental Dermatology. 2009;18:509–515.
[99] Becker W, Su B, Weisshart K, Holub O. FLIM and FCS detection in laser-scanning microscopes: increased efficiency by GaAsP hybrid detectors. Micr Res Tech. 2011;74:804–811.
[100] Studier H, Becker W. Megapixel FLIM. Proc SPIE 8948; 2014.
[101] Becker W, Bergmann A, Biskup C, Zimmer T, Klöcker N, Benndorf K. Multi-wavelength TCSPC lifetime imaging. Proc SPIE 4620; 2002. p. 79–84.
[102] Becker & Hickl GmbH. Modular FLIM systems for Zeiss LSM 710/780/880 family laser scanning microscopes. User handbook. 6th edition; 2015.
[103] Baggaley E, Gill MR, Green NH, Turton D, Sazanovich IV, Botchway SW, Smythe C, Haycock JW, Weinstein JA, Thomas JA. Dinuclear Ruthenium(II) complexes as two-photon, time-resolved emission microscopy probes for cellular DNA. Angew Chem Int Ed Engl. 2014;53:3367–3371.
[104] Baggaley E, Botchway SW, Haycock JW, Morris H, Sazanovich IV, Williams JAG, Weinstein JA. Long-lived metal complexes open up microsecond lifetime imaging microscopy under multiphoton excitation: from FLIM to PLIM and beyond. Chem Sci. 2014;5:879–886.
[105] Becker & Hickl GmbH. DCS-120 Confocal Scanning FLIM Systems. User handbook. 6th edition; 2015.

Arnd Krueger

4 Ultrafast lasers in biophotonics

Abstract: Ultrafast lasers, with pulse widths of a few picoseconds (10^{-12} s) or less, enable the time resolution required in laser-excited fluorescence lifetime spectroscopy and imaging techniques. Moreover, they deliver high pulse peak powers at high repetition rates and moderate average output powers. This unique characteristic of ultrafast pulses enables nonlinear optical interaction with tissue at low overall exposure, and drives applications such as deep-tissue multiphoton microscopy and high-precision eye surgery.

4.1 Introduction

In laser terminology the label "ultrafast" typically refers to laser pulses of a few picoseconds (1 ps = 10^{-12} s) or shorter. The typical average output power P_{ave} of ultrafast lasers used in biomedical applications is moderate (tens of milliwatts to a few watts), and the energy per pulse is rather low (microjoules or less). However, because of the extremely short duration the instantaneous peak power P_{peak} during each pulse is high. With τ and R indicating pulse width and repetition rate, P_{peak} is defined by:

$$P_{\mathrm{peak}} = \frac{1}{R\tau} P_{\mathrm{ave}}. \tag{4.1}$$

As a practical example, an ultrafast laser with a pulse width τ of 100 fs, and a repetition rate R of 80 MHz delivers with each pulse a peak power P_{peak} that is 5 orders of magnitude higher than its average power P_{ave}. The unique ability to deliver this peak power to tissue without exposing it to elevated pulse energies and average powers resulted in the development of a variety of biomedical applications that are not enabled by any other type of laser. Responding to end user requirements, laser manufacturers have engineered ultrafast lasers from complex laboratory systems into reliable and user-friendly devices. Consequently they have become ubiquitous in many commercial biomedical applications. In multiphoton microscopy, turn-key lasers delivering wavelength-tunable ultrafast pulses with energies in the nanojoule regime advance optical imaging to deeper tissue penetration while improving overall cell viability. In ophthalmology, ultrafast lasers with microjoule pulse energies are deployed for highly localized tissue ablation – facilitating high-precision surgical procedures by reducing heat deposition and collateral damage to the surrounding tissue. Before a more detailed summary of today's most important applications of ultrafast lasers in biophotonics is presented in Section 4.4, the technical aspects of ultrafast lasers will be reviewed: Section 4.2 will discuss some fundamentals of ultrafast laser pulses, including their generation through modelocking. In addition it will describe the effects

https://doi.org/10.1515/9783110429985-006

of dispersion on an ultrafast laser pulse propagating through an optical component (like a microscope objective), and potential compensation for these effects. Section 4.3 will continue with a review of a variety of different types of ultrafast lasers used in biomedical applications.

4.2 Fundamentals of ultrafast pulse generation and propagation

4.2.1 Modelocking

The generation of ultrafast laser pulses typically relies on continuous-wave (CW) modelocking, which was first demonstrated in 1964 [1]. Unless a laser is controlled to operate single-mode (or single-frequency), its output is a superposition of many longitudinal modes oscillating in the laser cavity (Fig. 4.1). Their wavelengths λ_S are defined by the standing wave condition, which for a linear cavity as in Fig. 4.1 is expressed by:

$$\lambda_S = \frac{2L}{s}, \tag{4.2a}$$

$$\nu_S = \frac{c}{\lambda_S} = s\frac{c}{2L} = s\nu_{cv}, \tag{4.2b}$$

where L is the cavity length of the laser, s an integer, and c the speed of light. ν_S represents the frequencies of the longitudinal modes, which are multiples of the cavity frequency $\nu_{cv} = c/(2L)$. The generation of ultrafast pulses requires a large spectral emission bandwidth with thousands of active modes. Typically they would behave like independent oscillators with randomly distributed relative phases. However, if in a way the phases are locked together so that all active modes oscillate synchronously, they interfere constructively at any given location within the cavity only once every round trip time T:

$$T = \frac{2L}{c}. \tag{4.3}$$

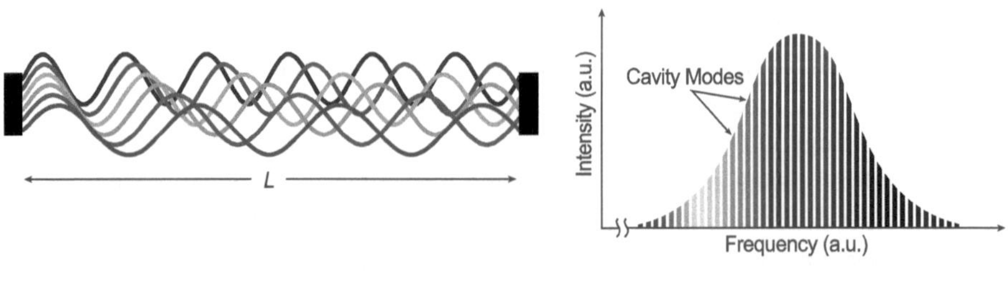

Fig. 4.1: Visualization of the standing wave condition (left). Unless a laser is controlled to operate single-mode (or single-frequency), its spectral output (right) is a superposition of many longitudinal modes oscillating in the laser cavity.

At any time in between, the amplitudes will cancel out. In other words, the active modes interfere in space and time to form a short pulse circulating around the cavity. Every time the circulating pulse reaches the output coupler, part of it escapes the cavity, resulting in a pulse train with a repetition rate equal to the cavity frequency:

$$R = \frac{1}{T} = \frac{c}{2L} = \nu_{cv}. \tag{4.4}$$

A typical repetition rate for a mode-locked laser is 80 MHz, which corresponds to a cavity length L of about 1.8 meters and a cavity roundtrip time (or period between two subsequent output pulses) T of about 12 nanoseconds.

Related to the Heisenberg uncertainty principle, the achievable pulse duration τ is inversely proportional to the spectral bandwidth $\Delta\nu$ of the laser:

$$\tau \propto \frac{1}{\Delta\nu}. \tag{4.5}$$

As an example, a 10 femtosecond laser pulse at a center wavelength of 800 nm has an associated spectrum spanning roughly 70 nanometers. Condition (4.5) implies that an ultrafast laser cavity needs to be designed to provide gain over a spectral range broad enough to support the pulse width for which it is intended. In particular, the bandwidth of the gain material's laser-active emission needs to be sufficiently wide.

4.2.1.1 Active and regenerative modelocking

Fig. 4.2 indicates the building blocks required in the design of an ultrafast laser. Besides standard cavity optics like the end mirrors and the gain material together with an efficient pump source, an ultrafast laser must integrate a modelocking mechanism and typically a device for controlling the optical dispersion within the resonator. Dispersion compensation will be discussed in Section 4.2.2, after a review of the modelocking techniques most commonly used in commercial ultrafast lasers.

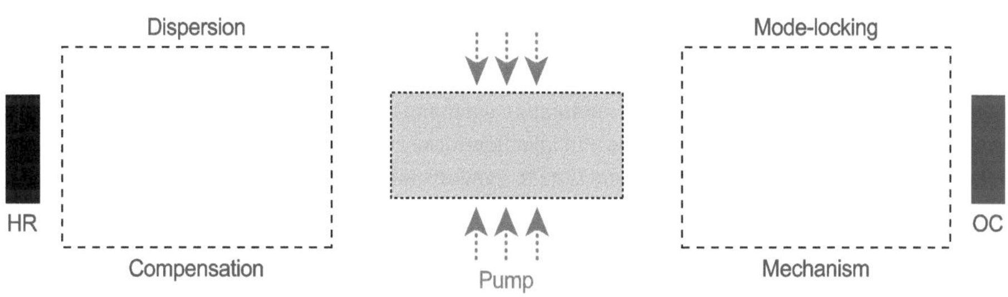

Fig. 4.2: Schematics of the basic building blocks typically required in an ultrafast laser. HR: high reflecting cavity end mirror, OC: output-coupling cavity end mirror.

Active modelocking typically employs an acousto-optic modulator (AOM) mounted close to one of the laser's end mirrors. A piezoelectric transducer driven by a radio frequency $\nu_{\mathrm{ML}} = \nu_{cv}/2$ generates a standing acoustic wave in the bulk of the AOM perpendicular to the intra-cavity laser beam. The induced pressure variations create a refractive index grating, which turns on and off at twice the radio frequency, i.e., at $2\nu_{\mathrm{ML}} = \nu_{cv}$. The AOM diffracts light away from the cavity beam when the grating is present and thus creates a modulated loss at the cavity frequency ν_{cv}. In the time domain, this loss is synchronized with the cavity roundtrips and does not affect a circulating pulse passing through the modulator when the loss is at its minimum – thus favoring pulsed, mode-locked operation over CW output. In the frequency domain, for each active longitudinal mode ν_{s} the periodic loss creates modulation sidebands at $\nu_{\mathrm{s}} \pm \nu_{cv}$, i.e., the frequencies of the neighboring modes ν_{s-1} and ν_{s+1}, respectively (Fig. 4.3). This process seeds and injection-locks all longitudinal modes to the same phase, so that they interfere in space and time to generate an ultrafast pulse.

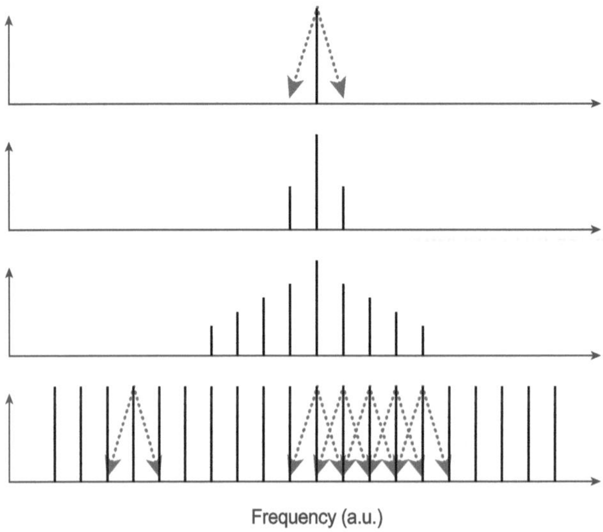

Fig. 4.3: Schematics of the principle of active modelocking. For each active longitudinal mode the periodic loss by the modulator creates modulation sidebands at precisely the frequencies of the neighboring modes. This process seeds and injection-locks all longitudinal modes to the same phase, so that they interfere in space and time to generate an ultrafast pulse.

As emphasized before, the loss modulation of the AOM needs to be precisely at the cavity frequency. This is automatically achieved by *regenerative modelocking*, removing the need for external adjustments. In regenerative modelocking the modulation frequency ν_{ML} for the AOM is derived directly from the laser cavity (Fig. 4.4). When the laser is running freely, i.e., without being mode-locked, mode beating will result in a weak modulation of the laser output at the cavity frequency υ_0. The mode beating

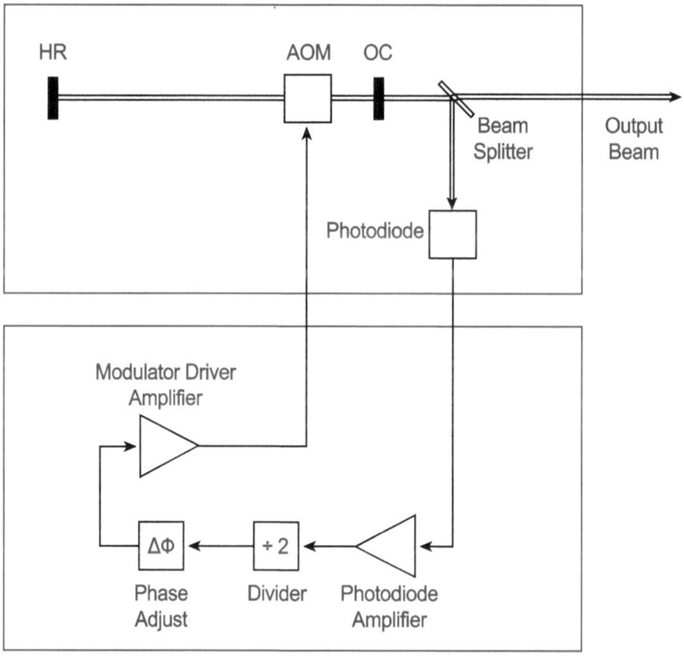

Fig. 4.4: Schematics of regenerative modelocking. The mode beating frequency is detected by a photodiode, divided by 2, and amplified to generate the drive signal for the modulator (AOM). As its frequency is derived directly from the laser cavity, the loss modulation is perfectly matched to the cavity frequency. HR: high reflecting cavity end mirror, OC: output-coupling cavity end mirror.

is detected by a photodiode and the filtered signal is divided by two and amplified to generate $\nu_{\mathrm{ML}} = \nu_{cv}/2$. As the radio frequency ν_{ML} is derived directly from the laser cavity, the loss modulation is always perfectly matched to the cavity frequency. Ultrafast pulsed operation is not only self-starting, but also reliably sustained. Regenerative modelocking is used in automated Ti:sapphire lasers, which deliver wavelength-tunable femtosecond pulses for applications like multiphoton microscopy.

4.2.1.2 Kerr-lens modelocking

Kerr-lens modelocking was discovered in 1991 [2] and became the first passive modelocking technique commercialized in solid state lasers. It is based on the optical Kerr effect, an instantaneous nonlinear interaction between the high-intensity ultrafast pulse circulating within the cavity and the gain crystal itself. As a practical approximation, the optical Kerr effect can be described by adding an intensity-dependent term to the refractive index n:

$$n'(\boldsymbol{r}, t) = n + n_2 I(\boldsymbol{r}, t), \tag{4.6}$$

where n_2 is the nonlinear refractive index and $I(\boldsymbol{r}, t)$ describes the intensity of the ultrafast pulse in space and time. For an ultrafast pulse circulating in a TEM_{00} cavity, the spatial intensity distribution $I(\boldsymbol{r})$ is symmetrical with respect to the axis of the resonator and highest in the center (Fig. 4.5). Provided n_2 is positive, the spatial refractive index variation induced by the Kerr effect acts like a lens, leading to *self-focusing* of the ultrafast pulse. The temporal index variation caused by the propagating pulse at any given location causes a time-dependent phase shift called *self-phase modulation*, which affects the spectral bandwidth of the pulse.

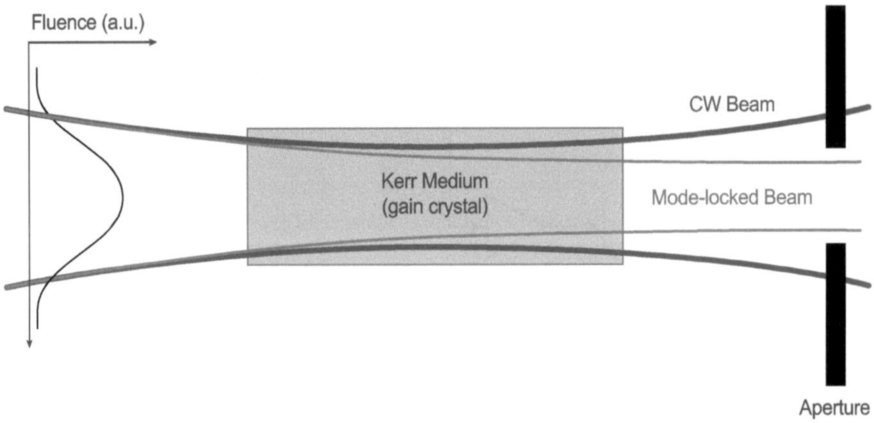

Fig. 4.5: Schematics of Kerr-lens modelocking. The self-focusing induced in the gain medium by the optical Kerr effect results in a lower cavity loss when the laser is mode-locked (pulsed), compared to CW operation.

Self-focusing is significant when the laser is modelocked and peak intensities are high, but it is negligible in CW operation. Kerr-lens modelocking relies on that difference. In *hard aperture Kerr-lens modelocking*, a pinhole in the laser cavity increases the loss for CW operation, while barely obstructing the modelocked beam due to its smaller radius at the aperture's location. In *soft aperture Kerr-lens modelocking*, the Kerr effect results in a better overlap of the intracavity beam with the pump beam in the crystal, resulting in higher gain for modelocked operation.

As a passive technique, Kerr-lens modelocking does not require drive electronics. It is self-sustaining. However, it is typically not self-starting and requires a starting mechanism, like an actuator that briefly moves one of the cavity mirrors. This disturbance of the cavity alignment induces fluctuations, which seed the circulating ultrafast pulse favored by the cavity design. Kerr-lens modelocking is mainly used in Ti:sapphire lasers, typically for generating ultrashort pulses (< 10 fs) at ≈ 800 nm, or for delivering wavelength-tunable femtosecond pulses.

4.2.1.3 Saturable absorber mirrors for modelocking

The use of a *semiconductor saturable absorber mirror* (SESAM) for modelocking a solid state laser was first reported in 1992 [3]. Today saturable absorber mirrors are integrated in many commercial ultrafast lasers as a robust modelocking mechanism. This section provides a short overview of the basic concept, detailed reviews can be found in [4, 5].

A saturable absorber mirror can be entirely made of layers of semiconducting materials and manufactured by semiconductor epitaxy using standard deposition techniques, such as metal-organic chemical vapor deposition (MOCVD). In its simplest form (sometimes called a saturable Bragg reflector or SBR), the bulk of the device is designed to form a Bragg reflector, a periodic structure of two alternating layers with different refractive indices (Fig. 4.6, based on [5]), each with the optical thickness of a quarter of the output wavelength for which the laser is designed. This structure, which is also referred to as a quarter-wave stack, forms a high-reflecting mirror. On its top a thin layer of semiconducting material is deposited, which acts as a saturable absorber for the targeted output wavelength. The whole device is typically covered by a dielectric anti-reflection coating. In a mode-locked laser the saturable absorber mirror replaces one of the cavity mirrors – in a standing wave cavity typically the high-reflecting end mirror. The saturable absorber introduces a significant cavity loss when exposed to the relatively low fluence in CW operation. However, the high peak fluence caused by an ultrafast pulse circulating in the cavity instantaneously and temporarily bleaches the absorber, minimizing the cavity loss for mode-locked operation (Fig. 4.7). Properly designed, modelocking is self-starting and self-sustaining. Because saturable absorber mirrors are optimized for operation at a certain wavelength, they are not utilized in broadly wavelength-tunable ultrafast lasers, such as Ti:sapphire oscillators. However, they are widely integrated in commercial mode-locked solid state or fiber oscillators utilizing rare earth-doped gain materials at output wavelengths at around 1050 nm (i.e. neodymium or ytterbium-doped) or 1550 nm (i.e. erbium-doped).

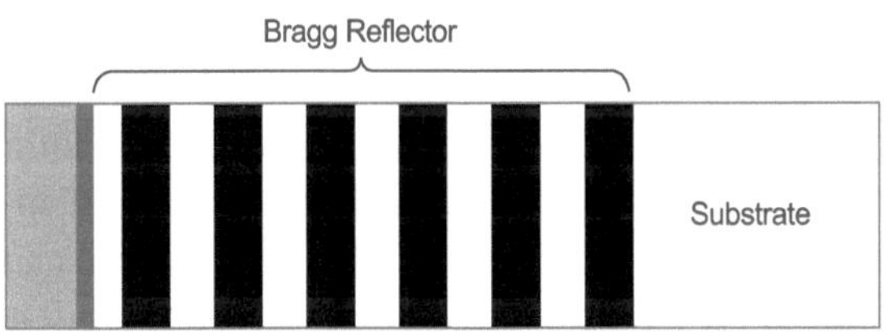

Fig. 4.6: Principle of a semiconductor saturable absorber mirror (schematic, based on [5]). The bulk consists of a quarter-wave stack of semiconductor layers, forming a high-reflecting Bragg mirror. A thin top layer (blue) forms a quantum well, which acts as a saturable absorber. A dielectric anti-reflection coating protects the top surface.

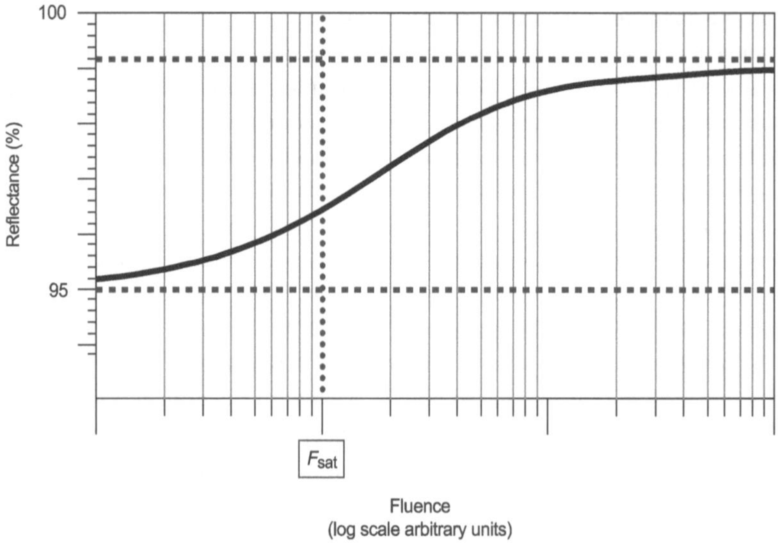

Fig. 4.7: The reflectance of a saturable absorber mirror changes with fluence (schematic). When integrated into a laser, the cavity loss is significant in CW operation (low fluence). However, the loss is low in mode-locked operation due to the high pulse peak intensities, which temporarily bleach the absorber.

Recently, single-walled carbon nanotubes (SWCNTs) have received considerable attention as saturable absorber material. Embedded in a polymer host surrounding a fiber taper, they have been used for modelocking ultrafast fiber lasers at 1 μm and 1.5 μm via evanescent field coupling [6, 7]. Compared to bulk saturable absorber mirrors this is an in-line device in fiber format, which facilitates integration into all-fiber-based ultrafast lasers [8].

4.2.2 The effect of dispersion on ultrafast pulses

As discussed in Section 4.2.1, an ultrafast pulse carries significant spectral bandwidth, which is inversely proportional to its pulse width. When propagating through an optical medium, the pulse is subjected to a frequency-dependent phase shift, which can dramatically alter the temporal profile of the pulse. Conventionally this phase shift is expanded into a Taylor series around the center spectral angular frequency $\omega_0 = 2\pi\nu_0$ of the spectral bandwidth:

$$\begin{aligned}\phi(\omega) &= \phi_0 + \left.\frac{\partial\phi}{\partial\omega}\right|_{\omega_0}(\omega-\omega_0) + \frac{1}{2}\left.\frac{\partial^2\phi}{\partial\omega^2}\right|_{\omega_0}(\omega-\omega_0)^2 + \frac{1}{6}\left.\frac{\partial^3\phi}{\partial\omega^3}\right|_{\omega_0}(\omega-\omega_0)^3 + \cdots \\ &\equiv \phi_0 + \phi_1(\omega_0)(\omega-\omega_0) + \frac{1}{2}\phi_2(\omega_0)(\omega-\omega_0)^2 + \frac{1}{6}\phi_3(\omega_0)(\omega-\omega_0)^3 + \cdots \end{aligned} \tag{4.7}$$

This approach allows a straightforward understanding of the effect of material dispersion on pulse propagation. The first term in (4.7) adds a constant value to the phase distribution, while the second one represents a phase shift that varies linearly and adds a delay to the pulse. Accordingly $T_g \equiv \phi_1$ is called the group delay. Both terms do not alter the temporal profile of the pulse. The third term introduces a frequency-dependent phase shift, which is proportional to the first derivative of the group delay $D_2 \equiv \phi_2 = (\partial T_g/\partial \omega)$. D_2 is measured in fs^2 and typically referred to as group delay dispersion or GDD [9]. GDD separates the frequency components of the pulse in time and causes pulse broadening. When D_2 is positive (called normal dispersion), spectral components with higher frequencies are delayed with respect to those with lower ones – the short ("*blue*") wavelengths within the pulse's optical spectrum trail the long ("*red*") ones, inducing a *positive linear chirp*. Considering a pulse with a Gaussian spectral distribution and no chirp, the pulse broadening caused by GDD can be calculated as [4, 10]:

$$\tau = \tau_0 \sqrt{1 + \left(4(\ln 2)\frac{D_2}{\tau_0^2}\right)^2}, \tag{4.8}$$

where τ_0 and τ are the pulse widths before and after propagation through an optical medium. Terms of higher order than GDD are of lesser value in equation (4.7), but can become more significant as the bandwidth increases. Only for very short ultrafast pulses well below 50 fs, the fourth term requires consideration when dealing with the effects of dispersive pulse broadening. As most ultrafast lasers utilized in biomedical applications have longer pulse width, this review will focus on compensating for pulse broadening caused by GDD and neglect the higher order terms.

Dispersion compensation is critical within ultrafast lasers as the pulse continuously circulates through the gain material and other dispersive intra-cavity components. Their intrinsic GDD (in combination with any self-phase modulation) induces a positive chirp on the pulse. This can only be ignored in some picosecond lasers such as mode-locked neodymium-doped vanadate lasers, where the effect is small enough ($|D_2/\tau_0^2| \ll 1$). Shorter pulse ultrafast lasers require intra-cavity dispersion compensation for stable pulse formation and pulse width optimization. This is accomplished by adding a source of negative GDD into the cavity, which cancels out the positive dispersion of the gain material and other intra-cavity components (Fig. 4.2). However, dispersion compensation is also critical extra-cavity in applications that require the ultrafast laser pulses to travel through a large amount of optically dispersive components before they reach their target. Here the addition of negative GDD to the beam path ensures that the pulses are as short as possible or necessary to induce the intended effects within the object of interest, such as a tissue sample. An important example will be discussed in Section 4.4.3.3.

4.2.2.1 Dispersion compensation

Proven methods for dispersion compensation involve optical components such as prisms, gratings, dispersive mirrors, or negative dispersion fibers. This chapter will review the most important ones utilized in and with ultrafast lasers in biophotonics.

A common method is based on the angular dispersion from wavelength-dependent refraction in optical prisms [11, 12], with the effect that the different spectral components of the ultrafast pulse travel on slightly different paths. The most convenient arrangement is the symmetrical four-prism sequence illustrated in Fig. 4.8, because it reconstructs the beam at the output by recombining the different wavelength components. Arranging the prisms close to Brewster's angle minimizes the optical losses and permits intra-cavity use. The negative GDD generated by the geometric effect depends on the distance between the prisms and is close to linear over a wide wavelength range. The prism glass itself adds positive dispersion. Therefore the overall negative GDD provided by the complete prism sequence can be fine-tuned by adjusting the prism spacing, or much more conveniently by translating the tips of the center prism pair (Pr2, Pr3) into the beam. As the overall intra-cavity GDD varies with the output wavelength, it is this flexibility that makes the prism compensator the perfect choice for wavelength-tunable femtosecond lasers. It is widely used for optimizing the pulse width of ultrafast Ti:sapphire lasers, either intra-cavity or externally in multiphoton microscopy applications. The prism sequence as illustrated in Fig. 4.8 can be simplified by replacing Pr3 and Pr4 with a back-reflecting mirror or periscope (positioned at the dashed line). Inside a laser cavity this is usually the high-reflecting end mirror. A periscope can be used in applications where input and output beam need to be spatially separated.

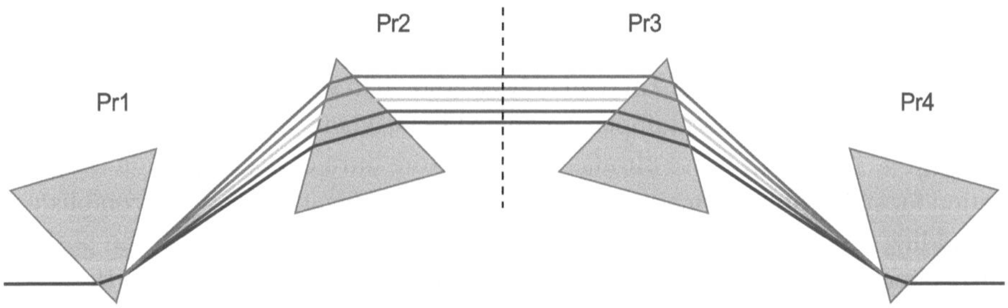

Fig. 4.8: Four-prism sequence for dispersion control. The optical path length *L* for longer (red) wavelengths is larger than for the shorter (blue) wavelengths. This geometric effect introduces negative GDD, which depends on the prism spacing. The overall negative GDD of the sequence can be fine-tuned most conveniently by translating the tips of the center prism pair (Pr2, Pr3) into the beam, which adds or removes positive GDD. The sequence can be reduced to two physical prisms by inserting a retro-reflector at the dashed line.

Spectral components are spatially separated between the prism Pr2 and Pr3, and travel on parallel paths. Translating a slit transversely through that region provides for convenient wavelength selection within a tunable femtosecond laser. This additional use of the dispersion compensator obviates the need for an additional intra-cavity component for wavelength tuning, such as a birefringent filter.

Another approach to dispersion compensation is based on the use of *dispersive* or *chirped mirrors* [13]. Typically, dielectric high-reflecting cavity mirrors are designed for maximum reflectivity at the center wavelength λ_0 of the laser output. They consist of a regular periodic structure of two alternating, $\lambda_0/4$ thick coatings, forming a Bragg reflector optimized for λ_0. In a dispersive mirror the thickness of the layers is varied across the stack of coatings, so that the shorter wavelengths within the bandwidth of a femtosecond pulse are reflected closer to the surface of the mirror, while longer wavelengths penetrate deeper into the structure before being redirected (Fig. 4.9). This leads to a wavelength-dependent phase shift, where the longer wavelength components of the pulse trail the shorter ones, thus creating a negative chirp. Used as cavity reflectors and appropriately designed, dispersive mirrors can provide sufficient negative GDD for stable pulse generation in ultrafast lasers. Because the amount of negative GDD is preset once the cavity is constructed, this approach does not support wide wavelength tunability. However, due to their better management of higher order phase distortions (mainly TOD), dispersive mirrors enable shorter pulses than can be achieved by using intra-cavity prisms for dispersion compensation. In fact, Ti:sapphire lasers with their dispersion controlled through chirped mirrors can generate pulse widths well below 10 fs [14]. Dispersive mirrors are also used extra-cavity for counteracting broadening in fixed wavelength biomedical applications [15, 16].

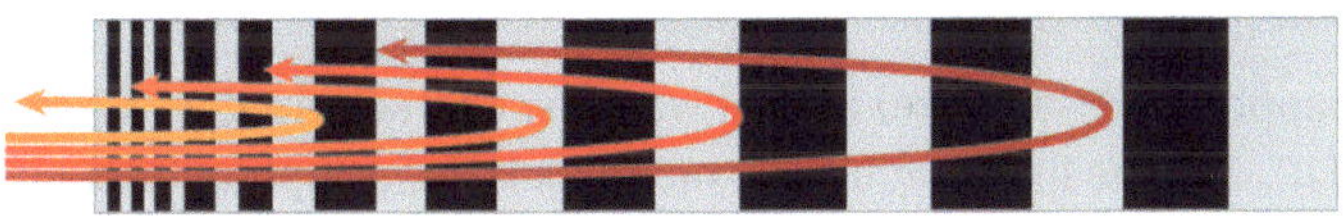

Fig. 4.9: Principle of a dispersive mirror (schematic). The thickness of the dielectric layers forming a high-reflecting mirror varies across the stack of coatings, so that the spectral components of a femtosecond pulse are reflected at different depths. If the longer wavelength components penetrate deeper and trail the shorter ones after reflection, the mirror induces a negative chirp.

The dispersion of the doped step index silica fibers, which are typically used as the gain material in ultrafast fiber lasers, is dominated by the material dispersion of fused silica glass: With a zero dispersion wavelength at 1.3 μm, dispersion is positive (normal) for shorter and negative (anomalous) for longer wavelengths. In compact ultrafast fiber lasers, dispersion compensation is preferably achieved through all-fiber-based techniques [17]. Chirped fiber Bragg gratings used in reflection provided the first practical means [18], and can be implemented directly into the gain fiber using phase mask writing techniques. Today's optical fiber technology offers other all-fiber-based options for designing ultrafast fiber lasers with properly balanced intra-cavity dispersion. In the normal dispersion regime of fused silica glass below 1.3 µm, microstructured fibers, such as photonic crystal or photonic bandgap fibers, can be constructed with negative (anomalous) dispersion [19] and integrated into a fiber laser for dispersion control. For wavelengths in the anomalous dispersion regime above 1.3 µm, such as the important range around 1.55 µm, optical fibers with large Germanium concentrations can be used for dispersion compensation [17].

4.3 Types of ultrafast lasers used in biophotonics

The earliest generation of ultrafast lasers utilized jets of organic dyes dissolved in viscous solvents as gain materials and were pumped by highly inefficient argon ion lasers. These dye laser systems tended to be large, complicated and too unreliable for commercial applications. As a result, their use was confined to research laboratories operated by laser experts. This has changed dramatically with the emergence of optically pumped ultrafast solid state and fiber lasers. Progress in this area was fueled by advances in high gain-bandwidth crystals and fibers, intra-cavity optics, and semiconductors lasers as optical pump sources.

The following paragraphs will review some of the ultrafast lasers used in biophotonics. Gain-switched picosecond diode lasers like those used in fluorescence lifetime spectroscopy and imaging are not included.

4.3.1 Femtosecond Ti:sapphire lasers

The titanium-doped sapphire (Ti:sapphire) crystal was first characterized as a laser gain material in 1982 [20, 21]. The first commercial mode-locked Ti:sapphire laser became available with the release of the Spectra-Physics *Tsunami*® oscillator in 1990. Since then Ti:sapphire lasers have become the workhorses in many ultrafast application areas including biophotonics, where today's fully automated and wavelength-tunable femtosecond Ti:sapphire lasers are indispensable tools in multiphoton microscopy.

Ti:sapphire crystals are produced by introducing Ti_2O_3 into a melt of Al_2O_3. A boule of single-crystalline material is grown from this melt where Ti^{3+} ions are substituted for a small percentage of the Al^{3+} ions. These Ti^{3+} ions are responsible for the emission of a Ti:sapphire laser. Absorption transitions occur over a broad wavelength range in the visible spectrum, from below 400 nm to about 650 nm (Fig. 4.10). Although the corresponding fluorescence band starts at around 600 nm, lasing is typically limited to wavelengths longer than 670 nm due to re-absorption losses within the crystal. Nonetheless, Ti:sapphire can provide spectral gain over an exceptionally wide wavelength range, enabling commercial mode-locked lasers with pulse width of

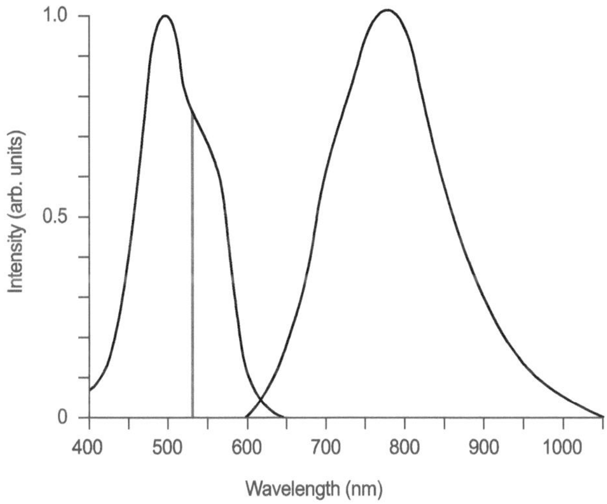

Fig. 4.10: Absorption (left) and emission (right) responsible for the output of a Ti:sapphire laser. The green line indicates the 532 nm wavelength of the solid state lasers used for optical pumping. The broad emission enables the generation of extremely short or widely wavelength-tunable femtosecond pulses.

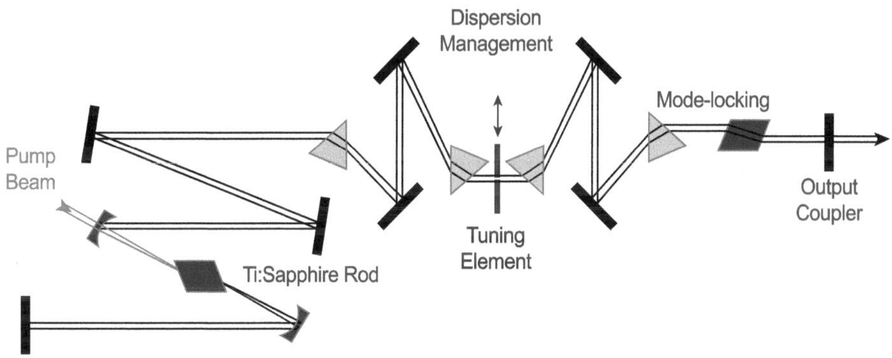

Fig. 4.11: Principle of a wavelength-tunable femtosecond Ti:sapphire laser (schematics of the Spectra-Physics *Tsunami*® cavity).

only a few femtoseconds at a center wavelength close to the peak of the spectral gain at around 800 nm. Alternatively, Ti:sapphire lasers can be designed for delivering longer pulses (i.e., tens of femtoseconds or more), which, due to their lower spectral bandwidth, can be wavelength-tuned over a spectral range as wide as 300 nm (Fig. 4.11).

One of the few drawbacks of Ti:sapphire as a gain material is the requirement for optical pumping in the visible wavelength range, where the availability of high-power diode lasers is limited. However, when commercial high-power, diode-pumped solid state green lasers emerged in 1996 [22, 23] and replaced the large and inefficient argon-ion pump lasers used previously, compact and user-friendly ultrafast Ti:sapphire lasers came within reach. These solid-state lasers are reliable and efficient, and their small footprint and low heat generation enable integration of pump laser and mode-locked Ti:sapphire cavity into a single, compact device. Consequently, in 1999 Spectra-Physics® (Santa Clara, CA, USA) released the *Mai Tai*® laser as the first fully automated and compact wavelength-tunable femtosecond oscillator (Fig. 4.12), followed two years later by Coherent® (Santa Clara, CA, USA) with the *Chameleon*® laser. In the meantime these products have increased their tuning range and average power and are widely integrated into commercial and home-built multiphoton microscopes, which benefit from the lasers' compact footprint and user-friendly operation. The latest development was the integration of automated dispersion compensation in 2007 to counteract dispersive pulse broadening in multiphoton microscopes [24, 25]. With dispersion compensation the output pulses from the laser are negatively chirped, so that the positive dispersion of the microscope will compress them close to their original pulse width (Fig. 4.13). As the negative chirp can be varied, the pulse width at the specimen can be optimized depending on the microscope objective and the wavelength used. Once calibrated, this optimization is fully automated – any wavelength change results in simultaneous adjustment of the negative chirp to the required amount.

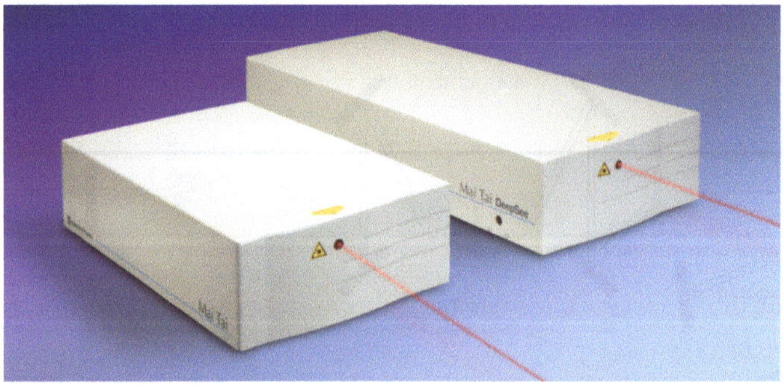

Fig. 4.12: Example of compact and fully automated, wavelength-tunable Ti:sapphire lasers, without (left) and with (right) dispersion compensation (Spectra-Physics *Mai Tai*® and *Mai Tai*® DeepSee™ series).

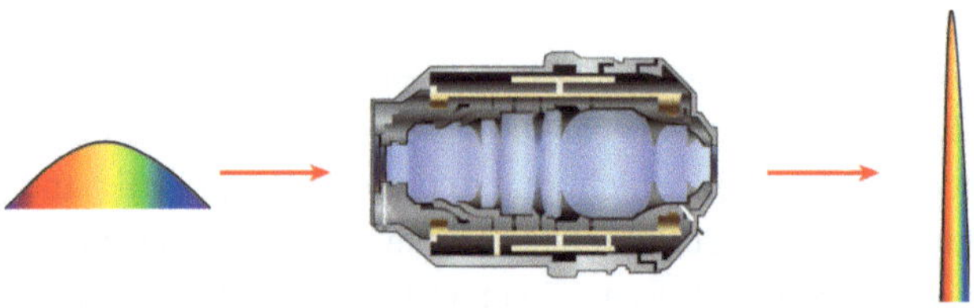

Fig. 4.13: Principle of application-specific dispersion compensation. After exiting the laser, the femtosecond pulses are negatively chirped. The positive dispersion of the application setup (symbolized by the microscope objective) recompresses the pulses close to the laser's original pulse width. This is important for applications that rely on peak power at the location of a specimen.

4.3.2 Other ultrafast diode-pumped solid state lasers

In many ways Ti:sapphire is an exceptional gain material for ultrafast lasers, explaining its tremendous commercial success. However, there are others that have recently become important, most notably host materials doped with the rare-earth ions of neodymium (Nd^{3+}) and ytterbium (Yb^{3+}). Due to their limited emission bandwidth, they neither enable widely wavelength-tunable femtosecond pulses nor pulse widths as short as those delivered by Ti:sapphire laser. However, they can be directly pumped by diode lasers, removing some of the complexity involved with Ti:sapphire. Because ultrafast lasers based on the aforementioned rare-earth dopants usually operate at a fixed center wavelength, modelocking is typically realized by the integration of a SESAM into the cavity.

Commercial Nd:vanadate (Nd:YVO_4) oscillators deliver picosecond pulses (about 8 ps) at average powers up to tens of watts at 1064 nm and repetition rates around 80 MHz [26]. The frequency-doubled output at 532 nm is used to synchronously pump optical parametric oscillators (OPOs) to generate the two synchronized and wavelength-tunable picosecond pulse trains required for Coherent Anti-Stokes Raman Scattering (CARS) [27, 28] and Stimulated Raman Scattering (SRS) microscopy [29].

Direct diode-pumped, ytterbium-doped tungstate crystals such as Yb:KGW and Yb:KYW have been successfully utilized in the design of very compact femtosecond lasers, emitting at around 1040 nm [30, 31]. Commercially available oscillators deliver multiwatt average powers with pulse widths down to below 200 fs [32]. Wavelength flexibility as required by many microscopy applications is achieved by synchronously pumping an OPO. Higher energy ytterbium-based laser sources such as the Spectra-Physics *femtoTrain*™ and *Spirit*® series are successfully deployed in high-precision eye surgery applications.

4.3.3 Ultrafast fiber lasers

Over the past decade fiber laser technology has developed rapidly – motivated by general inherent benefits of fiber-based laser architectures, together with rapid advances in the performance of critical components including single-mode, rare earth-doped silica gain fibers and the respective diode lasers for optical pumping.

In general, the most immediate differences between a fiber laser and a conventional solid state laser arise from the physical construction. The extended length and large outer surface area of optical fibers allow for efficient cooling of the gain material. The waveguide architecture enables high optical gain and wall-plug efficiencies. In addition, optical fibers can be coiled to allow for a compact footprint of the overall laser module.

When considering ultrafast oscillators in particular, the much smaller cross section (or mode area) of the active core in a fiber laser leads to higher peak power densities, when compared to crystal-based solid state lasers. Consequently, to avoid adverse nonlinear effects the intra-cavity peak power needs to be limited. In recent years many advances have been made to bridge that gap, mainly by increasing the mode area of the gain fiber. Adding diode-pumped chirped pulse fiber amplifiers provides another route to increase the pulse energies from all-fiber laser systems [17].

Commercial diode-pumped erbium- and ytterbium-doped fiber lasers are deployed in a variety of biophotonics applications. The fundamental (1550 nm) and frequency doubled (775 nm) output of erbium-doped femtosecond fiber lasers have been used for second- (SHG) and third-harmonic generation (THG) microscopy [33, 34].

Commercial all-fiber picosecond supercontinuum sources have been deployed as flexible direct excitation sources in fluorescence-based bioinstrumentation applications, including flow cytometry and confocal microscopy [35–37]. The supercontinuum is generated in a highly nonlinear fiber, such as a photonic crystal fiber (PCF) [38, 39]. As the generated supercontinuum has contributions in the visible wavelength range, optical or acousto-optical bandpass filters can be used to select appropriate spectral bands for flexible, selective excitation of fluorophores in cells or tissues. As the light transmitted to the specimen is in fact a train of ultrafast pulses with MHz repetition rates, this approach not only enables the (quasi-CW) detection of the amount of fluorescence, but also the measurement of the respective lifetimes in applications such as fluorescence lifetime imaging (FLIM).

An alternative approach combines spectral control of a fiber-based supercontinuum with second-harmonic generation to construct tunable ultrafast laser sources with high spectral purity. Delivering picosecond pulses and optimized for the visible spectral range, these sources are used for quasi-CW excitation of fluorescence and for measuring fluorescence lifetimes in confocal microscopy [40]. Although the peak powers generated are low compared to the tunable solid state lasers typically deployed for these applications, tunable ultrafast NIR and IR fiber sources of this kind have also been demonstrated in nonlinear cell and tissue imaging, including CARS [41, 42] and SRS microscopy [43].

4.3.4 Ultrafast optical parametric oscillators

An optical parametric oscillator (OPO) relies on a nonlinear process termed parametric down conversion [44]. In an appropriate nonlinear crystal, an input photon is split to produce two photons of lower energy, referred to as the signal and idler photons, where the sum of the two photons conserve the original photon energy (Fig. 4.14). By definition, the signal has the higher photon energy. Momentum must also be conserved, necessitating a fixed phase relationship between the input, signal and idler beams. This phase matching condition means that, for a given pump wavelength and a given crystal temperature and angle, only one signal (and idler) wavelength can be generated. Changing the angle and/or temperature of the crystal produces tunable output. The crystal is placed in an optical cavity which is resonant for at least one of the generated waves – typically the signal. Because the nonlinear process is instantaneous, i.e., without storage of energy within the crystal, optical parametric gain in a pulsed OPO requires the presence of a pump pulse coincident with the circulating signal (or idler) pulse. In the ultrafast time domain this is realized through synchronous pumping: The optical length of the OPO cavity is stabilized to match that of the pump laser, such that the circulating signal (or idler) pulse passes through the nonlinear crystal at the same time a pump pulse arrives.

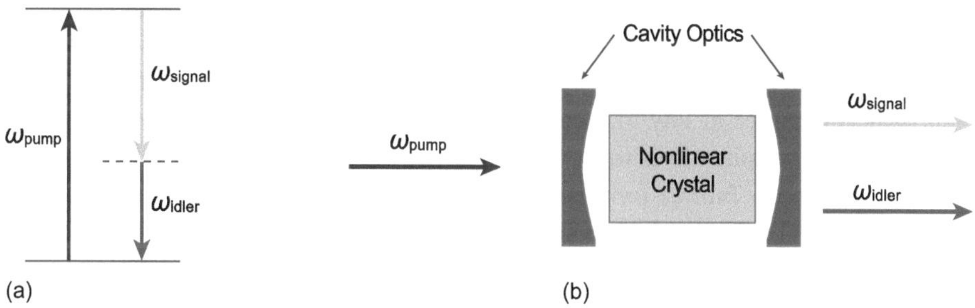

Fig. 4.14: Principle of an optical parametric oscillator (OPO). A photon is converted in a nonlinear crystal into a signal and an idler photon such that the energy is conserved (a). The crystal is placed into an optical cavity which is resonant for at least one of the generated waves, typically the signal (b).

In biophotonics, picosecond OPOs pumped by picosecond neodymium vanadate ($Nd:YVO_4$) DPSS lasers generate the two synchronized and wavelength-tunable pulse trains required for Coherent Anti-Stokes Raman Scattering (CARS) [27, 28] and Stimulated Raman Scattering (SRS) microscopy [29].

Femtosecond OPOs have been utilized to extend the spectral reach of ultrafast Ti:sapphire lasers deployed in multiphoton-excited fluorescence and harmonic generation microscopy to excitation wavelengths beyond 1100 nm [45–47]. It has been demonstrated that excitation at 1280 nm enables imaging about twice as deep into mouse brain compared to excitation at 775 nm [46]. This is the result of reduced light scatter, which is the major limitation to penetration depth at these wavelengths. Ti:sapphire laser-pumped femtosecond OPO systems have been used successfully in multimodal nonlinear imaging, enabling the simultaneous measurement of complementary biological information by combining two-photon excited fluorescence and/or second harmonic generation with CARS and/or sum frequency generation microscopy [48, 49]. The latest advance in this area is based on OPOs pumped by high-power Yb-based DPSS lasers. The Spectra-Physics *InSight®* oscillator was the first commercial product – a compact, fully automated one-box ultrafast source delivering from one output beam sub-120 femtosecond pulses, which are continuously wavelength-tunable from 680 nm to 1300 nm. It provides convenient access to the relative transparency window of tissue in vivo, including the important part beyond 1000 nm, where excitation photons are subjected to lower scatter and penetrate deeper. Together with an optional second synchronized femtosecond pulse train at 1041 nm, this source can be utilized for dual-wavelength and multimodal applications, including two-color excitation, and CARS or SRS microscopy [50].

4.4 Applications of ultrafast lasers in biophotonics

4.4.1 Overview

The unique optical properties of ultrafast lasers – generally their ability to deliver high peak intensities at low to moderate average powers and pulse energies – enable biomedical applications that are impossible or at least difficult to realize with other laser sources. Tab. 4.1 provides a snapshot: The applications listed in the top six columns can be divided into two categories: items (a) through (d) can be summarized as laser scanning microscopy techniques, while (e) and (f) are invasive procedures. The latter are not within the scope of this book. However, as femtosecond lasers have revolutionized surgical vision correction procedures, these applications deserve a short review in the next section. Beyond that, the application focus will be on the laser scanning microscopy applications, in particular on multiphoton imaging.

4.4.2 Femtosecond lasers in vision correction

Laser-assisted in situ keratomileusis (LASIK) is the most commonly performed interventional procedure to correct myopia (shortsightedness), hyperopia (farsightedness) and astigmatism. It relies on optimizing the refractive properties of the cornea by removing corneal tissue through photoablation using ultraviolet nanosecond output pulses of an excimer laser. LASIK involves the creation of a thin corneal flap to give the excimer laser pulses access to deeper corneal layers. The flap is closed back to its original position after photoablation, as the preservation of the surface epithelial layer accelerates recovery. Traditionally the flap has been created mechanically using specialized mechanical cutting tools called microkeratomes. However, in recent years these microkeratomes have been widely replaced by clinically approved "blade-less" devices, which utilize near-infrared (NIR) femtosecond laser pulses for the creation of the flap.

The cornea is transparent for NIR pulses. However, in the focus of the beam, the peak power density can be high enough to cause laser-induced optical breakdown (LIOB) and photodisruption of the tissue. By using femtosecond lasers together with a high numerical aperture objective the damage can be spatially confined to the focal volume. The highly localized deposition of energy generates a microplasma, which

Tab. 4.1: Ultrafast laser applications in biophotonics.

	Application	Details	Critical ultrafast laser parameter or capability	Typical ultrafast lasers used
(a)	Confocal microscopy (and other fluorescence-based instrumentation like flow cytometry)	Quasi-CW polychrome source for flexible selection of fluorophores and excitation wavelength(s)	Nonlinear frequency conversion (source-internal)	Polychrome fiber sources
(b)	Fluorescence lifetime imaging (FLIM)	Typically on a confocal microscope	Time resolution	Gain-switched diode lasers, polychrome fiber sources
(c)	Multiphoton microscopy	Two- and three-photon excited fluorescence, second- or third-harmonic generation, CARS, SRS, pump-probe, optogenetics/photostimulation	Localized nonlinear light-matter interaction with inherent optical sectioning	Tunable Ti:sapphire lasers, tunable OPOs
(d)	Pulsed stimulated emission depletion (STED) microscopy	Spatial resolution beyond optical diffraction limit ("super-resolution")	Pulse synchronization of two wavelengths	Gain-switched diode lasers, polychrome fiber sources, tunable Ti:sapphire lasers
(e)	Vision correction	Corneal flap creation for laser-assisted in situ keratomileusis (LASIK) Cataract surgery – Corneal incisions – Anterior capsulotomy – Lens fragmentation Other high-precision incisions	Nonlinear light-matter interaction, photodisruption	Fixed-wavelength ($\approx 1\,\mu m$) diode-pumped solid state or fiber lasers with pulse energies in the microjoule range
(f)	Other high-precision cell manipulation	Transfection, microdissection Cell/gene/tissue manipulation	Nonlinear light-matter interaction with ablation	Short pulse Ti:sapphire lasers
(g)	Others	Optical coherence tomography Diffuse optical tomography Terahertz generation	Bandwidth Time resolution Bandwidth, frequency conversion	Various

is followed by secondary side effects such as the emission of a shock wave and the generation of cavitation bubbles [51]. However, in contrast to nanosecond or even picosecond pulses, femtosecond laser photodisruption minimizes these side effects and thus the collateral damage. It requires much lower pulse energies, minimizing the range of the photodisruptive shock wave and the size of the cavitation bubbles [52]. The collateral thermal damage zone is extremely limited, because the thermal diffusion is in the sub-micron range [53].

The corneal flap for LASIK is created by scanning the focus of the femtosecond laser beam at the desired depth to form a dissection plane while leaving a hinge for the flap to remain connected with the cornea. Compared to a microkeratome, the femtosecond laser allows for flexibility with respect to the thickness of the flap, and – more importantly – enables much greater precision, which results in faster recovery times, improved visual results and greater safety [51, 52]. Today, there are a number of suppliers for femtosecond laser-based instruments, which have been clinically approved for cutting corneal flaps [54]. The lasers deployed in these devices are diode-pumped, rare-earth (Nd or Yb) doped solid state or fiber lasers delivering pulse widths of a few hundred femtoseconds at repetition rates in the high kilohertz to low megahertz regime. Pulse energies range up to a few microjoules, typically requiring pulse amplification [55].

Flap cutting for LASIK was the first femtosecond laser-based, clinically approved vision correction procedure, but others have been developed since then [56–58]. Currently there is significant focus on bringing devices to market that utilize femtosecond pulses for high-precision interventional processes involved in the transplant of intraocular lenses (IOLs) during cataract surgery. Approximately 18.9 million conventional (i.e. non-laser-based) cataract procedures were performed worldwide in 2010. The USA alone accounted for 3.3 million – more than three times the amount of LASIK procedures in the same period [59], indicating that there is a significant market opportunity in laser-based cataract surgery.

Because of their unique capability of removing controlled amounts of material with extremely low collateral thermal damage, ultrafast lasers have also been used for precise manipulation of tissue and individual cells in a wide area of biomedical research applications, including nanosurgery in developmental biology and the neurosciences [60], and the processing of teeth and other hard tissues [61].

4.4.3 Ultrafast lasers in microscopy

4.4.3.1 Confocal microscopy

In conventional confocal laser scanning microscopy, a tightly focused continuous-wave (CW) laser beam is used as the fluorescence excitation source. The laser focus is reimaged onto a pinhole called the confocal aperture, which serves to prevent fluorescence generated outside the focal plane from reaching the detector. By measuring the fluorescence intensity, a three-dimensional image is constructed point by point as the laser spot or the sample stage is sequentially scanned in all three dimensions. The spatial resolution is close to the diffraction limit of the excitation wavelength. Because of the multitude of fluorophores used and their different excitation wavelengths, flexible laser scanning confocal microscopes typically integrate several single-wavelength diode, or diode-pumped solid state lasers covering the visible spectrum. More recently, high repetition rate picosecond fiber-based laser sources have been deployed for quasi-CW excitation, enabling flexible selection of the excitation wavelength(s) in confocal microscopy (and other fluorescence-based bioinstrumentation like flow cytometers). These supercontinuum-based devices have been reviewed in Section 4.3.3. With their picosecond pulse width they can also be used for confocal fluorescence lifetime imaging (FLIM). Combined with time-correlated single-photon counting (TCSPC) in the detection channel, fluorescence lifetimes (typically a few nanoseconds) are measured pixel by pixel during the scan and displayed as a false-color image. Confocal FLIM provides additional information about the specimen, because the fluorescence lifetimes depend on the microenvironment of the fluorescent molecules.

4.4.3.2 Pulsed STED microscopy

Stimulated emission depletion (STED) microscopy was first conceptualized [62] and realized by Stefan Hell [63], and is one of the optical "super-resolution" microscopy or "nanoscopy" techniques that have been developed over the past few years [64]. Like a confocal laser scanning microscope, a STED microscope establishes an image of a biological specimen pixel by pixel by excitation and detection of fluorescence. However, in a STED microscope, a second, red-shifted laser beam is overlapped with the excitation beam. The wavelength of this so-called depletion or STED beam is tuned to a (red-shifted) emission line of the excited fluorophore, and the beam is shaped such that it forms a doughnut when focused. In a pulsed STED microscope both, the excitation and depletion beams, consist of synchronized trains of short pulses (typically tens to hundreds of picoseconds) at MHz repetition rates. Each excitation pulse is immediately followed by a short depletion pulse, which causes unidirectional stimulated emission. Fluorophores from within the doughnut-shaped focus of the depletion beam are quenched. They do not contribute to the spontaneous fluorescence, which is detected for image contrast. This effect reduces the effective spot size. The resulting higher lateral (x, y) image resolution increases with the intensity of the depletion beam and can be better than 10 times the diffraction limit, which constraints the resolution of the confocal (and any other conventional optical) microscope [65]. isoSTED, a combination of the STED approach with 4Pi microscopy, adds super-resolution also for the axial (z) direction [66].

The excitation source for pulsed STED can be a compact gain-switched picosecond diode laser. As the STED beam usually requires higher intensities, Ti:sapphire lasers have been utilized for depleting fluorophores with emission spectra in the infrared wavelength region. To avoid nonlinear effects during depletion their femtosecond pulses are typically stretched into the picosecond regime by passing them through dispersive optical fibers or bulk materials. The use of picosecond supercontinuum fiber sources for STED microscopy has been demonstrated [67], as has been the use of femtosecond pulses for the excitation of the fluorophores through two-photon absorption [68].

The use of pulsed sources is very effective because the depletion is synchronized to the fluorescent cycle of the fluorophore. However, the use of CW beams has also been shown viable [69] and reduces the complexity of the overall setup.

4.4.3.3 Multiphoton microscopy

Conventional confocal microscopes are essential tools in today's biomedical research laboratories. However, they have some limitations for in vivo and deep-tissue imaging: Out-of-focus excitation, which does not contribute to the construction of the image, increases the potential of damaging living specimens. In addition, the visible and ultraviolet wavelengths optimal for direct excitation do not penetrate deeply into tissue. And, reimaging the fluorescence generated in the focus of the laser beam through the laser scanner onto the confocal aperture generates losses and reduces the sensitivity.

In 1990 W. Denk, J. Strickler and W. Webb at Cornell University, Ithaca, NY, first showed that these limitations could be avoided by using femtosecond lasers for two-photon excitation of fluorophores [70] (Fig. 4.15 (a)). As any other nonlinear interaction, this process has a very low probability. However, the peak power of femtosecond pulses together with a high numerical aperture (NA) microscope objective generates high enough photon densities within the focal volume to drive two- (and even three-)photon absorption, while elsewhere in the sample the power density is too low. This inherent spatial sectioning capability obviates the need for a confocal pinhole to suppress out-of-focus fluorescence. The detector can be located close to the specimen, resulting in increased sensitivity and an improved signal-to-noise ratio. As there is virtually no out-of-focus excitation, phototoxicity is greatly reduced. Moreover, because multiphoton imaging uses longer wavelengths where there is less absorption and scattering, the two-photon technique can penetrate much farther into tissue (100 s compared to 10 s of micrometers).

Because of those advantages, two-photon excited fluorescence microscopy was quickly embraced by cell biologists and neuroscientists. Commercial system deployment was fueled by the advent of fully automated, and widely wavelength-tunable femtosecond lasers. Today these user-friendly sources routinely deliver pulses as short as 70 fs with peak powers as high as 450 kW. With a wavelength tuning range of more than 600 nm they enable the user to select the optimum two-photon excitation wavelengths for the most commonly used fluorophores.

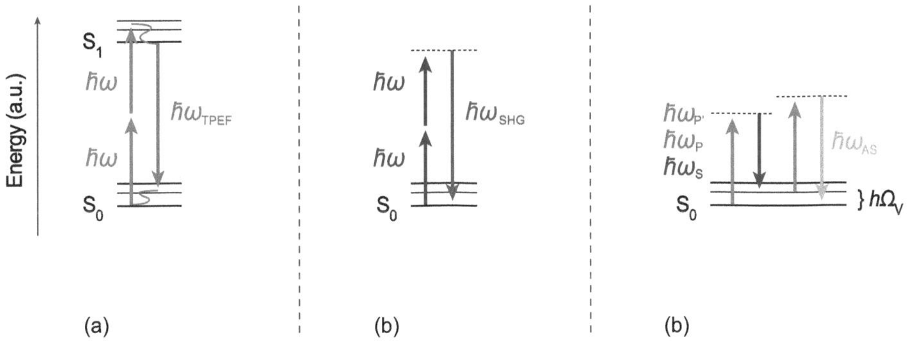

Fig. 4.15: Energy diagrams for selected multiphoton microscopy modalities: (a) two-photon excited fluorescence (TPEF), (b) second-harmonic generation (SHG), (c) coherent anti-Stokes Raman scattering (CARS).

The integration of automated dispersion compensation counteracts the pulse broadening caused by the positive dispersion of the microscope setup (Fig. 4.13). Depending on system configuration, the uncompensated dispersion of a commercial multiphoton microscope can stretch a 100 femtosecond pulse to several 100 femtoseconds before it interacts with the specimen. This unwanted effect reduces the nonlinear excitation efficiency and therefore the brightness of the image. In fact the number of photons absorbed per fluorophore per pulse, n_a, is inversely proportional to the pulse duration and can be expressed by:

$$n_a \approx \frac{P_{ave}^2}{\tau_p}, \tag{4.9}$$

where P_{ave} is the average laser power and τ_p is the pulse duration at the specimen [69]. One way to increase fluorescence and image brightness is raising the average power. However, that heightens the risk of tissue damage caused by linear absorption. Minimizing the pulse duration by compensating for the positive dispersion of the microscope not only optimizes image brightness and/or tissue penetration depth, but also improves the signal-to-noise ratio in multiphoton microscopy [71].

Fig. 4.16 demonstrates the positive effect of shorter pulses on image quality. In addition to two-photon generated fluorescence the images include signal resulting from second-harmonic generation (SHG), where two infrared photons are converted into one with twice the energy or half the wavelength (Fig. 4.15 (b)). SHG imaging does not require fluorescent labels, but spatially ordered structures of molecules with certain symmetry requirements [72]. As a second order nonlinear process it also benefits from higher peak power and thus shorter pulses at the specimen.

Fully automated widely-tunable femtosecond with integrated dispersion compensation have become ubiquitous in today's state-of-the-art two-photon microscope systems. Compact and easy to integrate, they automatically minimize the pulse width delivered to the biological specimen at any wavelength within their tuning range.

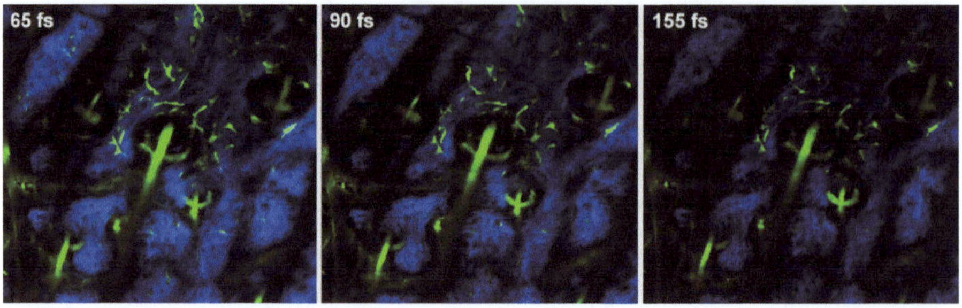

Fig. 4.16: Benefit of shorter pulses in two-photon excited fluorescence (TPEF) and second-harmonic generation (SHG) microscopy. Green fluorescence protein (TPEF, in green) and collagen (SHG, in blue) in mouse ear (in vivo image, 500 µm × 500 µm), pulse width at the sample was changed between 65 fs (left) and 155 fs (right) using a Spectra-Physics Mai Tai DeepSee laser. All other parameters were kept constant, including average power (20 mW at 870 nm) and detector sensitivity. Image courtesy of Dr. Claudio Vinegoni at Center for Systems Biology, Massachusetts General Hospital. Image reprinted from the October 2010 edition of *Laser Focus World*. Copyright 2010 by PennWell.

Two-photon excited fluorescence microscopy is presently the most commonly utilized multiphoton imaging modality. Second- (and third-) harmonic generation is generally available from the same equipment. However, there are other multiphoton techniques in use today that require more than one input wavelength, including Coherent Anti-Stokes Raman Scattering (CARS) [73], and Stimulated Raman Scattering (SRS) microscopy [29]. They offer the same inherent 3D optical sectioning capability. Both are label free, i.e., they do not require fluorophores, but rather utilize intrinsic properties of the specimen for image contrast. CARS is a four-wave mixing process of the so called Pump (P), Stokes (S), Probe (P′) and Anti-Stokes (AS) waves (Fig. 4.15 (c)). To simplify the setup, Pump and Probe waves are typically set at the same frequency (wavelength) and originate from the same laser beam ($\omega_P = \omega_{P'}$, $\lambda_P = \lambda_{P'}$). If for molecules within the focal volume the frequency difference between the Pump and Stokes waves matches a molecular vibration ($\Omega_V = \omega_P - \omega_S$), resonant enhancement creates a collimated, directional Anti-Stokes beam at higher energy ($\omega_{AS} = 2\omega_P - \omega_S$). As a third order process, CARS microscopy requires ultrafast laser pulses to deliver the necessary high peak intensities. Synchronized Pump and Stokes pulse trains are overlapped in space and time, and scanned together within the specimen. The Anti-Stokes signal generated in the focus of both beams is measured pixel by pixel and used to establish the CARS image.

Optical Parametric Oscillators (OPOs) are commonly used for CARS (and SRS) microscopy, as they deliver wavelength-tunable pulse trains that are inherently synchronized. For CARS microscopy generally pulse durations of a few picoseconds are optimal, as the narrower spectral bandwidth matches the spectral width of most Raman lines, thereby reducing the nonresonant background. However, some biologically significant vibrations, like the C–H_2 stretching modes in lipids at around 2850 cm^{-1}, have a broader spectral profile and enable good signal-to-background ratio when probed with femtosecond pulses [48]. One obvious advantage of using femtosecond pulses is multimodality – the ability to combine CARS microscopy with, for example, two-photon excited fluorescence and SHG imaging, both of which benefit from shorter pulse widths [48, 49]. In addition, a femtosecond OPO extends the spectral reach for two-photon excited fluorescence and harmonic generation microscopy to excitation wavelengths beyond 1000 nm, resulting in lower photon scatter and increased tissue penetration depth [45–47, 50]. Consequently, recently released compact, turnkey ytterbium-based OPOs have taken market share away from the Ti:sapphire lasers, since these new sources deliver femtosecond pulses that are tunable from 680 nm all the way to 1300 nm, all from one compact package that includes automated dispersion compensation.

As discussed in Section 4.3.3, ultrafast fiber lasers have recently been deployed for multiphoton microscopy, including CARS imaging [41, 42] and SRS imaging [43]. While being somewhat limited in their spectral reach and/or average power and thus their flexibility in the application space, these sources are far more compact and less costly.

4.5 Summary and outlook

Over the past 15 years ultrafast lasers have developed from rather complex systems requiring frequent user intervention to compact, reliable and easy-to-use devices. Fueled by these technological improvements, ultrafast lasers have quickly penetrated the biomedical field, where their unique capability of delivering high peak intensities at moderate average powers enabled new applications. Today, the biggest commercial markets are in ophthalmology and microscopy. Femtosecond lasers enable unprecedented precision and safety in LASIK vision correction procedures – resulting in reduced complications and faster patient recovery. The use of femtosecond lasers is poised to expand into other high-precision interventional procedures in ophthalmology, like cataract surgery. In microscopy, ultrafast lasers have enabled novel nonlinear imaging techniques, which provide new insights in biomedical research such as the neurosciences and the understanding of Alzheimer's and other diseases. Ultimately this will lead to better health care and higher quality of life.

Laser manufacturers will continue to extend the reach of existing ultrafast laser platforms and – more importantly – develop new ones. The result will be innovative, user-friendly, and affordable ultrafast lasers that will keep up with the changing requirements of established and novel biomedical applications. Solid state sources with bulk gain materials will maintain their dominant position in the market for the foreseeable future. But, as ultrafast fiber lasers advance in their performance characteristics, they will increasingly penetrate the application space as compact alternatives.

The integration of nonlinear processes, such as parametric wavelength conversion, into ultrafast laser sources will continue to gain momentum for accessing important wavelengths that are otherwise difficult to reach.

References

[1] Hargrove LE, Fork RL, Pollack MA, Locking of HeNe laser modes induced by synchronous intracavity modulation. Appl Phys Lett. 1964;5:4–5.

[2] Spence DE, Kean PN, Sibbett W. 60-fsec pulse generation from a self-modelocked Ti:sapphire laser. Opt Lett. 1991;16:42–44.

[3] Keller U, Miller DAB, Boyd GD, Chiu TH, Ferguson JF, Asom MT. Solid state low-loss intracavity saturable absorber for Nd:YLF lasers: an antiresonant semiconductor Fabry-Perot saturable absorber. Opt Lett. 1992;17:505–507.

[4] Paschotta R, Keller U. Passively mode-locked solid-state lasers. In: Sennaroglu A, editor. Solid-State Lasers and Applications. Boca Raton, FL: CRC Press; 2007. p. 259–318.

[5] Keller U. Ultrafast solid-state laser oscillators: a success story for the last 20 years with no end in sight. Appl Phys B. 2010;100:15–28.

[6] Kieu K, Mansuripur M. Femtosecond laser pulse generation with a fiber taper embedded in carbon nanotube/polymer composite. Opt Lett. 2007;32:2242–2244.

[7] Kieu K, Wise FW. All-fiber normal-dispersion femtosecond laser. Opt Express 2008;16: 11453–11458.

[8] Kieu K. Carbon nanotubes facilitate easy fabrication of mode-locked fiber lasers. SPIE Newsroom. 2010. doi: 10.1117/2.1201009.003241.

[9] Encyclopedia of Laser Physics and Technology [Internet]. Available at www.rp-photonics.com.

[10] Diels J, Rudolf W. Ultrashort laser pulse phenomena. 2nd edition. Burlington, MA: Academic Press; 2006.

[11] Martinez OE, Gordon JP, Fork RL. Negative group-velocity dispersion using refraction. J Opt Soc Am. 1984;B1:1003–1006.

[12] Fork RL, Martinez OE, Gordon JP. Negative dispersion using pair of prisms. Opt Lett. 1984;9:150–152.

[13] Szipöcs R, Ferencz K, Spielmann C, Krausz F. Chirped multilayer coatings for broadband dispersion control in femtosecond lasers. Opt Lett. 1994;19:201–203.

[14] Stingl A, Lenzner M, Spielmann C, Krausz F, Szipöcs R. Sub-10-fsec mirror-dispersion-controlled Ti:sapphire laser. Opt Lett. 1995;20:602–604.

[15] Uchugonova A, Müller J., Bückle R, Tempea G. Isemann A, Stingl A, König K. Negatively chirped laser enables nonlinear excitation and nanoprocessing with sub-20-fs pulses. Proc SPIE 6860; 2008. p. 686015.

[16] König K, Uchugonova A. Multiphoton imaging and nanoprocessing of human stem cells. In: Pavone FS, editor. Laser imaging and manipulation in cell biology. Weinheim: WILEY-VCH Verlag GmbH & Co. KGaA; 2010. p. 9–34.

[17] Fermann ME, Hartl I. Ultrafast fiber laser technology. IEEE J of Selected Topics in Quantum Electronics 2009;15:191–206.

[18] Farries M, Ragdale C, Reid D. Broadband chirped fibre Bragg filters for pump rejection and recycling in erbium doped fibre amplifiers. Electron Lett. 1992;28:487–489.

[19] Knight JC, Arriaga J, Birks T, Ortigosa-Blanc A, Wadsworth WJ, Russell PStJ. Anomalous dispersion in photonic crystal fiber. IEEE Photon Tech Lett. 2000;12:807–809.

[20] Moulton PF. Twelfth International Quantum Electronics Conference in Munich; 1982 June.

[21] Moulton PF. Spectroscopic and laser characteristics of Ti:Al_2O_3. J Opt Soc Am B. 1986;3:125–132.
[22] Nighan WL, Cole J. > 6 W of stable, 532-nm TEM_{00} output at 30 % efficiency from an intracavity-doubled diode-pumped multiaxial mode Nd:YVO_4 laser. OSA Proc. Advanced Solid-State Lasers, Postdeadline Paper PD-4; 1996.
[23] Nighan WL, Craig B DPSS lasers challenge water-cooled ion lasers. Laser Focus World. 1996;32.4:63–70.
[24] David D, Krueger A, Feru P. Dispersion compensation sharpens multiphoton microscopy. Laser Focus World. 2007;43.1:117–120.
[25] David V, Feru P. Seeing deeper with multiphoton microscopy. Biophotonics International. 2007;14(1):34–37.
[26] High Q Laser GmbH. picoTRAIN data sheet. Rankweil, Austria; Printed February 2010.
[27] Ganikhanov F, Carrasco S, Xie XS, Katz M, Seitz W, Kopf D. Broadly tunable dual-wavelength light source for coherent anti-Stokes Raman scattering microscopy. Opt Lett. 2006;31:1292–1294.
[28] Evans CL, Xu X, Kesari S, Xie XS, Wong STC, Young GS. Chemically-selective imaging of brain structures with CARS microscopy. Opt Express. 2007;15:12076–12087.
[29] Freudiger CW, Min W, Saar BG, Lu S, Holtom GR, He C, Tsai JC, Kang JX, Xie XS. Label-free biomedical imaging with high sensitivity by stimulated raman scattering microscopy. Science. 2008;322:1857–1861.
[30] Yamazoe S, Katou M, Kasamatsu T. ultra-compact laser diode-pumped femtosecond solid-state laser. Fujifilm Research & Development. 2009;54:48–52.
[31] Kim GH, Kang U, Heo D, Yashin VE, Kulik AV, Sall' EG, Chizhov SA. A compact femtosecond generator based on an Yb:KYW crystal with direct laser-diode pumping. J Opt Technol. 2010;77:225–229.
[32] High Q Laser GmbH. femtoTRAIN data sheet. Rankweil, Austria; Printed February 2010.
[33] Aviles-Espinosa R, Loza-Alvarez P, Brodschelm A, Kaenders W. Ultrafast lasers & THG microscopy for flexible in-vivo cell research. BioOptics World. 2010;3.5:18–21.
[34] Hellerer T. Fs-lasers for life sciences. Optik & Photonik. 2008;4:35–38.
[35] Clowes J. Next generation light sources for biomedical applications. Opt Photon. 2008;1:36–38.
[36] Hayes T. Fibre lasers stretch from cells to wind farms. Opto & Laser Europe. 2009;170:15–19.
[37] Telford WG, Subach FV, Verkhusha VV. Supercontinuum white light lasers for flow cytometry. Cytometry A. 2009;75(5):450–459.
[38] Nielsen FD, Jespersen KG, Pedersen MØ, Qian Y, Feuchter T, Andersen B, Mattsson K, Pedersen CF, Thomsen CL. High-power visible supercontinuum source. Proceedings Photonics West; 2006.
[39] Kudlinski A, George AK, Knight JC, Travers JC, Rulkov AB, Popov SV, Taylor JR. Zero-dispersion wavelength decreasing photonic crystal fibers for ultraviolet-extended supercontinuum generation. Opt Express. 2006;14:5715–5722.
[40] Lison F, Hellerer T. Fully tunable visible laser source is valuable for biophotonics. Laser Focus World. 2009;(45)11:53–57.
[41] Krauss G, Hanke T, Sell A, Träutlein D, Leitenstorfer A, Selm R, Winterhalder M, Zumbusch A. Compact coherent anti-Stokes Raman scattering microscope based on a picosecond two-color Er:fiber laser system. Opt Lett. 2009;34:2847–2849.
[42] Selm R, Winterhalder M, Zumbusch A, Krauss G, Hanke T, Sell A, Leitenstorfer A. Ultrabroadband background-free coherent anti-Stokes Raman scattering microscopy based on a compact Er:fiber laser system. Opt Lett. 2010;35:3282–3284.

[43] Gambetta A, Kumar V, Grancini G, Polli D, Ramponi R, Cerullo G, Marangoni M. Fiber-format stimulated-Raman-scattering microscopy from a single laser oscillator. Opt Lett. 2010;35:226–228.
[44] Boyd RW. Nonlinear optics. San Diego: Academic Press; 1992. p. 77ff.
[45] Balu M, Baldacchini T, Carter J, Krasieva TB, Zadoyan R, Tromberg BJ. Effect of excitation wavelength on penetration depth in nonlinear optical microscopy of turbid media. J Biomed Opt. 2009;14:0105081–0105083.
[46] Kobat D, Durst ME, Nishimura N, Wong AW, Schaffer CB, Xu C. Deep tissue multiphoton microscopy using longer wavelength excitation. Opt Express. 2009;17:13354–13364.
[47] Andresen V, Alexander S, Heupel WM, Hirschberg M, Hoffman RM, Friedl P. Infrared multiphoton microscopy: subcellular-resolved deep tissue imaging. Current Opinion in Biotechnology. 2009;20:54–62.
[48] Chen H, Wang H, Slipchenko MN, Jung Y, Shi Y, Zhu J, Buhman KK, Cheng JX. A multimodal platform for nonlinear optical microscopy and microspectroscopy. Opt Express. 2009;17:1282–1290.
[49] Smith PG. Coherent Anti-Stokes Raman Scattering (CARS) imaging complements other multimodal imaging techniques. Spectra-Physics Application Note DS-011116. 2011:1–4.
[50] Klein J, Smith PG. Multiphoton imaging takes new directions. Biophotonics. 2015;22(2):29–31.
[51] Juhasz T, Loesel FH, Kurtz RM, Horvath C, Bille JF, Mourou G. Corneal refractive surgery with femtosecond lasers. IEEE J of Selected Topics in Quantum Electron. 1999;5(4):902–910.
[52] Juhasz T, Spooner G, Sacks, Suárez C, Raksi F, Zadoyan R, Sarayba M, Kurtz RM. Ophthalmic applications of ultrashort pulsed lasers. Proc SPIE 5340; 2004. p. 66–75.
[53] Freebody M. Femtosecond lasers enable bladeless eye surgery. Biophotonics. 2011;18(2):18–21.
[54] Lachman M. Cataract and refractive surgery: hoping for a rebound. Medtech Insight. 2009;11(5).
[55] Strickland D, Mourou G. Compression of amplified chirped optical pulses. Opt. Communications. 1985;56:219–221.
[56] Lachman M. ESCRS 2009: The next wave in cataract and presbyopia devices. Medtech Insight. 2009;11(9).
[57] Lachman M. Laser cataract surgery comes into focus at ASCRS. Medtech Insight. 2010;12(5).
[58] Lachman M. Ophthalmology community focuses on laser cataract surgery. Medtech Insight. 2010;12(10).
[59] Mahdavi S. Laser cataract surgery: the next new thing in ophthalmology. Cataract & Refractive Surgery Today. 2011;11(3):83–87.
[60] Nishimura N, Schaffer CB, Kleinfeld D. In vivo manipulation of biological systems with femtosecond laser pulses. Proc SPIE 6261 High-Power Laser Ablation VI; 2006. p. 62611J.
[61] Strassl M, Wieger V, Brodoceanu D, Beer F, Moritz A, Wintner E. Ultra-short pulse laser ablation of biological hard tissue and biocompatibles. J of Laser Micro/Nanoengineering. 2008;3(1):30–40.
[62] Hell SW, Wichmann J Breaking the diffraction resolution limit by stimulated emission: stimulated-emission-depletion fluorescence microscopy. Opt Lett. 1994;19(11):780–782.
[63] Klar TA, Hell SW. Subdiffraction resolution in far-field fluorescence microscopy. Opt Lett. 1999;24(14):954–956.
[64] Hell SW. Far-Field Optical Nanoscopy. In Gräslund A, Rigler R, Widengren J, editors. Single molecule spectroscopy in chemistry, physics and biology. Berlin: Springer; 2009. p. 365–398
[65] Harke B, Keller J, Ullal CK, Westphal V, Schönle A, Hell SW. Resolution scaling in STED microscopy. Opt Express. 2008;16(6):4154–4162.

[66] Schmidt R, Wurm CA, Jakobs S, Engelhardt J, Egner A, Hell SW. Spherical nanosized focal spot unravels the interior of cells. Nature Meth. 2008;5(6):539–544.
[67] Wildanger D, Rittweger E, Kastrup L, Hell SW. STED microscopy with a supercontinuum laser source. Opt Express. 2008;16(13):9614–9621.
[68] Moneron G, Hell SW. Two-photon excitation STED microscopy. Opt Express. 2009;17(17): 14567–14573.
[69] Willig KI, Harke B, Medda R, Hell SW. STED microscopy with continuous wave beams. Nature Meth. 2007;4(11):915–918.
[70] Denk W, Strickler JH, Webb WW. Two-photon laser scanning fluorescence microscopy. Science. 1990;248:73–76.
[71] Field J, Carriles R, Sheetz KE, Chandler EV, Hoover EE, Tillo SE, Hughes TE, Sylvester AW, Kleinfeld D, Squier JA. Optimizing the fluorescent yield in two-photon laser scanning microscopy with dispersion compensation. Opt Express. 2010;18(13):13661–13672.
[72] Campagnola PJ, Loew LM. Second-harmonic imaging microscopy for visualizing biomolecular arrays in cells, tissues and organisms. Nat Biotech. 2003;21:1356–1360.
[73] Zumbusch A, Holtom GR, Xie XS. Three-dimensional vibrational imaging by coherent anti-Stokes Raman scattering. Rev Lett. 1999;82:4142–4145.

Part II: **Modern nonlinear microscopy of live cells**

Luca Lanzano, Giuseppe Vicidomini, Lorenzo Scipioni, Marco Castello, and Alberto Diaspro

5 STED microscopy: exploring fluorescence lifetime gradients for super-resolution at reduced illumination intensities

Abstract: The 21st century has opened with the development of several strategies to push the spatial resolution of far-field fluorescence microscopy beyond the diffraction limit. Among all these super-resolved techniques, stimulated emission depletion (STED) microscopy stands out for its general principles and its fast acquisition capability. However, the application of STED microscopy to live-cell imaging has been limited by the typically high illumination powers required by early implementations of the technique. One way to preserve the effective resolution of a STED microscope, but with a significant reduction of the illumination intensity, is by using the nanosecond fluorescence dynamics information contained in a time-resolved STED experiment. Two different implementations exploring this idea have been demonstrated so far, respectively the so-called gated-STED (g-STED) microscopy and Separation of Photons by Lifetime Tuning (SPLIT)-STED microscopy. Here we discuss the common principle behind the two methods and the benefits and major differences between the two implementations.

5.1 Introduction

Since its invention in 1908, far-field florescence microscopy has been an essential source of major discoveries in life sciences [1, 2]. Throughout the 20th century far-field microscopy has grown enormously and all its distinctive features, such as high sensitivity and specificity through fluorescent tagging, the possibility to investigate the interior of living organisms and the ability to extract quantitative data about molecular dynamics and organizations, were continuously refined. However, the fundamental barrier limiting the spatial resolution of early far-field fluorescent microscopy, was still intact at the end of the 20th century. Namely, the diffraction barrier precluded far-field microscopy access to the entire sub-cellular components whose structural complexities were below around half the wavelength of the light used for the investigation (i.e., ≈ 200 nm).

The 21st century opened with a fundamental microscopy revolution, which saw the introduction of new viable concepts able to effectively overcome the diffraction barrier and push the spatial resolution of a far-field fluorescence microscope to the double-digit nanometer scale [3–7]. These techniques are usually referred to as super-

https://doi.org/10.1515/9783110429985-007

resolved microscopies and their importance has been recently recognized by the award of the 2014 Nobel Prize in Chemistry to their inventors.

Even though super-resolved fluorescence microscopy currently includes many different techniques [8–14], with fundamental differences in implementation and the photophysical mechanism used to circumvent the diffraction barrier, all of them rely on the same basic principle [15]. All super-resolved techniques resolve features smaller than the diffraction limit by getting the fluorophores (defining the structures) transiently into two discernible (pseudo-)states, i.e., (pseudo-)states with different spectral or temporal or any other detectable responses to the illumination. The transition between distinguishable states ensures that the signal recorded stems from fluorophores within a region of the sample whose size is much smaller than the diffraction limit. Scanning this region across the sample leads to images with subdiffraction resolution.

Among all these super-resolved techniques, stimulated emission depletion (STED) microscopy [8] stands out for its generality and its fast recordings. Indeed, STED microscopy uses two fundamental processes to transiently transfer a fluorophore from a dark state to a bright state (the fluorophore emits a detectable signal): the excitation process transfers the fluorophore from the singlet ground state (dark) to the singlet excited state (bright) and the stimulated emission (SE) process does the opposite. The use of these two fundamental processes makes, at least in theory, STED microscopy compatible with all fluorescent probes. However, at the same time the use of these transitions imposes the use of high dose of illumination intensity to effectively drive the fluorophore into the dark state. In particular, in a typical STED microscope, a focused Gaussian beam excites all the fluorophores within a diffraction-limited spot. Successively, a focused doughnut-shaped beam, the STED beam, featuring a zero-intensity point coaligned with the peak of the Gaussian excitation spot and a wavelength able to induce stimulated emission, quenches all the excited fluorophores except those located in a tiny subdiffraction sized region around the zero-intensity point. In order to efficiently quench fluorescence, the stimulated emission process has to compete with spontaneous de-excitation, which normally occurs after few nanoseconds ($\tau_0 \approx 1\text{–}10\,\text{ns}$) from the excitation event. This short time window in which stimulating photons have to act, together with the low cross section of the stimulated emission process ($\sigma_{\text{STED}} \approx 10^{-17}\text{–}10^{-16}\,\text{cm}^2$) [16], demand a high dose of illumination intensity from the STED beam, i.e., a STED microscope normally operates with intensities of the order of 0.1–1 GW/cm^2. Such a range of intensities might introduce side effects such as photobleaching of the fluorophores and phototoxicity for the sample. As a consequence, STED microscopy was long thought to be incompatible with live-cell recording and long-term imaging and a prominent part of the developments in the field of STED microscopy has been focused on the mitigation of this drawback.

Two synergetic strategies have been followed. The first strategy aims at the development of new fluorescent probes with high photostability and/or with spectral fingerprint in the near-infrared region, where phototoxicity reduces [17, 18]. The second strategy aims at the reduction of illumination intensity without sacrificing the spatial resolution and without changing the basic mechanisms of STED microscopy [19, 20].

Within this latter strategy, it has been demonstrated that using the nanosecond fluorescence dynamics information of a time-resolved STED microscopy experiment it is possible to improve the effective resolution of STED microscopy without increasing the (peak) intensity of the STED beam or, equivalently, to reach a certain subdiffraction resolution with reduced (peak) intensity [19, 21]. Two different implementations, but exploring the same idea and working on similar data sets, have been demonstrated so far, respectively the so-called gated-STED (g-STED) microscopy [19, 22–24] and separation of photons by lifetime tuning STED (SPLIT-STED) microscopy [20].

Gated-STED and SPLIT-STED microscopy are based on the observation that since the stimulated emission process opens a new de-excitation pathway for the excited fluorophore, it also shortens the excited-state lifetime of the fluorophore, i.e., the time that the fluorophore spends on average in the excited state after the excitation event. In particular, if the STED beam is running in continuous wave (CW) the lifetime of the fluorophore decreases as $\tau_{S1} = 1/(k_0 + k_{STED})$, with

(i) $k_0 = 1/\tau_0$ the rate of spontaneous de-excitation, which is the inverse of the excited-state lifetime τ_0 in the absence of the STED beam;
(ii) $k_{STED} \propto I^*_{STED}$ the rate of stimulated emission, which is directly proportional to the instantaneous intensity I^*_{STED} of the STED beam [25].

Consequently, the doughnut-shape of the STED beam imposes a spatial signature for the excited-state lifetime of the fluorophores within the excitation spot, i.e., the excited-state lifetime of a fluorophore changes according to its position. The effective excited-state lifetime decreases away from the zero-intensity point, reaching a minimum in the proximity of the doughnut crest where there is maximum STED beam intensity.

As a matter of fact, by selecting only those photons that stem from the long-lived fluorophores it is possible to further reduce the volume from which the registered signal is generated, thus further increasing the resolution of the STED system, without increasing the intensity of the STED beam. It is important to highlight that, since the intensity of the beam does not increase, fluorescence quenching does not improve and the effective fluorescent volume does not change: it is only the region from which the fluorescence signal is recorded that decreases.

The gated-STED and the SPLIT-STED microscopes vary according to the method they use to select the photons from the long-lived fluorophores, which reside in the central part of the excitation volume. However, the two implementations share similar STED architecture. In particular, both need access to the so-called photon arrival time information, namely, for each fluorescent photon it is necessary to know the time passed from the excitation event that generated the photon. This is normally obtained using a pulsed laser which synchronizes the excitation events and a dedicated electronic that measures (with a precision of few tens of picoseconds) the arrival time of the photons.

Given the photon arrival times for each pixel the gated-STED microscope simply selects the photons within a certain time window (time-gated detection) [19], whilst the SPLIT-STED microscope performs a phasor-based analysis on the full photon arrival time histogram [20].

In this chapter we discuss the benefits and the major differences of the two implementations.

5.2 Gated- and SPLIT-STED theory

5.2.1 Temporal point spread function

Both gated-STED and SPLIT-STED microscopy decode extra spatial information from the nanosecond temporal dynamics of the fluorescence signal. In particular, both methods take advantage of the spatial signature of the excited-state lifetime induced by the SE process. Indeed, fluorescence photons are emitted with different temporal dynamics according to the position of the fluorophore inside the STED detection volume. Roughly speaking, photons emitted from the center of the detection volume are characterized by a slow dynamics, whilst photons emitted from the periphery follow fast dynamics. It is clear that both methods improve spatial resolution by highlighting photons that originate from the center and discarding photons that originate from the periphery.

The common starting point for a theoretical description of both approaches is the characterization of the temporal variation of the fluorescent detection volume of the STED system. Since we are mainly describing the detection volume in an imaging system, we can also refer to it with the term point spread function (PSF). For the sake of clarity we will limit the analysis to the lateral dimensions (bi-dimensional model). To obtain an analytical description of the temporal PSF we make the following assumptions:

- a parabolic approximation for the doughnut-shaped STED beam

$$I_{\mathrm{STED}}(r) = I_{\mathrm{STED}}(w)r^2/w^2,$$

  with $I_{\mathrm{STED}}(w)$ the STED beam intensity at position $r = w$;
- a Gaussian profile of the confocal PSF $h(r) = \exp(-2r^2/w^2)$, with w being the beam waist along the radial direction and $r^2 = x'^2 + y'^2$ the radial distance from the focal point ($x = 0, y = 0$);
- a single exponential decay rate for the unperturbed fluorophores $\gamma_0 = 1/\tau_0$, where τ_0 is the unperturbed excited-state lifetime;

– the fluorophores are excited at time zero ($t = 0$) and the STED beam is active during the entire time course of the experiments (i.e., the STED beam is running in CW).

The instantaneous probability of stimulated emission depends linearly on the STED beam intensity:

$$\gamma_{\mathrm{STED}}(r^2) = \gamma_0 \left[\frac{I_{\mathrm{STED}}(r)}{I_{\mathrm{SAT}}} \right], \tag{5.1}$$

where the constant I_{SAT}, usually called saturation intensity, represents the value of intensity for which $\gamma_{\mathrm{STED}} = \gamma_0$. The resulting radial distribution of the decay rates can be approximated by a parabolic function:

$$\gamma(r^2) = \gamma_0 + \gamma_{\mathrm{STED}}(r^2) = \gamma_0 + \gamma_0 \left[I_{\mathrm{STED}}(w) \frac{r^2/w^2}{I_{\mathrm{SAT}}} \right] = \gamma_0 + \gamma_0 k_{\mathrm{S}} r^2/w^2, \tag{5.2}$$

where we have defined:

$$k_{\mathrm{S}} = \frac{I_{\mathrm{STED}}(w)}{I_{\mathrm{SAT}}}. \tag{5.3}$$

According to this definition, k_{S} is the ratio between $I_{\mathrm{STED}}(w)$ and the saturation value I_{SAT} for which the probability of decay due to SE and spontaneous emission are equal.

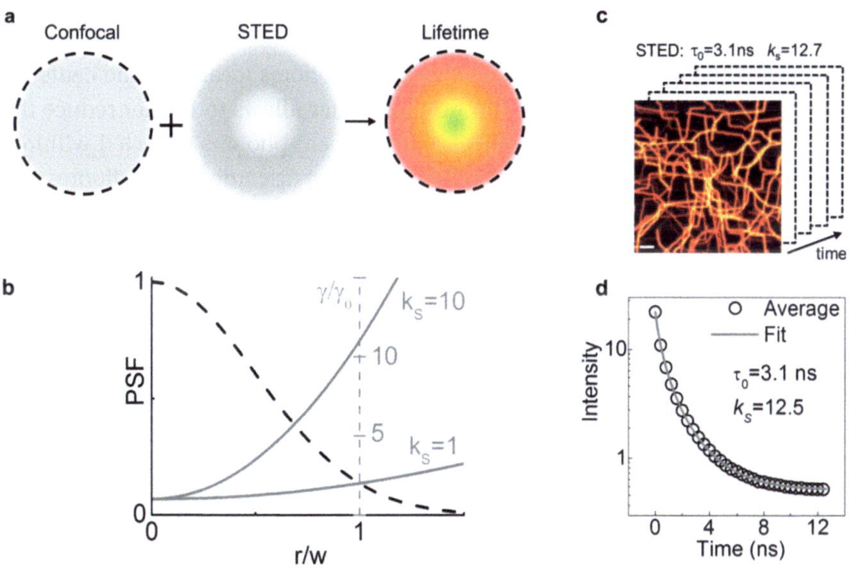

Fig. 5.1: (a) A doughnut-shaped STED beam coaligned with a confocal spot generates a continuous distribution of dynamics within the PSF. (b) The STED beam intensity determines the relative variation of decay rate γ/γ_0 (solid line) across the PSF (dashed line). (c) Simulated time-resolved STED image of filamentous structures resembling cytoskeletal networks. The input parameters were: $\tau_0 = 3.1$ ns, $k_{\mathrm{S}} = 12.7$, $S = 120$, $B = 0.5$. (d) The average STED decay is fitted to equation (5.11) to obtain the parameter k_{S} ($k_{\mathrm{S}} = 12.5$ from the fit in keeping with the input).

Importantly, the value of k_S quantifies the relative variation of decay rate values within the PSF of the CW-STED microscope [20] (Fig. 5.1 (b)).

The time-dependent fluorescence intensity $F(x, y, t)$ at each pixel can be expressed as:

$$F(x, y, t) = K \int e^{-\gamma(r^2)t} \rho(x', y')\, e^{-2r^2/w^2}\, dx'\, dy', \tag{5.4}$$

where K is a constant that depends on the quantum yield of the fluorophore, the maximum of the excitation intensity and the detection efficiency, $r^2 = (x' - x)^2 + (y' - y)^2$ and $\rho(x, y)$ is the density of fluorophores.

Since the point spread function of an optical system is defined as the image of an ideal point source, we can obtain the equation of the temporal PSF by substituting the density of fluorophores $\rho(x', y')$ with Dirac's δ function in equation (5.4). The temporal PSF reads:

$$h_{STED}(r, t) = K\, e^{-\gamma(r^2)t}\, e^{-2r^2/w^2}. \tag{5.5}$$

5.2.2 Gated-STED microscopy

The gated-STED microscope increases the spatial resolution of a CW-STED implementation by discarding early photons, most likely generated from short-lived fluorophores located in the periphery of the detection volume, and highlighting late photons most likely generated from long-lived fluorophores located in the center of the detection volume. In a nutshell, the temporal filter allows to further reduce the effective detection volume. For this purpose, fluorescent photons collected within a short time interval from the excitation events $[0, T_g]$ are discarded whilst fluorescent photons in the remaining part of the excitation pulse period $[T_g, T]$ generate the final image. The point spread function of this imaging modality can be derived from the temporal integration of equation (5.5) in the interval of collection $[T_g, T]$. If we assume that the pulse interval is much longer than the excited-state lifetime of the fluorophores ($T \gg \tau_0$), the PSF of a gated CW-STED microscope reads:

$$\begin{aligned} h_{g,CW\text{-}STED}(r) &= \int_{T_g}^{T} h_{STED}(r, t)\, dt \\ &= K\, e^{-2r^2/w^2} \frac{1}{\gamma_0 + \gamma_0 k_S r^2/w^2} e^{-(\gamma_0 + \gamma_0 k_S r^2/w^2)T_g} \\ &= K\, e^{-\gamma_0 T_g}\, e^{-2r^2/w^2} \frac{1}{\gamma_0 + \gamma_0 k_S r^2/w^2} e^{-T_g \gamma_0 k_S r^2/w^2}. \end{aligned} \tag{5.6}$$

Notably, the effective PSF can be divided into four terms:

(i) a constant term that denotes the amplitude of the PSF, clearly indicating that the signal decreases exponentially as the time-gating value T_g increases;

(ii) a Gaussian function that denotes the confocal PSF;

(iii) a Lorentzian function that reflects the PSF of a non-gated ($T_g = 0$) CW-STED microscope;
(iv) a Gaussian term that reflects the reduction of the effective PSF for increasing T_g.

The PSF (equation (5.6)) immediately highlights the positive and negative aspects of the gCW-STED implementation. For increasing T_g the effective PSF shrinks in size, furthermore the Gaussian term dominates over the Lorentzian term removing the PSF tails which drastically decrease the contrast, thus the effective resolution in the conventional CW-STED microscope. The resolution of a gated CW-STED microscope theoretically increases to infinity for increasing T_g, but the reduction of the PSF's amplitude poses an effective limit. Indeed, the reduction of the PSF's amplitude translates into a reduction of the signal-to-noise ratio (SNR) of the gated images. It is clear that a strong reduction of the SNR can cancel out the benefits of the time-gating. In real life, the value T_g is settled to half of the excited-state lifetime τ_0, in this case the image contrast effectively improves, but the PSF's amplitude is reduced only by a factor of 0.4. However, recently it has been shown that the SNR ratio in gated-STED implementation can be recovered by combining the usually discarded early photons with the conventional gated-STED image (late photons) through a multi-image deconvolution algorithm [26, 27].

5.2.3 SPLIT-STED microscopy

The goal of SPLIT is to achieve super-resolution by an explicit separation of the different components of the fluorescent signals according to their "temporal decays". In particular, SPLIT microscopy isolates the slow component that is most likely induced by fluorophores in the center of the effective detection volume of a CW-STED system. Thus, the maximum achievable spatial resolution will be ultimately determined by the ability to distinguish between different temporal dynamics. In this framework the problem of resolving spatial features is translated into the spectroscopy problem of resolving the stimulated emission-induced variations of the fluorescence lifetime. First, we will find an explicit expression for the temporal decay of fluorescence emitted from the center of the detection volume in a CW-STED microscope. Second, we will describe a fit-free analysis of the fluorescence decays aimed at quantifying the relative contribution of this component at each pixel of the image.

5.2.3.1 Explicit formulas for the fluorescence decays in CW-STED microscopy

Starting from equation (5.4), we conveniently switch to a system of polar coordinates centered on the pixel (x, y) and integrate along ϕ':

$$F(x, y, t) = K \int_0^\infty C(r^2)\, \mathrm{d}r^2\, \mathrm{e}^{-\gamma(r^2)t}\, \mathrm{e}^{-2r^2/w^2}, \qquad (5.7)$$

with

$$C(r^2)\,\mathrm{d}r^2 = \int_0^{2\pi} \rho(r', \phi')r\,\mathrm{d}r\,\mathrm{d}\phi. \tag{5.8}$$

With this definition, $C(r)$ describes the effective concentration of fluorophores in a concentric ring of radius r around the pixel position. The formula shows that the temporal dynamics of $F(x, y, t)$ encodes nanoscale spatial information in the distribution of exponential decay components. This distribution is fully described in this model by the parameters γ_0 and k_S in equation (5.2).

The parameter $\tau_0 = 1/\gamma_0$ depends on the specific fluorophore and can be measured from the sample with the same instrumentation by setting the STED beam power to zero. The parameter $k_S = I_{STED}(w)/I_{SAT}$ is proportional to the STED beam power and its precise value depends on the optical configuration (in particular the shape of the excitation and the STED beams) and on the properties of the sample. We can show that, using our analytical model of SE-induced lifetime variations, we can estimate the value of k_S from the very same image $F(x, y, t)$ by considering the average time-resolved decay $I(t)$ of all the pixels of an image:

$$I(t) = \iint_{x,y} F(x, y, t)\,\mathrm{d}x\,\mathrm{d}y. \tag{5.9}$$

For simplicity, consider the case of an image obtained by scanning around a point-like object, for which $C(r^2) \approx \delta(r^2)$:

$$\begin{aligned}\iint_{x,y} F(x, y, t)\,\mathrm{d}x\,\mathrm{d}y &= K \iint_{x,y} \mathrm{d}x\,\mathrm{d}y \int_0^{\infty} C(r^2)\,\mathrm{d}r^2\,\mathrm{e}^{-\gamma(r^2)t}\,\mathrm{e}^{-2r^2/w^2} \\ &\approx K \int_0^{\infty} \mathrm{d}r^2\,\mathrm{e}^{-\gamma(r^2)t}\,\mathrm{e}^{-2r^2/w^2} \\ &\propto \mathrm{e}^{-\gamma_0 t}\frac{1}{1 + k_S\gamma_0 t/2}.\end{aligned} \tag{5.10}$$

If we take into account the presence of uncorrelated background we obtain the following functional form for the average fluorescence decay under CW-stimulated emission [20]:

$$I(t) = A\,\mathrm{e}^{-\gamma_0 t}\frac{1}{1 + k_S\gamma_0 t/2} + B. \tag{5.11}$$

In order to assess the generality of this formula for the imaging of non-point-like structures, we simulated time-resolved STED images of filamentous structures similar to those found in cytoskeletal networks (Fig. 5.1 (c)). The images were obtained by assuming an unperturbed fluorescence lifetime value $\tau_0 = 3.1$ ns whereas the STED intensity level was such that the variations across the PSF could be modeled by a value $k_S = 12.7$ according to equation (5.2). The resulting time-resolved STED image can be seen as the convolution of the object with a spatio-temporal PSF (equation (5.4)). As expected, when we look at the average temporal decay of all the pixels of the image (Fig. 5.1 (d)), this is not a single exponential. Here, the nonexponential behavior is not due to the presence of several fluorescent species (we are assuming a single fluorescent species with a specific τ_0) but rather to the continuous distribution of decay rates induced by the STED beam. The nonexponential temporal trend derived in our model and described by equation (15.1) can be a good approximation of the average STED decay (Fig. 5.1 (d)). If the average STED decay is fitted to equation (5.11) we obtain the parameters $\tau_0 = 3.1$ ns and $k_S = 12.5$ in keeping with the input. It is worth mentioning that with this analysis we are able to estimate experimentally the parameters τ_0 and k_S of the spatio-temporal PSF without needing to know the exact value of the waist w. This is because in our model the parameter k_S has been defined as an indicator of the relative variation of decay rate across the confocal PSF.

Now that we have a quantitative model for describing the SE-induced lifetime variations under the CW-STED microscopy architecture, we need a method to isolate the longer lifetime components of the distribution (Fig. 5.1 (a)) in order to produce a higher resolution image. It is convenient to approximate the continuous distribution of decays to a discrete number n of decay components. Consider for instance the simplest case $n = 2$ (Fig. 5.2 (a)). In order to find an explicit expression for the associated temporal decays (Fig. 5.2 (b)), we split the integral in equation (5.10) into $n = 2$ parts:

$$F(x, y, t) = K \int_0^{r_1} C(r^2)\,\mathrm{d}r^2\, \mathrm{e}^{-\gamma(r^2)t}\, \mathrm{e}^{-2r^2/w^2} + K \int_{r_1}^{\infty} C(r^2)\,\mathrm{d}r^2\, \mathrm{e}^{-\gamma(r^2)t}\, \mathrm{e}^{-2r^2/w^2}. \tag{5.12}$$

Within this approximation, we need to define two components which do not depend on the function $C(r^2)$. By approximating, $C(r^2) \approx C_0^i$, we can write, for $i = 1, 2$:

$$\int_{r_{i-1}}^{r_i} C(r^2)\,\mathrm{d}r^2\, \mathrm{e}^{-\gamma(r^2)t}\, \mathrm{e}^{-2r^2/w^2} \approx C_0^i \int_{r_{i-1}}^{r_i} \mathrm{d}r^2\, \mathrm{e}^{-\gamma(r^2)t}\, \mathrm{e}^{-2r^2/w^2}, \tag{5.13}$$

where $r_j = 0, r_1, \infty$, for $j = 0, 1, 2$, respectively. We are approximating $C(r^2)$ inside (r_{i-1}^2, r_i^2) with its average value within this interval and ignoring its variations inside (r_{i-1}^2, r_i^2). The boundary r_1 of the two subdiffraction volumes can be arbitrarily chosen in such a way that:

$$\int_0^T I_1(t)\,\mathrm{d}t = \int_0^T I_2(t)\,\mathrm{d}t. \tag{5.14}$$

Then we define the time-dependent decay of the i-th component as:

$$\begin{aligned} I_i(t) &= \int_{r_{i-1}}^{r_i} \mathrm{d}r^2\, \mathrm{e}^{-\gamma(r^2)t}\, \mathrm{e}^{-2r^2/w^2} \\ &\propto \mathrm{e}^{-\gamma_0 t} \frac{1}{1 + k_S\gamma_0 t/2} \left(\mathrm{e}^{-(1+k_S\gamma_0 t/2)2r_{i-1}^2/w^2} - \mathrm{e}^{-(1+k_S\gamma_0 t/2)2r_i^2/w^2}\right). \end{aligned} \tag{5.15}$$

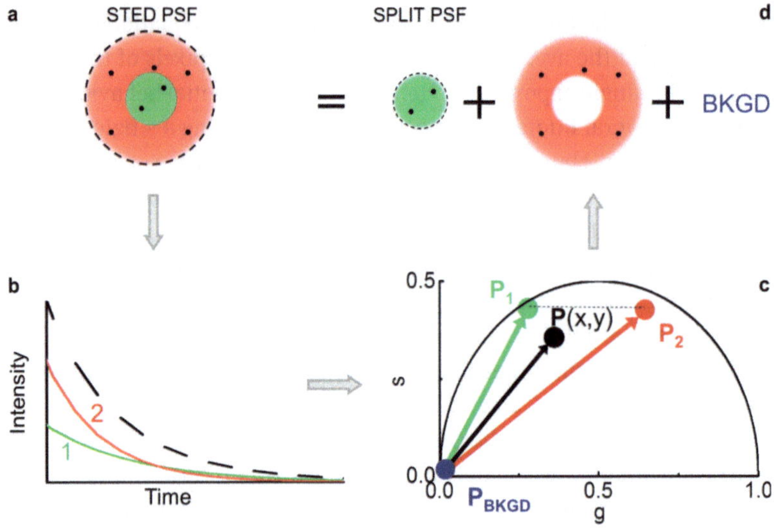

Fig. 5.2: (a) The STED-induced continuous distribution of decay rates across the PSF is approximated by only 2 spatial components. (b) The measured fluorescence decay (dashed line) is expressed as the linear combination of the slower decay component associated to the inner part of the PSF (1) and the faster decay component associated to the outer part of the PSF (2). (c) The decays are represented as vectors in the phasor plot: the experimental phasor $\boldsymbol{P}(x, y)$ is assumed to be a linear combination of the phasors $\boldsymbol{P}_1$ and $\boldsymbol{P}_2$ plus the phasor of the uncorrelated background ($\boldsymbol{P}_{BKGD}$). (d) The fractional contribution of each component ($\boldsymbol{P}_1$, $\boldsymbol{P}_2$, $\boldsymbol{P}_{BKGD}$) is calculated and the photons are sorted into three separate images. The SPLIT image associated to the slower component has higher spatial resolution than the STED image and is separated from the background.

Then we can write the intensity as a linear combination of components:

$$F(x, y, t) \approx K \lfloor C_0^1(x, y) I_1(t) + C_0^2(x, y) I_2(t) \rfloor \tag{5.16}$$

The intensity decay at each pixel (Fig. 5.2 (b), dashed line) can be expressed as the sum of a slow (Fig. 5.2 (b), green line) and a fast (Fig. 5.2 (b), red line) decay component, associated respectively to the fluorophores located on the inner and outer part of the PSF (Fig. 5.2 (a)).

5.2.3.2 Phasor analysis in SPLIT-STED

Now, we will describe a fit-free analysis of the fluorescence decays aimed at quantifying the relative contribution of a specific component at each pixel of an image. This analysis is based on the phasor approach to FLIM [28–30], a methodology that has found application in FRET [31–35], spectral fingerprinting [36, 37], correlation spectroscopy [38, 39] and super-resolution [20]. An important outcome of the whole SPLIT-STED approach is that any contribution to the fluorescence signal that is uncorrelated on the nanosecond timescale is separated from the final super-resolved image.

For example, the anti-Stokes fluorescence emission induced by the STED beam. Notably, this source of background will be uncorrelated since the STED beam is running in CW. Anti-Stokes fluorescence emission is a common source of background in STED microscopy, especially when the STED beam's wavelength is closer to the emission maximum of the fluorophore [40, 41].

In the phasor analysis any decay $J(t)$ is converted via a Fast Fourier Transform (FFT) into a pair of phasor coordinates (g, s) defined as:

$$\begin{aligned} g &= \frac{\int_0^T J(t)\cos(2\pi t/T)\,\mathrm{d}t}{\int_0^T J(t)\,\mathrm{d}t}, \\ s &= \frac{\int_0^T J(t)\sin(2\pi t/T)\,\mathrm{d}t}{\int_0^T J(t)\,\mathrm{d}t}, \end{aligned} \tag{5.17}$$

where T is the pulse interval. Here we assume that $T \gg \tau_0$, thus the function $J(t)$ has already decayed to background [30]. The slow and fast decays (Fig. 5.2 (b)) can be now represented as two vectors $\boldsymbol{P}_1 = (g_1, s_1)$ and $\boldsymbol{P}_2 = (g_2, s_2)$ in the phasor plot (Fig. 5.2 (c)).

We can write the intensity at one pixel as the sum of two components plus a term representing any uncorrelated background $F(x, y, t) = A_1(x, y)I_1(t) + A_2(x, y)I_2(t) + I_{\mathrm{BKGD}}(x, y)$. Thus, the phasor of each pixel $\boldsymbol{P}(x, y) = (g_1(x, y), s_1(x, y))$ can be expressed as a linear combination of the phasors of the two components and the phasor of the background

$$\begin{aligned} g(x, y) &= \frac{\int_0^T [A_1(x, y)I_1(t) + A_2(x, y)I_2(t) + I_{\mathrm{BKGD}}(x, y)]\cos(2\pi t/T)\,\mathrm{d}t}{\int_0^T [A_1(x, y)I_1(t) + A_2(x, y)I_2(t) + I_{\mathrm{BKGD}}(x, y)]\,\mathrm{d}t} \\ &= \frac{N_1(x, y)}{N(x, y)}g_1 + \frac{N_2(x, y)}{N(x, y)}g_2, \\ s(x, y) &= \frac{\int_0^T [A_1(x, y)I_1(t) + A_2(x, y)I_2(t) + I_{\mathrm{BKGD}}(x, y)]\sin(2\pi t/T)\,\mathrm{d}t}{\int_0^T [A_1(x, y)I_1(t) + A_2(x, y)I_2(t) + I_{\mathrm{BKGD}}(x, y)]\,\mathrm{d}t} \\ &= \frac{N_1(x, y)}{N(x, y)}s_1 + \frac{N_2(x, y)}{N(x, y)}s_2, \end{aligned} \tag{5.18}$$

where the total number of photons $N(x, y)$ detected at one pixel is the sum of the photons originating in the two spatial components plus the uncorrelated background $N(x, y) = N_1(x, y) + N_2(x, y) + N_{\mathrm{BKGD}}(x, y)$. Since the uncorrelated background is independent of t, its phasor coordinates are $\boldsymbol{P}_{\mathrm{BKGD}} = (0, 0)$ (Fig. 5.2 (c)). Using vector notation we can rewrite:

$$\boldsymbol{P}(x, y) = \frac{N_1(x, y)\boldsymbol{P}_1 + N_2(x, y)\boldsymbol{P}_2}{N(x, y)} = f_1(x, y)\boldsymbol{P}_1 + f_2(x, y)\boldsymbol{P}_2.$$

This linear system of equations can be written in the form $\boldsymbol{P} = \boldsymbol{Mf}$, where $\boldsymbol{f} = (f_1, f_2)$ is the vector of the unknown fractional components and M_{ij} is the matrix

$$\boldsymbol{M} = \begin{pmatrix} g_1 & g_2 \\ s_1 & s_2 \end{pmatrix}, \tag{5.19}$$

which describes the two different temporal dynamics in the phasor domain. The solution of this system is given by $\boldsymbol{f} = \boldsymbol{M}^{-1}\boldsymbol{P}$. The images $N_i(x, y)$ $(i = 1, 2)$ of the photons emitted in each of the two subdiffraction volumes and the image $N_{\mathrm{BKGD}}(x, y)$ of the background can be calculated as $N_i(x, y) = f_i(x, y)N(x, y)$ and

$$N_{\mathrm{BKGD}}(x, y) = \left[1 - \sum_{i=1}^{n} f_i(x, y)\right] N(x, y), \tag{5.20}$$

respectively (Fig. 5.2 (d)). By using a model for the STED-induced variations of fluorescence lifetime and the phasor analysis of lifetime data we have obtained a super-resolved image $N_1(x, y)$ characterized by a higher effective resolution than the CW-STED counterpart (Fig. 5.2 (d)).

Extending this procedure to n components we can state that the original image $N(x, y)$ can be split into $n + 1$ images based on the assumption that we are able to distinguish, within our observation volume, n different dynamics, associated with n linearly independent vectors in the n-dimensional phasor space. If we take noise into account, we find that the larger the number n of components, the higher will be the noise propagated to the final SPLIT images. Indeed, since the SPLIT image is calculated as: $N_1(x, y) = f_1(x, y)N(x, y)$, it is affected by the noise in the STED image $N(x, y)$ and, in addition, by any noise in the factor $f_1(x, y)$. The factor $f_1(x, y)$ is calculated at each pixel from the measurement of $g(x, y)$ and $s(x, y)$ which contain experimental noise. The noise in the measurement of g and s (or higher harmonics components) is propagated to the fraction f_1 through the linear system $\boldsymbol{P} = \boldsymbol{Mf}$ depending on the matrix $\boldsymbol{M}$. One way to quantify the error propagation through this linear system is by considering the condition number k_{cond} defined as the product of the norms of the matrices $\boldsymbol{M}$ and its inverse $\boldsymbol{M}^{-1}$, $k_{\mathrm{cond}} = \|\boldsymbol{M}\| \, \|\boldsymbol{M}^{-1}\|$. This parameter decreases with higher values of k_S and increases exponentially with the value of n [20]. Even if, theoretically, unlimited spatial resolution could be foreseen by considering higher values of n, we must conclude that, for practical purposes, noise will restrict the use of the SPLIT algorithm only to low values of n.

5.3 Gated- and SPLIT-STED comparison

The operational difference between gated-STED and SPLIT-STED is that, for each pixel, the gated-STED microscope simply selects the photons within a certain time window (time-gated detection), whereas the SPLIT-STED microscope performs a phasor-based analysis on the full photon arrival time histogram. In both cases this operation is aimed at isolating the slower component corresponding to the inner part of the PSF. However, there is also a strong conceptual difference between these two approaches: whilst the SPLIT-STED approach attempts to perform a true separation of the dynamics components, the gated-STED approach selects the fluorescence photons that most likely belong to the slow dynamics (the longer T_g is, the higher is the probability that the registered photons belong to the slow dynamics and thus are generated from the center of the detection volume). On the other hand, time-gating is straightforward: it does not require the data analysis involved in the SPLIT framework and is simple to implement. In particular, since it does not need the full photon arrival time histogram, the expensive time-correlated single-photon counting based hardware can be substituted by cheaper time-gated electronics [19, 23, 42].

The difference between the two approaches can be more easily understood by looking at a specific experimental case. Fig. 5.3 (a) and (c) report the time-resolved confocal and STED images of an area of a fixed cell where microtubules have been immunostained with the dye Oregon Green 488. The phasor plot of the confocal image (Fig. 5.3 (b)) shows a single spot in agreement with the fact that all the pixels of the image have the same lifetime. The phasor plot of the STED image (Fig. 5.3 (d)) shows a significant spreading, indicating that the pixels of the image may have different lifetimes. Remarkably, there is a significant bending of the phasor toward the

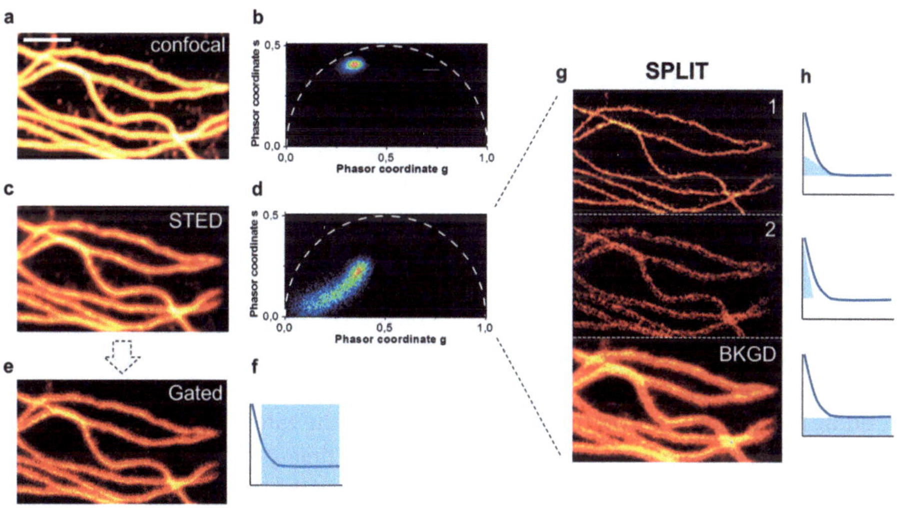

Fig. 5.3: Time-gating versus SPLIT analysis in time-resolved STED. Time-resolved confocal (a) and STED (c) images of tubulin in fixed HeLa cells immunostained with Oregon Green 488 and associated phasor plots (b, d). (e) Time-gated image (T_g = 1 ns) and (f) sketch of the operation of time-gating in the time domain. (g) SPLIT image series showing the images associated respectively to the slow component (1), fast component (2) and the uncorrelated background (BKGD). (h) Sketch of the operation of SPLIT in the time domain. The STED beam power was set to 40 mW. Scale bar 2 µm.

origin of the plot indicating the presence of a strong uncorrelated background in most of the pixels. The presence of uncorrelated background is explained by the anti-Stokes fluorescence emission induced by the CW-STED beam at $\lambda = 560$ nm on the dye Oregon Green 488 [20, 43]. In this case, when we perform time-gating, we efficiently reject the fast temporal component but the signal is still degraded by the contribution from the background (Fig. 5.3 (f) and (g)). Since the background is not separated from the high-resolution component, the final time-gated image does not show a significant improvement in spatial resolution (Fig. 5.3 (f)). On the other hand, by performing the SPLIT analysis, we are able to generate three separate images (Fig. 5.3 (g)) corresponding to the slower temporal component (1), the faster temporal component (2) and the background (BKGD). In this case the super-resolved SPLIT image is not degraded by the presence of the strong uncorrelated background signal (Fig. 5.3 (g)). This is because the SPLIT image represents the exclusive contribution of each temporal component to the total signal in a pixel (Fig. 5.3 (h)). In other words, only with the SPLIT approach are we performing a true separation of the dynamics components at each pixel of the image. It is worth noting that different approaches exist to remove the anti-Stokes emission background from gated-STED. In TCSPC-based gated-STED implementation, the anti-Stokes emission background can be subtracted/removed by estimating it from the photon arrival times histogram [27, 43], similar to SPLIT-STED. More generally, anti-Stokes emission background can be subtracted by using lock-in (synchronous) detection approaches [40].

5.4 Discussion and conclusions

We have presented two methods, gated-STED and SPLIT-STED, that provide a significant reduction of illumination intensity, without sacrificing spatial resolution and without changing the basic mechanisms of STED microscopy. Both methods exploit the nanosecond fluorescence dynamics information of a time-resolved STED microscope to improve the effective spatial resolution without increasing the (peak) intensity of the STED beam or, equivalently, to reach a certain subdiffraction resolution with reduced (peak) intensity. The main difference between the two approaches resides on the strategy they use to select the photons from the long-lived fluorophores residing in the central part of the fluorescent volume: the gated-STED microscope simply selects the photons within a certain time window (time-gated detection), while the SPLIT-STED microscope performs an explicit analysis of the temporal dynamics of fluorescence.

It is worth noting that this strategy of improving spatial resolution by using differences in fluorescence dynamics is common to other techniques that are not based on stimulated emission. For instance DSOM [44], which exploits a singlet-triplet state transition [45], and methods based on the switching dynamics of reversibly switchable fluorophores [46], also take advantage of the additional spatial information encoded in fluorescence dynamics. However, the timescale on which the dynamics is recorded depends on the photophysical process exploited by the technique. For instance, in the aforementioned methods it ranges from microseconds (DSOM) to milliseconds (reversible switching), which is much slower than what can be achieved in SPLIT-STED. Since in the latter case the dynamics is recorded in the nanosecond range, the maximum achievable temporal resolution of SPLIT-STED is not compromised and this approach is also compatible with those techniques, like fluorescence correlation spectroscopy (FCS) [47], that require temporal resolution down to the microsecond scale.

As a matter of fact, FCS has become one of the methods of choice to probe the motion of proteins and other molecules with minimal perturbation on specific subcellular compartments [48–50]. An efficient combination of FCS with super-resolution microscopy would represent the opportunity to probe molecular diffusion at the nanoscale inside cells. STED-FCS has been successfully applied to study the dynamics of membrane lipids at the nanoscale [51, 52]. In the context of live-cell experiments, the use of strategies aimed at reducing illumination intensity, without sacrificing spatial resolution, is of crucial importance. For this reason we expect that the principles of gated-STED and SPLIT-STED will be exploited in the near future also in this field [53].

References

[1] Diaspro A. Circumventing the diffraction limit. Il Nuovo Saggiatore. 2014;30:45–51.

[2] Cella Zanacchi F, Bianchini P, Vicidomini G. Fluorescence microscopy in the spotlight. Microscopy Research and Technique. 2014;77(7):479–482.

[3] Hell SW. Toward fluorescence nanoscopy. Nature Biotechnology. 2013;21(11):1347–1355.

[4] Hell SW. Microscopy and its focal switch. Nature Methods. 2009;6(1):24–32.

[5] Diaspro A, editor. Nanoscopy and multidimensional optical fluorescence microscopy. Chapman & Hall; 2009.

[6] Huang B, Babcock H, Zhuang X. Breaking the diffraction barrier: super-resolution imaging of cells. Cell. 2010;143(7):1047–1058.

[7] Hell SW, Sahl SJ, Bates M, Zhuang X, Heintzmann R, Booth MJ, Bewersdorf J, Shtengel G, Hess H, Tinnefeld P, Honigmann A, Jakobs S, Testa I, Cognet L, Lounis B, Ewers H, Davis SJ, Eggeling C, Klenerman D, Willig KI, Vicidomini G, Castello M, Diaspro A, Cordes T. The 2015 super-resolution microscopy roadmap. J Phys D: Appl Phys. 2015;48(44):443001–443001.

[8] Hell SW, Wichmann J. Breaking the diffraction resolution limit by stimulated emission: stimulated-emission-depletion fluorescence microscopy. Opt Lett. 1994;19(11):780–782.

[9] Hofmann M, Eggeling C, Jakobs S, Hell SW. Breaking the diffraction barrier in fluorescence microscopy at low light intensities by using reversibly photoswitchable proteins. Proceedings of the National Academy of Sciences of the United States of America. 2005;102(49):17565–17569.

[10] Gustafsson MG. Nonlinear structured-illumination microscopy: wide-field fluorescence imaging with theoretically unlimited resolution. Proceedings of the National Academy of Sciences of the United States of America. 2005;102(37):13081–13086.

[11] Hess ST, Girirajan TP, Mason MD. Ultra-high resolution imaging by fluorescence photoactivation localization microscopy. Biophysical Journal. 2006;91(11):4258–4272.

[12] Rust MJ, Bates M, Zhuang X. Sub-diffraction-limit imaging by stochastic optical reconstruction microscopy (STORM). Nature Methods. 2006;3(10):793–795.
[13] Betzig E, Patterson GH, Sougrat R, Lindwasser OW, Olenych S, Bonifacino JS, Davidson MW, Lippincott-Schwartz J, Hess HF. Imaging intracellular fluorescent proteins at nanometer resolution. Science. 2006;313(5793):1642–1645.
[14] Dertinger T, Colyer R, Iyer G, Weiss S, Enderlein J. Fast, background-free, 3D super-resolution optical fluctuation imaging (SOFI). Proceedings of the National Academy of Sciences of the United States of America. 2009;106(52):22287–22292.
[15] Hell SW. Far-field optical nanoscopy. Science. 2007;316(5828):1153–1158.
[16] Kastrup L, Hell SW. Absolute optical cross section of individual fluorescent molecules. Angew Chem Int Ed Engl. 2004;43(48):6646–6649.
[17] Lukinavicius G, Umezawa K, Olivier N, Honigmann A, Yang G, Plass T, Mueller V, Reymond L, Correa IR Jr, Luo ZG, Schultz C, Lemke EA, Heppenstall P, Eggeling C, Manley S, Johnsson K. A near-infrared fluorophore for live-cell super-resolution microscopy of cellular proteins. Nat Chem. 2013;5(2):132–139.
[18] Lukinavicius G, Blaukopf C, Pershagen E, Schena A, Reymond L, Derivery E, Gonzalez-Gaitan M, D'Este E, Hell SW, Gerlich DW, Johnsson K. SiR-Hoechst is a far-red DNA stain for live-cell nanoscopy. Nature Communications. 2015;6:8497.
[19] Vicidomini G, Moneron G, Han KY, Westphal V, Ta H, Reuss M, Engelhardt J, Eggeling C, Hell SW. Sharper low-power STED nanoscopy by time gating. Nature Methods. 2011;8(7):571–573.
[20] Lanzano L, Coto Hernandez I, Castello M, Gratton E, Diaspro A, Vicidomini G. Encoding and decoding spatio-temporal information for super-resolution microscopy. Nature Communications. 2015;6:6701.
[21] Moffitt JR, Osseforth C, Michaelis J. Time-gating improves the spatial resolution of STED microscopy. Optics Express. 2011;19(5):4242–4254.
[22] Vicidomini G, Hernandez IC, d'Amora M, Zanacchi FC, Bianchini P, Diaspro A. Gated CW-STED microscopy: a versatile tool for biological nanometer scale investigation. Methods. 2014;66(2):124–130.
[23] Coto Hernàndez I, Buttafava M, Boso G, Diaspro A, Tosi A, Vicidomini G. Gated STED microscopy with time-gated single-photon avalanche diode. Biomed Opt Express. 2015;6(6):2258–2267.
[24] Vicidomini G, Coto Hernàndez I, Diaspro A, Galiani S, Eggeling C. The Importance of Photon Arrival Times in STED Microscopy. In: Kapusta P, Wahl M, Erdmann R, editors. Advanced Photon Counting, vol. 15. Springer International Publishing; 2015. p. 283–301.
[25] Vicidomini G, Schonle A, Ta H, Han KY, Moneron G, Eggeling C, Hell SW. STED nanoscopy with time-gated detection: theoretical and experimental aspects. PLoS One. 2013;8(1):e54421.
[26] Castello M, Diaspro A, Vicidomini G. Multi-images deconvolution improves signal-to-noise ratio on gated stimulated emission depletion microscopy. Appl Phys Lett. 2014;105(23):234106.
[27] Coto Hernández I, Castello M, Lanzanò L, d'Amora M, Bianchini P, Diaspro A, Vicidomini G. Two-photon excitation STED microscopy with time-gated detection. Scientific Reports. 2016; 6:19419–19419 EP.
[28] Clayton AH, Hanley QS, Verveer PJ. Graphical representation and multicomponent analysis of single-frequency fluorescence lifetime imaging microscopy data. J Microsc. 2004;213(Pt 1):1–5.
[29] Redford GI, Clegg RM. Polar plot representation for frequency-domain analysis of fluorescence lifetimes. J Fluoresc. 2005;15(5):805–815.
[30] Digman MA, Caiolfa VR, Zamai M, Gratton E. The phasor approach to fluorescence lifetime imaging analysis. Biophysical Journal. 2008;94(2):L14–L16.

[31] Barreiro O, Zamai M, Yanez-Mo M, Tejera E, Lopez-Romero P, Monk PN, Gratton E, Caiolfa VR, Sanchez-Madrid F. Endothelial adhesion receptors are recruited to adherent leukocytes by inclusion in preformed tetraspanin nanoplatforms. J Cell Biol. 2008;183(3):527–542.
[32] Blaine J, Lanzano L, Giral H, Caldas Y, Levi M, Gratton E, Moldovan R, Lei T. Dynamic imaging of the sodium phosphate cotransporters. Adv Chronic Kidney Dis. 2011;18(2):145–150.
[33] Giral H, Cranston D, Lanzano L, Caldas Y, Sutherland E, Rachelson J, Dobrinskikh E, Weinman EJ, Doctor RB, Gratton E, Levi M. NHE3 regulatory factor 1 (NHERF1) modulates intestinal sodium-dependent phosphate transporter (NaPi-2b) expression in apical microvilli. J Biol Chem 2012;287(42):35047–35056.
[34] Giral H, Lanzano L, Caldas Y, Blaine J, Verlander JW, Lei T, Gratton E, Levi M. Role of PDZK1 protein in apical membrane expression of renal sodium-coupled phosphate transporters. J Biol Chem. 2011;286(17):15032–15042.
[35] Dobrinskikh E, Lanzano L, Rachelson J, Cranston D, Moldovan R, Lei T, Gratton E, Doctor RB. Shank2 contributes to the apical retention and intracellular redistribution of NaPiIIa in OK cells. Am J Physiol Cell Physiol. 2013;304(6):C561–C573.
[36] Cutrale F, Salih A, Gratton E. Spectral Phasor approach for fingerprinting of photo-activatable fluorescent proteins Dronpa, Kaede and KikGR. Methods Appl Fluoresc. 2013;1(3):35001.
[37] Fereidouni F, Bader AN, Colonna A, Gerritsen HC. Phasor analysis of multiphoton spectral images distinguishes autofluorescence components of in vivo human skin. J Biophotonics. 2014;7(8):589–596.
[38] Ranjit S, Lanzano L, Gratton E. Mapping diffusion in a living cell via the phasor approach. Biophysical Journal. 2014;107(12):2775–2785.
[39] Scipioni L, Gratton E, Diaspro A, Lanzanò L. Phasor analysis of local ICS detects heterogeneity in size and number of intracellular vesicles. Biophysical Journal. 2016;111:619–629.
[40] Vicidomini G, Moneron G, Eggeling C, Rittweger E, Hell SW. STED with wavelengths closer to the emission maximum. Optics express. 2012;20(5):5225–5236.
[41] Coto Hernàndez I, d'Amora M, Diaspro A, Vicidomini G. Influence of laser intensity noise on gated CW-STED microscopy. Laser Phys Lett. 2014;11(9):095603.
[42] Wu X, Toro L, Stefani E, Wu Y. Ultrafast photon counting applied to resonant scanning STED microscopy. J Microsc. 2015;257(1):31–38.
[43] Coto Hernandez I, Peres C, Cella Zanacchi F, d'Amora M, Christodoulou S, Bianchini P, Diaspro A, Vicidomini G. A new filtering technique for removing anti-Stokes emission background in gated CW-STED microscopy. J Biophotonics. 2014;7(6):376–380.
[44] Enderlein J. Breaking the diffraction limit with dynamic saturation optical microscopy. Applied Physics Letters. 2005;87(9)094105.
[45] Humpolickova J, Benda A, Machan R, Enderlein J, Hof M. Dynamic saturation optical microscopy: employing dark-state formation kinetics for resolution enhancement. Phys Chem Chem Phys. 2010;12(39):12457–12465.
[46] Yao J, Shcherbakova DM, Li C, Krumholz A, Lorca RA, Reinl E, England SK, Verkhusha VV, Wang LV. Reversibly switchable fluorescence microscopy with enhanced resolution and image contrast. Journal of Biomedical Optics. 2014;19(8):086018.
[47] Elson EL. Fluorescence correlation spectroscopy: past, present, future. Biophysical Journal. 2011;101(12):2855–2870.
[48] Lanzano L, Lei T, Okamura K, Giral H, Caldas Y, Masihzadeh O, Gratton E, Levi M, Blaine J. Differential modulation of the molecular dynamics of the type IIa and IIc sodium phosphate cotransporters by parathyroid hormone. Am J Physiol Cell Physiol. 2011;301(4):C850–C861.
[49] Cardarelli F, Lanzano L, Gratton E. Fluorescence correlation spectroscopy of intact nuclear pore complexes. Biophysical Journal. 2011;101(4):L27–L29.

[50] Cardarelli F, Lanzano L, Gratton E. Capturing directed molecular motion in the nuclear pore complex of live cells. Proceedings of the National Academy of Sciences of the United States of America. 2012;109(25):9863–9868.
[51] Eggeling C, Ringemann C, Medda R, Schwarzmann G, Sandhoff K, Polyakova S, Belov VN, Hein B, von Middendorff C, Schonle A, Hell SW. Direct observation of the nanoscale dynamics of membrane lipids in a living cell. Nature. 2009;457(7233):1159–1162.
[52] Vicidomini G, Ta H, Honigmann A, Mueller V, Clausen MP, Waithe D, Galiani S, Sezgin E, Diaspro A, Hell SW, Eggeling C. STED-FLCS: An Advanced Tool to Reveal Spatiotemporal Heterogeneity of Molecular Membrane Dynamics. Nano Lett. 2015;15(9):5912–5918.
[53] Lanzanò L, Scipioni L, Di Bona M, Bianchini P, Bizzarri R, Cardarelli F, Diaspro A, Vicidomini G. Measurement of nanoscale three-dimensional diffusion in the interior of living cells by STED-FCS. Nature Communications. 2017;8:65.

Peter T. C. So, Heejin Choi, Elijah Yew, and Christopher Rowlands

6 Principles and applications of temporal-focusing wide-field two-photon microscopy

Abstract: Temporal focusing allows for rapid optically sectioned two-photon wide-field microscopy. Depth sectioning is provided in a wide-field manner, without spatial focusing, by controlling the temporal width of femtosecond laser pulses near the focal plane. This spatial control of the temporal pulse width is achieved by diffracting the light off a grating resulting in spectral component separation and temporal broadening; these spectral components are only recombined at the focal plane to reproduce short, femtosecond pulses. Applications include (i) high speed functional imaging in the brain, (ii) fast FLIM and PLIM, (iii) cell-selective optogenetic excitation, and (iv) temporal focusing photodynamic therapy that may allow selective killing of cancer cells.

6.1 Introduction

The two-photon absorption process was first predicted by Dr. Göppert-Mayer in 1931 [6] and was developed by Sheppard and Webb groups into a powerful microscopic biomedical imaging technique [8–10]. In addition to fluorescence, second (SHG) and higher harmonic generation processes may also be produced via optical interaction with biomolecules with substantial polarizability [11–17]. Multiphoton microscopy has important advantages in 3D depth-resolved image acquisition. First, the quadratic dependence on photon flux results in inherent depth discrimination where over 80 % of fluorescent signal is localized to within a 1 μm thick region. Second, photodamage and photobleaching is localized to the same small volume. Third, the multiphoton excitation wavelength lies in the infrared region resulting in significantly lower tissue scattering and absorption and deeper imaging. Fourth, the wide separation of the excitation spectrum and the emission spectra allows more precise spectroscopic measurements.

While lacking depth resolution, conventional fluorescence microscopy, illuminating a sample using wide-field geometry allows 2D images to be acquired at high frame rate. For 3D optical microscopy, such as conventional two-photon laser scanning microscopy (TPLSM), excitation light is focused to a diffraction-limited spot that results in optical sectioning. However, sequential raster scanning of the excitation volume is required to produce a 3D image and data acquisition rate decreases as sample size or resolution increases. Finite fluorophore lifetime and fluorophore concentration ultimately limit fluorescence photon generation rate even with very fast scanning mechanisms, such as an acousto-optical deflectors [23, 24]. Therefore, the most effective way

https://doi.org/10.1515/9783110429985-008

to improve imaging speed is to parallelize the excitation process [25]. For example, multifocal multiphoton microscopy (MMM) is a successful parallelization approach [26–29]. This chapter focuses on a relatively new multiphoton technique, temporal focusing microscopy (TFM), which can be considered as a method of even higher degree of parallelization (by a factor of 10^2 to 10^4) [30–34]. Fig. 6.1 demonstrates the underlying concepts behind spatial focusing and temporal focusing. In Fig. 6.1 (a), an optical pulse is focused laterally in the spatial dimension, traveling along axial direction and its temporal pulse width is constant. The intensity at the focal spot reaches a maximum. Nonlinear optical processes such as two-photon absorption are proportional to the square of the pulse peak power, resulting in optical sectioning. For high numerical aperture (NA) objective, submicron lateral and axial resolution can be achieved. With temporal focusing, the optical pulse travels in axial direction without changing its lateral dimension, unlike spatial focusing (Fig. 6.1 (b)). However, the instantaneous intensity is maximized at the focal plane by manipulating the temporal width of the optical pulse such that it is minimized at the focal plane and is broader in out-of-plane locations. This approach allows wide-field imaging with depth discrimination. TFM controls the spectral dispersion of the ultrafast pulse along the optical path to manipulate pulse width to produce this optical sectioning effect.

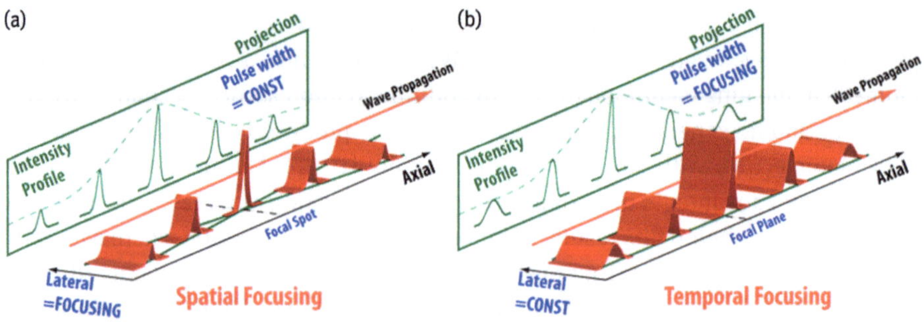

Fig. 6.1: (a) Spatial vs (b) temporal focusing multiphoton microscopy.

In this chapter, we will cover the basic principle of TFM and its typical implementation. We will provide both paraxial and vectorial formulation describing image formation for temporal focusing; the limited axial resolution of this approach will be explained and two methods for remedying this limitation will be described. A characterization of penetration depth of TFM for imaging into different tissues and organs will be presented. Finally, this chapter will close with examining several applications where TFM have demonstrated successes. Clearly, it is impossible to cover all application areas that have used TFM and our selection of specific application areas is inherently subjective. However, we believe that TFM excels in application spaces that are characterized by the need for simultaneous, possibly patterned, excitation over

a large area, requiring only moderate penetration depth, and needing high imaging speed. With these criteria in mind, we will cover four application areas including:
(1) functional imaging of neuronal network,
(2) fluorescence and phosphorescence lifetime imaging,
(3) cell-selective optogenetics, and
(4) cell-selective photodynamic therapy.

6.2 Invention of temporal focusing two-photon microscopy and basic operating principle

Temporally focusing wide-field two-photon microscopy was introduced almost simultaneously by Oron and Silberberg [30] and by Zhu and Xu [33] about a decade ago. Temporal focusing two-photon microscopy is, in essence, simply a 4-f pulse shaper proposed by Froehly, Colombeau, and Vampouille [35] and Martinez, Gordon, and Fork [36] with the exception that there is no second grating to recombine the pulse. Temporal focusing multiphoton microscopy works because of the time-bandwidth product which states that the product of the pulse width and the spectral content is a constant. In a typical experimental setup (Fig. 6.2), an ultrafast pulse is incident upon a grating. This approach separates the pulse into its constituent wavelengths, each diffracting at an angle governed by the grating equation. A focusing lens converts the wavelength-dependent angular separation of the spectrum into a wavelength-dependent positional offset at the back-focal plane of the objective. The objective collimates and recombines the various spectral components only within the focus of the objective. It is clear that with the exception of the grating surface and the focal plane the pulse

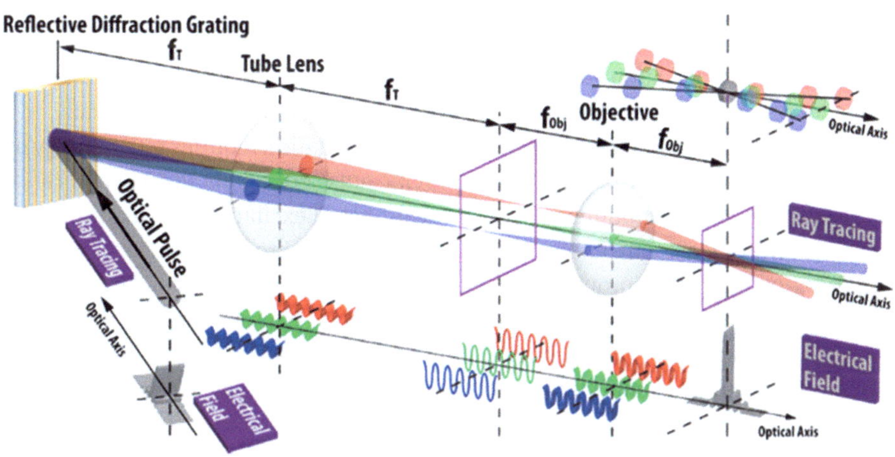

Fig. 6.2: Implementation of temporal focusing multiphoton microscopy.

width is broadened substantially everywhere in between as a result of the constancy of the time-bandwidth product. This reduces the photon flux everywhere between the grating and the focus. It is only at the focal plane where the spectral components of the pulse overlap and are recombined. As a result of this recombination, the time-bandwidth product dictates that the pulse width is once again minimum everywhere across the plane and photon flux is high enough for nonlinear optical processes to occur.

In terms of instrument design it is clear that a temporal focusing microscope is significantly simpler than that of single focus or multifoci scanning system for two reasons. First, since the whole 2D field is imaged simultaneously, raster scanning mechanism and the associated control electronics are not needed. Second, unlike high sensitivity photomultiplier tubes used in TPLSM requiring custom electronics interface, especially multianode photomultiplier tubes used in MMM [27], temporal focusing system uses commercially readily available CMOS or CCD cameras that significantly simplify data acquisition electronics design. However, for acquiring a 3D data cube, it is still necessary to scan the excitation plane in the axial direction. This axial scan can be accomplished by translating the objective or the specimen as in standard scanning microscopes. Alternatively, given the similarities between a temporal focusing microscope and a 4-f pulse shaper, both positive and negative dispersion may be achieved [36]. This property has been exploited in temporal focusing to shift the position of the focal plane along the z-axis effectively creating an axial scan through optical means [37, 38]. This is done by deliberately up-chirping or down-chirping the beam. This approach is very effective for using a temporal focusing microscope for patterning type applications, but is less effective for imaging type applications since the imaging range is defined by the position of the objective that typically has a relatively short depth of field.

Another important consideration in temporal focusing microscope instrument design lies on the choice of excitation lasers. While temporal focusing systems were first implemented with titanium-sapphire (Ti-Sa) oscillators, it is clear that these low pulse energy oscillators are not optimal for temporal focusing applications. Typically, in a single-point focusing system, efficient two-photon excitation requires pulse energy on the order of 0.1–1 nJ. These oscillators can provide pulse energy about 50 nJ and can at best excite 500 spots simultaneously. Since two-photon excitation is a quadratic process, it is always more efficient for high speed imaging to provide sufficient energy at each location to bring fluorophores to near excitation saturation instead of distributing the energy into more locations. Clearly, high peak power lasers are needed for temporal focusing multiphoton microscopy. Chan and Dong groups first introduced the use of regenerative amplifier as temporal focusing [39]. A typical regenerative amplifier delivers mJ level energy pulses allowing in principle simultaneous excitation of 10^6 to 10^7 locations. However, it should be noted that regenerative amplifiers do not provide 10^6 to 10^7 level imaging speed improvement because these amplifiers deliver comparable average power as Ti-Sa oscillators (on the order of 10 W level) and the corresponding pulse train repetition rate is often reduced to kHz to 10 kHz range. One can readily show that imaging speed improvement effectively scales with average power delivered into the specimen, which is limited mostly by specimen photodamage. The typical damage threshold for point scanning is on the order of 10 mW, while the temporal focusing system has delivered power up to 10 W over approximately a 1 mm^2 area. Therefore, we expect about two orders of magnitude speed increase when comparing point scanning vs temporal focusing systems. Recent introduction of ultra-high power fiber amplifier with power up to several hundred Watts may increase imaging speed further but the limit imposed by the specimen damage threshold of these lasers remains to be established. While these amplifiers have pulse train profiles very suitable for temporal focusing applications, they are also more limited than Ti-Sa oscillators in terms of fluorophore selectivity. These amplifiers typically have emission wavelengths at about 1050 nm with very narrow or no tuning range. While these wavelengths can excite a broad class of organic fluorophores, they are not well matched to most commonly used fluorescent proteins (with the exception of a few red variants). The conversion of these μJ or mJ pulse lasers to the appropriate wavelengths around 900–1000 nm will require optical parametric amplification and frequency doubling. Both of these processes have low efficiencies that greatly reduce the average power output of the final pulse train negating the imaging speed advantage of the temporal focusing approach. While it is likely that multiple hundred Watt amplifiers may be unsuitable for biomedical imaging applications due to specimen damage, these amplifiers may be combined with parametric amplifiers or parametric oscillators to provide much higher average power pulse train in the wavelength range suitable for fluorescent protein excitation that will greatly broaden the biomedical utility of temporal focusing imaging.

6.3 Image formation theory for temporal focusing microscopes

Image formation in a TFM has been previously analyzed [33, 37, 40, 41]. Most of the prior work assumes that the spectrum of the pulse is taken to be narrow such that $k \approx k_O$, where k is the wave number of the spectral components of the pulse while k_O is the mean wave number. These works further make the paraxial approximation where a pencil of rays is assumed and the sines of the angles involved are small and may be approximated to the angle itself. This is clearly not the case for ultrafast pulses and high NA objectives. We have recently published a theoretical extension on the theory of image formation for a temporal focusing microscope using the paraxial and vectorial approaches and compare the validity of both approaches through experiment [21]. We have also determined the 3D optical transfer function (OTF) from numerical calculation and compared with experiment. In this work, we make the assumption that $k \neq k_O$, whereas the diffracted angle can be assumed to be small. This is more consistent with most experimental design while being in contrast to some previous publications [33, 37]. For the vector case, by recasting the problem as a projected pupil function, we were able to apply standard FFT algorithms and the chirp z-transform to minimize the computational process.

Fig. 6.3 illustrates the difference between the FWHM of the intensity temporal focusing axial point spread function (PSF) as calculated from paraxial and vectorial theory. It is clear that the paraxial form and vector theory give similar results, although the drop-off is slower for the vector theory. Paraxial theory gives a smaller axial FWHM of 6.2 µm compared to 7.6 µm from vector theory. This may be attributable to the vector theory accounting for factors such as the apodization, that the spectral content of the ultrafast pulse has a Gaussian intensity profile, and that the wavefront after the objective is spherical and not parabolic. The closeness of the results can be explained if we note that even for 100 fs pulses, the angle of diffraction at the $1/e^2$ wavelength is small enough such that $\cos\theta_d \approx 1$. However, the slower fall off of the axial PSF has implications in poor axial sectioning of this approach.

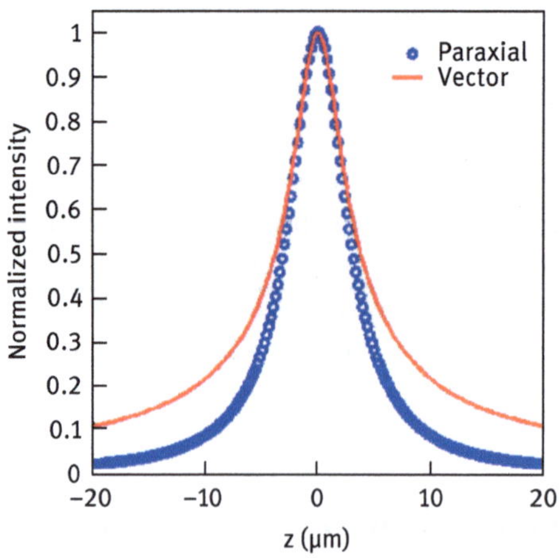

Fig. 6.3: The envelope of the temporal focus axial PSF for the paraxial approximation (open blue circles) and that calculated from vector theory (red line). The FWHM for the paraxial approach is 6.2 µm whereas for the vector approach it is 7.6 µm. [21]

We now consider the OTF of TFP. Fig. 6.4 shows the cross sections of a cross section of the 3D OTF along k_x–k_z plane of the 1P wide-field microscope, the 1p confocal microscope with a point detector, the 2P microscope, and the temporal focusing 2P microscope. In Fig. 6.4 (a), it is observed that the 1P wide-field microscope exhibits the well-known 'missing cone', indicating that 1P wide-field microscopy is without optical sectioning capability. In Fig. 6.4 (b) and (c), the inclusion of a confocal pinhole or the use of 2P excitation results in optical sectioning. The OTF for a TFM can be seen as a mix between a wide-field and a 2P microscope. The transverse frequency support is similar to that of a 2P microscope. It also misses axial frequency support for the low lateral frequency region but it does not have a true "missing cone" as shown in Fig. 6.5.

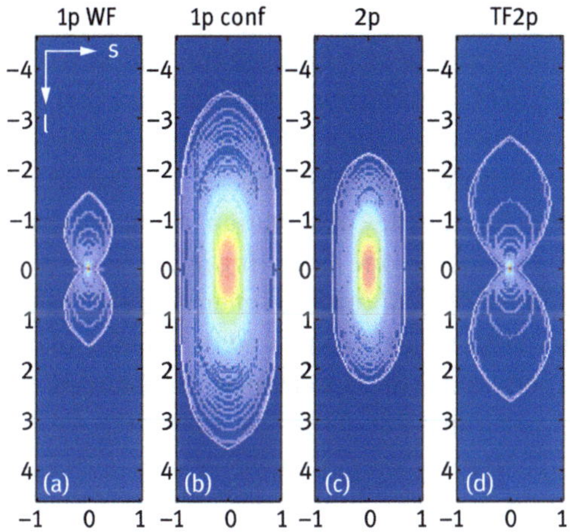

Fig. 6.4: (a) Slice view of the OTF for a 1p fluorescence wide-field microscope. The missing cone indicates that no axial information is transmitted. On the other hand, (b) is the slice of the 1P confocal microscope. The missing cone has been filled in and has optical sectioning. A similar OTF is found for (c) a conventional 2P microscope. (d) is the slice OTF of a temporal focusing 2P microscope. The approximate extent of the OTF has been outlined in white as a visual aid [21].

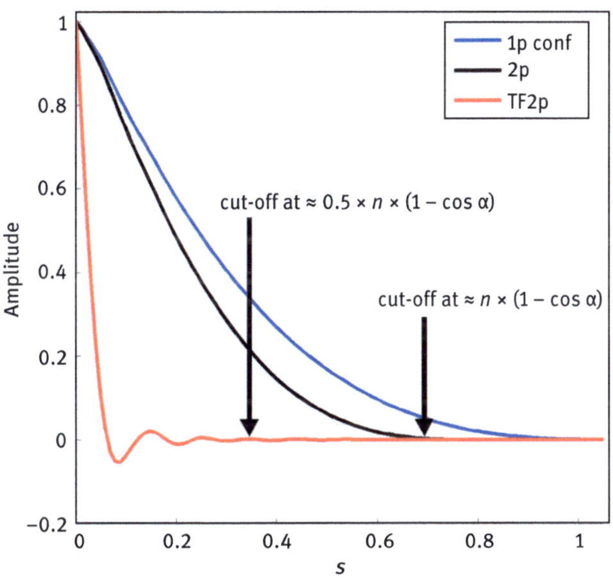

Fig. 6.5: Plotting the transfer function along the k_z-axis indicates that the frequency support for temporal focusing 2P is close to half that of the conventional 2P transfer function. The 1P confocal has a frequency support twice that of conventional 2P. The cut-off frequency for the 2P microscope is about twice as large as that of a temporal focusing 2P microscope. In all cases, the excitation wavelength was 790 nm and the emission wavelength of 395 nm [21].

Fig. 6.5 shows a plot along k_z-axis from the 3D OTF. The axial frequency support for a 2P microscope is given by $n(1 - \cos\alpha)$. Given that the point spread function of a temporal focusing 2P microscope has an FWHM of 1.8 µm, which is about twice that of a conventional 2P microscope at 0.8 µm, we expect the support for a temporal focusing 2P microscope to be close to half that of the conventional 2P transfer function in the axial direction and is supported by Fig. 6.5. Another key observation is that the drop-off is much quicker for a temporal focusing 2P microscope indicating much faster axial information loss as compared to conventional 2P or 1P confocal microscopes. The axial resolution is therefore much poorer for TFM. We next compared the validity of our high aperture formulation of TFM to experimental measurements. We obtained the 3D OTF (Fig. 6.6 (b)) by taking the 3D Fourier transform of the PSF image (Fig. 6.6 (a)) of a single bead. Since the beads have a peak emission of fluorescence at 485 nm we normalized the spatial frequencies by 485 nm for both cases. It is clear that the theoretically derived PSF and OTF are validated by experiment. A two-photon microscope is worth the complexity and expense because it is able to deliver good optically sectioned images. A TFM has optical sectioning effect that is poorer than that of a 1P confocal or a 2P microscope. We imaged a 0.2 µm fluorescent bead and summed the image over a region equivalent to 2 Airy Units (AU). We found that the FWHM of the imaged bead is 2.1 µm compared to 1.8 µm from theory, while Fig. 6.6 shows the fit.

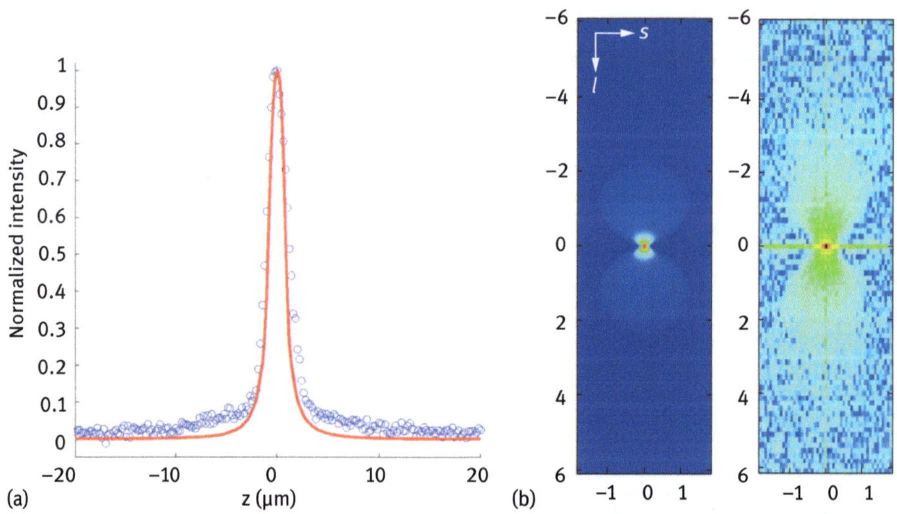

Fig. 6.6: Comparison between the numerically calculated (a) PSF and (b) OTF. (a) Experimental data is represented as blue circles while theory is plot as a red line. (b) Left is OTF from theory and right is experimental. In both cases, the excitation wavelength was 790 nm and the emission wavelength was taken as 485 nm for (a) [21].

Although the image of a single fluorescent bead provides a good estimate of the imaging properties of a microscope, a more common scenario found in imaging is to discriminate between signal that emanates from the in-focus plane and out-of-focus planes. For this purpose, the optical sectioning has been defined to be the response of the microscope to a plane of fluorescence with no transverse variations (i.e., no lateral spatial frequencies). Alternatively, it is the steepness of the gradient of the response due to a sea of fluorescence scanned through the focus. The signal of such an experiment was recorded and graphed in Fig. 6.7 (a). The first derivative of the recorded signal, a measure of the optical sectioning capability of a microscope, was obtained and is also presented in the same figure. From the data, we find that the FWHM of the optical response to a sea of fluorescence (given by the dotted red curve) to be 7.6 μm. What this means is that we are only able to locate the interface of an axially extended object with no lateral variation to within 7.6 μm. We further compared the optical response to a sea of fluorescence by measuring the optical sectioning effect, which is obtained by scanning a thin film of fluorescence through the focus. Fig. 6.7 (b) illustrates the response of the temporal focusing 2P microscope due to a thin layer of fluorescence. The measured response has an FWHM of 8 μm and is in good agreement with the derived optical sectioning response from a sea of fluorescence as well as the response calculated numerically.

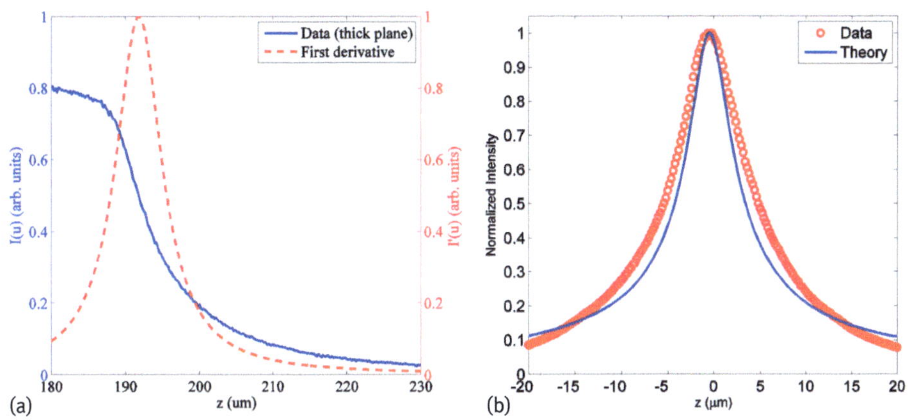

Fig. 6.7: Estimation of optical section of temporal focusing 2P microscope. (a) A thick layer of Rhodamine was scanned through the temporal focus and the recorded signal (blue, solid line) is plotted. The first derivative of the signal was obtained and a curve fitted to it (red, dashed line). (b) A thin layer of fluorescence (red open circles) was scanned through the temporal focus and the captured image was summed over an area corresponding to around 2 Airy Units (AU). This experiment is equivalent to taking the first derivative of the optical response to a sea of fluorescence. The data is compared to vector theory (blue line) and a good fit is obtained between experiment and theory [21].

6.4 Remedying the poor axial resolution of temporal focusing 2P microscopy

TFM suffers from low axial resolution when compared to the standard TPLSM that is partly caused by the under-utilization of the numerical aperture (NA) of the objective [18]. Another problem of TFM, that uses imaging detectors such as CCDs and CMOSs, is that its resolution is based on telecentric mapping of emission photons from the specimen plane to the image plane. For deep imaging, the scattering of emission photons degrades the image contrast obscuring the fine features of the specimen. In contrast, the scattered emission photons of TPLSM are integrated with a large area detector and the spatial structure and resolution of images depend only on the temporal sequence of the scanning process and the excitation point spread function [42]. In addition, it has been reported that the axial extent of the excitation volume is progressively broadened in a scattering medium as the imaging depth increases, which further adds more background noise [43]. Today, besides implementing temporal focusing in a line-scan geometry [31], there are two successful implementations that overcome the resolution limit of standard wide-field temporal focusing 2P microscopy.

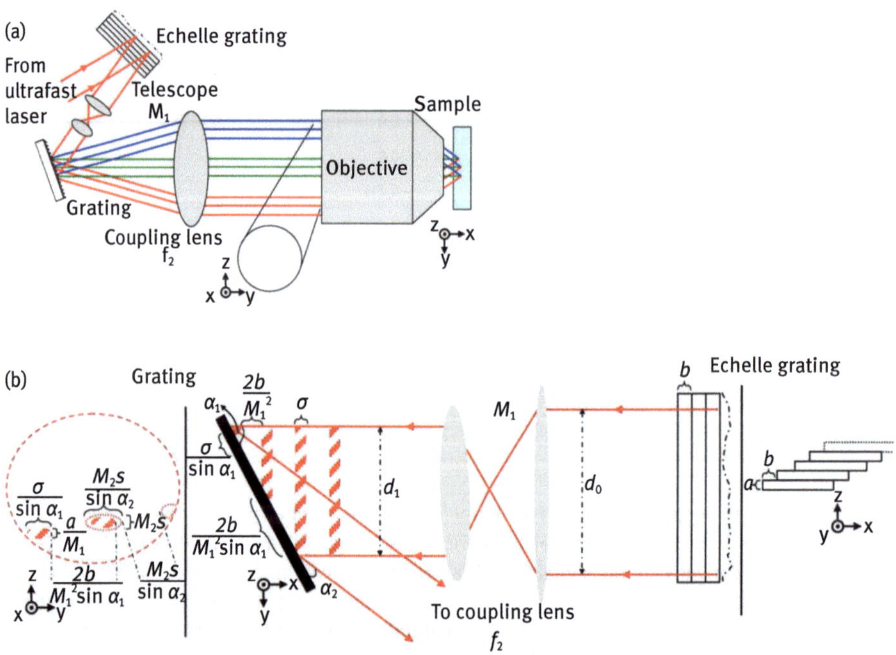

Fig. 6.8: (a) Implementation of multifocal temporal focusing 2P microscope. (b) Pulse splitting and propagation between Echelle and normal grating [18].

Vaziri and co-workers have shown that 0.85 μm FWHM axial resolution can be achieved at the expense of increasing complexity of the system by properly filling the back aperture of the objective [18]. They have termed this technique Multifocal Temporal Focusing (Fig. 6.8). Their implementation is similar to that of Tal and co-workers [31] in implementing line-scan temporal focusing. In the line-scan arrangement, a cylindrical lens first focuses the light in one dimension onto the grating. The tube lens expands the line focus to cover the objective back aperture in one dimension as in other line-focusing scanning microscopes while the dispersion of the grating expands the different spectral components of the ultrafast pulse along the orthogonal direction covering the full back aperture resulting in good axial confinement. The issue with this line-scan approach is that slow mechanical scanning is needed in one direction. Multifocal temporal focusing overcomes this limitation. Instead of using a cylindrical lens to focus light into the grating, an Echelle grating is inserted to reflect the ultrafast light pulse to the normal grating via a 4-f geometry. It is ensured that the grating period of the Echelle grating projected onto the sample via the two 4-f relay is small enough such that it is diffraction limited, thus ensuring the back aperture of the objective is filled along this direction. The grating step size is chosen such that the ultrafast pulses reflected by each step are sufficiently temporally delayed and do not interfere with each other. A more detailed description of this pulse propagation geometry is presented in Fig. 6.8 (b).

Multifocal temporal focusing has very successfully improved the axial confinement of TFM. It formed a 7 μm diameter spot in the specimen with an excellent aspect ratio featuring 0.85 μm confinement in the axial direction. This demonstrated that axial confinement similar to that of the line-scan temporal focusing system can be achieved and thus is comparable to that of traditional TPLSM (Fig. 6.9). The approach has also successfully been applied to improve cellular and tissue imaging resolution (Fig. 6.10).

We have recently demonstrated an alternative approach to overcome the poor axial resolution of temporal focusing by utilizing structured light illumination (SLI) in TFM to effectively reject background scattered emission photons and thereby improve image contrast when imaging in a turbid medium [19]. Equally importantly is that the axial resolution of this system is dictated by the spatial frequency of the structured light thereby allowing an axial resolution that is significantly better than TPLSM even in the absence of scattering. The core idea of this approach is that SLI acts as a virtual pinhole and the low frequency out-of-focus scattered photons beyond the depth of field of the objective can be removed computationally.

A class of depth-resolved imaging techniques based on SLI have been proposed to select a particular imaging plane and to reject out-of-focus background for standard wide-field single-photon microscopy [44, 45] and it has been shown that the axial resolution of these techniques is comparable to that of the confocal microscope [46, 47]. Of these methods, one effective approach we adapted is termed 'HiLo microscopy' that combines the in-focus high frequency content extracted from the uniformly illumi-

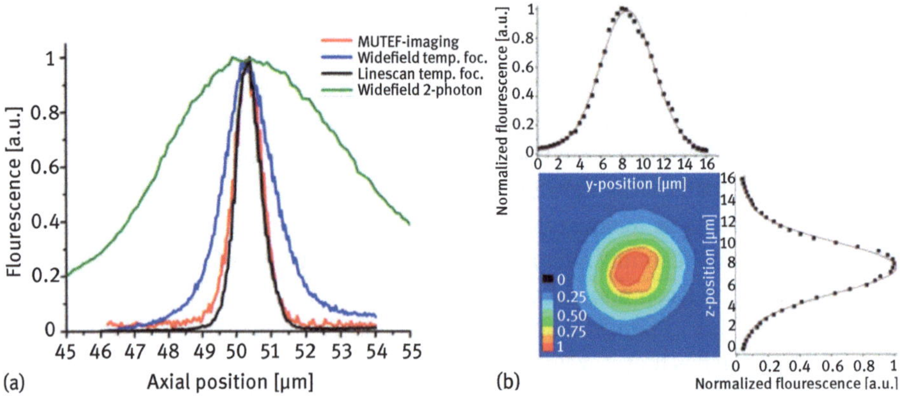

Fig. 6.9: (a) Axial confinement of multifocal temporal focusing 2P microscope (red line) is comparable with line temporal focusing system (black line). (b) Lateral light distribution corresponding to the measured axial confinement [18].

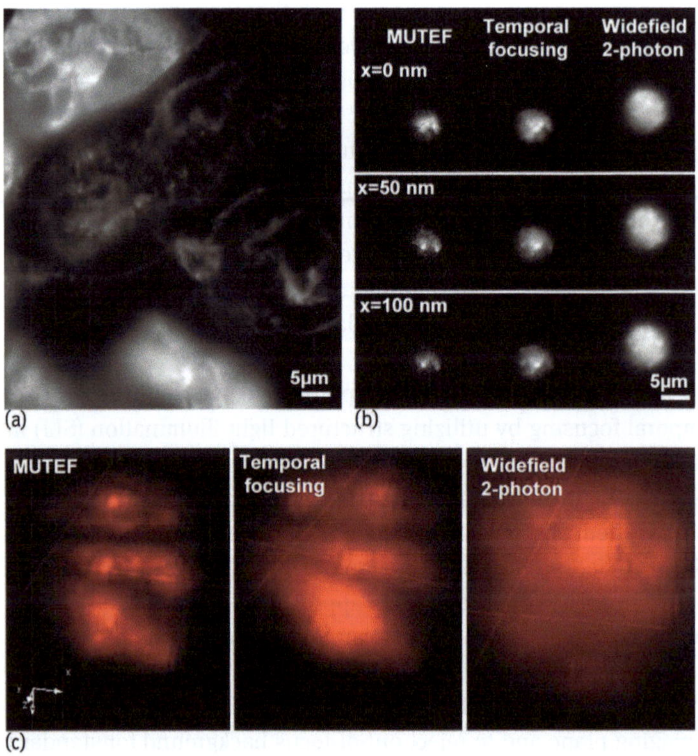

Fig. 6.10: Comparing multifocal temporal focusing 2P with temporal focusing 2P and wide-field 2P for imaging mouse kidney cells [18].

nated image (UI) and the in-focus low frequency content extracted from the structured light illuminated image (SI) to generate an optically sectioned image [45].

An implementation of structured light TFM is shown in Fig. 6.11 where either a phase grating (generated by a Michelson interferometer) or an intensity granting (generate by a physical grid) can be used. The theoretical axial resolution of SLI can be estimated using the defocused 2D optical transfer function (OTF) derived by Stokseth [48]. Fig. 6.12 (a) shows the plot of contrast at the fringe periods of $T_g = 3.42\,\mu m$, 1.71 µm, 0.85 µm, 0.43 µm, which correspond to the normalized fringe frequencies of 0.13, 0.26, 0.52, 1.04, respectively. For the fringe period of 1.71 µm and higher, FWHM of the axial resolution is expected to be better than that of normal temporal focusing systems reported in the literature [49, 50]. Furthermore, theoretically, SLI has the potential of attaining depth resolution better than TPLSM when the normalized fringe frequency is higher than 0.3 as shown in Fig. 6.12 (b). Ideally, the best optical sectioning is achieved when the normalized fringe frequency comes close to 1 but with the trade-off of the reduced signal-to-noise ratio [51]. For this comparison, the intensity of SLI is obtained by normalizing the defocused 2D OTF and the total fluorescence intensity of TPLSM generated at a given z-plane for a uniform specimen is calculated by integrating the two-photon intensity in each z-section [52]. The axial resolution improvement was demonstrated experimentally with a thin layer of Rhodamine solution both in the absence and in the presence of scattering medium. Experimental results show that axial sectioning similar to standard TPLSM can be achieved (Fig. 6.12). We further performed the measurement with a 15 µm thick mouse kidney sample under two scattering conditions. First, the experiments were performed without any scattering medium (0 % Lipofundin) and then undertaken with 2 % Lipofundin as an immersion medium. From Fig. 6.13, it is evident that SLI improves the axial resolution of

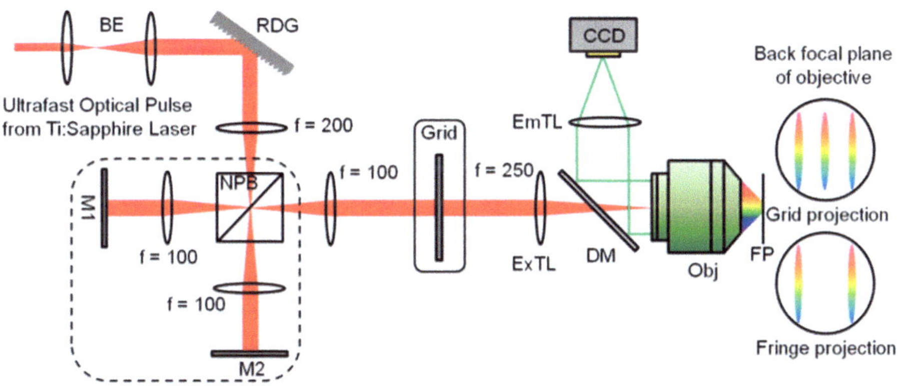

Fig. 6.11: Schematic diagram of the temporal focusing structured light illumination setup. Either interferometer or grid is used for the fringe projection or the grid projection. BE: beam expander, RDG: reflective diffraction grating, NPB: nonpolarizing beam splitter, M: mirror, ExTL: excitation tube lens, EmTL: emission tube lens, DM: dichroic mirror, Obj: objective, FP: focal plane [19].

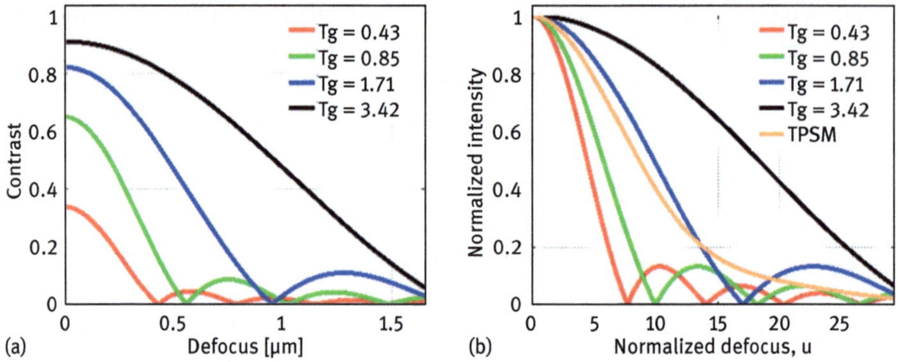

Fig. 6.12: (a) Contrast decay of fringe pattern of the spatial period of T_g = 0.43, 0.85, 1.71, 3.42 μm as a function of the defocus. (b) Normalized intensity of SLI and TPSM as a function of its distance from the focal plane [19].

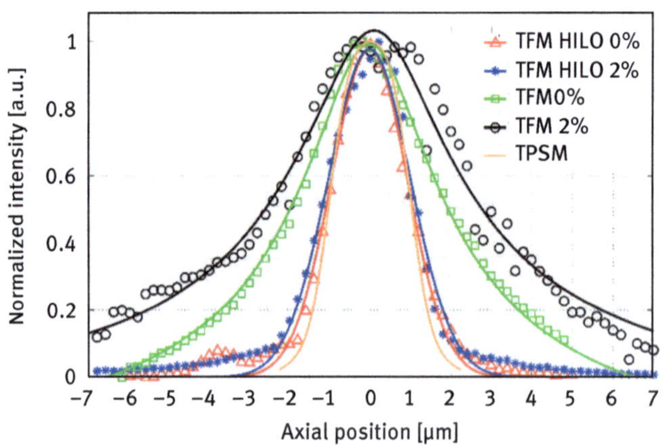

Fig. 6.13: Axial resolution of structured light temporal focusing 2P microscope measured with thin layer of Rhodamine solution in the absence and presence of scattering medium. Fringe period of SLI is 1.71 μm. (red) TFM HiLo at 0 % intralipid, (blue) TFM HiLo at 2 % intralipid, (green) TFM at 0 % intralipid, (black) TFM at 2 % intralipid, (orange) TPSM in Fig. 6.2 (b) convoluted with 2 μm thick Rhodamine solution [19].

TFM significantly. At 0 % Lipofundin, TFM's FWHM is 3.98 μm and TFM HiLo's FWHM is 2.22 μm. At 2 % intralipid, they are 5.21 μm and 2.42 μm, respectively. This broadening of the axial resolution has been reported in the literature [43]. It is interesting to note that HiLo processed data is not much affected by the scattering medium, which implies that HiLo-based SLI works robustly even in a highly scattering medium. The axial resolution improvement is further demonstrated with a prepared slide of sectioned mouse kidney. Fig. 6.14 shows the *xz*-section view of a TFM image without SLI and HiLo processed TFM images with fringe period of 3.42 μm, 1.71 μm, 0.85 μm respectively. As the fringe frequency increases, the sectioning capability of SLI increases and finer features of the specimen becomes visible more clearly. Particularly, the cross-sectional intensity plot along the depth direction indicated by the arrow is also shown on the right side. The weak intensity object in the middle of the sample, which is hidden by the two high intensity objects above and below, is discernible with SLI at $T_g = 0.85\,\mu m$. This effect is more clearly visible in the normalized intensity plot where the intensity is normalized with the peak intensity value.

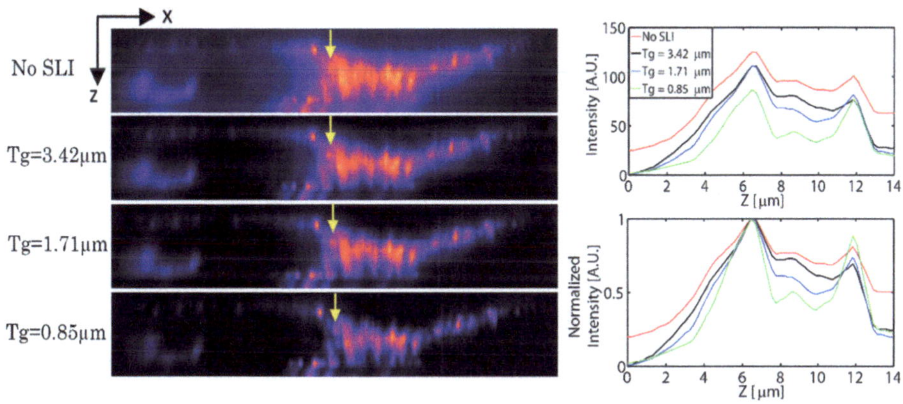

Fig. 6.14: *xz*-sections of the glomeruli and convoluted tubules structure in a mouse kidney sample acquired with TFM without SLI, HiLo processed TFM with fringe period of 3.42 μm, 1.71 μm, 0.85 μm, respectively. The thickness of the imaged portion is 14 μm. Intensity increases from purple to red. The cross-sectional intensity plot along the line indicated by the yellow arrow is also shown on the right-hand side [19].

In summary, both multifoci temporal focusing and structured light TPM can improve axial resolution of images to an extent at least comparable to TPLSMs. The multifoci temporal focusing approach has the advantage that axial confinement improvement is achieved on the excitation side, so it is applicable in patterning type applications and minimizes specimen photodamage. These advantages are traded off with the need for a custom-made Echelle grating. Structured light temporal focusing systems have advantages in simplicity of implementation, potentially improving resolution beyond point scanning, and affording rejection of scattered light background. These advantages are traded off with loss of out-of-plane fluorescence photons and the associated photodamage. Potentially this drawback may be partly alleviated by performing maximum likelihood photon reassignment [53].

6.5 Characterizing performance of temporal focusing 2P microscope for deep tissue imaging

It is expected that TFM can excite samples as deep as, if not deeper than conventional TPSM [54], but because conventional TPSM can make use of the scattered emission photons in the sample (since all the emitted photons are assumed to come from the focal volume), unlike temporal focusing which can only use the ballistic and weakly-scattered photons to form the image, it is expected that in scattering samples, such as biological tissue samples, performance will be worse. We have characterized the extent of this performance degradation in several different tissues [22]. We analyzed the ability of the imaging system to extract information by measuring the modulation transfer function (MTF) of tissue images at different depths. MTF is a property of an imaging system that describes the contrast of a given spatial frequency, placed perfectly in focus in the field of view and imaged through an optical system. As the spatial frequency increases, the contrast (the intensity ratio between the brightest and darkest points) drops. At the point where the MTF drops to zero, that spatial frequency can no longer be resolved; any features with this spatial frequency cannot be seen. Consequently, this implies that 'penetration depth' is user-specific; users who wish to see large features with low spatial frequencies will likely be able to see further into the sample compared to users who wish to see samples with finer features of interest. It is therefore helpful to try to estimate the MTF as a function of penetration depth. In order to ensure that the features to be discerned in different tissues are comparable, we stained cell nuclei by injecting Hoechst 34580 intraperitoneally and intravenously into a male FVB/NJ mouse (Jackson Laboratories) prior to sacrifice. We chose to target nuclei because they are morphologically similar in tissues tested. Further, these nuclei are small and often have fine features, which provide a high frequency support in the sample. By taking the Fourier transform of an image and taking a 'radial average' (i.e., combining all the 2D frequency components with the same frequency, regardless of angle), it is possible to estimate the MTF, assuming that the sample consists of spatial white noise. Clearly this is only a weak approximation, since the images of the stained nuclei will have a particular spatial frequency content, but the estimate is acceptable for determining the maximum spatial frequency that can be observed in the sample; by determining where the MTF reaches the background level, it is possible to define the maximum spatial frequency, and therefore determine the maximum penetration depth for a given desired spatial frequency.

After the animal was sacrificed, organs (kidney, liver, heart, spleen, lung, and white adipose tissue) were extracted and fixed by paraformaldehyde and mounted in a glass-bottomed dish for imaging. We then performed both TFM and TPSM of these organ specimens (Fig. 6.15). Excitation power levels were adjusted for different tissues and different image depths to ensure that the image brightness was comparable for different depths and for different tissues, ensuring that the differences in MTF were not simply due to shot noise or reduced signal. An image processing algorithm was

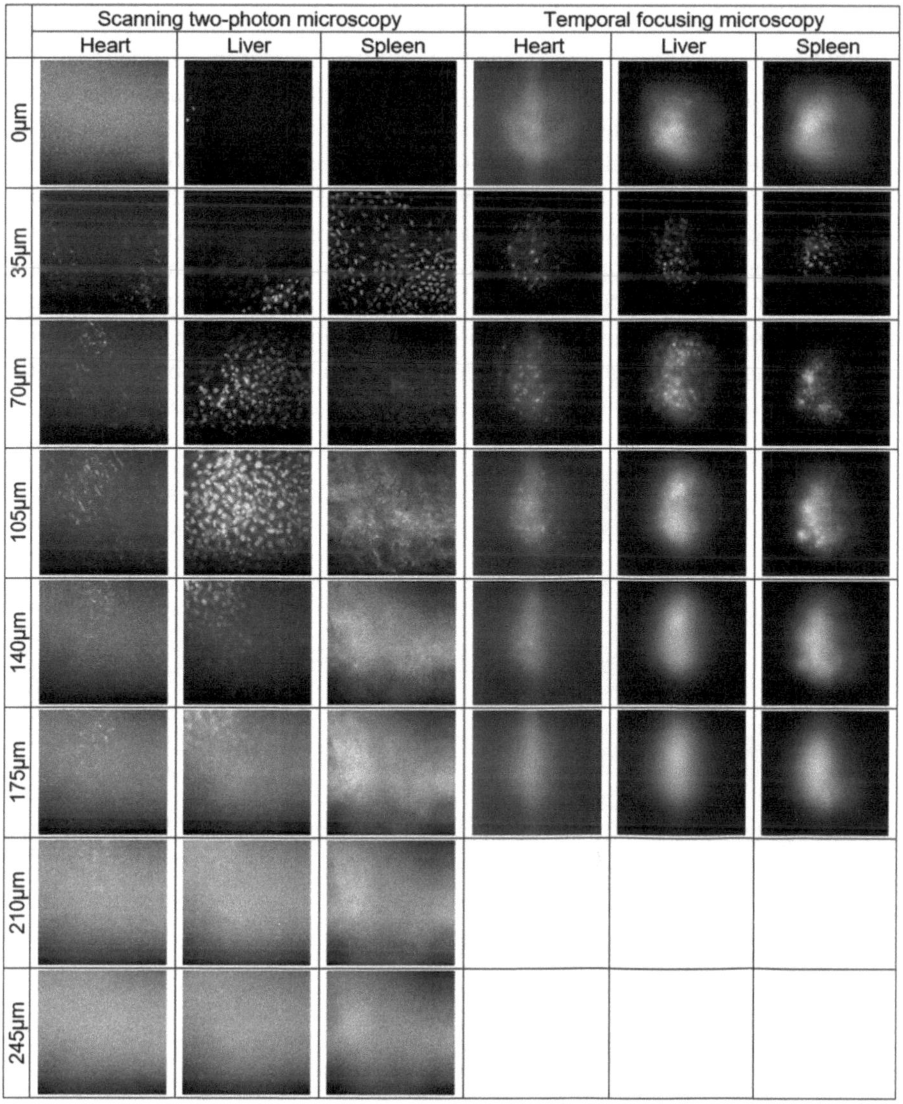

Fig. 6.15: Representative images acquired by both TFM and TPSM [22].

developed to automatically estimate MTF from these image stacks. First, the image is cropped so that it is square. The 2D FFT is then taken, and the image reshaped so that the low-frequency components lie at the center of the image and the high frequency components are at the edge. A 'radial average' is then performed, whereby a ring centered on the zero-frequency component is constructed, with a thickness of one FFT pixel. The ring is normalized such that the sum of all pixels in the ring is equal to 1, and then multiplied elementwise with the FFT image; the sum over the whole resulting image is then taken. This summed value is a crude estimate of the MTF at a given spatial frequency. The ring is then enlarged to cover a different spatial frequency, and the process repeated until the ring is larger than the image. The resulting plot of estimated MTF vs spatial frequency should have an approximately monotonic decay. The background level is determined by starting from the highest spatial frequency values and working towards the low frequency values. At each new spatial frequency value, it is compared with the mean and standard deviation of all the previous (higher frequency) values, which are all assumed to be equal to the baseline plus an individual noise term. If the value of the new point deviates from the mean by more than three times the standard deviation, it is assumed that it is not a noise term and therefore is the start of the image information; its value represents the highest observable spatial frequency in that image. The entire algorithm is illustrated in Fig. 6.16.

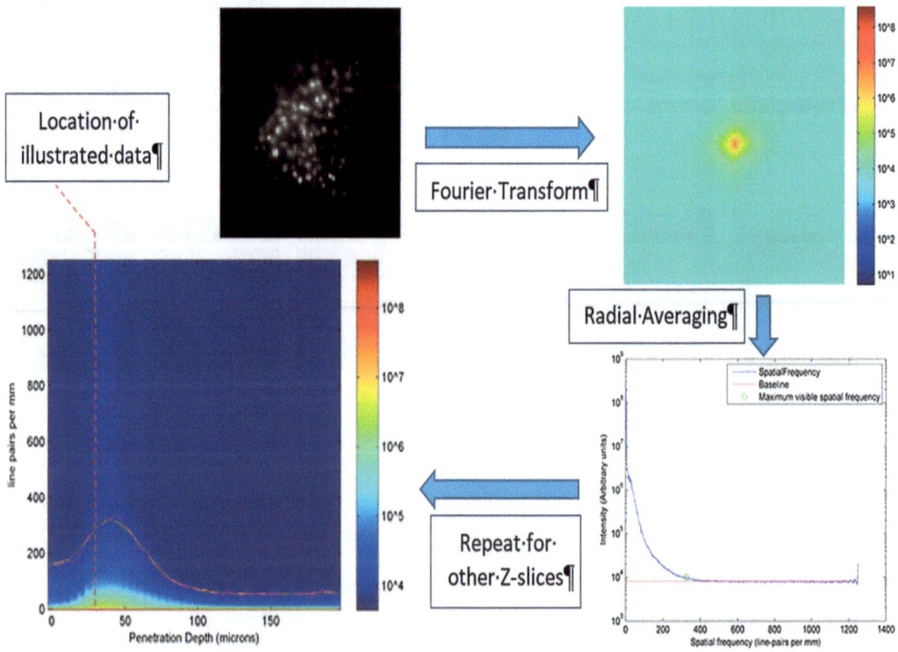

Fig. 6.16: Image analysis algorithm for extracting tissue imaging depth for TPM and TPSM [22].

Accepting that there are limits to the technique in terms of differences in stain penetration as a function of depth and organ type, it is clear from the data that, depending on the sample, TPSM can achieve approximately twice the penetration depth of TFM when subject to practical issues such as sample damage (Fig. 6.17). It was also possible to compare organs in terms of the achievable penetration depth; the heart consistently demonstrated the lowest penetration depth, around 50 µm for temporal focusing and 120 µm for scanning two-photon microscopy. The lungs, liver and kidneys were all very similar, with penetration depths of approximately 90 μm for temporal focusing and 150 μm for scanning two-photon microscopy. The spleen is slightly easier to penetrate, at around 100 μm for temporal focusing and 180 μm for scanning two-photon microscopy, but the low absorption and scattering due to the large lipid droplets in white adipose tissue meant that penetration depths of nearly 200 μm for temporal focusing and over 500 μm for scanning two-photon microscopy were possible. The especially large error bars were due to the large variation in measured scanning two-photon penetration depths; values of 190 μm, 635 μm and 695 μm were recorded, and the 190 μm result strongly skewed the results.

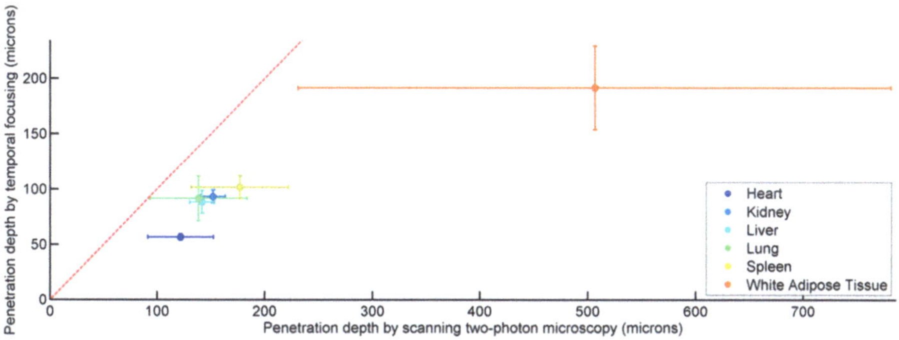

Fig. 6.17: A comparison of achievable imaging depth for different tissues imaged using TFM and TPSM [22].

6.6 Application 1: Functional imaging of neuronal network using TFM

One may argue that functional imaging of the neuronal network in a small animal brain in vivo is one of the most suitable applications of TFM, since functional imaging of neuronal communication requires high speed imaging. For the study of calcium signals, the time constant for sampling of the relevant network structure is on the order of 100 ms. While routine voltage imaging of neuronal network still awaits the development of better voltage sensors, it is clear that sampling a neuronal network with a time constant of several milliseconds is very well suited for TFM-type approaches.

One of the first defining applications of TFM in neuronal network imaging was performed by the Vaziri group; they successfully imaged the calcium dynamics of all the neurons within the head of *C. elegans* [1]. The design of the instrument is fairly straightforward (Fig. 6.18). The neurons in *C. elegans* expressed nuclear-localized GCaMP5K, which simplified the segmentation of functional signals from individual neurons. They acquired data from an image cube with approximately 60 μm lateral dimension and 30 μm axial dimension. Taking an axial step of 2 μm, they achieved

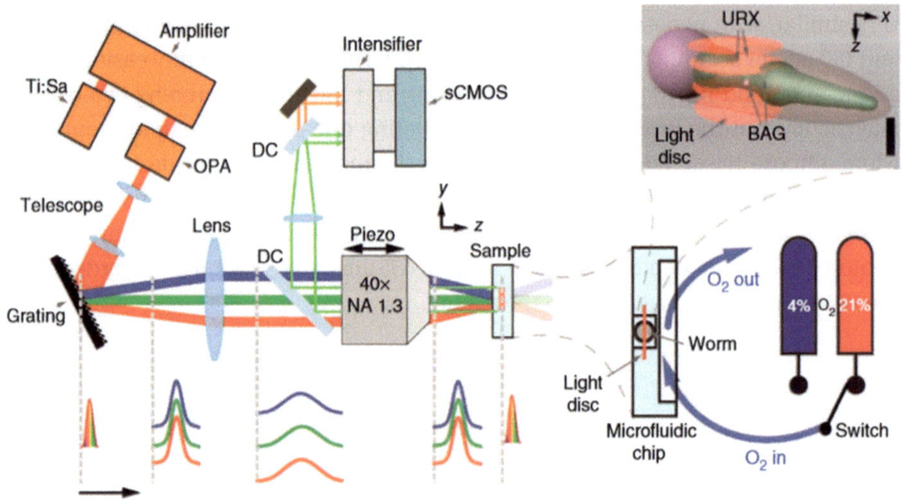

Fig. 6.18: TFM for imaging neuronal network in *C. elegans* [1].

a volumetric frame rate of about 13 Hz, although the GCaMP5K signal-to-noise level limited the actual imaging speed to about 4–6 Hz. The changes in neuronal calcium levels as a function of ambient oxygen concentration were studied across the whole network in the head region (Fig. 6.19). Importantly, the high speed imaging afforded by TPM allowed many neurons to be simultaneously imaged within the relevant time constant, potentially allowing neuronal communication networks to be deduced by examining the correlation matrix (Fig. 6.19 (a)).

While not strictly speaking wide-field imaging, the implementation of temporal focusing in line-focusing mode is an important extension (Fig. 6.20), removing the need for expensive and complex regenerative amplifiers in order to image large areas. The main limitation of using TFM for neuronal network imaging, especially for green sensors like GCaMPs, is limited laser power. Since commercially available regenerative amplifiers currently only output at around 800 nm, and since the excitation of GCaMP requires excitation in the 900–1000 nm range, the use of optical parametric amplification and frequency doubling is required, resulting in severe power loss as previously discussed. This limitation can be overcome by line-focusing, where power is not distributed over a plane, but only over one line, which is then scanned to acquire the image. Dana, Shoham and co-workers used a modified 150 kHz regenerative amplifier operating at 905 nm with 150 mW power, which proved sufficient for line-focusing application [3, 4]. Importantly, even with the laser operating near its performance margin, they significantly extended the imaging area up to about 500 μm laterally, albeit with a fairly modest lateral and axial resolution of about 2.6 μm and 25 μm respectively. Imaging neuronal cell cultures at a frame rate of about 10 frames per second, they were able to record calcium signals from neurons expressing GCaMP3. Importantly, they were able to simultaneously ontogenetically stimulate selected neurons using ChR2-mCherry via holographic beam forming (Fig. 6.21).

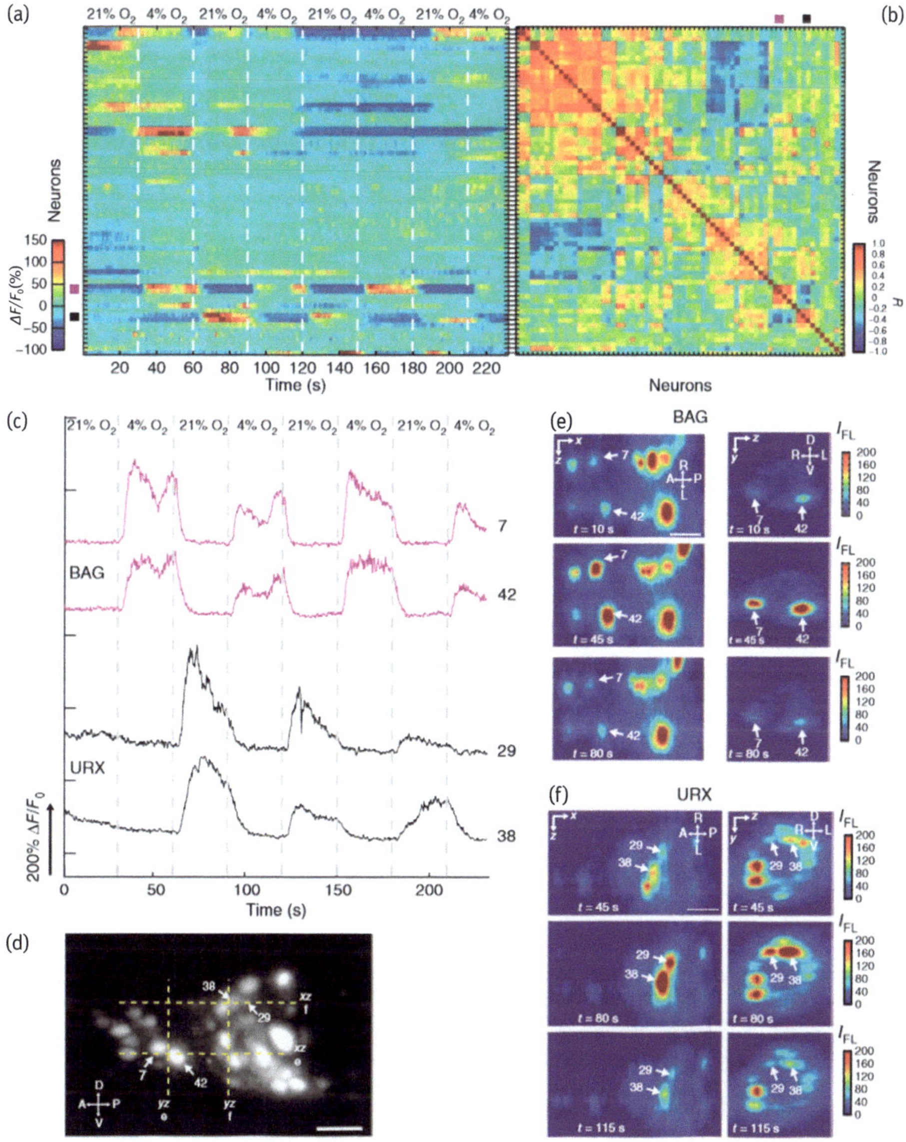

Fig. 6.19: (a) Correlation of calcium signal across different neurons. (b)–(d) Network calcium signals as a function of ambient oxygen levels [1].

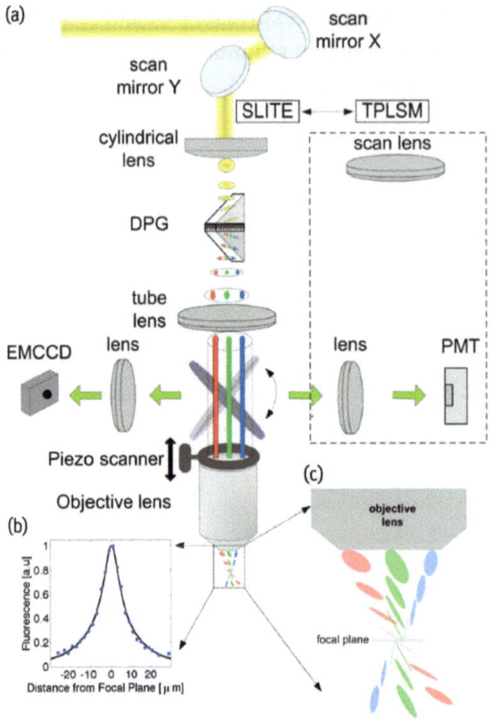

Fig. 6.20: Line-scan TPM developed for large area, high throughput neuronal network imaging [4].

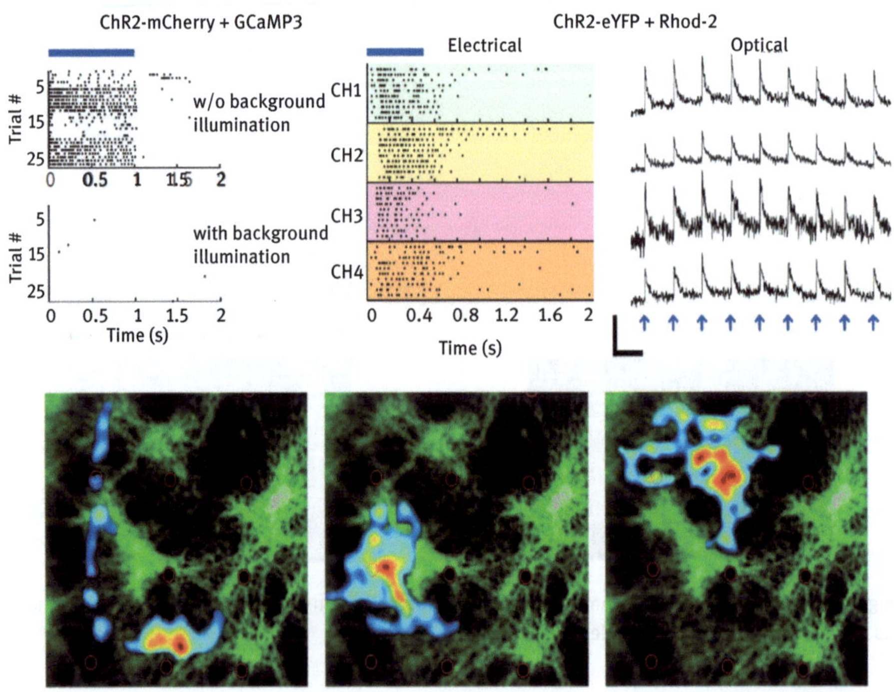

6.7 Application 2: Fluorescence and phosphorescence lifetime imaging using TFM

Fluorescence (FLIM) and phosphorescence (PLIM) lifetime microscopies are information-rich optical spectroscopic techniques [55–61]. A particularly important application of FLIM is in the quantification of fluorescence resonance energy transfer (FRET), the preferred method of quantifying intracellular protein-protein interactions in vivo. For example, FRET measurements have been used to measure intramolecular distances [62], and to observe dynamic conformational changes in proteins [63] and RNA [64]. FLIM has been also applied in disease diagnosis. Koenig and co-workers found that normal skin and cutaneous melanoma can be differentiated by their morphological appearances in combination with their fluorescence lifetime spectroscopic signatures [14]. A clinical trial across multiple centers in Europe is underway to test the utility of multiphoton FLIM in the minimally invasive diagnosis of melanoma [65].

While PLIM is not as widely used as FLIM, this methodology nonetheless has several important potential biomedical applications, due to the availability of phosphorescence-based oxygen sensors. Quenching of phosphorescence by oxygen affects the phosphorescence lifetime of such sensors, which then enables measurement of oxygen partial pressure in vivo in tissues or thick biological samples with high temporal and spatial resolution [66–68]. PLIM-based partial oxygen measurements can be used to quantify the degree of hypoxia in tissues or tumors, a critical physiological parameter of solid tumors that determines tumor growth, gene expression [69], metastatic potential [70], metabolism, prognosis [71–73], and response to therapy [74, 75]. The use of PLIM-based oxygen sensors has also enabled the quantification of oxygen supply and consumption in the brain, which is critical for understanding brain metabolism and cognitive function [76, 77]. Today, PLIM is not widely used mostly due to the associated long lifetime that entails typical image frame rates on the order of minutes to hours.

The implementations of FLIM and PLIM measurements in microscopy have different challenges. For FLIM, the need for picosecond level timing resolution of photon arrival time requires fast electronics; for PLIM, the long lifetime (up to milliseconds) significantly lengthens the required pixel residence time and slows the image frame rate. In general, FLIM and PLIM measurements can be performed either in the time domain or in the frequency domain. In the time domain, a light pulse of short duration excites the sample and the time delay of the emitted photon is measured. A histogram of these time delays provides a direct measurement of the fluorophore's lifetime. Typical fluorophores exhibit single exponential decay kinetics in a homogeneous microenvironment. However, in practice, even for fluorophores with single exponential decay dynamics, the experimentally measured time delay histogram is a convolution of an

◂ **Fig. 6.21:** Combining line-focus TPM imaging with optogenetic excitation [3, 4].

unknown instrumentation response and the exponential decay of the fluorophore. In the presence of more complex environments, multiple exponential decays must be resolved. The frequency domain approach is, mathematically, the Fourier equivalent of the time domain. In the frequency domain, the excitation light is modulated at a frequency ω. The intrinsic fluorescence temporal response acts as a low-pass filter and the emitted fluorescent light is phase-shifted and modulated and modified as a function of the underlying lifetime. In order to obtain accurate phase shift and demodulation measurements, the excitation light modulation frequency ω has to be of the same order of magnitude as the inverse of the lifetime (i.e., up to 10^8 Hz for FLIM). Phase shift and demodulation are often measured indirectly by either homodyne or heterodyne approaches. In the homodyne approach, measuring the steady state amplitudes at different phases allows the recovery of the waveform. In the case of heterodyne measurement, the signal is detected by a detector whose gain is modulated sinusoidally at a slightly increased frequency $\omega + \Delta\omega$. This electronic mixing process results in translating the phase and demodulation information to an electronic signal at the cross correlation frequency $\Delta\omega$ that can be readily isolated by low-pass filtering. The relative merits of time domain vs frequency domain approaches and homodyne vs heterodyne detection schemes have been discussed extensively in the literature [78, 79].

We have developed a very fast 3D FLIM and PLIM imaging system based on combining two complementary technologies:
(1) temporal focusing wide-field (TFWF) two-photon microscopy, a method for efficiently exciting a single 3D resolved plane in a translucent specimen, and
(2) camera-based heterodyne frequency-domain lifetime measurement, a method for highly parallelized wide-field imaging with picosecond lifetime resolution [20].

Fig. 6.22 shows the optical design of the temporal focusing wide-field FLIM/PLIM system.

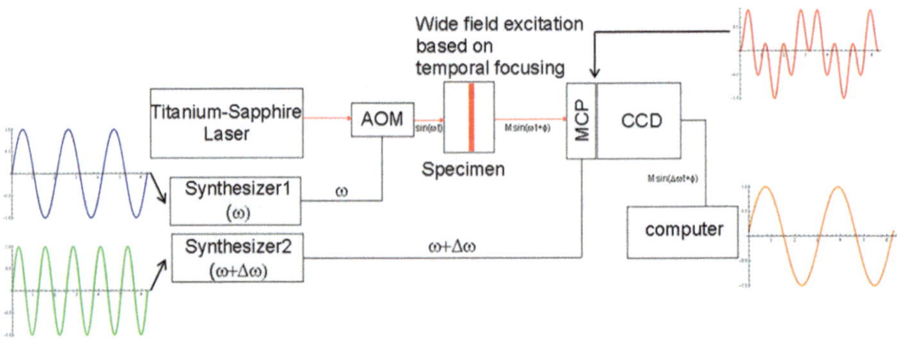

Fig. 6.22: Temporal focusing wide-field FLIM/PLIM design. Electronic subsystem: frequency domain lifetime measurement via heterodyne detection [20].

The accuracy of the developed wide-field 3D lifetime imaging system was evaluated by measuring the fluorescence lifetimes of Rhodamine B solution in different solvents that have been carefully quantified in literature [7]. Fitting the measurements to a single exponential decay model allows the extraction of lifetimes tabulated in Fig. 6.23 (left). The lifetime measurements obtained by the temporal focusing FLIM were in good agreement with literature values within about 0.05–0.1 ns. Note that the two ways to estimate the fluorescence lifetime (either from phase data or from modulation data) provide very similar results, which supports the choice of a single exponential decay model to fit Rhodamine B lifetime data. When plotting the lifetime measurements of each pixel into a single polar plot (a sine and cosine transform of temporal decay information into the spectral domain [80, 81]) the lifetimes of individual pixels cluster around a single location located on the universal circle as expected for single exponential decay processes (Fig. 6.23, right). The locations of these four distributions are consistent with the tabulated lifetimes obtained from averaging over the whole image.

	Tau (Ph)	Tau (Mod)	Tau (Magde)
FL	4.01	4.18	4.00
RhB/ EtOH	2.83	2.84	2.93
RhB/ MeOH	2.22	2.36	2.46
RhB/ H_2O	1.75	1.79	1.68

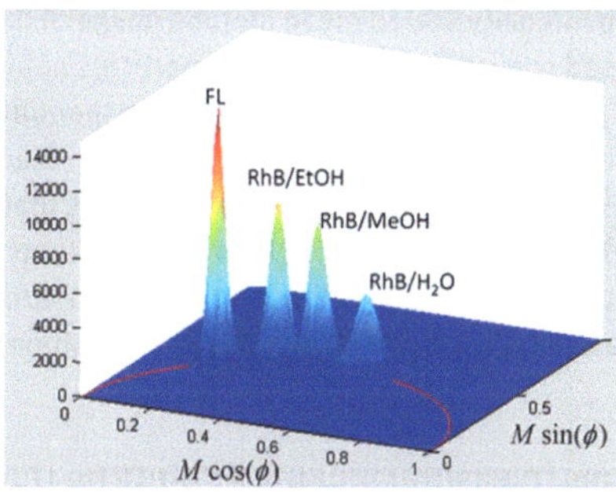

Fig. 6.23: Demonstration of accurate measurement of fluorescence lifetime of Rhodamine B solutions in different solvents by TFWF FLIM. Fluorescence in water was used as a reference. (Left) Tabulated results of estimated lifetime values τ extracted from either phase (Ph) or modulation (Mod) measurements. Literature values are also included as a reference [7]. (Right) Lifetime resolved data for each pixel from the fluorescein and Rhodamine solution images shown in polar plot format [20].

The most important advantage of temporal focusing for FLIM imaging is probably in throughput enhancement. The high data acquisition speed of the developed instrument is demonstrated by FLIM imaging of fixed fibroblasts loaded with conjugate polymer nanoparticles (CPN) of high two-photon cross section (previously measured in excess of 15 000 GM [82]). A representative intensity scaled lifetime image and the associated polar plot of pixel lifetime distribution of the specimen is shown in Fig. 6.24. The nonsymmetric, off universal circle distribution of pixel values in the polar plot indicates nonsingle exponential decay of CPNs. This is consistent with the different mean lifetimes measured from phase and modulation data. It is important to note that consistent pixel fluorescence lifetime measurements can be obtained with integration

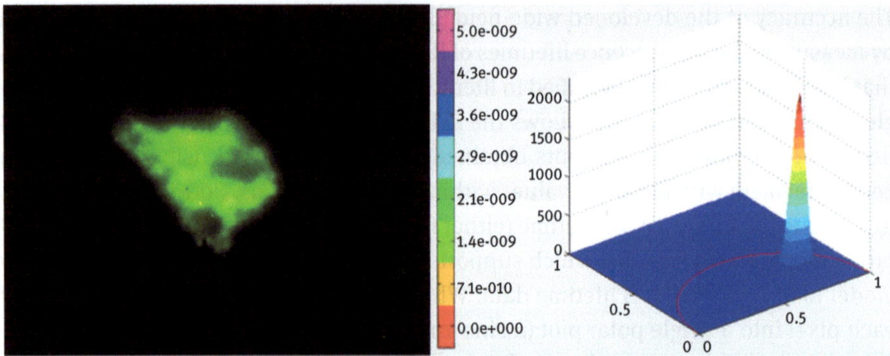

Integration Time	Tau. Phase (ns)	Tau. Mod (ns)
20	1.9±0.1	1.2±0.1
10	1.8±0.1	1.3±0.1
5	1.7±0.2	1.3±0.1
2	1.7±0.2	1.7±0.2

Fig. 6.24: (Top, left) An intensity scaled mean lifetime image of fixed fibroblasts with vacuoles loaded with endocytosed conjugated polymer nanoparticles of high two-photon absorption cross section. Color scale represents pixel lifetime values corresponding to the color bar with units of seconds. Image brightness represents pixel intensity values. Black regions are ignored in analysis corresponding to locations with intensity below 500 photons that are mostly outside the boundary of this cell. (Top, right) Representative polar plot of pixel lifetime values for 10 ms data acquisition time. (Bottom) Tabulated mean lifetime values and their standard deviations are estimated from the modulation or the phase data for four different image acquisition rates [20].

times as short as 5 ms (measurement uncertainty on the order of 0.1–0.2 ns). When the integration time is further reduced to 2 ms, phase-based lifetime measurement still provides a reasonable lifetime estimate (with less than 0.5 ns error) while modulation-based lifetime estimates start to deviate from values obtained with longer average time (> 0.5 ns).

The impact of parallelism inherent in TFM is probably more important for phosphorescence measurements. Phosphorescence lifetime measurement has many uses in biomedical imaging and spectroscopic analysis. One of the most important biomedical uses of molecular phosphorescence is monitoring oxygen concentration in biological systems. Minimally invasive measurements of partial oxygen pressure can be used, for example, to quantify hypoxia in tumors, an important determinant of tumor physiology [69, 70] and its response to therapy [75, 83]. However, 3D PLIM using point scanning confocal or multiphoton excitation is always slow since the required pixel residence time must be substantially longer than the phosphorescence lifetime (as long as a fraction of a millisecond). The estimated necessary duration for high-resolution 3D mapping of oxygen distribution in a tumor can exceed days using palladium-based probes (lifetime approaches one millisecond). In order to demonstrate the accuracy of the developed PLIM TFM system, we quantified oxygen concentration in solutions of 1 mM Tris(2,2′-bipyridyl) dichlororuthenium(II) hexahydrate in PBS equilibrated with several independently calibrated oxygen/nitrogen gas mixtures of O_2 mass fraction of 0, 4, 8, and 21 %. The phosphorescence lifetime of the ruthenium solution equilibrated with each mixture was found to be single exponential as indicated by the distribution of the estimated pixel lifetime that lies on the universal circle of the polar plot (Fig. 6.25, left). The measured phosphorescence lifetime ranges from about 400 to 600 ns. There is an inverse dependence of the phosphorescence lifetime on oxygen partial pressure in agreement with a Stern–Volmer relationship (Fig. 6.25, middle, right). We have demonstrated fast sequential 3D-resolved FLIM and PLIM TFM imaging of rhodamine labeled fibroblast in collagen scaffold containing ruthenium for in situ environment oxygen monitoring (Fig. 6.26).

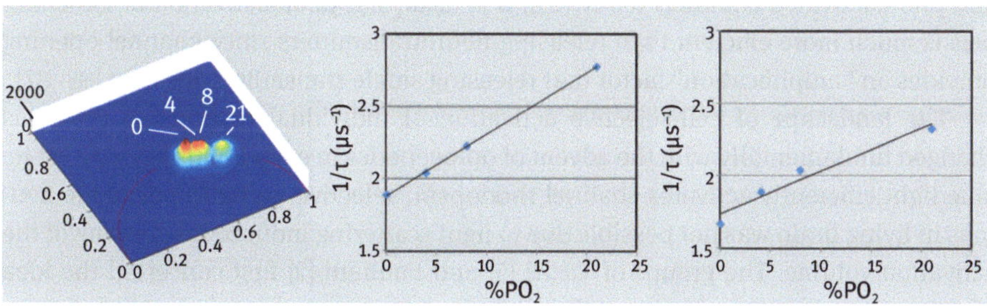

Fig. 6.25: Fast measurement of partial oxygen pressure by TFWF phosphorescence lifetime imaging. (Left) Polar plot of phosphorescence lifetime of 1 mM Tris(2,2′-bipyridyl) dichlororuthenium(II) hexahydrate solutions equilibrated with 0, 4, 8, and 21 % PO_2 gas mixtures. Inverse phosphorescence lifetimes (middle, from modulation data; right, from phase data) are plotted against O_2 concentration demonstrating Stern–Volmer dependence with R^2 values of 0.99 and 0.95 respectively for linear regression [20].

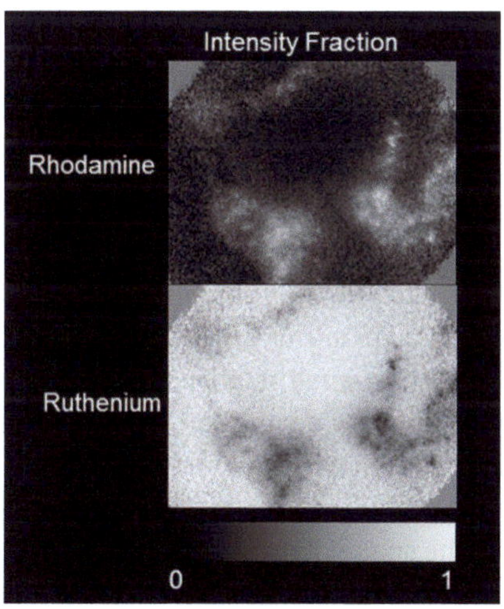

Fig. 6.26: Fast sequential 3D-resolved TFWF FLIM and PLIM imaging of human dermal fibroblasts seeded in a collagen scaffold. Fluorescence intensity fraction due to Rhodamine (top) and phosphorescence intensity fraction due to ruthenium (bottom) are resolved based on lifetimes [20].

6.8 Application 3: Cell-selective optogenetics using TFM

While TPSM has recently become the method of choice for high resolution in vivo tissue imaging, the importance of multiphoton excitation for inducing localized chemical reaction has been recognized by Denk, Webb and co-workers from almost the very beginning [10]. Several years after demonstrating TPSM, Denk further developed scanning photochemical microscopy to induce localized activation of neuron by using caged neurotransmitters [84]. While their work represented an important milestone in developing technologies to control neuronal circuits, it did not result in any major biological impact for two reasons. First, molecular cages at that time had very low two-photon cross section and the laser power required for uncaging often induced nonspecific photochemical reactions in cells. Second, with the development of optogenetics and photoactivated channels today [85], it is clear that photoactivation of ion channels is much more efficient than releasing neurotransmitters since channel opening provides an "amplification" factor that releasing single transmitter does not [85–87].

The landscape of cell-selective activation of individual neurons in vivo was changed fundamentally with the advent of optogenetics in early 2000s [85–87]. While blue light efficiently activates channel rhodopsin, selective excitation of single neurons in living brain was not possible due to light scattering induced broadening of the activation volume. The groups of Vaziri [5] and Emiliani [2] first conceived the idea of using two-photon excitation to localize the optogenetic process in 2010. Importantly, both of these works utilized temporal focusing geometry for neuron activation for several important reasons. First, it has been observed that the efficient activation of a single neuron requires activating many ion channels and activation with a diffraction-limited point is inefficient for this propose. Second, the study of neuronal circuits often requires simultaneous activation of multiple neurons and patterned excitation can be readily accomplished using temporal focused geometry.

Varizi and co-workers recently demonstrated a high aspect ratio two-photon excitation volume spanning 15 μm in diameter with less than 2 μm thickness enabling single cell excitation in tissues. They further investigated various parameters such as depth discrimination, excitation spot size, power, and duration (Fig. 6.27).

Emiliani and co-workers took temporal focusing-based two-photon optogenetics to the next step by introducing patterned excitation (Fig. 6.28). Since the grating surface is imaged onto the imaging plane of a TPM, pattern generation can be readily accomplished by projecting an appropriate intensity distribution on the grating. This can be accomplished with a photomask or an MEMS mirror. However, these approaches are light inefficient if the illumination area is small relative to the whole field of view. Emiliani and co-workers proposed an approach based on the principle of generalized phase contrast. In this approach, a phase-based spatial light modulator projects a phase pattern via a 4-f relay onto the grating via a phase-contrast filter positioned at the Fourier plane. In the ideal case, the spatial light modulator is a pure phase element and generating a phase-only pattern does not result in light losses,

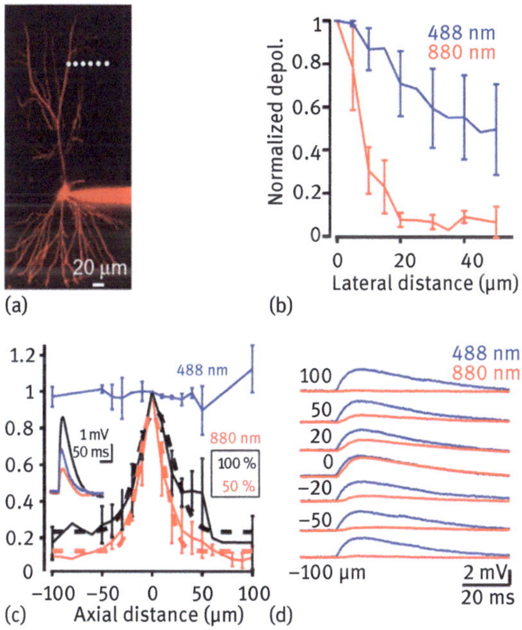

Fig. 6.27: Quantifying spatial resolution of temporal focusing optogenetics [5].

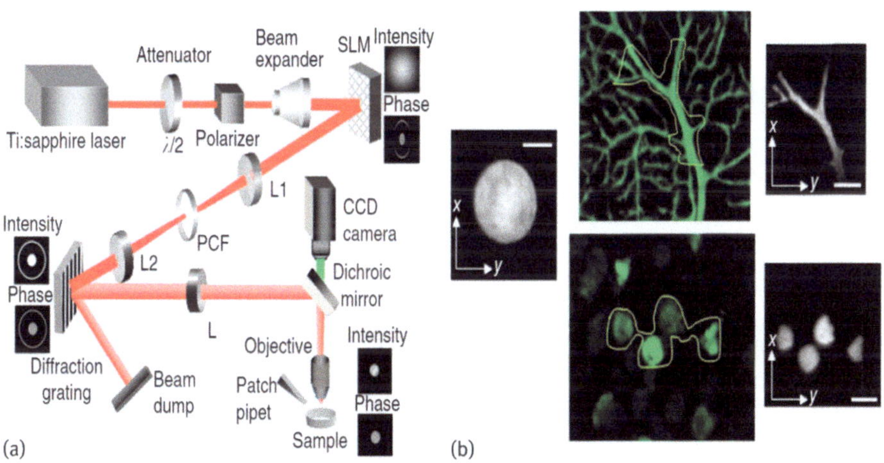

Fig. 6.28: Temporal focusing optogenetic design with patterning element based on generalized phase contrast [2].

although spatial light modulators suffer significant losses due to design limitations. The phase-contrast filter at the Fourier plane phase retards the low frequency components of the phase image relative to the high frequency components. The interference between the low frequency components with the high frequency components at the grating plane effectively converts the phase-only image at the SLM into an image with intensity contrast at the grating. It can be shown that by careful design of the phase contrast filter, one can maximize the image fidelity and contrast while transmitting higher total intensity than with masks or MEMS mirrors. Emiliani and co-workers have provided a thorough characterization of system parameters in temporal focusing patterned optogenetics and demonstrated simultaneous selective multiple cell activation in brain slices (Fig. 6.29).

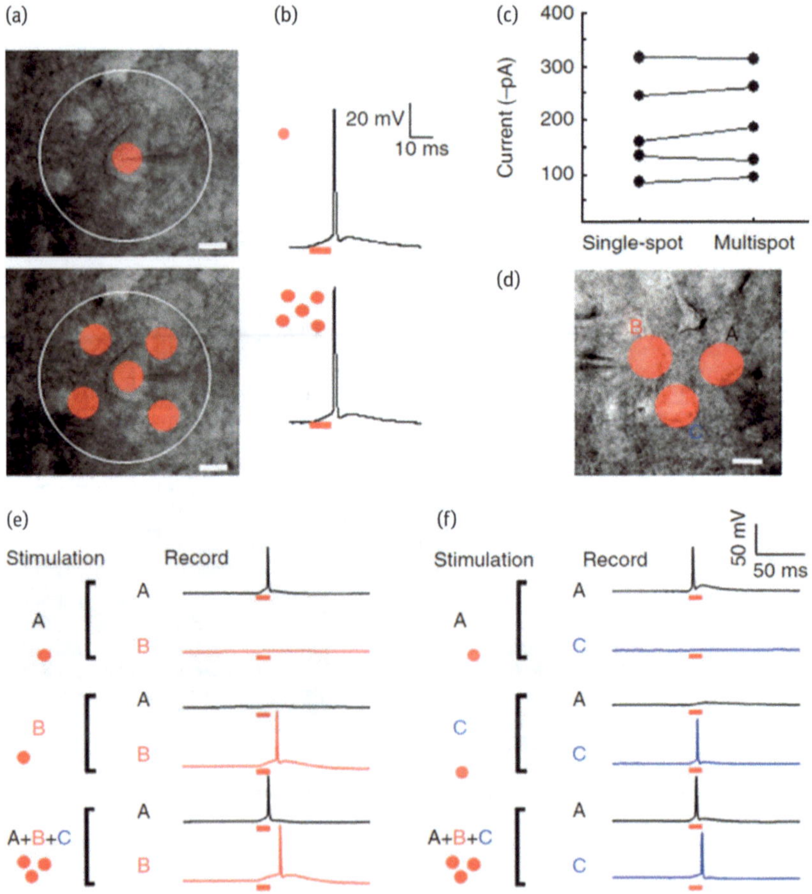

Fig. 6.29: Demonstrating selective multiple cell activation [2].

6.9 Application 4: Cell selective photodynamic therapy using TPM

It is broadly accepted that cancer is one of the most important medical challenges in the developed world. One of the primary treatments, chemotherapy, is extremely taxing on the patient due to the toxicity and low selectivity of commonly-used anticancer agents. While cancer cells are preferentially killed, a number of other biochemical systems in the body are negatively impacted, leading to side effects such as cognitive dysfunction, hair loss, and organ damage [88]. Photodynamic therapy (PDT) is a far more selective treatment option, wherein light is used to localize the antineoplastic effect and thereby avoid many of the downsides associated with more systemic treatment. In PDT, a chromophore called a photosensitizer is illuminated with specific wavelengths of light in order to create reactive radical species that generate downstream effects, including apoptosis and necrosis in populations of cells. By controlling the administration of light to select regions of tissue, PDT can be carried out in just the tissue of interest, sparing critical nerves and organs, and preserving functionality where conventional surgery and chemotherapy would cause irreparable damage. In the ultimate limit, light administration could be confined purely to selected cells, avoiding the surrounding tissue entirely. In this case, the photodynamic agent need only be optimized for factors like quantum yield, clearance rate and dark (un-illuminated) toxicity. Such control would enable precise treatment of individual cells within intact tissue for pinpoint anticancer therapy, such as, for example, when tumor cells have extensively infiltrated sensitive organs such as the eye.

The use of multiphoton excitation for photodynamic therapy is not new; Fisher et al. demonstrated two-photon excitation of some psoralen derivatives in 1997 [89], and in 1999, König, Riemann and Fischer described their work on two-photon excitation of photofrin and 5-ALA labeled Chinese Hamster Ovary (CHO) cells [90]. Demonstration of two-photon absorption in verteporfin and photofrin in vitro was performed by Khurana et al. in 2007 [91], and in the same year, two-photon occlusion of blood vessels in a chicken embryo was performed by Samkoe et al. using verteporfin [92]. Blood-vessel occlusion was demonstrated in mice by Collins et al., using a selection of different photosensitizers that were optimized for a high two-photon cross section [93]. The use of different novel photosensitizers was also pursued by Starkey et al., who argued that xenograft tumors could be treated using two-photon excitation from a regenerative amplifier focused through approximately 2 cm of tissue [94, 95].

Though this work is promising, the techniques used above all required the femtosecond laser focus to be scanned in the tissue, a serial process that is inherently slow. We have developed a temporal focusing PDT method by exciting millions of points in parallel, using temporal focusing. Our instrument is designed to project a temporal focusing plane approximately 700 μm × 700 μm onto a sample. Exposure times are on the order of 10 s to 30 s, which is approaching values that are clinically relevant. As an example application, this exposure duration and illumination area could be suitable for the treatment of residual cancer cells in a resection margin.

OVCAR-5 ovarian cancer cells (Fox Chase Cancer Institute) either grown in 2D on a collagen-coated 35 mm glass-bottomed dish (MatTek, P35GCOL-1.5-14-C) or grown in 3D model tumor nodules [96] were incubated with verteporfin solution for two hours and exposed to light using a TFM system. Patterning was accomplished by projecting the temporal focusing plane onto an intermediate focal plane containing a photomask, before being refocused onto the sample. The efficacy of PDT was evaluated after approximately 8–10 hours using live-dead label consisting of cytoplasmic green Calcein AM (live stain) and nucleic orange Ethidium Homodimer-1 (dead stain).

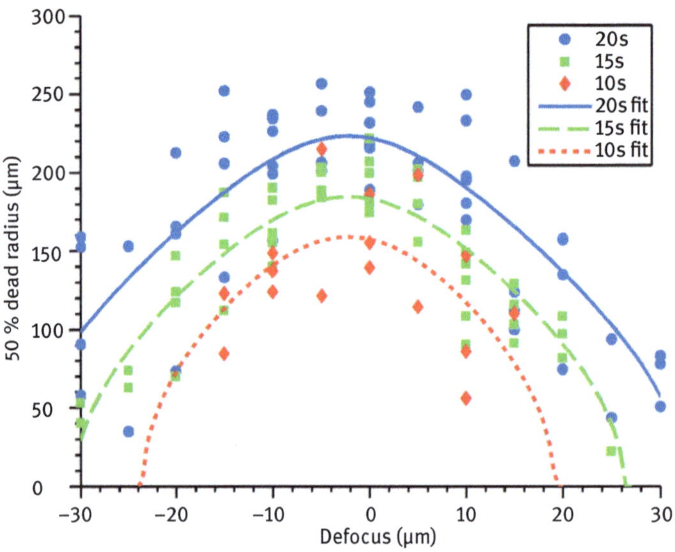

Fig. 6.30: 50 % cell death radius as a function of defocus for different exposure values.

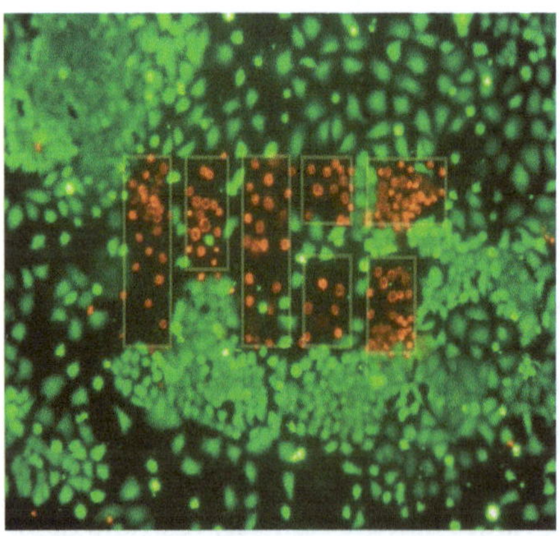

Fig. 6.31: Demonstrating patterning of the temporal focusing plane using the MIT logo via a live-dead stain. White lines illustrating the mask boundary are added as a guide to the eye and are not present in the image itself

We have quantified PDT parameters such as exposure duration and verteporfin concentration. The depth sectioning capability of patterned cell killing was demonstrated, and a depth resolution on the order of 20–30 μm was achieved for an exposure area approximately 300 μm in diameter (Fig. 6.30). We have not pushed the axial resolution limit further as PDT does not require axial resolution beyond the dimensions of a typical cell, which is on the order of several tens of microns. We have further demonstrated patterned cell killing in 2D and 3D cultures (Fig. 6.31).

6.10 Conclusion

In this chapter, the underlying physical principles for TFM are explained in detail. We have further evaluated the depth discrimination capability of TFM, in showing the technique to have relatively poor axial sectioning capability, and have described two more advanced TFM implementations that overcome this limitation. We have provided a survey of the usability of TFM in different tissues and organs. We show that, in general, the achievable penetration depth of TFM is approximately half that of depth as TPSM; the limitation stems from the fact that while infrared excitation light for TFM has similar penetration depth as TPSM, the use of a camera as a detector results in image degradation, due to the scattering of blue-shifted emission photons, resulting in a loss of image signal-to-background ratio. Finally, we describe four typical application areas, divided into two classes, where TFM has found important utility. In the class of rapid imaging, the first application relates to high speed functional imaging of cells and tissues, especially in the brain where rapid sampling of large areas is important. The second application of TFM enables fastest fluorescence and phosphorescence lifetime imaging in cells and tissues due to the high degree of parallelization. It is probably most important in the phosphorescence case where the lifetime is long and complete spectroscopic analysis at each spatial location may take almost one millisecond. The second class of applications utilizes the ability of TFM for depth resolved patterning in thick biological tissues. The most important application in this class is probably cell-selective optogenetic excitation that promises to offer a better understanding of neuronal circuit communications. We have further described an application in temporal focusing photodynamic therapy that may allow selective killing of cancer cells. We foresee applications of patterned temporal focusing as probably the more important direction, as temporal focusing-based imaging is always hampered by the acquired image being blurred by scattered emission photons.

Acknowledgment: We acknowledge support from National Institute of Health 5-P41-EB015871-28, 2R01EY017656-06, R21-NS091982-01, 1-U01-NS090438-01, 1U01CA202177-01, and 1R01HL121386-01A1, Singapore-MIT Alliance for Science and Technology, Connecticut Children's Medical Center, Hamamatsu Corp., and Samsung Advanced Institute of Technology. CJR further acknowledges support from the Wellcome Trust 093831/Z/10/Z.

References

[1] Schrodel T, et al. Brain-wide 3D imaging of neuronal activity in Caenorhabditis elegans with sculpted light. Nat Methods. 2013;10(10):1013–1020.

[2] Papagiakoumou E, et al. Scanless two-photon excitation of channelrhodopsin-2. Nature Methods. 2010;7(10)848–854.

[3] Paluch-Siegler S, et al. All-optical bidirectional neural interfacing using hybrid multiphoton holographic optogenetic stimulation. Neurophotonics. 2015; 2(3):031208.

[4] Dana H, et al. Hybrid multiphoton volumetric functional imaging of large-scale bioengineered neuronal networks. Nat Commun. 2014;5:3997.

[5] Andrasfalvy B, et al. Two-photon single-cell optogenetic control of neuronal activity by sculpted light. Proc Natl Acad Sci USA. 2010;107:11981–11986.

[6] Göppert-Mayer M. Über Elementarakte mit zwei Quantensprüngen. Ann Phys (Leipzig). 1931;5:273–294.

[7] Magde D, Rojas GE, Seybold PG. Solvent dependence of the fluorescence lifetimes of xanthene dyes. Photochem Photobiol. 1999;70(5):737–744.

[8] Sheppard CJR, et al. The scanning harmonic optical microscope. IEEE/OSA Conference on Laser Engineering and Applications. Washington; 1977.

[9] Gannaway JN, Sheppard CJR. Second harmonic imaging in the scanning optical microscope. Optics and Quantum Electronics. 1978;10:435–439.

[10] Denk W, Strickler JH, Webb WW. Two-photon laser scanning fluorescence microscopy. Science. 1990;248(4951):73–76.

[11] Campagnola PJ, et al. Second-harmonic imaging microscopy of living cells. J Biomed Opt. 2001;6(3):277–286.

[12] Campagnola PJ, et al. Three-dimensional high-resolution second-harmonic generation imaging of endogenous structural proteins in biological tissues. Biophys J. 2002;82(1 Pt1):493–508.

[13] Campagnola PJ, et al. High-resolution nonlinear optical imaging of live cells by second harmonic generation. Biophys J. 1999;77(6):3341–3349.

[14] Konig K, et al. Clinical two-photon microendoscopy. Microsc Res Tech. 2007;70(5):398–402.

[15] Laiho LH, et al. Two-photon 3-D mapping of ex vivo human skin endogenous fluorescence species based on fluorescence emission spectra. J Biomed Opt. 2005;10(2):024016.

[16] Mohler W, Millard AC, Campagnola PJ. Second harmonic generation imaging of endogenous structural proteins. Methods. 2003;29(1):97–109.

[17] Stoller P, et al. Polarization-dependent optical second-harmonic imaging of a rat-tail tendon. J Biomed Opt. 2002;7(2):205–214.

[18] Vaziri A Shank CV. Ultrafast widefield optical sectioning microscopy by multifocal temporal focusing. Optics Express. 2010;18(19):19645–19655.

[19] Choi H, et al. Improvement of axial resolution and contrast in temporally focused wide-field two-photon microscopy with structured light illumination. Biomed Opt Express. 2013;4(7):995–1005.

[20] Choi H, et al. 3D-resolved fluorescence and phosphorescence lifetime imaging using temporal focusing wide-field two-photon excitation. Optics Express. 2012;20(24):26219–26235.

[21] Yew EY, Sheppard CJ, So PT. Temporally focused wide-field two-photon microscopy: paraxial to vectorial. Opt Express. 2013;21(10):12951–12963.

[22] Rowlands CJ, et al. Objective, comparative assessment of the penetration depth of temporal-focusing microscopy for imaging various organs. J Biomed Opt. 2015;20(6):61107.

[23] Reddy GD, Saggau P. Fast three-dimensional laser scanning scheme using acousto-optic deflectors. J Biomed Opt. 2005;10(6):064038.

[24] Duemani Reddy G. et al. Three-dimensional random access multiphoton microscopy for functional imaging of neuronal activity. Nat Neurosci. 2008;11(6):713–720.
[25] So PT, Yew EY, Rowlands C. High-throughput nonlinear optical microscopy. Biophys J. 2013; 105(12):2641–2654.
[26] Bahlmann K, et al. Multifocal multiphoton microscopy (MMM) at a frame rate beyond 600 Hz. Optics Express. 2007;15(17):10991–10998.
[27] Kim K, et al. Multifocal multiphoton microscopy based on multianode photomultiplier tubes. Optics Express. 2007;15(18):11658–11678.
[28] Bewersdorf J, Pick R, Hell SW. Mulitfocal multiphoton microscopy. Opt. Lett. 1998;23:655–657.
[29] Buist AH, et al. Real time two-photon absorption microscopy using multipoint excitation. J Microscopy. 1998;192:217–226.
[30] Oron D, Tal E, Silberberg Y. Scanningless depth-resolved microscopy. Optics Express. 2005;13(5):1468–1476.
[31] Tal E, Oron D, Silberberg Y. Improved depth resolution in video-rate line-scanning multiphoton microscopy using temporal focusing. Optics Letters. 2005;30(13):1686–1688.
[32] Durst ME, Zhu G, Xu C. Simultaneous spatial and temporal focusing for axial scanning. Optics Express. 2006;14(25):12243–12254.
[33] Zhu GH, et al. Simultaneous spatial and temporal focusing of femtosecond pulses. Optics Express. 2005;13(6):2153–2159.
[34] Durst, ME, Straub AA, Xu C. Enhanced axial confinement of sum-frequency generation in a temporal focusing setup. Opt Lett. 2009;34(12):1786–1788.
[35] Froehly C, Colombeau B, Vampouille M. II Shaping and analysis of picosecond light pulses. In: Wolf E, editor. Progress in optics. Elsevier; 1983. p. 63–153.
[36] Martinez OE, Gordon JP, Fork RL. Negative group-velocity dispersion using refraction. J Opt Soc Am A. 1984;1(10):1003–1006.
[37] Durst ME, Zhu GH, Xu C. Simultaneous spatial and temporal focusing for axial scanning. Optics Express. 2006;14(25):12243–12254.
[38] Suchowski H, Oron D, Silberberg Y. Generation of a dark nonlinear focus by spatio-temporal coherent control. Optics Communications. 2006;264:6.
[39] Cheng LC, et al. Spatiotemporal focusing-based widefield multiphoton microscopy for fast optical sectioning. Opt Express. 2012;20(8):8939–8948.
[40] Wefers MM, Nelson KA. Space-time profiles of shaped utrafast optical waveforms. IEEE Journal of Quantum Electronics. 1996;32(1):12.
[41] Danailov MB, Christov IP. Time-space shaping of light-pulses by Fourier optical-processing. Journal of Modern Optics. 1989;36(6):725–731.
[42] Kim KH, et al. Multifocal multiphoton microscopy based on multianode photomultiplier tubes. Optics Express. 2007;15(18):11658–11678.
[43] Dana H Shoham S. Numerical evaluation of temporal focusing characteristics in transparent and scattering media. Optics Express. 2011;19(6):4937–4948.
[44] Neil MAA, Juskaitis R, Wilson T. Method of obtaining optical sectioning by using structured light in a conventional microscope. Optics Letters. 1997;22(24):1905–1907.
[45] Lim D, Chu KK, Mertz JJ. Wide-field fluorescence sectioning with hybrid speckle and uniform-illumination microscopy. Optics Letters. 2008;33(16):1819–1821.
[46] Weigel A, Schild D, Zeug A. Resolution in the ApoTome and the confocal laser scanning microscope: comparison. Journal of Biomedical Optics. 2009;14(1):014022.
[47] Lim D, et al. Optically sectioned in vivo imaging with speckle illumination HiLo microscopy. Journal of Biomedical Optics. 2011;16(1):016014.
[48] Stokseth PA. Properties of a defocused optical system. Journal of the Optical Society of America. 1969;59(10):1314–1321.

[49] Vaziri A, et al. Multilayer three-dimensional super resolution imaging of thick biological samples. Proceedings of the National Academy of Sciences of the United States of America. 2008;105(51):20221–20226.
[50] Kim D, So PTC. High-throughput three-dimensional lithographic microfabrication. Optics Letters. 2010;35(10):1602–1604.
[51] Karadaglic D, Wilson T. Image formation in structured illumination wide-field fluorescence microscopy. Micron. 2008;39(7):808–818.
[52] Sheppard CJR, Gu M. Image-formation in 2-photon fluorescence microscopy. Optik. 1990;86(3): 104–106.
[53] Singh VR, et al. Improving signal-to-noise ratio of structured light microscopy based on photon reassignment. Biomed Opt Express. 2012;3(1):206–214.
[54] Dana H, et al. Line temporal focusing characteristics in transparent and scattering media. Opt Express. 2013;21(5):5677–5687.
[55] Bastiaens PI, et al. Time-resolved fluorescence spectroscopy of NADPH-cytochrome P-450 reductase: demonstration of energy transfer between the two prosthetic groups. Biochemistry. 1989;28(21):8416–8425.
[56] Becker W, et al. Fluorescence lifetime imaging by time-correlated single-photon counting. Microsc Res Tech. 2004;63(1):58–66.
[57] Gerritsen HC, Vroom JM, de Grauw CJ. Combining two-photon excitation with fluorescence lifetime imaging. IEEE Engineering in Medicine & Biology Magazine. 1999;18(5):31–36.
[58] Gratton E, et al. Fluorescence lifetime imaging for the two-photon microscope: time-domain and frequency-domain methods. J Biomed Opt. 2003;8(3):381–390.
[59] Gryczynski I, Razynska A, Lakowicz JR. Two-photon induced fluorescence of linear alkanes; a possible intrinsic lipid probe. Biophys Chem. 1996;57(2-3):291–295.
[60] Konig K, et al. Two-photon excited lifetime imaging of autofluorescence in cells during UVA and NIR photostress. J Microsc. 1996;183(Pt 3):197–204.
[61] Lakowicz JR. Emerging applications of fluorescence spectroscopy to cellular imaging: lifetime imaging, metal-ligand probes, multi-photon excitation and light quenching. Scanning Microsc Suppl. 1996;10:213–224.
[62] Clegg RM, Murchie AI, Lilley DM. The four-way DNA junction: a fluorescence resonance energy transfer study. Braz J Med Biol Res. 1993;26(4):405–416.
[63] Deniz AA, et al. Single-molecule protein folding: diffusion fluorescence resonance energy transfer studies of the denaturation of chymotrypsin inhibitor 2. Proc Natl Acad Sci USA. 2000;97(10):5179–5184.
[64] Zhuang X, Rief M. Single-molecule folding. Curr Opin Struct Biol. 2003;13(1):88–97.
[65] Dimitrow E, et al. Spectral fluorescence lifetime detection and selective melanin imaging by multiphoton laser tomography for melanoma diagnosis. Exp Dermatol. 2009;18(6):509–515.
[66] Helmlinger G, et al. Interstitial pH and pO2 gradients in solid tumors in vivo: high-resolution measurements reveal a lack of correlation. Nat Med. 1997;3(2):177–182.
[67] Torres Filho IP, et al. Noninvasive measurement of microvascular and interstitial oxygen profiles in a human tumor in SCID mice. Proc Natl Acad Sci USA. 1994;91(6):2081–2085.
[68] Ziemer LS, et al. Oxygen distribution in murine tumors: characterization using oxygen-dependent quenching of phosphorescence. J Appl Physiol. 2005;98(4):1503–1510.
[69] Harris AL. Hypoxia – a key regulatory factor in tumour growth. Nat Rev Cancer. 2002;2(1): 38–47.
[70] Rofstad EK. Microenvironment-induced cancer metastasis. Int J Radiat Biol. 2000;76(5): 589–605.
[71] Evans SM, Koch CJ. Prognostic significance of tumor oxygenation in humans. Cancer Lett. 2003;195(1):1–16.

[72] Koukourakis MI, et al. Enhanced expression of SPARC/osteonectin in the tumor-associated stroma of non-small cell lung cancer is correlated with markers of hypoxia/acidity and with poor prognosis of patients. Cancer Res. 2003;63(17):5376–5380.
[73] Weinmann M, Belka C, Plasswilm L. Tumour hypoxia: impact on biology, prognosis and treatment of solid malignant tumours. Onkologie. 2004;27(1):83–90.
[74] Brown JM, Giaccia AJ. The unique physiology of solid tumors: opportunities (and problems) for cancer therapy. Cancer Res. 1998;58(7):1408–1416.
[75] Tannock IF, Rotin D. Acid pH in tumors and its potential for therapeutic exploitation. Cancer Res. 1989;49(16):4373–4384.
[76] Sakadzic S, et al. Two-photon high-resolution measurement of partial pressure of oxygen in cerebral vasculature and tissue. Nat Methods. 2010;7(9):755–759.
[77] Wang Y, et al. In vivo integrated photoacoustic and confocal microscopy of hemoglobin oxygen saturation and oxygen partial pressure. Opt Lett. 2011;36(7):1029–1031.
[78] Becker W. Advanced Time-Correlated Single Photon Counting Techniques. Berlin: Springer; 2015.
[79] Periasamy A, Clegg RM, editors. FLIM microscopy in biology and medicine. Boca Raton: Chapman and Hall; 2009.
[80] Redford GI, Clegg RM. Polar plot representation for frequency-domain analysis of fluorescence lifetimes. J Fluoresc. 2005;15(5):805–815.
[81] Digman MA, et al. The phasor approach to fluorescence lifetime imaging analysis. Biophys J. 2008;94(2):L14–L16.
[82] Rahim NAA, et al. Conjugated polymer nanoparticles for two-photon imaging of endothelial cells in a tissue model. Advanced Materials. 2009;21(34):3492–3496.
[83] Brown JM. Tumor microenvironment and the response to anticancer therapy. Cancer Biol Ther. 2002;1(5):453–458.
[84] Denk W. Two-photon scanning photochemical microscopy: mapping ligand-gated ion channel distributions. Proc Natl Acad Sci USA. 1994;91(14):6629–6633.
[85] Boyden ES, et al. Millisecond-timescale, genetically targeted optical control of neural activity. Nat Neurosci. 2005;8(9):1263–1268.
[86] Zemelman BV, et al. Selective photostimulation of genetically chARGed neurons. Neuron. 2002;33(1):15–22.
[87] Nagel G, et al. Channelrhodopsin-2, a directly light-gated cation-selective membrane channel. Proc Natl Acad Sci USA. 2003;100(24):13940–13945.
[88] Partridge AH, Burstein HJ, Winer EP. Side effects of chemotherapy and combined chemohormonal therapy in women with early-stage breast cancer. J Natl Cancer Inst Monogr. 2001;30:135–142.
[89] Fisher WG, et al. Simultaneous two-photon activation of type-I photodynamic therapy agents. Photochem. Photobiol. 1997;66(2):141–155.
[90] Koenig K, Riemann I, Fischer P. Photodynamic therapy by nonresonant two-photon excitation. Proc SPIE 3592; 1999.
[91] Khurana M, et al. Quantitative in vitro demonstration of two-photon photodynamic therapy using photofrin and visudyne. Photochem Photobiol. 2007;83(6):1441–1448.
[92] Samkoe KS, et al. Complete blood vessel occlusion in the chick chorioallantoic membrane using two-photon excitation photodynamic therapy: implications for treatment of wet age-related macular degeneration. J Biomed Opt. 2007;12(3):034025.
[93] Collins HA, et al. Blood-vessel closure using photosensitizers engineered for two-photon excitation. Nat. Photonics. 2008;2:420–424.

[94] Starkey JR, et al. New two-photon activated photodynamic therapy sensitizers induce xenograft tumor regressions after near-IR laser treatment through the body of the host mouse. Clin Cancer Res. 2008;14(20):6564–6573.
[95] Starkey JR, et al. Vascular targeting to the SST2 receptor improves the therapeutic response to near-IR two-photon activated PDT for deep-tissue cancer treatment. Biochim Biophys Acta. 2013;1830(10):4594–4603.
[96] Canny J. A computational approach to edge detection. IEEE Trans Pattern Anal Mach Intell. 1986;8(6):679–698.

Shagufta Rehman Alam, Meghan J. O. Melia, Horst Wallrabe, Zdenek Svindrych, Dhyan Chandra, Suchitra Joshi, Jaideep Kapur, and Ammasi Periasamy

7 FLIM-FRET microscopy

Abstract: Fluorescence Lifetime Imaging (FLIM) offers a non-invasive approach to assess the metabolic state of cells and tissues under different pathophysiological conditions. In cancer, cellular metabolism is deranged. Increased glycolysis and defective mitochondrial oxidative phosphorylation (OXPHOS) is reported in several cancers. Also, mitochondrial dysfunction has been implicated in different neurodegenerative diseases. In this study, we have used the FLIM-FRET approach to assess metabolic activity by following the NAD(P)H and Trp lifetimes in in vitro models of prostate cancer and chemically induced seizure. Two prostate cancers and one normal cell line were assessed for their metabolic response upon stimulation with glucose and upon treatment with doxorubicin, a commonly used anticancer drug which induces apoptosis. Furthermore, we have used in vitro mouse brain slices to assess metabolic activity upon chemically inducing seizure with KCl in real-time. Mitochondrial regions of interest were taken for FLIM data analysis. We observed an increase in metabolic activity as assessed by an increase in NAD(P)H-a_2%, E% and NADPH : NADH ratio upon stimulation with glucose in prostate cancer cells and a decrease in these parameters in the in vitro seizure model, respectively. Interesting, we also observed an increase in mitochondrial activity as assessed by an increase in NAD(P)H-a_2% and NADPH : NADH ratio in prostate cancer cells upon treatment with doxorubicin with accompanied changes in cellular morphology. Correction of mitochondrial OXPHOS activity and induction of apoptosis is a promising strategy in cancer treatment. Nevertheless, our results show that NAD(P)H and Trp lifetimes can be used as biomarkers to understand metabolic activity in different pathophysiological states.

7.1 Introduction

7.1.1 Tissue autofluorescence as a biomarker

Imaging of tissue or cellular autofluorescence provides a non-invasive approach and a valuable tool to assess the metabolic state of living cells and tissues. Autofluorescence lifetime imaging has a wide application in disease diagnosis and in assessing the disease state by quantitative analysis of cellular metabolism [1, 2]. There has been active research to understand the metabolic activity by following the tissue or cellular metabolic signatures through the endogenous fluorophores under different physiological and pathological conditions. These endogenous fluorophores include amino

https://doi.org/10.1515/9783110429985-009

acids like phenylalanine, tyrosine and tryptophan; structural proteins like collagen and elastin; and coenzymes involved in cellular metabolism like nicotinamide adenine dinucleotide phosphate (NAD(P)H), flavin mononucleotide (FMN) and flavin adenine dinucleotide (FAD). Tab. 7.1 provides a summary of the fluorescent properties of these endogenous fluorophores [3–10].

7.1.2 Metabolic coenzymes in cellular metabolism

Adenosine triphosphate (ATP) is the high-energy molecule that stores energy and is essential to all the physiological processes. All processes that require energy for cellular functions, obtain it directly from the stored ATP. It is the energy currency of life and therefore could provide a direct measure of the metabolic state of cells and tissues. But, using the fluorescence signature of ATP/ADP to understand the metabolic state is technically not feasible because excitation ($\approx$ 259 nm) and emission ($\approx$ 370 nm) of ATP is masked by tryptophan-containing proteins [2]. However, the coenzymes like NAD(P)H and FMN/FAD involved in the metabolic pathways are autofluorescent, spectrally separate and therefore can be used for metabolic profiling studies [11] (Fig. 7.1 and Tab. 7.1).

Fig. 7.1 provides a snapshot of the major metabolic pathways where the coenzymes like NAD(P)H and FMN/FAD are produced. During glycolysis, 2 molecules of NADH are produced in the cytoplasm. In order for glycolysis to take place the cell maintains low levels of NADH within the cytoplasm. The cell does so by transporting cytoplasmic NADH to mitochondria by malate-aspartate shuttle [2] where NADH is oxidized to NAD+ producing malate; malate is transported to mitochondrial matrix, mitochondrial NAD+ is reduced to NADH which in turn is oxidized to NAD+ by the Electron Transport Chain (ETC) in the respiratory chain in the mitochondrial matrix [2, 12]. In the mitochondria, NADH is also produced by citric acid cycle or Krebs cycle. The energy stored in mitochondrial NADH is transferred to an electrochemical proton gradient that is essential to drive phosphorylation of ADP to ATP by F0F1 ATP synthase, (Fig. 7.1) to produce ATP, the energy currency of the cell.

The coenzymes FMN/ FAD are derived from riboflavin (vitamin B_2) and are covalently bound to flavoproteins in enzymatic reactions of the Krebs cycle and ETC which take place in the mitochondria [2]. The ETC complex II of succinate dehydrogenase contains bound FAD cofactor which is reduced to FADH2. FADH2 is then oxidized to FAD by coenzyme Q [2, 12]. Only the oxidized form of the flavins, FpOx (FAD or FMN) are fluorescent, whereas the reduced forms Fp (FADH2 or FMNH2), are nonfluorescent. On the other hand, NAD(P)H is fluorescent, whereas the oxidized form NAD(P)+ is nonfluorescent [2, 5].

Tab. 7.1: Fluorescence properties of endogenous fluorophores used as biomarkers. Table shows fluorescent properties, excitation wavelength, emission wavelength, and fluorescent lifetimes for endogenous fluorophores used as biomarkers.

Fluorophores	Excitation wavelength (nm)	Emission wavelength (nm)	Conditions to measure the lifetime (ns)	References
Proteins/amino acids				
Phenylalanine	240–270; 265 (max.)	280 (max.)	(1) 7.5 ns in solution in PBS pH 7.4, 0.5 mg/ml, (Fluka > 99 % purity).	(1) McGuinness et al., 2006 [3]
			(2) 6.8 ns in solution in water at neutral pH	(2) Lakowicz et al., 2006 [5]
Tyrosine	260–280; 275 (max.)	280–340; 300 (max.)	(1) 2.4 ns calf thymus nucleosome core particles; (each has 30 Tyrosine) in 3 M NaCl + 0.2 mM EDTA, pH 6.5.	(1) Ashikawa et al., 1982 [4]
			(2) 3.6 ns in solution in water at neutral pH	(2) Lakowicz et al., 2006 [5]
Tryptophan	240–300; 295 (max.)	300–400; 350 (max.)	3.1 ns in solution in water at neutral pH	Alcala et al., 1987 [6]; Lakowicz et al., 2006 [5]
Structural proteins				
Collagen	250–360; 325, 360 (max.)	320–480; 400, 405 (max.)	2.5–3.5 ns in ex vivo human atherosclerotic plaque	Phipps et al., 2012 [7]
Coenzymes				
NADH	300–400; 340 (max.)	400–500; 460 (max.)	(1) 0.4 ns (free) NADH (Sigma) dissolved in MOPs buffer (100 mM, pH 7.0)	(1) Lakowicz et al., 1992 [8]
			(2) 1–5 ns (bound) NADH dissolved in Tris buffer (20 mM, pH 8.0)	(2) Lakowicz et al., 2006 [5]
			(3) NADH (Sigma) dissolved in Tris buffer (100 mM, pH 8.5)	(3) Yu and Heikal, 2009 [10]
FMN/FAD	300–500; 390, 450 (max.)	500–600; 535 (max.)	4.89 ns for FMN; 3.38 ns and 0.12 ns for FAD in solution at pH 7.5 and at 3 °C	Lakowicz et al., 2006 [5]

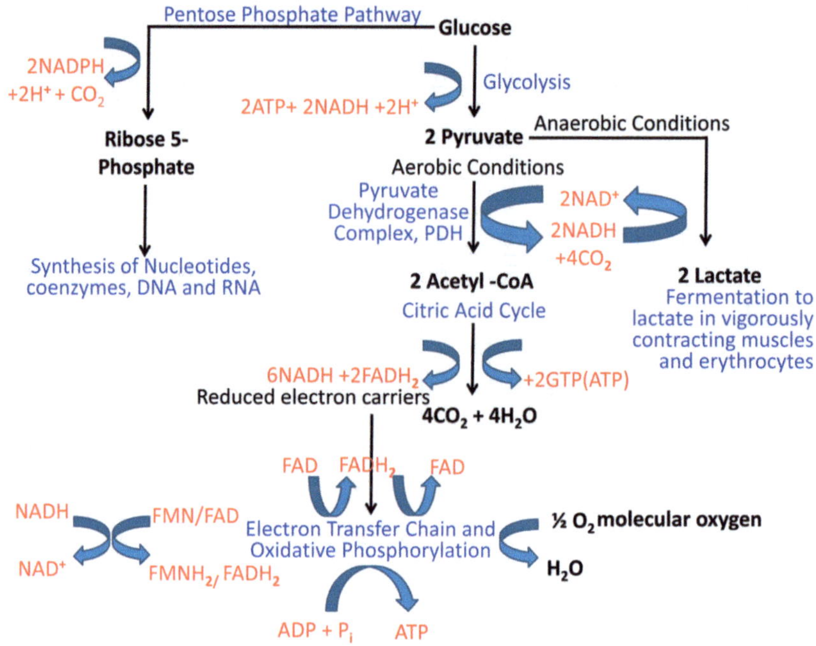

Fig. 7.1: Illustration of different metabolic pathways where NADH, NADPH and FMN/FAD coenzymes are produced [12].

It has been observed that during respiration, the oxidation of NADH to NAD+ through the ETC results in a loss of NADH fluorescence signal while flavoprotein oxidation results in an increase in FpOx fluorescence. After treatment with mitochondrial inhibitors like cyanide used to block the ETC, an increase in NADH fluorescence signal and a decrease in FpOx fluorescence signal has been observed [2, 13].

The coenzyme NADPH is recognized by enzymes that are involved primarily in anabolic (reductive) reactions like lipid, cholesterol and fatty acid synthesis. NADH is recognized by enzymes that are involved in catabolic reactions of glycolysis and by enzymes of mitochondrial ETC. Therefore, the ratio of NAD(P)H:NAD(P) could reflect the overall redox state of the cell [14]. Additionally, A NAD(P)H (reduced), and FAD (oxidized) have been used as natural biomarkers of cellular redox state, energy metabolism and mitochondrial abnormalities under different pathophysiological conditions [15–19].

In certain conditions like cancer, glycolysis is deranged. It has been observed that glucose uptake and glycolysis proceed faster in cancer than in noncancerous tissues [12]. Tumors also depend on anaerobic glycolysis for ATP production due to lack of a good capillary system for continued supply with oxygen. The smaller numbers of mitochondria in tumor cells could also account for less ATP being generated from ETC-oxidative phosphorylation in mitochondria and hence, more ATP is needed from glycolysis resulting in increased glycolytic rate in cancer. Also, in some tumor cells several glycolytic enzymes are overproduced, for example isozyme of hexokinase which is insensitive to feedback inhibition by glucose 6-phosphate [12]. Therefore, monitoring changes in cytosolic NADH could provide insight on the glycolytic rate under different pathophysiological conditions [20].

7.1.3 Tryptophan

Tryptophan (Trp) is an essential amino acid and humans are incapable of synthesizing Trp; therefore, it must be supplemented through dietary intake. Tryptophan is a precursor for serotonin, a neurotransmitter and indoleacetate (a plant growth factor), and for nicotinate (niacin) (Fig. 7.2), which in turn is a precursor of coenzymes NAD and NADP involved in metabolic pathways [12]. Only 3 out of the 20 amino acids, tyrosine (Try or Y), phenylalanine (Phe or F) and tryptophan (Trp or W), all aromatic, are autofluorescent, absorbing and emitting in the UV region. Their fluorescence intensity and lifetime mainly provide information on the composition and structure of the protein they are part of and changes in overall cellular microenvironment [8, 21]. Of these 3 amino acids, Trp is moderately fluorescent with a quantum yield of 0.1–0.5 and has the highest absorption coefficient, therefore the highest brightness and the longest emission wavelength of 350 nm [21]. Alterations in the tryptophan metabolism could be a common pathological pathway for clinically distinct neurodegenerative disorders, like Parkinson's, Huntington's and Alzheimer's diseases [22].

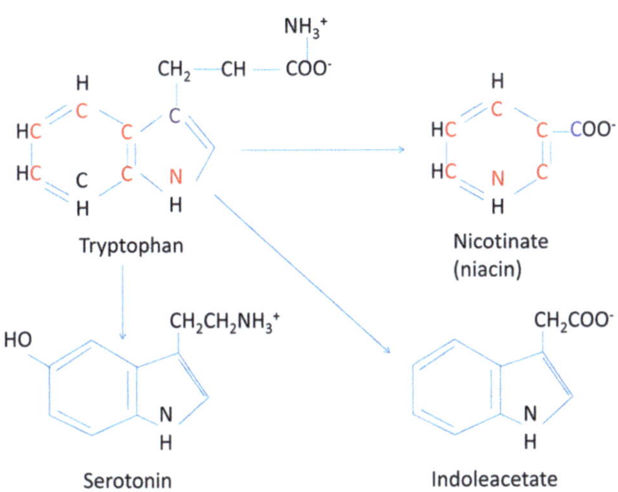

Fig. 7.2: Illustration of tryptophan as a precursor of serotonin, indoleacetate and for nicotinate which in turn is a precursor of NAD and NADP. Colored atoms trace the source from the tryptophan [12].

7.2 Examples of some applications of NAD(P)H and tryptophan lifetime and intensity imaging to diseases

Changes in the metabolic state of the cell or the tissue due to changes in the physiological processes or due to the onset of a pathological condition, result in modifications in the abundance of endogenous fluorophores, their distribution and biochemical properties of their environment and therefore lead to differences in the intensity and lifetime distributions of the endogenous fluorophores [11]. Endogenous fluorophores, which serve as biomarkers, can hence be exploited for the development and advancement of non-invasive detection systems for different pathophysiological conditions. For example, significant efforts are being made towards the development of non-invasive methods for the early detection of cancer where the therapeutic interventions are highly effective [23]. Understanding cell metabolism through time-resolved autofluorescence microscopy has gained impetus but is also challenging at the same time. Development of methods for the detection of optical signatures has shown promising results and there has been rising interest in studying NADH and Trp signals in cancers and other disease conditions using single- or multiphoton excitation approaches to assess their intensity and/or lifetime distribution as metabolic signatures. For example, Trp and NADH/FAD optical markers have been used to differentiate between normal and Human Papillomavirus (HPV) immortalized epithelial cells by quantitative analysis of their autofluorescence signal. Trp autofluorescence was used in this study as a sensitive indicator of general protein expression in cells. Morphological differences and an increase in Trp mean intensity were seen when normal and human papillomavirus immortalized epithelial cells were compared, and an increase in the NADH and FAD mean intensities with overall decrease in the redox state of the HPV immortalized cells was observed [23]. The autofluorescence signal from Trp, NADH and FAD along with collagen and porphyrins have been used to understand and differentiate the neoplastic changes after induction of squamous cell carcinoma in mice models using an Excitation Emission Matrix (EEM) based analysis, where excitation was achieved from 280–460 nm at 10 nm intervals and the fluorescence emission was recorded from 300–750 nm [24]. Also, in another study the ratio of NADH to the tryptophan signal has been shown to serve as an indicator of cellular metabolism to distinguish between cancerous and normal cervical cells [13]. Tryptophan metabolism and indoleamine 2,3-dioxygenase activity are linked to cancer development and progression. It has been demonstrated in the literature that women with breast cancer had lower levels of tryptophan compared to women without breast cancer [25]. In our recent publications we used multiphoton excitation to study NADH and Trp lifetimes and their interactions by resonance energy transfer in nontumorigenic (MCF10A) and tumorigenic (HeLa) cells before and after glucose stimulation to understand the glycolytic activity in nontumorigenic versus tumorigenic cells [26] and we show that bound:free NADH ratios increase whereas Trp lifetime quenches (degradation of Trp) resulting in an increase in the efficiency of energy transfer (E%) after glucose stimulation. The magnitude of these outcomes was seen to be significantly higher in tumorigenic (HeLa) versus nontumorigenic (MCF10A) cells. The quenching of Trp fluorescence intensities and lifetime by protein-bound NADH in solutions using pig lactate dehydrogenase and malate dehydrogenase using a spectrofluorometer was described way back in 1979. It was demonstrated that most, if not all, of the tryptophan quenching by NADH was due to Trp-NADH resonance energy transfer [27]. In the current study we used two pathophysiological model systems at [A] cellular level human prostate cancer cells and at [B] tissue level in vitro seizure model using mouse hippocampal brain slices induced with KCl; and applied non-invasive multiphoton imaging to understand the metabolic changes by measuring NAD(P)H and Trp intensity, lifetime distribution and Efficiency of Energy Transfer ($E\%$) under different conditions and treatments. We have used two prostate cancer cell lines (African American and Caucasian prostate cancer cell lines, see Methodology section) for glucose stimulation studies and before and after treatment with doxorubicin as an attempt to assess these metabolic signatures under these conditions and to understand the efficacy of anticancer drug treatment. Also, we have used the FLIM measurements to understand the metabolic activity in real time in an in vitro seizure model.

7.2.1 Prostate cancer and treatment

Prostate cancer is one of the most common cancers in men in the USA. More than 2 million people are prostate cancer survivors alone in America [28]. Early diagnosis of cancer and cancer follow-up post-treatment is a challenging problem. As mentioned above, tremendous efforts are being made towards developing non-invasive methods of early cancer detection when the cancer treatment is most effective [23]. Non-invasive methods can be used to image the metabolic signatures to understand the metabolic profile in cancer, changes in metabolic profile during cancer progression and post-cancer treatment. In the current study we have used two prostate cancer cell lines; African American (AA) and Caucasian (CA). The occurrence of prostate cancer in AA men is higher than in CA men, and prostate cancer is more aggressive in AA men. It has been shown recently that differential content of mitochondrial DNA (mtDNA) can be linked to prostate tumors. The higher incidence and aggressive form of prostate cancer seen in AA men as compared to CA men can be attributed to the lower level mtDNA content in AA men [29]. A number of chemotherapeutic drugs for prostate cancer are currently being used for its treatment [28]. In the present study we used the anticancer drug doxorubicin and analyzed NAD(P)H and Trp lifetimes before and after the treatment as a way to understand the efficacy of doxorubicin drug treatment by following the cellular metabolic signatures.

7.2.2 Doxorubicin

Doxorubicin is an anthracycline chemotherapeutic commonly used for cancer treatment and is regarded as the most potent chemotherapeutic by the Food and Drug Administration, USA [30]. Its potency comes from the fact that there is rapid uptake of doxorubicin by cells in 3–5 min after intravenous injection and its ability to attack rapidly dividing cells, regardless of which phase of the cell cycle the cell is in, and slowing disease progression, which has been known for a long time. Doxorubicin enters the cells by passive diffusion and binds to proteasome's 20S subunit. Doxorubicin then uses proteasome's 20S complex to shuttle into the nucleus through the nuclear pore. Since doxorubicin has a higher affinity for nuclear DNA it dissociates with proteasome's 20S to bind to nuclear DNA. Doxorubicin's primary action is to inhibit enzymes topoisomerase I and II, and to interfere with DNA replication and transcription by intercalating into DNA. Upon binding to topoisomerase I and II it induces a range of cytotoxic events, like DNA damage. When there is a failure to repair these breaks in DNA, the cellular growth cycle is inhibited at checkpoints G1 and G2 and apoptosis is triggered [31, 32]. Other actions of doxorubicin includes free radical generation which further induces DNA damage and its ability to not only intercalate nuclear DNA but mitochondrial DNA as well [33].

7.3 Multiphoton NAD(P)H and Trp FLIM-FRET microscopy in prostate cancer

7.3.1 Cell culture

The African American cell lines were grown in high glucose (25 mM) Dulbecco's Modified Eagle Medium (Life Technologies) supplemented with 10 % cosmic calf serum (Hyclone), 1 % Penicillin-Streptomycin (Life Technologies), and 4 mM Sodium Pyruvate (Life Technologies) as per the recommendations. The Caucasian prostate cancer cell lines, LNCaP, were maintained in RPMI 1640 (Life Technologies) supplemented with 10 % cosmic calf serum (Hyclone) and 1 % Penicillin-Streptomycin (Life Technologies). Normal human prostate epithelial cells were obtained from the American Type Culture Collection and were maintained in Prostate Epithelial Growth Medium (Lonza). All cells were maintained in the cell culture incubator, at 37 °C with 5 % CO_2. The media was changed every 48 hours, and the cells were passaged once they reached 80 % confluence. For imaging, each cell type was plated onto 25 mm No. 1.5 glass coverslips (Thermo Scientific), in their recommended growth medium. The cells were allowed to proliferate in the incubator until they had reached 70–80 % confluence.

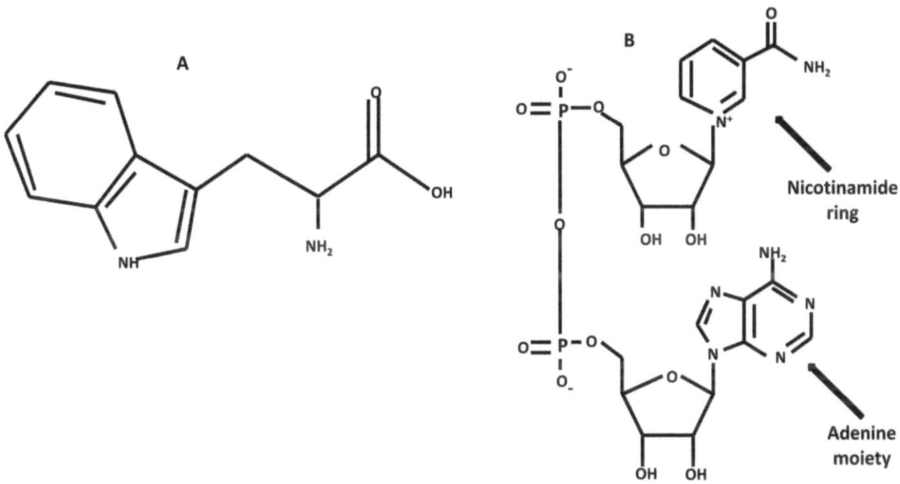

Fig. 7.3: Chemical structures of (A) tryptophan and (B) reduced nicotinamide adenine dinucleotide (NADH). The fluorescence of the nicotinamide ring is quenched by the adenine moiety in the same molecule resulting in shorter lifetime whereas NADH in a bound-to-enzymes state acquires a more stretched conformation where its fluorescence is no longer quenched by the adenine moiety, hence gets significantly enhanced, with a much longer lifetime [5, 20].

7.3.2 Basics of FRET

Förster Resonance Energy Transfer (FRET) is a distance-dependent phenomenon in which energy transfer takes place from donor molecule in the excited state to another neighboring acceptor molecule in the ground state, via long-range dipole-dipole coupling [34]. FRET is described as the nonradiative transfer of energy which occurs when the emission spectrum of the donor fluorophore overlaps with the excitation spectrum of the acceptor fluorophore in close proximity. FRET can occur only when the conditions of FRET are satisfied. There are three main criteria for FRET to occur: spectral overlap of at least 30 % or more between emission spectrum of the donor and the absorption spectrum of the acceptor fluorophores, the molecular distance should be within 1–10 nm between donor and acceptor molecules and the matching dipole orientation (K^2) between donor and acceptor fluorophores, meaning that the emission dipole of the donor and the absorption dipole of the acceptor must be oriented to each other and not perpendicular to each other [5]. Due to FRET there is quenching of the donor fluorescence and sensitized emission from the acceptor fluorophore. Therefore, FRET provides spatial-temporal resolution of the dynamics of interactions between different biomolecules. The efficiency of energy transfer is inversely dependent upon the sixth power of the distance (r) separating the donor and acceptor molecule subject to the Förster distance (R_0), which is the distance between the donor and acceptor fluorophores at which the energy transfer efficiency is 50 %. Since the efficiency of the energy transfer varies with the sixth power of the distance separating the donor and acceptor molecules, the molecular distance at which FRET can occur is limited to 1–10 nm. Therefore, FRET acts as a spectroscopic ruler and provides a non-invasive method to visualize protein-protein interactions [35, 36], DNA-protein [37] and RNA-protein [38] interactions.

Basics on FRET are well described in the following references [5, 36, 39–45]. One of the major applications of fluorescence lifetime imaging is the measurement of FRET. In FLIM-FRET measurements, FRET events are identified if there is a reduction in donor lifetimes as a result of quenching of fluorescence in the presence of the acceptor. The efficiency of energy transfer or $E\%$ can hence be calculated from the following equation $E = 1 - \tau_{DA}/\tau_D$, where τ_D is the unquenchenced donor lifetime (donor alone) and τ_{DA} is the quenched donor lifetime in the presence of acceptor [46]. Since only donor lifetimes are measured in FLIM-FRET, there is no need for bleed-through corrections, which are required in intensity-based FRET measurements. Since lifetime measurements are independent of the concentration of the fluorophores and excitation laser intensity but are sensitive to environmental changes, FLIM-FRET is a very sensitive and accurate way of detecting molecular interactions. In this study we have used FLIM-FRET to understand the metabolic activity in prostate cancer cells, efficacy of anticancer drug treatment and a way to understand metabolic activity in an in vitro seizure model.

7.3.3 Multiphoton FLIM-FRET microscopy

Ultraviolet radiation is used to illuminate NAD(P)H (340 nm) and Trp (280 nm) and it is phototoxic to the cells. Multiphoton microscopy for NAD(P)H and Trp can be used to understand cell metabolism in different pathophysiological conditions without damaging the cells. Multiphoton FLIM-FRET microscopy provides spatio-temporal resolution required to distinguish between different physiological states. Also, NADH signals coming from different compartments: cytoplasmic, mitochondrial and nuclear can be delineated. NADH signals come from cell cytoplasm (from glycolysis, fermentation, pyruvate dehydrogenase complex), mitochondria (from citric acid cycle/Krebs cycle, electron transfer chain and oxidative phosphorylation) and nucleus (by gene expression). The NADH signal that comes from the nucleus plays a role in gene expression but not in metabolic activity [13, 47, 48]. During the lifetime measurement it is observed that the fluorescence of the nicotinamide ring is quenched by the adenine moiety in the same molecule (Fig. 7.3), resulting in a typical average lifetime of 0.4 ns of the free NADH [5]. When NADH is bound to various enzymes it acquires a more stretched conformation where its fluorescence is no longer quenched, hence gets significantly enhanced, and has a much longer lifetime, typically around 2.5 ns but can also go up to 5 ns (Tab. 7.1) [5]. Due to the conformational heterogeneity of the various enzymes to which NADH is bound, it can have complex lifetime distributions with multiexponential decay [49]. It has been reported that the maximum emission of free NADH is around 470 nm, whereas bound NADH is blue-shifted to 440 nm and has been used to distinguish normal from oral squamous carcinoma cells [16]. On the other hand, the Trp amino acid has a lifetime of around 3.02 ns and has a single lifetime component [6].

7.3.3.1 Experimental setup

The FLIM imaging system (Fig. 7.4) consists of a Zeiss 780 confocal/multiphoton (MP) laser scanning system coupled to the Zeiss inverted epi-fluorescence microscope and controlled with the Zen software (Carl Zeiss Inc.). An ultrafast (150 fs) tunable 10 W-pumped pulsed Chameleon laser, operating at 80 MHz and at 740 nm was used for the multiphoton excitation of NAD(P)H and Trp (Coherent, Inc.). The laser was coupled to the Zeiss 780 unit to scan the specimen via an *XY* raster scanning mechanism using galvomirrors.

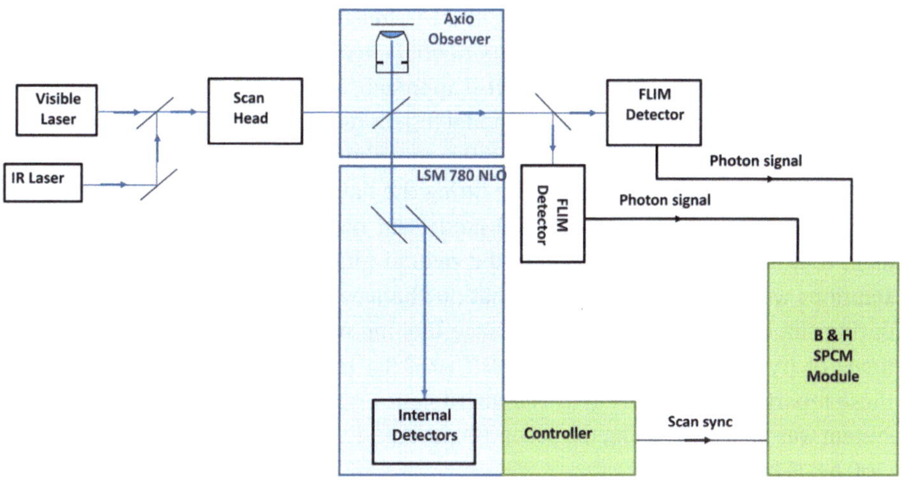

Fig. 7.4: Schematic of the experimental setup of the two-three-photon excitation FLIM imaging system.

An ET650sp-3P filter (Chroma Technology Corp.) was used below the nose piece to transmit the emission light to the external (non-descanned) detection unit. The non-descanned output port is coupled to the two T-adaptor in series to couple three FLIM hybrid PMT detectors with the spectral response of 300–600 nm and a full width at half maximum response time of ≈ 150 ps (Becker & Hickl GmbH). The non-descanned configuration provides 80 % transmission at 340 nm.

The fluorescence decay per pixel was measured using SPC-150 TCSPC board (Becker & Hickl GmbH) and data acquisition was controlled by SPCM software (v8.91), set at 256 time bins in each excitation detection period (39 ps/bin). The TCSPC device synchronizes the lifetime detector to the excitation pulse and the scanning clock, and records both the arrival time (relative to the excitation pulse) and the spatial (*X*, *Y*, *Z*) information for each detected photon [50]. A reference signal from the MP laser is fed to the TCSPC device. Repeated excitation-emission cycles result in a photon count histogram (often termed a fluorescence decay profile) recorded for each pixel of an image.

A Zeiss 40× 1.1 NA W (C-Apochromat UV-VIS-IR; UV transmission is 60 % at 340 nm) or 40× 1.3 NA oil, (EC Plan-Neofluar, UV transmission is 60 % at 340 nm) objective lens was used to focus the light on the sample and collect the emission. The "donor" Trp and "acceptor" NAD(P)H photons were collected using HQ360/40-2p (#230753, Chroma Technologies Corp.) and ET480/40-2p (#232292, Chroma Technologies Corp.) band-pass filters, respectively. To maintain cell viability, the average excitation power at the specimen plane was kept at average power 2 mW with the collection time of at least about 30 s to collect acceptable photon counts. Photobleaching at this imaging condition was less than 10 %, negligible for FLIM analysis.

The decay data was analyzed by the SPCImage software v5.1 (Becker & Hickl GmbH), which allows single/multi-exponential curve fitting on a pixel-by-pixel basis using a weighted least-squares numerical approach. Cellular NAD(P)H lifetime images were analyzed using bi-exponential model functions to calculate the lifetime components of free and bound NADH, and relative contribution of each to the decay. Cellular Trp lifetime images were obtained by fitting the data to a single-exponential model function, since a double-exponential model did not yield any significant improvement of fitting. At the completion of the cellular imaging experiments, the standard solutions were re-imaged to ensure that no changes to the instrument had occurred during the experiment (i.e., fluorescence lifetime or intensity). The measured lifetimes of tryptophan (3.3 ns) and NADH (0.67 ns) in solution were consistent with those reported in literature [51]. The instrument response function (IRF) of the FLIM system was measured using the second-harmonic generated signals from urea crystal (300 ps) [52].

7.3.4 Image processing and analysis

The florescence lifetime data were fitted in SPCImage (v5.1, Becker & Hickl GmbH) according to our standardized procedure (tryptophan: binning 5, single-component fit; NAD(P)H: binning 2, two-component fit; fixed 'shift' parameter). SPCImage was used to process each image in both NAD(P)H and Trp channels, generating photon images, t_1, t_2, $a_1\%$, $a_2\%$, and chi squared for each pixel of each channel. For the Trp channel t_2 and $a_2\%$ were not calculated as Trp has a single lifetime. The exported results from multiple samples and FOVs were loaded into ImageJ as a multidimensional hyperstack using a simple macro, and inspected for excessive scatter (or second harmonic signal) and χ^2 values. The photon images were used to create 2-by-2 pixels regions of interest for analysis using the pFRET ROI [53] selection tool in ImageJ and the relevant fitting parameters extracted from individual ROIs were saved as text files. The saved data was imported into Excel with a custom VBA (Visual Basics for Applications) macro and automatically processed using a set of Excel template spreadsheets. The template spreadsheets were elaborated to calculate and graph histograms of lifetimes (individual components as well as amplitude-weighed means), relative amplitudes of fit components, fit quality (χ^2 values) and collect and present the results from all experiments. Minimal manual intervention was required to collect and compare control vs treated samples. The tryptophan quenching efficiency was calculated according to $E = 1 - \tau/\tau_0$, where $\tau_0 = 3.1$ ns. The reduced dinucleotide phosphate concentration is calculated according to NADPH% $= 100 \times (\tau_2 - \tau_b)/(\tau_a - \tau_b)$, where τ_2 is the lifetime of the longer component of the NAD(P)H fit (corresponding to the bound NAD(P)H), $\tau_a = 4.4$ ns and $\tau_b = 1.5$ ns [54].

7.4 Research applications

7.4.1 Effect of glucose stimulation on metabolic activity in normal and prostate cancer cells

In cancer, glycolysis is known to get deranged. As mentioned above, glucose uptake and glycolysis proceed faster in cancer than in noncancerous tissues [12]. Therefore, we first wanted to assess the glycolytic activity in the prostate cancer cell lines in comparison to Normal Human Prostate Epithelial (NHPE) primary cells using NAD(P)H and Trp lifetime distribution, NAD(P)H-a_2% and efficiency of energy transfer (E%) between Trp and NADH as readouts. For NHPE primary cells the basal level of glucose in growth medium was 5 mM whereas for African American (E006AAT1B) and Caucasian (LNCaP) prostate cancer cell lines it was 25 mM and 11 mM respectively. Since the African American (E006AAT1B) cell line had a higher basal glucose level, for the glucose stimulation study it was dropped down to 5 mM to match the Normal Human Prostate Epithelial (NHPE) primary cells, and cultured overnight. Whereas for Caucasian (LNCaP) prostate cancer cell lines it was left at 11 mM. In general, for imaging, the coverslip with cells plated onto them was placed into a stainless steel chamber with 500 μl of the growth medium. This chamber was then placed in the Zeiss 780 microscope's heated stage with temperature maintained at 37 °C throughout the imaging process. Blood/gas mixture (20.9 % O_2, 4.99 % CO_2 and rest balanced with N_2) was bubbled through a water bath and to the chamber to ensure proper humidity and live cell imaging conditions. The image acquisition time was 90 seconds in order to collect sufficient photons in the tryptophan channel.

Each coverslip was imaged at 3–5 fields of view (FOV) as a 0 min time point control. After the control slide was imaged, the growth media was replaced with growth media containing glucose at a final concentration of 25 mM. Glucose stimulation was done for 30 min on stage (stage incubator, Carl Zeiss, PeCon Tempcontrol 37-2 digital) using the same coverslip of cells used for 0 min time point control. The NAD(P)H and Trp lifetime images were then acquired. An increase in NAD(P)H-a_2% as an indicator of increased metabolic activity after glucose stimulation was clearly seen for the prostate cancer cells (E006AAT1B and LNCaP) (Fig. 7.5, adapted from SPIE Proceedings, paper number 9712-23, 2016). The magnitude of shift to the right was observed to be the highest for African American (E006AAT1B), which was 50 %, followed by the Caucasian (LNCaP) at 27.3 % and hardly any appreciable change for the normal prostate cells (NHPE). Prostate cancer in African American men is described to be the most aggressive form [29] and therefore the glycolytic activity in these cells could probably be higher compared to the Caucasian and normal prostate cancer cells to keep up with their metabolic demand and continued energy production in the form of ATP.

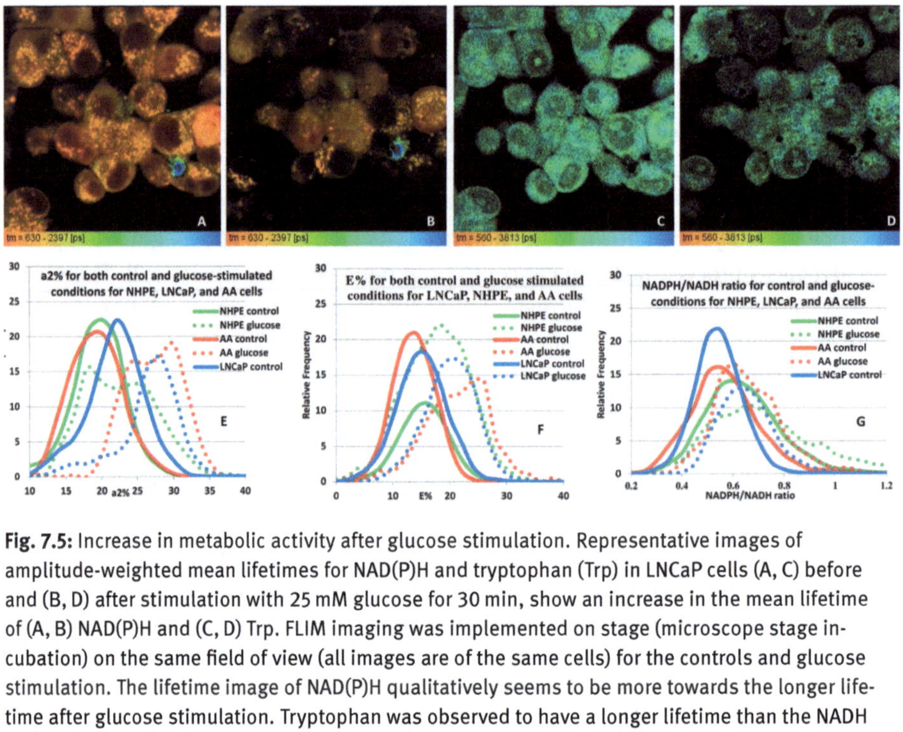

Fig. 7.5: Increase in metabolic activity after glucose stimulation. Representative images of amplitude-weighted mean lifetimes for NAD(P)H and tryptophan (Trp) in LNCaP cells (A, C) before and (B, D) after stimulation with 25 mM glucose for 30 min, show an increase in the mean lifetime of (A, B) NAD(P)H and (C, D) Trp. FLIM imaging was implemented on stage (microscope stage incubation) on the same field of view (all images are of the same cells) for the controls and glucose stimulation. The lifetime image of NAD(P)H qualitatively seems to be more towards the longer lifetime after glucose stimulation. Tryptophan was observed to have a longer lifetime than the NADH (C) before and (D) after glucose stimulation as can be seen in the color spectrum at the bottom of the image. Increase in the (E) a_2%, (F) E%, and (G) NADPH/NADH ratio histograms for normal (NHPE) and prostate cancer (LNCaP and E006AT1B) cells. In each graph the solid line represents control cells and the dotted line represents cells stimulated with 25 mM glucose. Glucose stimulation causes a right shift in (E) a_2% and (F) E%. Additionally, note that the shift is greater in the E006AAT1B cells, followed by LNCaP cells, and NHPE cells, demonstrating an increase in cellular metabolism in the E006AAT1B cells compared to the two other cell types.

The observed higher NAD(P)H-a_2% for the African American (E006AAT1B) cells in the present study underscores this phenomenon. On the other hand, NHPE cells grow very slowly as compared to the prostate cancer cells which divide very fast, and therefore, 30 min of glucose stimulation may not be sufficient to see a glycolytic response in the normal prostate primary epithelial cells. The observed higher NAD(P)H-a_2% means that there is an increase in the enzyme-bound fraction of NAD(P)H as a result of which the Trp lifetime quenches due to the increase in the energy transfer (E%) between Trp and NADH after glucose stimulation as previously reported by us [26, 52]. The magnitude of shift to the right for the E% distribution was observed to be the highest for African American (E006AAT1B) at 78.6 %, followed by the Caucasian (LNCaP) at 42.9 % and only a marginal increase of 12.5 % was seen in the normal prostate cells (NHPE). We further wanted to delineate the ratio of NADPH : NADH as a measure of the anabolic to catabolic ratio based on the formula described elsewhere [54]. We observed modest increases of 9.1 %, for the African American (E006AAT1B), 18.2 % for the Caucasian (LNCaP) and 8.3 % for the normal prostate cells (NHPE) after 30 min of glucose stimulation.

7.4.2 Effect of doxorubicin treatment on metabolic activity in prostate cancer cells

Doxorubicin, an anthracycline chemotherapeutic is commonly used for cancer treatment because it is most potent, is rapidly taken up by the cells (3–5 min) and has the ability to attack rapidly dividing cells [30, 31]. Doxorubicin inhibits topoisomerase I and II, inhibits DNA replication and transcription by intercalating into DNA, arrests cellular growth cycle at checkpoints G1 and G2 and triggers apoptosis [31, 32]. Doxorubicin treatment at 0.5 μM final concentration was done in Caucasian LNCaP prostate cells. NAD(P)H and Trp lifetime images were acquired for 0 min time point control and at 60 min of doxorubicin treatment of the same FOVs and same cells. At 60 min treatment it was clearly seen that these cells had rounded up and lost their normal healthy cellular morphology. These characteristics indicate cellular stress and the beginning of cell death. Change in cellular morphology accompanied with the 60 % increase in NAD(P)H-a_2% and 20 % in NADPH : NADH ratio was observed at 60 min of treatment (Fig. 7.6). We did not observe any change in E% at 60 min. Defective OXPHOS activity is reported in prostate cancer; therefore, correction of mitochondrial OXPHOS activity and induction of apoptosis is a promising strategy in cancer treatment [28]. Our recent observations using FLIM-FRET approach has been that treatment with doxorubicin could correct this defect as we observe an increase in the OXPHOS activity before induction of apoptosis [55].

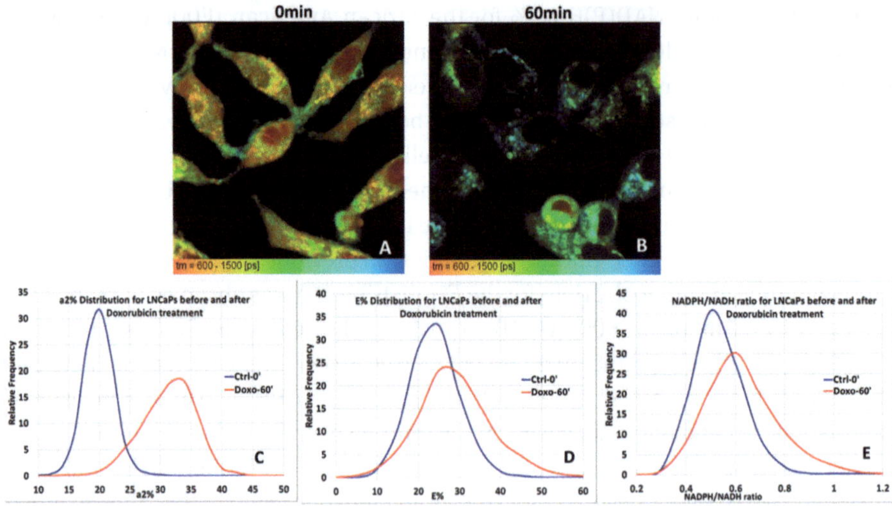

Fig. 7.6: Increase in mitochondrial metabolic activity after treatment with anticancer drug doxorubicin. Representative images of amplitude-weighted mean lifetimes for NAD(P)H for LNCaP cells (A) at 0 min control before, and (B) after treatment with 0.5 µM doxorubicin at 60 min for the same field of view and of the same cells. The lifetime of NAD(P)H shifts towards the longer lifetime upon doxorubicin treatment. At 60 min treatment it is clearly seen that these cells had rounded up and lost their normal healthy cellular morphology, which indicates cellular stress and the beginning of cell death. The graphs show the (C) $a_2\%$, (D) $E\%$, and (E) NADPH/NADH ratio for LNCaP cells before (blue), and at 60 min (red) after doxorubicin treatment. We saw an increase in the $a_2\%$ and NADPH/NADH ratio with treatment, indicating an increase in mitochondrial metabolic activity before cell death.

7.5 Multiphoton NAD(P)H and Trp FLIM-FRET microscopy of hippocampal tissue in an in vitro KCl induced seizure model

7.5.1 FLIM-imaging of hippocampal tissue in an in vitro KCl induced seizure model

We have further tried to understand the metabolic changes in a KCl induced seizure mouse model. For this we used the adult male C57Bl6 mouse hippocampal slices (from Dr. Jaideep Kapur) and tried to assess the differences in NAD(P)H and Trp lifetimes and abundances under conditions to chemically induce seizure using KCl [56]. About 90 % of the energy generated in the brain in form of ATP comes from oxidative phosphorylation within mitochondria [57]. Mitochondrial dysfunction has been implicated in different neurodegenerative states [58–60]. Also, the excessive metabolic demand during a seizure could result in damage to the mitochondria and cell death due to excitotoxic mechanisms [61, 62]. A relation between changes in mitochondrial functional states and control of synaptic activity has been described at peripheral synapses [63]. To understand the metabolic changes in a KCl induced seizure model, we first looked at the distribution of mitochondria in the CA1 region of hippocampal slices (Fig. 7.7 (A) and (B)). Hippocampal slices 300 μm thick from mouse brain slices were maintained in oxygenated aCSF medium containing (in mM): 127 NaCl, 2 KCl, 1.5 $CaCl_2$, 1.5 $MgSO_4$, 25.7 $NaHCO_3$, 1.1 KH_2PO_4, and 10 glucose and was bubbled with (95 % O_2, 5 % CO_2). Mitochondria was stained with live cell stain MitoTracker Red FM (Invitrogen, Molecular Probes) 200 nM/45 min and with Dapi (blue) in aCSF 33 °C with bubbled (95 % O_2, 5 % CO_2). The slices were washed in aCSF before imaging. Imaging was done with 2p 1060 nm excitation at 29 % with emission filters 411–553 nm for Dapi and 615–695 nm (for MitoTracker Red FM).

For NAD(P)H and Trp FLIM imaging, mouse hippocampal slices were imaged in an inverted multiphoton microscope (Zeiss Axio Observer, Zeiss LSM 780, Coherent Chameleon Ultra II, ZEN software) in a glass-bottom dish using 40× water immersion objective. Oxygenated (95 % O_2, 5 % CO_2) aCSF medium was perfused at a rate of 1 ml/min and maintained at 33 °C. Control samples were compared with samples treated with 5 mM and 10 mM KCl on stage in real time. Two-photon excitation of NAD(P)H and three-photon excitation of tryptophan was achieved at the same time using 740 nm femtosecond pulses (about 150 fs at 80 MHz repetition rate); the power was set to 2 % in the ZEN software (corresponding to about 10 mW in the focal plane). Two-channel lifetime images (340–380 nm for tryptophan, 460–500 nm for NAD(P)H) were collected using a time-correlated single-photon counting hardware (Becker and Hickl SPC-150 TCSPC boards, two HPM-100 GaAs PMTs in non-descanned configuration, SPCM software) as described above. Images were taken from a depth of about 30 μm from the sample surface to reduce the absorption of tryptophan fluorescence and the collection time was 90 seconds. Both the basal and apical regions around CA1 neurons were covered by 2-by-2 pixels ROIs for further FLIM analysis.

7.5.2 Effect of KCl treatment on metabolic activity in mouse hippocampal slices

In vitro brain slices were used to generate and propagate seizure activity by using KCl to induce depolarization in brain slices and mimic neuronal firing during seizure [56]. We used 10 mM and 5 mM KCl to induce depolarization in mouse hippocampal slices to assess the effect of induced depolarization on metabolic activity. We looked at the CA1 regions of the mouse hippocampus as described earlier [56]. We first did the labeling of hippocampal slices with MitoTracker Red FM to better understand the distribution of the mitochondria in the CA1 neurons since 90 % of the energy generated in the brain in form of ATP comes from oxidative phosphorylation within mitochondria [57, 64]. Both the apical and basal regions of CA1 neurons were abundant in mitochondria (Fig. 7.7 (B)). Therefore, we took these regions for NAD(P)H and Trp FLIM analyses as described in the above sections. Changes in the NAD(P)H mean lifetime was seen within 10 min of 10 mM (Fig. 7.7 (D)) and 5 mM (not shown) KCl treatment done on stage and as compared to the untreated, 0 min time point control (Fig. 7.7 (C)). Shift to shorter lifetime values for NAD(P)H was seen after KCl treatment. The $a_2\%$ is used as an indicator of metabolic activity and interestingly, we saw a 14.28 % decrease in the $a_2\%$ post-KCl treatment (Fig. 7.7 (E)) suggesting there could be some damage to the mitochondria or even cell death due to excitability induced by KCl treatment. We did not see drastic differences in $E\%$ distribution pre- and post-KCl treatment (Fig. 7.7 (F)) as there was overlap in the standard deviations and there was no difference in Trp mean lifetimes (not shown). Also, we saw a decrease of 8.36 % in the NADPH : NADH ratio. Taken together, the decrease in the $a_2\%$ and NADPH : NADH ratio suggests an overall decrease in metabolic activity.

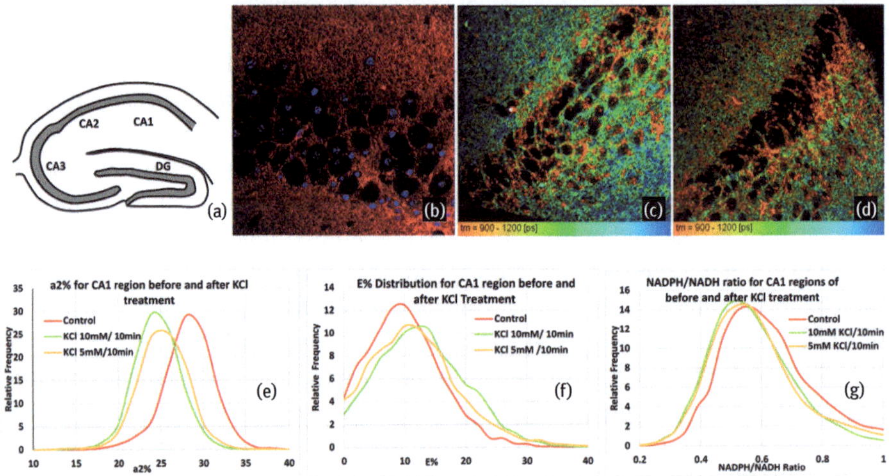

Fig. 7.7: Multiphoton FLIM imaging and analysis of the mouse hippocampus brain slice. (A) Diagrammatic illustration of the hippocampus showing the CA1, CA2, CA3 and DG regions [64]. Apical and basal regions of CA1 neurons were taken for FLIM imaging and analysis. (B) Mitochondrial distribution in CA1 neurons of the mouse hippocampus slice as can be seen by live tissue labeling with MitoTracker Red FM and nucleoli stained with Dapi. Representative images of amplitude-weighted mean lifetimes for NAD(P)H in CA1 neurons (C) before and (D) after treatment with 10 mM KCl. The graphs show the (E) $a_2\%$, (F) $E\%$, and (G) NADPH/NADH ratio for apical and basal regions of CA1 neurons before (red) and after 5 mM KCl (yellow) and 10 mM KCl (green) treatments. Note that the $a_2\%$ and NADPH/NADH ratio shifts to the left with treatment, indicating a decrease in cellular metabolic activity after treatment with KCl.

7.6 Conclusion

Our results taken together show that NAD(P)H and Trp lifetimes, and their abundances, could provide valuable information on the metabolic state of the cells and tissues under study. One could follow the NAD(P)H-$a_2\%$ increase or decrease in response to different stimuli or insult. Additionally, one could also use these metabolic signatures to understand seizure activity. Through this work we conclude that metabolic lifetime imaging can be used in assessing the disease state and efficacy of drug treatment by quantitative analyses of cellular metabolism.

Acknowledgment: We acknowledge funding from the National Institutes of Health (NIH) EB020843, and University of Virginia. African American (E006AAT1B) and a Caucasian (LNCap) prostate cancer cell lines were obtained from Dr. Dhyan Chandra, Roswell Park Cancer Institute.

References

[1] Chance B, Legallais V, Schoener B. Metabolically linked changes in fluorescence emission spectra of cortex of rat brain, kidney and adrenal gland. Nature. 1962;195:1073–1075.

[2] Nichols M, Ward K, Zholudeva V, Smith J, Hallworth R. Autofluorescence Lifetime Imaging. In: Ghukasyan V, Heikal A, editors. Natural Biomarkers for Cellular Metabolism. Boca Raton, FL: CRC Press, Taylor & Francis Group; 2014. p. 77–105.

[3] McGuinness CD, Macmillan AM, Sagoo K, McLoskey D, Birch DJS. Excitation of fluorescence decay using a 265 nm pulsed light-emitting diode: Evidence for aqueous phenylalanine rotamers. Appl Phys Lett. 2006;89:063901.

[4] Ashikawa I, Nishimura Y, Tsuboi M, Watanabe K, Iso K. Lifetime of tyrosine fluorescence in nucleosome core particles. J Biochem. 1982;91:2047–2055.

[5] Lakowicz JR. Principles of fluorescence spectroscopy. 3rd edition. Springer US; 2006.

[6] Alcala JR, Gratton E, Prendergast FG. Resolvability of fluorescence lifetime distributions using phase fluorometry. Biophys J. 1987;51:587–596.

[7] Phipps JE, Sun Y, Fishbein MC, Marcu L. A fluorescence lifetime imaging classification method to investigate the collagen to lipid ratio in fibrous caps of atherosclerotic plaque. Lasers Surg Med. 2012;44:564–571.

[8] Lakowicz JR, Szmacinski H, Nowaczyk K, Johnson ML. Fluorescence lifetime imaging of free and protein-bound NADH. Proc Natl Acad Sci USA. 1992;89:1271–1275.

[9] König K. Clinical multiphoton tomography. J. Biophotonics. 2008;1:13–23.

[10] Yu Q, Heikal AA. Two-photon autofluorescence dynamics imaging reveals sensitivity of intracellular NADH concentration and conformation to cell physiology at the single-cell level. J Photochem Photobiol B Biol. 2009;95:46–57.

[11] Richards-Kortum R, Sevick-Muraca E. Quantitative optical spectroscopy for tissue diagnosis. Ann Rev Phys Chem. 1996;47:555–606.

[12] Nelson D, Cox M. Lehninger Principles of Biochemistry. 4th edition. W. H. Freeman; 2004.

[13] Li D, Zheng W, Qu JY. Two-photon autofluorescence microscopy of multicolor excitation. Opt Lett. 2009;34:202–204.

[14] Kilfoil PJ, Tipparaju SM, Barski O a., Bhatnagar A. Regulation of ion channels by pyridine nucleotides. Circ Res. 2013;112:721–741.

[15] Benson RC, Meyer RA, Zaruba ME, McKhann GM. Cellular autofluorescence–is it due to flavins? J Histochem Cytochem. 1979;27:44–48.

[16] Rück A, Hauser C, Mosch S, Kalinina S. Spectrally resolved fluorescence lifetime imaging to investigate cell metabolism in malignant and nonmalignant oral mucosa cells. J Biomed Opt. 2014;19:096005.

[17] Pires L, Nogueira MS, Pratavieira S, Moriyama LT, Kurachi C. Time-resolved fluorescence lifetime for cutaneous melanoma detection. Biomed Opt Express. 2014;5:3080–3089.

[18] Skala MC, Riching KM, Gendron-Fitzpatrick A, et al. In vivo multiphoton microscopy of NADH and FAD redox states, fluorescence lifetimes, and cellular morphology in precancerous epithelia. Proc Natl Acad Sci USA. 2007;104:19494–19499.
[19] Skala MC, Riching KM, Bird DK, et al. In vivo multiphoton fluorescence lifetime imaging of protein-bound and free nadh in normal and pre-cancerous epithelia. J Biomed Opt. 2007;12:24014.
[20] Heikal AA. Intracellular coenzymes as natural biomarkers for metabolic activities and mitochondrial anomalies. Biomark Med. 2010;4:241–263.
[21] Valeur B, Berberan-Santos MN. Autofluorescence and fluorescence labeling in biology and medicine. In: Molecular fluorescence: principles and applications. Wiley-VCH Verlag GmbH & Co. KGaA; 2012. p. 479–505.
[22] Szalardy L, Klivenyi P, Zadori D, Fulop F, Toldi J, Vecsei L. Mitochondrial disturbances, tryptophan metabolites and neurodegeneration: medicinal chemistry aspects. Curr Med Chem. 2012;19:1899–1920.
[23] Mujat C, Greiner C, Baldwin A, et al. Endogenous optical biomarkers of normal and human papillomavirus immortalized epithelial cells. Int J Cancer. 2008;122:363–371.
[24] Diagaradjane P, Yaseen MA, Yu J, Wong MS, Anvari B. Autofluorescence characterization for the early diagnosis of neoplastic changes in DMBA/TPA-induced mouse skin carcinogenesis. Lasers Surg Med. 2005;37:382–395.
[25] Lyon DE, Walter JM, Starkweather AR, Schubert CM, McCain NL. Tryptophan degradation in women with breast cancer: a pilot study. BMC Res Notes. 2011;4:156.
[26] Jyothikumar V, Sun Y, Periasamy A. Investigation of tryptophan-NADH interactions in live human cells using three-photon fluorescence lifetime imaging and Förster resonance energy transfer microscopy. J Biomed Opt. 2013;18:060501.
[27] Torikata T, Forster LS, Johnson RE, Rupley JA. Lifetimes and NADH quenching of tryptophan fluorescence in pig heart cytoplasmic malate dehydrogenase. J Biol Chem. 1979;254:3516–3520.
[28] American Cancer Society. Prostate Cancer [Internet]; 2015 [cited 2015 Nov 12]. Available from: www.cancer.org/cancer/prostatecancer/.
[29] Koochekpour S, Marlowe T, Singh KK, Attwood K, Chandra D. Reduced Mitochondrial DNA Content Associates with Poor Prognosis of Prostate Cancer in African American Men. PLoS One. 2013;8:e74688.
[30] Carvalho C, Santos RX, Cardoso S, et al. Doxorubicin: the good, the bad and the ugly effect. Curr Med Chem. 2009;16:3267–3285.
[31] Tacar O, Sriamornsak P, Dass CR. Doxorubicin: An update on anticancer molecular action, toxicity and novel drug delivery systems. J Pharm Pharmacol. 2013;65:157–170.
[32] Tehranian N, Sepehri H, Mehdipour P, et al. Combination effect of PectaSol and Doxorubicin on viability, cell cycle arrest and apoptosis in DU-145 and LNCaP prostate cancer cell lines. Cell Biol Int. 2012;36:601–610.
[33] Ashley N, Poulton J. Mitochondrial DNA is a direct target of anti-cancer anthracycline drugs. Biochem Biophys Res Commun. 2009;378:450–455.
[34] Sun Y, Periasamy A. Localizing protein–protein interactions in living cells using fluorescence lifetime imaging microscopy. In: Verveer PJ, editor. Advanced fluorescence microscopy SE – 6. New York: Springer; 2015. p. 83–107.
[35] Day RN, Periasamy A, Schaufele F. Fluorescence resonance energy transfer microscopy of localized protein interactions in the living cell nucleus. Methods. 2001;25:4–18.
[36] Periasamy A, Day RN. Molecular Imaging: FRET microscopy and spectroscopy. New York: Oxford University Press; 2005.

[37] Parkhurst LJ, Parkhurst KM, Powell R, Wu J, Williams S. Time-resolved fluorescence resonance energy transfer studies of DNA bending in double-stranded oligonucleotides and in DNA-protein complexes. Biopolymers. 2002;61:180–200.
[38] Rehman S, Gladman JT, Periasamy A, Sun Y, Mahadevan MS. Development of an AP-FRET based analysis for characterizing RNA-protein interactions in myotonic dystrophy (DM1). PLoS One. 2014;9:e95957.
[39] Clegg RM. The history of FRET: From conception through the labors of birth. Rev Fluoresc. 2006:1–45.
[40] Förster T. Energy migration and fluorescence. J Biomed Opt. 2012;17:11002.
[41] Förster T. Intermolecular energy migration and fluorescence. Energetics. 1948;148:55–75.
[42] Periasamy A. Advanced light microscopy. Methods. 2014;66:121–123.
[43] Sun Y, Wallrabe H, Seo SA, Periasamy A. FRET microscopy in 2010: The legacy of Theodor Förster on the 100th anniversary of his birth. ChemPhysChem. 2011;12:462–474.
[44] Wallrabe H, Periasamy A. Imaging protein molecules using FRET and FLIM microscopy. Curr Opin Biotechnol. 2005;16:19-27.
[45] Periasamy A, Wallrabe H, Chen Y, Barroso M. Quantitation of protein–protein interactions: confocal FRET microscopy. In: Correia J, Detrich HW, editors. Biophysical tools for biologists, volume two: in vivo techniques. Academic Press; 2008. p. 569–598.
[46] Clegg RM, Periasamy A, editors. FLIM microscopy in biology and medicine. Boca Raton, FL: Chapman and Hall/CRC; 2009.
[47] Zhang Q, Piston DW, Goodman RH. Regulation of corepressor function by nuclear NADH. Science. 2002;295:895–897.
[48] Li D, Zheng W, Qu JY. Time-resolved spectroscopic imaging reveals the fundamentals of cellular NADH fluorescence. Opt Lett. 2008;33:2365–2367.
[49] Stringari C, Cinquin A, Cinquin O, Digman MA, Donovan PJ, Gratton E. Phasor approach to fluorescence lifetime microscopy distinguishes different metabolic states of germ cells in a live tissue. Proc Natl Acad Sci USA. 2011;108:13582–13587.
[50] Becker W. TCSPC Handbook 6th edition [Internet]; 2015 [cited 2015 Nov 12]. Available from: www.becker-hickl.de/literature.htm#handb.
[51] Pradhan A, Pal P, Durocher G, et al. Steady state and time-resolved fluorescence properties of metastatic and nonmetastatic malignant cells from different species. J Photochem Photobiol B. 1995;31:101–112.
[52] Jyothikumar V, Sun Y, Periasamy A. Tryptophan as an alternative biomarker for cellular energy metabolism. In: Ghukasyan VV, Heikal AA, editors. Natural biomarkers for cellular metabolism. Boca Raton, FL: CRC Press, Taylor & Francis Group; 2015. p. 157–177.
[53] Sun Y, Rombola C, Jyothikumar V, Periasamy A. Förster resonance energy transfer microscopy and spectroscopy for localizing protein-protein interactions in living cells. Cytometry A. 2013;83:780–793.
[54] Blacker TS, Mann ZF, Gale JE, et al. Separating NADH and NADPH fluorescence in live cells and tissues using FLIM. Nat Commun. 2014;5:3936.
[55] Alam SR, Wallrabe H, Svindrych Z, Chaudhary AK, Christopher KG, Chandra D, Periasamy A. Investigation of mitochondrial metabolic response to doxorubicin in prostate cancer cells: An NADH, FAD and tryptophan FLIM assay. Sci Rep. 2017;7:10451.
[56] Traynelis SF, Dingledine R. Potassium-induced spontaneous electrographic seizures in the rat hippocampal slice. J Neurophysiol. 1988;59:259–276.
[57] Gunter TE, Gunter KK, Sheu SS, Gavin CE. Mitochondrial calcium transport: physiological and pathological relevance. Am J Physiol. 1994;267:C313–C339.
[58] Ankarcrona M, Dypbukt JM, Bonfoco E, et al. Glutamate-induced neuronal death: a succession of necrosis or apoptosis depending on mitochondrial function. Neuron. 1995;15(4):961–973.

[59] SchinderAF, Olson EC, Spitzer NC, Montal M. Mitochondrial dysfunction is a primary event in glutamate neurotoxicity. J Neurosci. 1996;16:6125–6133.
[60] White RJ, Reynolds IJ. Mitochondrial depolarization in glutamate-stimulated neurons: an early signal specific to excitotoxin exposure. J Neurosci. 1996;16:5688–5697.
[61] Meldrum BS. Cell damage in epilepsy and the role of calcium in cytotoxicity. Adv Neurol. 1986;44:849–855.
[62] Olney JW, Collins RC, Sloviter RS. Excitotoxic mechanisms of epileptic brain damage. Adv Neurol. 1986;44:857–877.
[63] Tang YG, Zucker RS. Mitochondrial involvement in post-tetanic potentiation of synaptic transmission. Neuron. 1997;18:483–491.
[64] Bindokas VP, Lee CC, Colmers WF, Miller RJ. Changes in mitochondrial function resulting from synaptic activity in the rat hippocampal slice. J Neurosci. 1998;18:4570–4587.

Angelika Rück, Jasmin Breymayer, and Sviatlana Kalinina

8 TCSPC FLIM and PLIM for metabolic imaging and oxygen sensing

Abstract: Correlated imaging of phosphorescence and fluorescence lifetime parameters of metabolic markers is a challenge for direct investigating mechanisms related to cell metabolism and oxygen tension. A large variety of clinical phenotypes is associated with mitochondrial defects accomplished with changes in cell metabolism. In many cases the hypoxic microenvironment of cancer cells shifts metabolism from oxidative phosphorylation (OXPHOS) to anaerobic or aerobic glycolysis. Also during stem cell differentiation a switch in cell metabolism is observed. Mitochondrial dysfunction associated with hypoxia has been invoked in many complex disorders such as type 2 diabetes, Alzheimer's disease, cardiac ischemia/reperfusion injury, tissue inflammation and cancer.

Cellular responses to oxygen tension have been studied extensively; optical imaging techniques based on time-correlated single photon counting (TCSPC) to detect oxygen concentration and distribution will be discussed in this chapter. By inspecting the fluorescence decay characteristics of intrinsic coenzymes different anaerobic and aerobic metabolic pathways can be imaged. In addition, oxygen tension can be imaged by considering the phosphorescence lifetime of a phosphorescent probe. As will be shown, the combination of both fluorescence lifetime imaging (FLIM) of coenzymes like NAD(P)H and FAD and phosphorescence lifetime (PLIM) of phosphorescent dyes could provide valuable information about the correlation of metabolic pathways and oxygen tension.

8.1 Cellular energy metabolism in different environments

To maintain its metabolism, every cell needs to produce energy in the form of adenosine triphosphate (ATP). Glycolysis, Krebs cycle, and oxidative phosphorylation (OXPHOS) enable cells to gain this energy. In glycolysis, one molecule of glucose is processed to two molecules of pyruvate. Two molecules of ATP are produced and two molecules of oxidized nicotinamide adenine dinucleotide (NAD^+) are reduced to $NADH + H^+$. Under anaerobic conditions, the resulting pyruvate is processed to lactate by oxidation of $NADH + H^+$ and, therefore, NAD^+ is again available for the glycolysis enzyme glyceraldehyde 3-phosphate dehydrogenase. These processes take place in the cytoplasm. Under aerobic conditions, pyruvate decarboxylation occurs in mitochondria and two molecules of pyruvate are converted to acetyl-CoA while two molecules of NAD^+ are reduced to $NADH + H^+$. Subsequently, two molecules of acetyl-CoA enter the Krebs cycle in mitochondria and generate six molecules of $NADH + H^+$,

https://doi.org/10.1515/9783110429985-010

two molecules of reduced flavin adenine dinucleotide ($FADH_2$), and two molecules of guanosine triphosphate (GTP). During OXPHOS, electrons from $NADH + H^+$ and $FADH_2$ are transferred on oxidizers via redox reactions in the inner mitochondrial membrane. In this electron transport chain, complexes I to IV, ubiquinone and cytochrome c are involved. Changes in the conformation of complexes I, III and IV enable the transport of protons from $NADH + H^+$ and $FADH_2$ to the intermembrane space. ATP synthase uses energy from the proton gradient in the intermembrane space to generate ATP. In OXPHOS, $NADH + H^+$ is reduced to NAD^+. The main reactions in glycolysis and OXPHOS are schematically demonstrated in Fig. 8.1.

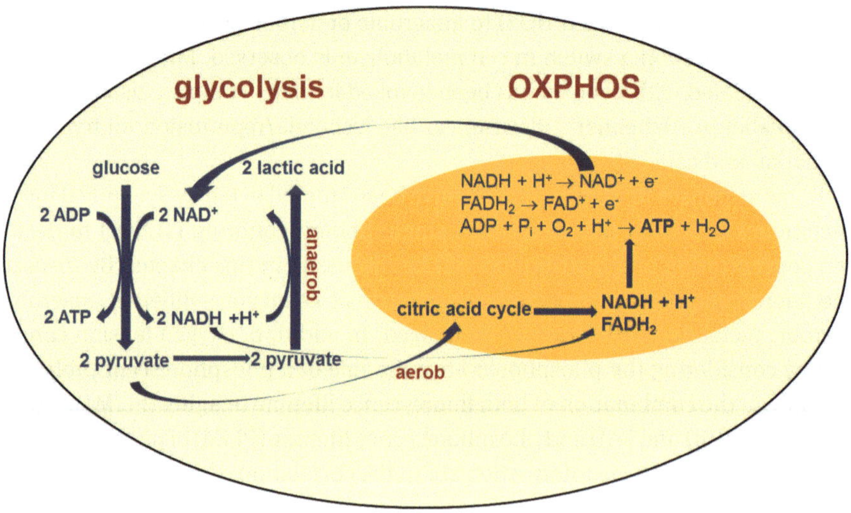

Fig. 8.1: Main reactions involved in glycolysis and OXPHOS in a living cell to maintain energy metabolism.

Even in the presence of oxygen, cancer cells remodel their energy metabolism from OXPHOS to anaerobic glycolysis with an increased uptake of glucose, a process known as aerobic glycolysis or the "Warburg effect" [1, 2]. Recent research shows that this is not true for all types of cancer cells. Whether they gain their energy from OXPHOS or glycolysis or both depends on the tumor microenvironment (hypoxia), tumor size or activated oncogenes [3, 4].

Switches in metabolism are also present in stem cells during tissue development. In pluripotent embryonic stem cells and induced pluripotent stem cells that are rapidly proliferating, metabolism corresponds to aerobic glycolysis. In differentiating embryonic stem cells, glycolysis decreases and OXPHOS increases. Quiescent adult stem cells want to avoid damage from reactive oxygen species (ROS) to ensure life-long tissue renewal capacity and to reside in a hypoxic environment and, therefore, gain energy from glycolysis. Proliferating adult stem/progenitor cells for tissue homeostasis and renewal show a metabolism with different combinations of glycolysis and OXPHOS [5–7]. Oxygen tension therefore seems to be important as a metabolic regulator.

8.2 Cellular energy metabolism and FLIM of autofluorescent coenzymes

Identifying and characterizing the complex cellular and molecular mechanisms of energy metabolism and monitoring metabolic alterations could be useful as a biomarker for characterizing disease progression or determining the efficacy of novel therapies. Demand is high to develop robust tools for monitoring metabolic activity with high spatial and temporal resolutions. Whereas spatial resolution of the widespread clinically used imaging techniques of positron-emission tomography (PET) and functional magnetic resonance imaging (fMRI) is low, optical techniques based on time-correlated single photon counting (TCSPC) allow high resolution on a cellular level [8]. Additionally, the development of 2-photon (2P) microscopy-based techniques profit from the spatially confined nonlinear excitation effect and deep penetrating nature of near-infrared excitation light [9–12]. 2P microscopy was coupled with time-resolved fluorescence imaging microcopy (FLIM), using TCSPC methods [13, 14]. By inspecting fluorescence decay characteristics of intrinsic coenzymes, 2P FLIM offers the possibility to determine redox states of cells and to directly image different metabolic pathways such as glycolysis and OXPHOS, which drive ATP production in the cytosol and in the mitochondria as demonstrated in Fig. 8.1, or key paths of antioxidant defense [15, 16].

Observation of cell metabolism through time-resolved autofluorescence imaging is a new and challenging procedure (for comprehensive reviews on FLIM, see [17]). It is based on the detection of the fluorescence lifetime of the metabolic coenzymes NAD(P)H (nicotinamide adenine dinucleotide (phosphate)) and FAD^+ (flavin adenine dinucleotide). These enzymes can reflect the redox state and metabolic changes during cell carcinogenesis and differentiation by the redox ratio, which is defined as the ratio of the fluorescence intensity of FAD^+ and NAD(P)H [18]. A change in the optical redox ratio causes a change in the fluorescence lifetimes of NAD(P)H and FAD^+ [19, 20]. NAD(P)H is located in the mitochondria of living cells as well as the cytoplasm. NAD(P)H not bound to proteins ("free" NAD(P)H) typically possesses a short fluorescence lifetime around 400 ps because of quenching of the reduced nicotinamide by the adenine group. If it is bound to proteins, the lifetime is much longer (1–6.5 ns, depending on the target to which the cofactor binds) [16, 21, 22]. Due to conformational heterogeneity of the different enzymes, bound NAD(P)H can have complex lifetime distributions with more than one exponential component [23–25]. In [25] the researchers tried to resolve NAD(P)H fluorescence in a "quasi-global" 4-component multi-exponential fit. The lifetime of bound NAD(P)H also varies due to the existence of the two redox couples NAD+/NADH and NADP+/NADPH in the cells [16]. The maximum emission of free NAD(P)H is around 470 nm, whereas the maximum is blue-shifted towards 440 nm when bound to proteins [26–29]. Fig. 8.2 demonstrates how to separate bound and free NAD(P)H by combined spectral and lifetime detection. In this example 2P-FLIM was studied in OKF6/TERT-2, SCC-4 and SCC-25 cells. The mean lifetime, τ_{mean}, is calculated by a two exponential fitting procedure

$$I(t) = a_1\, e^{-t/\tau}{}_1 + a_2\, e^{-t/\tau}{}_2, \tag{8.1}$$

where $I(t)$ is the fluorescence intensity at time t, τ_1 and τ_2 are the lifetimes of the free and bound fluorescence components, a_1 and a_2 their relative contributions, and the distribution histogram for the two spectral regions around 436 nm and 470 nm is shown for all cells. The lifetime is represented in false colors. Free NAD(P)H was found in a higher amount at 470 nm, which correlates well with the shorter lifetime. NAD(P)H found in the cell nucleus was mostly free, it plays a role in gene expression but not in metabolic activity and the fluorescence lifetime of the bound fraction was shorter than in the mitochondria and cytosol [16, 30, 81, 32].

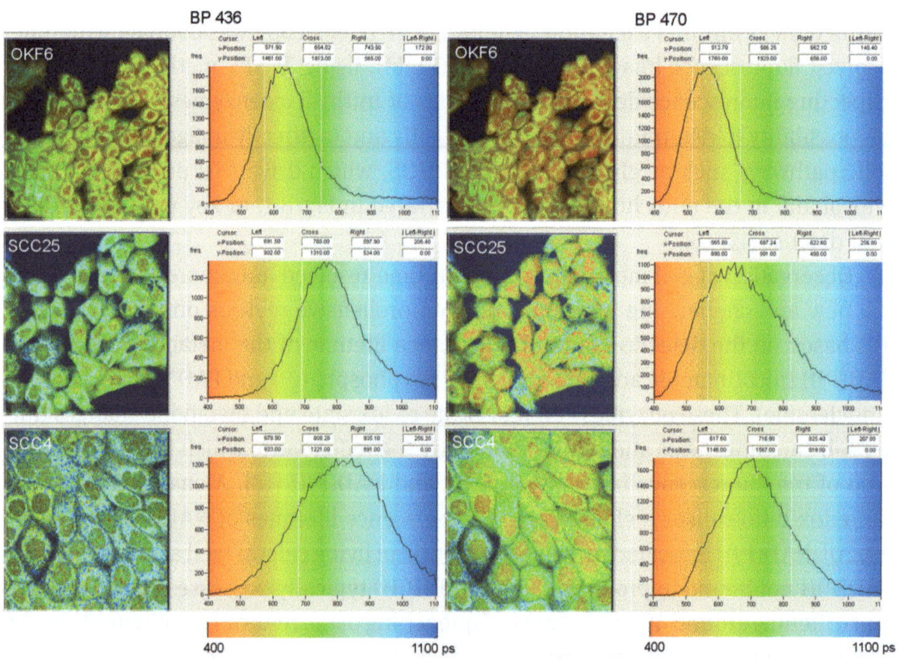

Fig. 8.2: 2P-FLIM and distribution histogram of the mean lifetime of NAD(P)H with BP (436 ± 10) nm (higher contribution of bound NAD(P)H) and BP (470 ± 10) nm (higher contribution of free NAD(P)H) of OKF6/TERT-2, SCC-25 and SCC-4 cells after excitation with 720 nm. The mean lifetime, τ_m, was calculated by a two exponential fitting procedure. The histograms show the distribution of the τ_m-value within the range 400–1100 ps. Image from [15]. Reproduced with permission from SPIE.

There are many reports in the literature which confirm a relation between metabolic phenotypes and fluorescence lifetimes of NAD(P)H [33, 34]. During adipogenic differentiation of human salivary gland stem cells the mean fluorescence lifetime of NAD(P)H and FAD^+ was longer for differentiated cells [35]. This was correlated with an increase in oxygen consumption and an increase in aerobic cell metabolism. In the *Caenorhabditis elegans* germ line different metabolic states of stem cells could be distinguished by the phasor approach to fluorescence lifetime imaging and correlated with the redox state of the cells [23]. During carcinogenesis a shortening of the lifetime of NAD(P)H was reported for low-grade and high-grade precancerous tissue compared to normal tissue whereas the lifetime of FAD^+ increased [19]. This was correlated with a shift from oxidative phosphorylation to glycolysis. In another report, the metabolic state of intestinal stem cells in a living small intestine was characterized by the phasor approach by a high ratio of free/bound NAD(P)H and a short lifetime of NAD(P)H, indicating glycolysis, as mostly expected in highly proliferative stem cells and cancer cells [36]. In cerebral tissue, in vivo 2P-NAD(P)H-FLIM showed multiple decaying exponentials, representing different enzyme-bound formulations [25]. Here, the metabolic activity of neurons, astrocytes, vascular endothelial cells, and others were monitored during periods of anoxia.

It is evident that observation of cell metabolism by autofluorescence FLIM could be a straightforward tool for fluorescence guided diagnosis. The method is even discussed as being used in the clinic for imaging of brain tumours [37, 38]. However, in contrast to most reports the lifetime of NAD(P)H was found to be longer in the tumor (glioblastoma multiforme) than in the normal tissue (normal cortex). For a comprehensive review on fluorescence lifetime techniques in medical applications see [39]. A short NAD(P)H lifetime correlated with higher concentration of free NAD(P)H and glycolytic switch, is, therefore, not an exclusive rule for tumors and depends on the type of cancer [15, 37, 38, 40, 41]. In a recent publication the mean lifetime of NAD(P)H of a variety of malignant breast cancer cells was increased over that of a noncancerous mammary epithelium cell line, whereas for other types a decrease was observed [41]. Therefore, the situation is complex and more information is needed to conclude on metabolic pathways. Separation of NADH and NADPH fluorescence lifetimes as described in [16] is one possibility, as bound NADPH plays a role especially in antioxidant defense and could be critical to oxygen tension [42]. The involvement of both, NAD(P)H and FAD^+ fluorescence lifetimes and intensities, as described for optical metabolic imaging (OMI) is important to quantify heterogeneous cell populations and to characterize the genetic and phenotypic heterogeneity of cancers [43, 44]. Substantial phenotypic variations might be induced by hypoxic stress, the mechanisms are not yet consistently known [45]. It seems that oxygen tension significantly influences metabolic pathways (see Section 8.1). Imaging methods to investigate oxygen tension are therefore of main interest (see Section 8.4).

8.3 FLIM of NADH and NAD(P)H

The redox state plays a central role in the regulation of energy production and other metabolic reactions in cells. The two redox couples NAD^+/NADH and $NADP^+$/NADPH are the most important determinants of the redox state and they are engaged in different metabolic pathways [46]. Free radical formation is determined by the redox state of NAD, while the NADP redox state is involved in antioxidant defense within glutathione reductase, in superoxide production during phagocytosis in NADPH oxidase [47] or in other enzymes as mitochondrial transhydrogenase [48]. Therefore, many different metabolic reactions are driven by NAD(P)H. Investigating the biological functions of NAD and NADP is important to understand fundamental properties of living cells. New strategies are needed to distinguish the redox couples. As NADP is phosphorylated at a special site of the molecule, the fluorescence properties of the nicotinamide ring of NADPH are identical to those of NADH [49]. Spectral separation is therefore not possible. However, as demonstrated [16], FLIM differentiates quantitatively between the two cofactors. By using genetic and pharmacologic approaches it was shown that the relative amount of NADPH versus NADH can be determined from the lifetime τ_2 of the enzyme bound NAD(P)H, which is significantly increased to approx. 3.6 ns for the phosphorylated molecule. To distinguish NADH and NADPH, therefore, requires free fitting of τ_2 in equation (8.1) to resolve unknown responses of NAD versus NADP pools. Due to the existence of the two redox couples NAD^+/NADH and $NADP^+$/NADPH in the cells, the lifetime of bound NAD(P)H can vary during experiments. As bound NADPH plays a role in antioxidant defense, τ_2 could be critical to oxygen tension [42]. In an interesting work by Niesner et al. [44] activation of NADPH oxidase (NOX2) during phagocytosis was demonstrated by FLIM. The τ_2-distribution in each region of increased fluorescence lifetime in polymorphonuclear cells (PMNs) treated with phorbol 12-myristate 13-acetate (PMA) was evaluated in order to identify the specific fluorescence lifetime peak of NADPH bound to NADPH oxidase, which was around (3670 ± 170) ps. This is shown in Fig. 8.3.

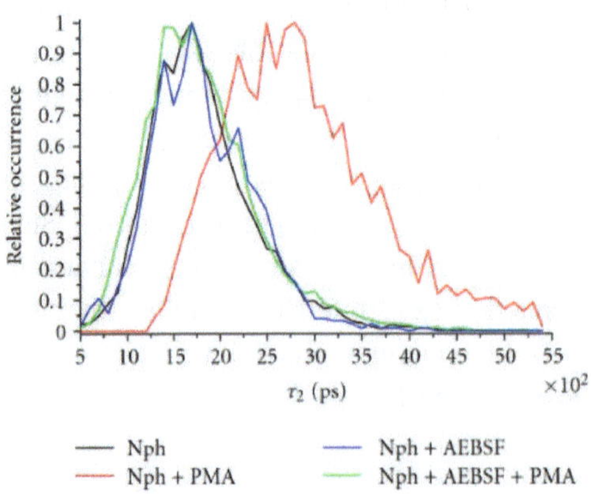

Fig. 8.3: The histogram of τ_2 in each region of increased fluorescence lifetime in PMNs treated with PMA identifies the specific fluorescence lifetime peak of NADPH bound to NOX2. Image from [50].

8.4 PLIM of oxygen sensors

As a terminal electron acceptor in the mitochondrial respiratory chain oxygen plays a central role in cellular energy production of mammalian organisms. Disturbance of oxygen supply induces a number of metabolic changes and can lead to cell death [51]. In fact, hypoxia-associated defective mitochondrial function has been invoked in such complex disorders as type 2 diabetes, Alzheimer's disease, cardiac ischemia/reperfusion injury, tissue inflammation and cancer [52, 53]. Therefore, measurements of oxygen levels in tissue are helpful for the investigation of metabolic activity under pathological conditions. Different technologies including microelectrodes [54, 55], electron paramagnetic resonance (EPR) systems [56, 57], nitroimidazole staining [58, 59], and Raman microspectroscopy [60] have been developed to measure oxygen in vivo and are used to monitor oxygen consumption rate and tissue oxygenation (concentration of oxygen in situ). The level of oxygen in tissue is usually evaluated as partial oxygen tension (oxygen pressure, pO_2) and consequently measured in mmHg (kPa or Torr) or in μM.

Technologies based on phosphorescence quenching by oxygen have given rise to a generation of non-invasive optical methods for oxygen tension measurement that can achieve extremely high spatial and temporal resolution. Having a triplet state as ground state oxygen is a very effective quenching agent for fluorophores in the triplet state. Generally, due to collision with an oxygen molecule the energy of phosphorescent molecules in the triplet state is transferred to oxygen instead of phosphorescence emission. The resultant decrease in intensity and lifetime of phosphorescence is described by the Stern–Volmer equation for luminescence quenching:

$$I_0/I = \tau_0/\tau = 1 + K_{SV}[Q], \tag{8.2}$$

where I and τ are luminescence intensity and lifetime, respectively, in the presence of a quenching agent, I_0 and τ_0 are luminescence intensity and lifetime in the absence of a quenching agent, K_{SV} is the Stern–Volmer constant, which quantifies the quenching efficiency and $[Q]$ is a quencher concentration. The luminophore concentration itself is not mentioned in the equation which suggests that quenching is independent of luminophore concentration. Oxygen tension therefore can be determined by just measuring the lifetime changes.

Porphyrins [61] and metalloporphyrins [62–64], complexes of ruthenium [65, 66] and various other rare-earth metals [67, 68] are widely used as oxygen-sensing probes to measure average pO_2 and to perform a detailed mapping in biological samples. The choice and modification of the probes are both determined by the aim of the measurement. It seems that the most popular probes are Pt- and Pd-porphyrins which can be used in nanoparticles as well as conjugated with cell-penetrating peptides to provide easy delivering into the cells or conjugated with macromolecules like PEG for oxygen control in extracellular space [69–73]. Unlike metalloporphyrins, ruthenium complexes exhibit relatively short lifetimes of oxygen-dependent luminescence

(< 5 µs) with good photostability [66, 74, 75]. Of particular interest is the endogenous protoporphyrin IX, whose delayed fluorescence provides a way to directly measure mitochondrial oxygen tension in cultured cells and in tissues without loading the samples with any additional probe [76, 77].

Lifetime measurements of oxygen-sensitive probes can be performed by time-resolved techniques usually by using time-gated or TCSPC methods. Application of TCSPC to confocal laser scanning microscopes provides a proper spatial resolution for detailed mapping of biological samples loaded with pO_2-sensitive probes [62, 65, 74]. Actually, the preferred method for phosphorescence lifetime imaging (PLIM) is TCSPC in combination with confocal or multiphoton laser scanning [8, 70, 78]. Moreover, the technique of simultaneous FLIM and PLIM can be used for correlative imaging of the fluorescence lifetime of metabolic coenzymes like NAD(P)H and pO_2-sensitive phosphorescence [8, 74].

8.5 TCSPC FLIM and PLIM

TCSPC is based on the measurement of the arrival time of the first emitted photon relative to the excitation pulse. In combination with fast photomultipliers for detecting single photons, this technique achieves the highest accuracy [14]. The modern implementation of TCSPC is multidimensional, when for every photon not only the time in the single period is determined but also other parameters [8]. The recording process builds up a photon distribution over the scan coordinates and the arrival times of the photons after the excitation pulses [8, 14, 79]. The technique features excellent time resolution, near-ideal photon efficiency [80], and dissemination of multi-exponential fluorescence decay profiles into their decay components.

The method of multidimensional TCSPC is not directly applicable to PLIM because of the much longer lifetimes that are typical for phosphorescence. To apply TCSPC with a titanium-sapphire laser, whose pulse period is far too short to observe decay functions of phosphorescence, a technique based on an additional on-off modulation of a high-frequency pulsed laser can be used [8, 81, 82].

We recently presented results on simultaneous NAD(P)H-FLIM and PLIM based on ruthenium tris-(2,2′-bipyridyl) dichloride ($Ru(BPY)_3$) [74, 83, 84]. We used modulated high-frequency pulsed multiphoton laser scanning in combination with an advanced multidimensional TCSPC technique. Fluorescence was recorded during the on-phase of the laser, phosphorescence during the off-phase. Laser modulation was achieved by controlling the opto-acoustic modulator (AOM) of the laser scanning microscope by a signal generated in the TCSPC system (Becker & Hickl GmbH, Berlin, Germany). This system was coupled to the NDD port of an LSM 710 (Zeiss, Germany). The basic setup of the two-channel detection system is shown in Fig. 8.4.

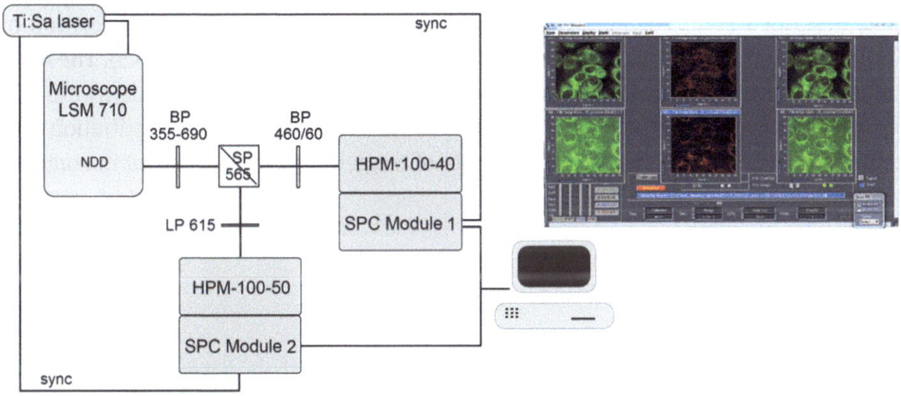

Fig. 8.4: System for simultaneous NAD(P)H-FLIM and PLIM based on ruthenium tris-(2,2′-bipyridyl) dichloride. An advanced two-channel multidimensional TCSPC technique is used with a modulated high-frequency pulsed two-photon titanium-sapphire laser. Image from [83].

The difference between fluorescence and phosphorescence decay times was used for their separation during recording. The fluorescence and the phosphorescence signals were also separated spectrally (see Fig. 8.4). A two-photon excitation at 780 nm induces the phosphorescence of $Ru(BPY)_3$ as well as the fluorescence of intracellular NAD(P)H. A band-pass filter 460/60 nm selected the spectral range for NAD(P)H lifetime detection; a phosphorescent signal was not detected within this spectral range. Because of a lack of other phosphorescent molecules within the cells, it was not necessary to use narrow band-pass filters for imaging of $Ru(BPY)_3$ and the applied long-pass filter LP 615 was sufficient.

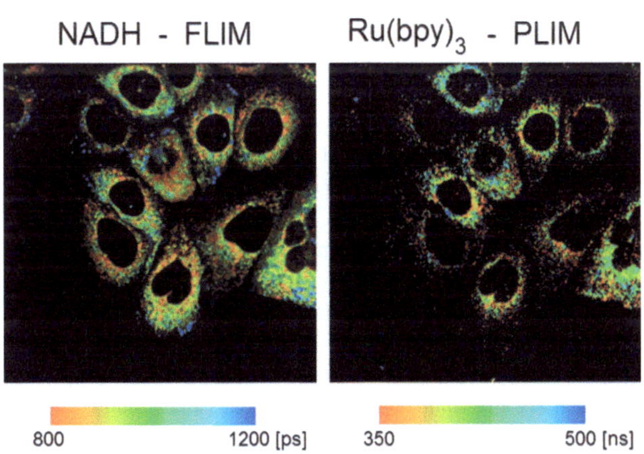

Fig. 8.5: Mean lifetime images of NAD(P)H fluorescence and $Ru(BPY)_3$ phosphorescence measured simultaneously.

The described setup for FLIM/PLIM imaging enables simultaneous investigation of intracellular NAD(P)H and $Ru(BPY)_3$ for living cells in 2D cultures (Fig. 8.5). The technique provides valuable information about the correlation of metabolic pathways and oxygen tension. It can be used to analyze the intracellular oxygen distribution and helps to deepen understanding of the role of oxygen in cell metabolism in complex in vivo systems.

References

[1] Gatenby RA, Gillies RJ. Why do cancers have high aerobic glycolysis? Nat Rev Cancer. 2004;4:891–899.
[2] Warburg O. On the origin of cancer cells. Science. 1956;123:309–314.
[3] Jose C, Bellance N, Rossignol R. Choosing between glycolysis and oxidative phosphorylation: A tumor's dilemma? Biochim Biophys Acta. 2011;1807:552–561.
[4] Garcia-Heredia JM, Carnero A. Decoding Warburg's hypothesis: tumor-related mutations in the mitochondrial respiratory chain. Oncotarget. 2015;6:41582–41599.
[5] Shyh-Chang N, Daley GQ, Cantley LC. Stem cell metabolism in tissue development and aging. Development. 2013;140:2535–2547.
[6] De Miguel MP, Alcaina Y, de la Maza DS, Lopez-Iglesias P. Cell metabolism under microenvironmental low oxygen tension levels in stemness, proliferation and pluripotency. Curr Mol Med. 2015;15:343–359.
[7] Mohyeldin A, Garzon-Muvdi T, Quinones-Hinojosa A. Oxygen in stem cell biology: A critical component of the stem cell niche. Cell Stem Cell. 2010;7:150–161.
[8] Becker W. Advanced time-correlated single photon counting applications. Springer Series in Chemical Physics Volume 111. Cham, CH: Springer International Publishing; 2015.
[9] Zipfel WR, Williams RM, Christie R, Nikitin AY, Hyman BT, Webb WW. Live tissue intrinsic emission microscopy using multiphoton-excited fluorescence and second harmonic generation. Proc Natl Acad Sci USA. 2003;100:7075–7080.
[10] Svoboda K, Yasuda R. Principles of two-photon excitation microscopy and its applications to neuroscience. Neuron. 2006;50:823–839.
[11] Huang S, Heikal AA, Webb WW. Two-photon fluorescence spectroscopy and microscopy of NAD(P)H and flavoprotein. Biophys J. 2002;82:2811–2825.
[12] Schenke-Layland K, Riemann I, Stock UA, König K. Imaging of cardiovascular structures using near-infrared femtosecond multiphoton laser scanning microscopy. J Biomed Opt. 2005;10:240171–240175.
[13] Ehlers A, Riemann I, Stark M, König K. Multiphoton fluorescence lifetime imaging of human hair. Microsc Res Tech. 2007;70:154–161.
[14] Becker W. Advanced time-correlated single photon counting techniques. Springer Series in Chemical Physics Volume 81. Berlin, Heidelberg: Springer-Verlag; 2005.
[15] Rück A, Hauser C, Mosch S, Kalinina S. Spectrally resolved fluorescence lifetime imaging to investigate cell metabolism in malignant and nonmalignant oral mucosa cells. J Biomed Opt. 2014;19:96005.
[16] Blacker TS, Mann ZF, Gale JE, et al. Separating NADH and NADPH fluorescence in live cells and tissues using FLIM. Nat Commun. 2014;5:3936.
[17] Periasamy A, Clegg RM. FLIM microscopy in biology and medicine. New York: CRC Press; 2009.

[18] Chance B, Schoener B, Oshino R, Itshak F, Nakase Y. Oxidation-reduction ratio studies of mitochondria in freeze-trapped samples. NADH and flavoprotein fluorescence signals. J Biol Chem. 1979;254:4764–4771.
[19] Skala MC, Riching KM, Gendron-Fitzpatrick A, et al. In vivo multiphoton microscopy of NADH and FAD redox states, fluorescence lifetimes, and cellular morphology in precancerous epithelia. Proc Natl Acad Sci USA. 2007;104:19494–19499.
[20] Skala MC, Ramanujam N. Multiphoton redox ratio imaging for metabolic monitoring in vivo. Methods Mol Biol. 2010;594:155–162.
[21] Lakowicz JR, Szmacinski H, Nowaczyk K, Johnson ML. Fluorescence lifetime imaging of free and protein-bound NADH. Proc Natl Acad Sci. 1992;89:1271–1275.
[22] Wu Y, Zheng W, Qu JY. Sensing cell metabolism by time-resolved autofluorescence. Opt Lett. 2006;31:3122–3124.
[23] Stringari C, Cinquin A, Cinquin O, Digman MA, Donovan PJ, Gratton E. Phasor approach to fluorescence lifetime microscopy distinguishes different metabolic states of germ cells in a live tissue. Proc Natl Acad Sci USA. 2011;108:13582–13587.
[24] Chorvatova A, Mateasik A, Chorvat DJr. Spectral decomposition of NAD(P)H fluorescence components recorded by multi-wavelength fluorescence lifetime spectroscopy in living cardiac cells. Laser Phys Lett. 2013;10:125703.
[25] Yaseen MA, Sakadzic S, Wu W, Becker W, Kasischke KA, Boas DA. In vivo imaging of cerebral energy metabolism with two-photon fluorescence lifetime microscopy of NADH. Biomed Opt Express. 2013;4:307–321.
[26] König K, Schneckenburger H. Laser-induced autofluorescence for medical diagnosis. J Fluoresc. 1994:4:17–40.
[27] Lakowicz JR. Principles of fluorescence spectroscopy. 2nd ed. New York: Kluwer/Academic Plenum Publishers; 1999.
[28] Huber R, Buchner M, Li H, Schlieter M, Sperfeld AD, Riepe MW. Protein binding of NADH on chemical preconditioning. J Neurochem. 2000:75:329–335.
[29] Vekshin NL. Photonics of biopolymers. Berlin, Heidelberg: Springer-Verlag; 2002.
[30] Li D, Zheng W, Qu JY. Time-resolved spectroscopic imaging reveals the fundamentals of cellular NADH fluorescence. Opt Lett. 2008;33:2365–2367.
[31] Zhang Q, Piston DW, Goodman RH. Regulation of corepressor function by nuclear NADH. Science. 2002;295:1895–1897.
[32] Fjeld CC, Birdsong WT, Goodman RH. Differential binding of NAD+ and NADH allows the transcriptional corepressor carboxyl-terminal binding protein to serve as a metabolic sensor. Proc Natl Acad Sci USA. 2003;100:9202–9207.
[33] Heikal AA. Intracellular coenzymes as natural biomarkers for metabolic activities and mitochondrial anomalies. Biomark Med. 2010;4:241–263.
[34] Ghukasyan VV, Kao FJ. Monitoring cellular metabolism with fluorescence lifetime of reduced nicotinamide adenine dinucleotide. J Phys Chem C. 2009;113:11532–11540.
[35] König K, Uchugonova A, Gorjup E. Multiphoton fluorescence lifetime imaging of 3D-stem cell spheroids during differentiation. Microsc Res Tech. 2011;74:9–17.
[36] Stringari C, Edwards RA, Pate KT, Waterman ML, Donovan PJ, Gratton E. Metabolic trajectory of cellular differentiation in small intestine by phasor fluorescence lifetime microscopy of NADH. Sci Rep. 2012;2:568.
[37] Sun Y, Hatami N, Yee M, et al. Fluorescence lifetime imaging microscopy for brain tumor image-guided surgery. J Biomed Opt 2010;15:56022.
[38] Leppert J, Krajewski J, Kantelhardt SR, et al. Multiphoton excitation of autofluorescence for microscopy of glioma tissue. Neurosurgery. 2006;58:759–767.

[39] Marcu L. Fluorescence lifetime techniques in medical applications. Ann Biomed Eng. 2012;40:304–331.
[40] McGinty J, Galletly NP, Dunsby C, et al. Wide-field fluorescence lifetime imaging of cancer. Biomed Opt Express. 2010;1:627–640.
[41] Walsh AJ, Cook RS, Manning HC, et al. Optical metabolic imaging identifies glycolytic levels, subtypes, and early-treatment response in breast cancer. Cancer Res. 2013;73:6164–6174.
[42] Chorvatova A, Aneba S, Mateasik A, Chorvat D, Comte B. Time-resolved fluorescence spectroscopy investigation of the effect of 4-hydroxynonenal on endogenous NAD(P)H in living cardiac myocytes. J Biomed Opt. 2013;18:67009.
[43] Walsh AJ, Skala MC. Optical metabolic imaging quantifies heterogeneous cell populations. Biomed Opt Express. 2015;6:559–573.
[44] Walsh AJ, Cook RS, Sanders ME, et al. Quantitative optical imaging of primary tumor organoid metabolism predicts drug response in breast cancer. Cancer Res. 2014;74: 5184–5194.
[45] Solaini G, Baracca A, Lenaz G, Sgarbi G. Hypoxia and mitochondrial oxidative metabolism. Biochim Biophys Acta. 2010;1797:1171–1177.
[46] Ying W. NAD+/NADH and NADP+/NADPH in cellular functions and cell death: regulation and biological consequences. Antioxid Redox Signal. 2008;10:179–206.
[47] Segal AW. The function of the NADPH oxidase of phagocytes and its relationship to other NOXs in plants, invertebrates, and mammals. Int J Biochem Cell Biol. 2008;40:604–618.
[48] Rydström J. Mitochondrial NADPH, transhydrogenase and disease. Biochim Biophys Acta. 2006;1757:721–726.
[49] De Ruyck J, Famerée M, Wouters J, Perpète EA, Preat J, Jacquemin D. Towards the understanding of the absorption spectra of NAD(P)H/NAD(P)+ as a common indicator of dehydrogenase enzymatic activity. Chem Phys Lett. 2007;450:119–122.
[50] Niesner R, Narang P, Spiecker H, Andresen V, Gericke KH, Gunzer M. Selective detection of NADPH oxidase in polymorphonuclear cells by means of NAD(P)H-based fluorescence lifetime imaging. J Biophys. 2008,doi:10.1155/2008/602639.
[51] Brahimi-Horn MC, Pouyssegur J. Oxygen, a source of life and stress. FEBS Lett. 2007;581:3582–3591.
[52] Mayevsky A. Mitochondrial function in vivo evaluated by NADH fluorescence. Cham, CH: Springer International Publishing; 2015.
[53] Solaini G, Baracca A, Lenaz G, Sgarbi G. Hypoxia and mitochondrial oxidative metabolism. Biochim Biophys Acta. 2010;1797:1171–1177.
[54] Buerk DG. Measuring tissue PO2 with microelectrodes. Methods Enzymol. 2004;381:665–690.
[55] Braun RD, Lanzen JL, Snyder SA, Dewhirst MW. Comparison of tumor and normal tissue oxygen tension measurements using OxyLite or microelectrodes in rodents. Am J Physiol Heart Circ Physiol. 2001;280:H2533– H2544.
[56] Bobko AA, Dhimitruka I, Eubank TD, Marsh CB, Zweier JL, Khramtsov VV. Trityl-based EPR probe with enhanced sensitivity to oxygen. Free Radic Biol Med. 2009;47:654–658.
[57] Williams BB, Khan N, Zaki B, Hartford A, Ernstoff MS, Swartz HM. Clinical electron paramagnetic resonance (EPR) oximetry using India ink. Adv Exp Med Biol. 2010;662:149–156.
[58] Arteel GE, Thurman RG, Yates JM, Raleigh JA. Evidence that hypoxia markers detect oxygen gradients in liver: pimonidazole and retrograde perfusion of rat liver. Br J Cancer. 1995;72:889–895.
[59] Varia MA, Calkins-Adams DP, Rinker LH, et al. Pimonidazole: a novel hypoxia marker for complementary study of tumor hypoxia and cell proliferation in cervical carcinoma. Gynecol Oncol. 1998;71:270–277.

[60] Torres Filho IP, Terner J, Pittman RN, Somera LG 3rd, Ward KR. Hemoglobin oxygen saturation measurements using resonance Raman intravital microscopy. Am J Physiol Heart Circ Physiol. 2005;289:H488–H495.

[61] Okura I. Photosensitization of Porphyrins and Phthalocyanines. Amsterdam, NL, Gordon & Breach, 2001.

[62] Kurokawa H, Ito H, Inoue M, et al. High resolution imaging of intracellular oxygen concentration by phosphorescence lifetime. Sci Rep. 2015;5:1–13.

[63] Lebedev AY, Cheprakov AV, Sakadzić S, Boas DA, Wilson DF, Vinogradov SA. Dendritic phosphorescent probes for oxygen imaging in biological systems. ACS Appl Mater Interfaces. 2009;1:1292–1304.

[64] Sakadzić S, Roussakis E, Yaseen MA, et al. Two-photon high-resolution measurement of partial pressure of oxygen in cerebral vasculature and tissue. Nat Methods. 2010;7:755–759.

[65] Hosny NA, Lee DA, Knight MM. Single photon counting fluorescence lifetime detection of pericellular oxygen concentrations. J Biomed Opt. 2012;17:016007.

[66] Coogan MP, Court JB, Gray VL, et al. Probing intracellular oxygen by quenched phosphorescence lifetimes of nanoparticles containing polyacrylamide-embedded [Ru(dpp(SO3Na)2)3]Cl2. Photochem Photobiol Sci. 2010;9:103–109.

[67] Papkovsky DB, Zhdanov AV, Fercher A, Dmitriev RI, Hynes J. Phosphorescent oxygen-sensitive probes. Basel, CH: Springer-Verlag; 2012.

[68] Papkovsky DB, Dmitriev RI. Biological detection by optical oxygen sensing. Chem Soc Rev. 2013;42:8700–8732.

[69] Dmitriev RI, Zhdanov AV, Ponomarev GV, Yashunski DV, Papkovsky DB. Intracellular oxygen-sensitive phosphorescent probes based on cell-penetrating peptides. Anal Biochem. 2010;398:24–33.

[70] Dmitriev RI, Borisov SM, Kondrashina AV, et al. Imaging oxygen in neural cell and tissue models by means of anionic cell-permeable phosphorescent nanoparticles. Cell Mol Life Sci. 2015;72:367–381.

[71] Golub AS, Popel AS, Zheng L, Pittman RN. Analysis of phosphorescence in heterogeneous systems using distributions of quencher concentration. Biophys J. 1997;73:452–465.

[72] Ferchner A, Borisov SM, Zhdanov AV, Klimant I, Papkovsky DB. Intracellular O2 sensing probe based on cell-penetrating phosphorescent nanoparticles. ACS Nano. 2011;5:5499–5508.

[73] Wang XD, Gorris HH, Stolwijk JA, et al. Self-referenced RGB colour imaging of intracellular oxygen. Chem Sci. 2011;2:901–906.

[74] Kalinina S, Bisinger D, Breymayer J, Rück A. Cell metabolism, FLIM and PLIM and applications. Proc SPIE 9329; 2015. doi: 10.1117/12.2079166.

[75] Lo KK, Lee TK, Lau JS, Poon WL, Cheng SH. Luminescent biological probes derived from ruthenium(II) estradiol polypyridine complexes. Inorg Chem. 2008;47:200–208.

[76] Mik EG, Stap J, Sinaasappel M, et al. Mitochondrial PO2 measured by delayed fluorescence of endogenous protoporphyrin IX. Nat Methods. 2006;3:939–945.

[77] Mik EG, Johannes T, Zuurbier CJ, et al. In vivo mitochondrial oxygen tension measured by a delayed fluorescence lifetime technique. Biophys J. 2008;95:3977–3990.

[78] Becker W, Bergmann A, Hink MA, König K, Benndorf K, Biskup C. Fluorescence lifetime imaging by time-correlated single-photon counting. Microsc Res Tech. 2004;63:58–66.

[79] Becker W, Su B, Bergmann A, Weisshart K, Holub O. Simultaneous fluorescence and phosphorescence lifetime imaging. Proc SPIE 7903; 2011. doi:10.1117/12.875204.

[80] Gerritsen HC, Asselbergs MA, Agronskaia AV, Van Sark WG. Fluorescence lifetime imaging in scanning microscopes: acquisition speed, photon economy and lifetime resolution. J Microsc. 2002;206:218–224.

[81] Becker W, Su B, Weisshart K, Holub O. FLIM and FCS detection in laser-scanning microscopes: increased efficiency by GaAsP hybrid detectors. Microsc Res Tech. 2011;74:804–811.
[82] Becker W. The bh TCSPC Handbook. 6th ed. Berlin: Becker & Hickl GmbH; 2015.
[83] Kalinina S, Breymayer J, Schäfer P, et al. Correlative NAD(P)H-FLIM and oxygen sensing-PLIM for metabolic mapping. J Biophotonics. 2016. doi: 10.1002.
[84] Kalinina S, Rück A. FLIM and PLIM in biomedical research – An innovative way to combine autofluorescence and oxygen measurements. Photon Laser Med. 2016;5:257–266.

Karsten König

9 Laser tweezers are sources of two-photon effects

Abstract: Laser tweezers or optical traps are established laser tools for optical non-contact manipulation of micron/submicron sized objects in liquids such as non-adherent biological cells in medium. Typical laser traps are based on optical gradient forces generated with high numerical aperture near-infrared (NIR) continuous wave (cw) laser microscopes. The laser-cell interaction is determined by a change of the momentum due to the beam direction being altered by refraction. In order to avoid laser absorption, NIR cw lasers such as the Nd:YAG laser at 1064 nm, the frequency doubled erbium:YAG fiber laser at 760 nm, the tunable cw Ti:sapphire ring laser, and laser diodes at wavelengths < 800 nm are employed. They are considered to be safe tweezer sources. However, two-photon absorption effects may occur due to the generation of high MW/cm^2 laser intensities when using tightly focused cw laser beams at a power of 100 mW or more. These nonlinear effects can be used for two-photon excited fluorescence spectroscopy of trapped objects. However, when using low-wavelength NIR (< 800 nm) traps, potential "UV-like" photodamaging effects have to be considered during cell manipulation. The use of the Nd:YAG laser at 1064 nm is recommended when using laser tweezers for optical sperm transport for laser-assisted in vitro fertilization (IVF).

9.1 Introduction

Light exerts pressure that can be used to accelerate or buoy nanometer-sized and micrometer-sized particles against gravity. Furthermore, a tightly focused laser beam can pull transparent particles into the focal volume of a high NA objective.

If the particle is greater than the wavelength (Mie particle), the trapping phenomenon can be explained by refraction. The laser beam transmits the object and leaves it in another direction, thereby generating a momentum change and subsequently a force. The net force results in the positioning of the particle within the focal volume (Fig. 9.1).

Arthur Ashkin pioneered laser manipulation of micrometer-sized objects, such as microspheres and bacteria, based on optically-induced forces in the piconewton range [1]. His first publication on a stable laser trap based on two opposite beams holding a single microsphere dates back to January 1970. In his abstract he wrote:

> Micron-sized particles have been accelerated and trapped in stable optical potential wells using only the force of radiation pressure from a continuous laser. It is hypothesized that similar accelerations and trapping are possible with atoms and molecules using laser light tuned to specific optical transitions. The implications for isotope separation and other applications of physical interest are discussed [2].

https://doi.org/10.1515/9783110429985-011

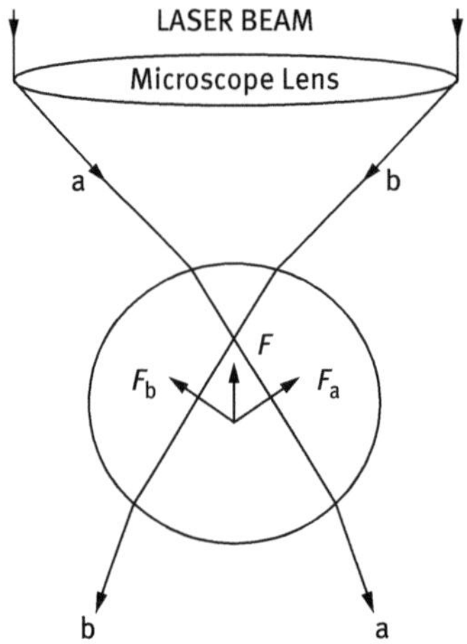

Fig. 9.1: Principle of optical trapping using ray optics.

In 1986, he published with his new co-worker Steven Chu and others on optical trapping and cooling of atoms [3, 4]. They used a single laser beam instead of two beams and realized the first stable 3D atom trap. Later on, Chu further developed the method of optical manipulation of atoms [5]. In 1997, Steven Chu, Claude Cohen-Tanneudji, and William D. Philip were awarded with the Nobel Prize for their work on the accuracy of atomic clocks and acceleration measurements based on trapped cooled atoms.

Starting from 1987, Ashkin used the single beam optical trap to manipulate biological objects, such as viruses, bacteria, and intracellular organelles [6]. Interestingly, the group faced problems when using the green argon ion laser as trapping source. A laser power of 100 mW damaged trapped bacteria, likely due to absorption and subsequent heating effects. This optical damage limits the use of the green laser trap in biomedical sciences. The problem could be solved when switching to a NIR laser source [7].

Biological objects do not significantly absorb red and NIR light up to 1200 nm. There are some skin cells and erythrocytes that contain the weakly NIR-light-absorbing pigments melanin and hemoglobin, respectively. However, pigment-free biological cells appear nearly NIR transparent. Also, scattering in NIR is negligible compared to UV or visible light. Therefore, the red and NIR spectral range between 600 nm and 1200 nm is termed the “optical window of cells and tissues”. The heating effect in NIR trapped living pigment-free cells such as CHO (Chinese hamster ovary) cells and sperm cells was measured to be less than 2 K at 100 mW laser power [8].

When Ashkin and his coworkers employed an Nd:YAG laser at 1064 nm, bacteria and yeast cells survived the trap for some hours when using a mean power of 80 mW and started to reproduce even during confinement in the laser trap. Also, red blood cells could be trapped without damaging the membrane. However, a significant lower laser power was required [7].

Later on, laser traps (laser tweezers) became a valuable tool in cell biology. Applications include the determination of motor protein forces, DNA molecule stretching, cell sorting, and spectroscopy of motile cells such as algae and sperm [9–20].

The motor proteins kinesin, dynein, and myosin are responsible for the active molecule transport in cells. ATP consumption results in conformation changes that are used to move binding partners along the cytoskeleton. Laser tweezers were used to understand these chemomechanical processes on a single molecule level, to determine the exact pN forces exerted by the motor proteins for organelle transportation within living cells and to calculate the transportation velocity [9–12]. The DNA molecule became a further biological object of interest for laser trapping. By attaching microbeads to a single DNA molecule, the DNA could be stretched and its elasticity determined [13, 20, 21]. Cell sorting by the use of laser tweezers, a microfluidic chip, and a target cell recognition system has been realized, for example, to sort human embryonic stem cells [26].

Laser tweezers are also an interesting tool to study highly motile cells, such as certain bacteria, algae, and sperm. By keeping them in the trap, sensitive spectroscopic measurements, such as optical metabolic imaging by fluorescence lifetime imaging, can be performed [15, 23]. Furthermore, the motility forces can be determined. We performed measurements and calculations to determine the force of a healthy human sperm cell to be 44 pN [18]. Furthermore, our group has demonstrated that the motility force did not decrease significantly after freezing and thawing. The frozen sperm of a healthy donor will still be a "high quality sperm".

Since NIR laser tweezers were considered to be perfectly safe manipulation tools, the clinical application of optical traps was considered. Trapping of sperm is of special interest for clinicians working in the field of male infertility treatment. Conventional IVF (in vitro fertilization) techniques rely on mechanical intracytoplasmic sperm microinjection (ICSI) directly into the egg cell (oocyte) and subzonal insemination (SUZI) by less invasive injection of sperm between the egg cell membrane and the outer zona pellucida layer. Yona Tadir and other IVF experts [27, 28] came up with the idea to use laser tweezers as a non-contact optical tool to transport single sperm cells of low motility directly to the egg cell. Furthermore, it was suggested to use an additional pulsed laser to drill a hole through the zona pellucida (LZD: laser zona drilling) and the oocytes membrane as well as to support hatching (LAH: laser assisted hatching) by zona opening of the embryo. Furthermore, the microsurgery laser could be used to cut off the flagellum from the sperm in order to maneuver the optically-trapped sperm head more easily (Fig. 9.2).

In order to realize laser-assisted IVF with cw laser tweezers and pulsed lasers for cell surgery, we performed laser safety studies on trapped human sperm using a tunable cw Ti:sapphire ring laser. We were astonished to find that trapping sperm with 760 nm laser tweezers for one minute resulted in a stop of flagellar motion and finally cell death. That means, cw NIR laser microbeams used as laser tweezers may cause cell damage. The reason is nonlinear absorption based on non-resonant absorption

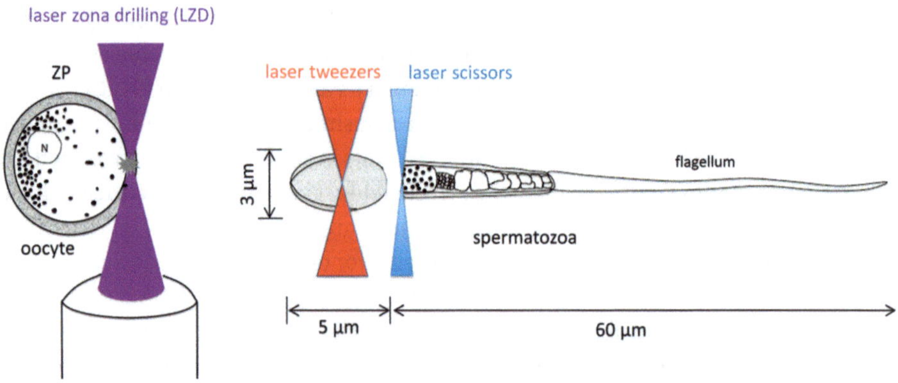

Fig. 9.2: Laser-assisted in vitro fertilization. "Bad quality" sperm of low motility can be transported to and into the oocyte by laser tweezers. To assist the transport into the oocyte and the "fusion", the outer membrane of the egg cell, the zona pellucida, can be perforated by non-invasive intracellular surgery with a second pulsed laser beam (laser zona drilling). Furthermore, the flagellum can be cut off for easier optical trapping. Finally, laser-assisted hatching of the embryo can be performed by a further opening of the surrounding zona. N: nucleus, ZP: zona pellucida.

of two NIR trapping photons [14, 16, 17, 24]. The required high MW/cm^2 intensity is provided by the cw laser beam of 100 mW or more power when focusing with high NA objectives. This chapter describes the experimental setup and the methods leading to the discovery of laser tweezers as sources of two-photon effects and their possible damage potential.

9.2 Experimental setup

Optical trapping of sperm and laser surgery of oocytes combined with fluorescence imaging were performed with a modified inverted fluorescence microscope (Axiovert 135M, ZEISS, Germany) equipped with galvoscanners for fast beam scanning and a joystick controlled motorized stage for additional stage scanning. The trapping beam was provided by an argon-ion laser pumped cw Ti:sapphire ring laser (899-01, Coherent Inc., USA, beam waist diameter 0.6 mm, divergence 1.7 mrad). The microsurgery laser was a frequency-doubled (KDP) Q-switched Nd:YAG laser with 4–6 ns pulse length (Surelite 1, Continuum, USA). Both beams filled the back aperture of a NA 1.30 ZEISS Neofluoar objective (100×, 440480, WD: 0.24 mm, immersion oil 518C with refractive index $n = 1.51$). Power regulation was performed with a polarizing beam attenuator (Karl Lambrecht Corp., USA). A 100 W mercury lamp (HBO) provided 365 nm and 405 nm fluorescence excitation light. The fluorescence was detected with a slow-scan cooled CCD camera (ZVS-47DEC, ZEISS, Germany) (Fig. 9.3). In addition, the microscope was equipped with a halogen lamp as white light source for brightfield microscopy. Further information can be obtained from [24].

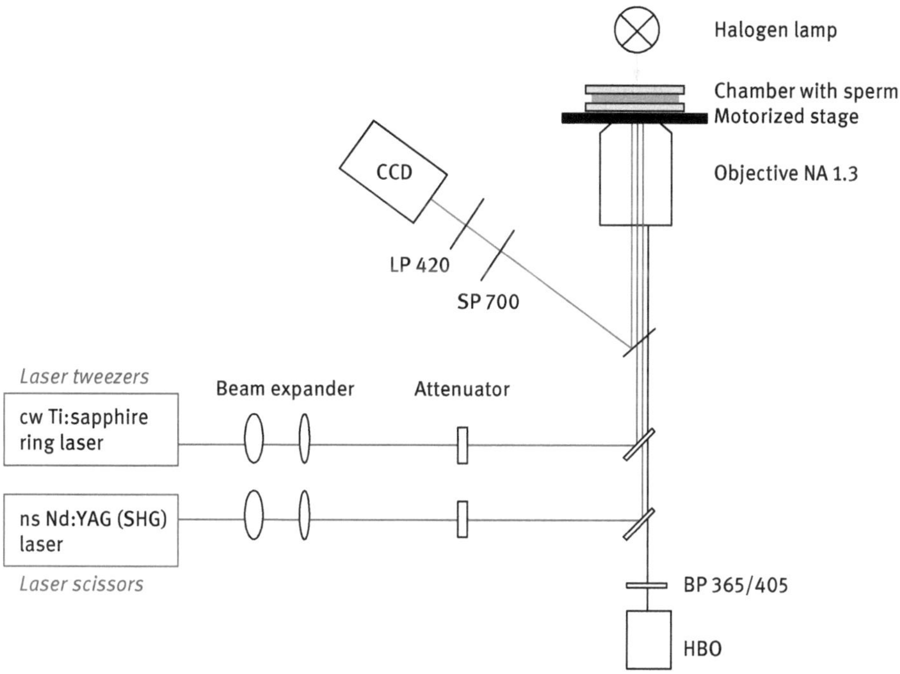

Fig. 9.3: Laser tweezers and laser scissors were realized with a cw Ti:sapphire ring laser and an ns frequency doubled Nd:YAG laser. Both lasers were coupled to a microscope equipped with high NA objectives, a motorized specimen stage and a cooled sensitive CCD camera as detection system.

To determine the optical forces generated by the optical trap on the confined specimen, accurate *in situ* power measurements are required. Often measurements are performed in air with a power meter equipped with a limited sensor size after laser transmission through the objective. However, this does not allow precise measurements of highly convergent beams as in the case of high NA objectives. The measured values are, therefore, different from the real powers at the trapping sample (*in situ* power). In order to determine the correction factors, a sandwich system consisting of identical opposite ZEISS brightfield objectives and the microchamber (with a thin 0.5 mm spacer and filled with PBS) between, was employed. The objectives were aligned in such a way that both had the same focal spot within the chamber resulting in incoming and leaving parallel beams. In the case of nonexpanded laser beams, a total system transmission of 33.9 % was measured (at 800 nm). That means, that each set consisting of the objective, the drop of oil, the 0.16 mm thick coverslip, and a 0.25 mm PBS layer, has a transmittance of about $(0.339)^{0.5} = 0.58$. Further measurements revealed a correction factor of 1.5 for the particular power meter [18].

9.3 Materials and methods

Human spermatozoa of three donors with normal semen parameters were obtained according to the World Health Organization guidelines. Semen was layered on a discontinuous isotonic Percoll gradient (Pharmacia, Sweden) and centrifuged for 15 min at 200 g. The bottom layer was washed with HEPES buffered fresh human tubal fluid (HTF, Irvine Scientific, USA) and centrifuged again for 10 min at 100 g. Finally, the pellet containing the sperm was diluted in HEPES buffered isotonic saline solution containing 1 % human serum albumin (HSA). The sperm was injected into microchambers consisting of two 0.16 mm thick glass windows separated by a typical 3 mm thick silicon layer as spacer (calibration experiments: 0.5 mm thick spacer). The injection was performed through the silicon layer. Laser trapping was performed with the microchamber on the motorized stage through the glass window [24].

9.4 Determining the trapping force

The net trapping force F of the optical trap depends on the trapped object, the medium, and the quality of the trap (e.g., beam alignment) represented by the trapping parameter Q as well as on the laser power P and can be calculated as:

$$F = QP/c,$$

where c is the velocity of light in medium. The trapping parameter can be derived from the Stokes equation by calculating the drag force F_{drag} that is exerted when the trapped sample is moved with the maximum possible velocity v_{max} before dropping off from the trap:

$$F_{drag} = 6\pi\mu R_{eq}v_{max},$$

where μ and R_{eq} are the viscosity of the medium and the length parameter of the sample, respectively. Laser tweezers interact with the sperm head and not with the midpiece nor flagellum. In order to use the Stokes formula also for the ellipsoidal shaped sperm head ($3 \times 5\,\mu m^2$), we developed a hydrodynamic model and calculated the length parameter R_{eq} of human sperm. Then, we excised the 60 µm long flagellum of 10 sperm samples with the 532 nm surgery microbeam and trapped the sperm heads (Fig. 9.3). By changing the stage speed we determined the drop-off velocity v_{max}. Based on these data, we could finally determine a trapping parameter $Q = 0.12$ for human spermatozoa in our NIR Ti:sapphire trap for an 800 nm laser wavelength [18].

9.5 Determining the motility force

In a next step, 100 healthy spermatozoa were trapped at 800 nm. We started with a laser power of 150 mW and slowly reduced the power. The drop-off power was determined for each individual sperm. The most motile sperm required a minimum power of 142 mW to confine it in the trap. The lowest power value for a motile sperm was found to be 29 mW.

With a mean drop-off power of (82 ± 38) mW, the calculated trapping parameter $Q = 0.12$, and the assumption of a nearly linear swim motion, the typical motility force of healthy human sperm can be determined to be (44 ± 22) pN.

When switching the laser wavelength from 800 nm down to 760 nm, we noticed an interesting effect on trapped spermatozoa. The motility force dropped within one minute trapping time. The trapped spermatozoa finally became paralyzed. A minimum (drop-off) trapping power as low as 0.3 mW was determined for immobile sperm [18].

9.6 Fluorescence imaging of trapped spermatozoa

The trapped cells can be imaged in the visible range above 420 nm with the cooled slow-scan CCD camera. Autofluorescence images were taken with UV exposure at 365 nm using the mercury lamp. As seen in Fig. 9.4, spermatozoa consist of

(i) a head that contains the non-fluorescent DNA,
(ii) a midpiece that contains mitochondria with autofluorescent coenzymes, and
(iii) the flagellum for motion.

Long UVA exposure paralyzed the sperm and sometimes a weak autofluorescence on the front part of the sperm head was monitored.

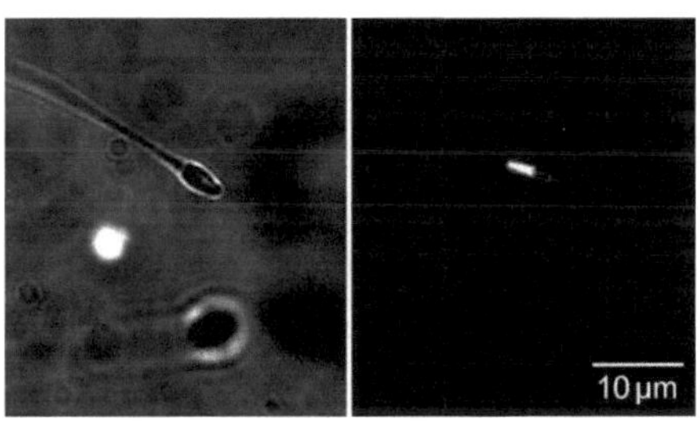

Fig. 9.4: White-field image and autofluorescence image of a single trapped sperm [35].

In order to study possible phototoxic effects of NIR trapping beams, a Live/Dead Assay based on intracellular fluorescence was employed. The sperm heads were stained with the green-fluorescent live cell indicator SYBR14 and the red fluorescent dead cell indicator propidium iodide (Molecular Probes, USA). The fluorescence was excited with the mercury lamp for short ms exposure times and detected with the cooled CCD camera. When using 100 mW laser traps at 800 nm or higher wavelength, the spermatozoa remained alive up to 10 minutes in the optical trap. However, when using a shorter wavelength, such as 760 nm, the sperm died as monitored by color change of the fluorescent sperm head from green to red [24].

9.7 Trap-induced two-photon fluorescence

Surprisingly, when switching off the mercury lamp the sensitive cooled CCD camera could still detect a tiny fluorescent spot in the labeled sperm head. The position was identically with the intracellular trapping beam position [14].

In order to prove whether the fluorescence was caused by a two-photon effect, the power was enhanced. Indeed, the fluorescence intensity increased nonlinearly with a P^2 dependence as indicated in Fig. 9.5. This function is typical for two-photon effects.

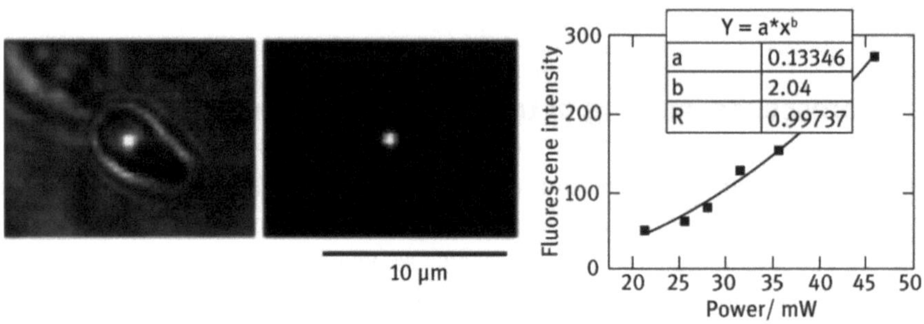

Fig. 9.5: Trap-induced two-photon excited fluorescence in the sperm head. The fluorescence intensity depends on the laser power squared [35].

Trapping beams at 760 nm (2.62×10^{-19} J photon energy) can be focused down to a spot size with a diameter of $0.61\lambda/\mathrm{NA} = 357$ nm and an exposure area of 0.1 µm², respectively, when using an NA = 1.3 objective. This results in a laser intensity of 100 MW/cm² at 100 mW laser power and a photon flux density of more than 4×10^{26} photons $\mathrm{cm}^{-2}\,\mathrm{s}^{-1}$. This high concentration pf photons in space and time is sufficient to induce nonlinear effects, in particular to excite fluorophores with a high fluorescence quantum yield via non-resonant two-photon absorption.

9.8 Trap-induced nonlinear phototoxic effects

That finally means that cw laser tweezers are sources of two-photon effects. Two-photon absorption may result in the generation of heat, fluorescence, and photochemical reactions as indicated in Fig. 9.6. In principle, 760 nm traps can induce similar photodamage effects as 760 nm/2 = 380 nm UV effects. UV light at 380 nm may induce reactive oxygen species (ROS) that can kill a cell.

We determined the phototoxic effect of the cw Ti:sapphire ring laser trap (100 mW) in dependency on wavelength. No phototoxic effects could be found for trapping wavelengths of 800 nm or higher. Cells trapped at 780 nm could be held alive for 8 min. The most damaging effect was found for 760 nm traps. In that case, sperm survived only one minute ((65 ± 20) s).

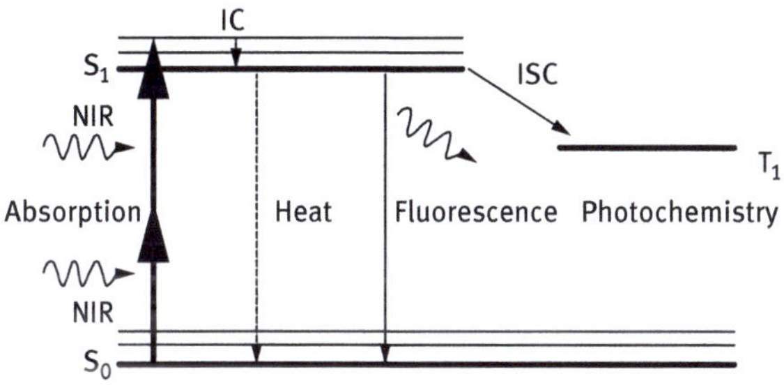

Fig. 9.6: Two-photon absorption of NIR trap photons may induce heat, fluorescence, and destructive photochemical reactions in the trapped specimen. IC: internal conversion, ISC: intersystem crossing, S: singlet state, T: triplet state.

A damage parameter (DP) was defined as the reciprocal value of the mean time when red fluorescence appeared, indicating cell death [24]. The DP values as a function of the trapping wavelength are depicted in Fig. 9.7.

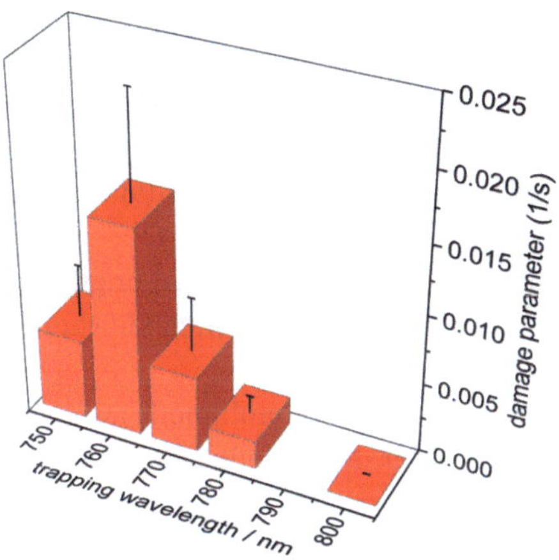

Fig. 9.7: Trap-induced killing of human sperm using a tunable cw Ti:sapphire ring laser [24].

9.9 Photodamage effects as a result of mode-beating phenomena

Under normal conditions, the cw Ti:sapphire ring laser operates in the multimode regime, that means, a variety of longitudinal modes is present. Superposition of these modes can cause mode-beating effects resulting in power fluctuations.

In our cw laser trap we measured such power fluctuations based on the occurrence of unstable pulses with a repetition frequency f as a multiple of the 180 MHz base frequency. This base frequency corresponded to the cavity length L of 1.70 m according to the formula $f = c/L$ with c as the velocity of light.

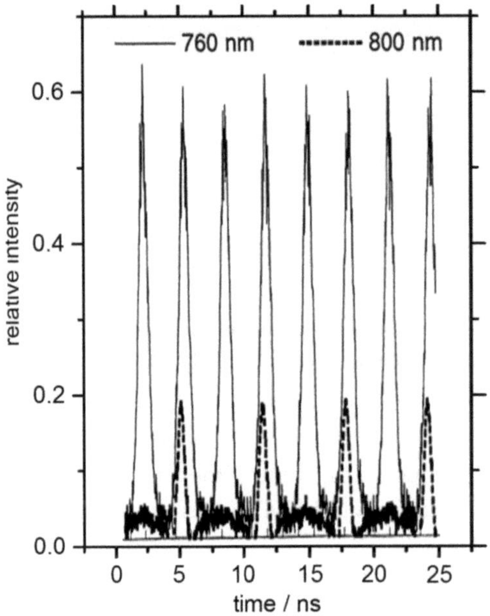

Fig. 9.8: Unstable laser pulses may occur in multimode cw laser due to mode-beating phenomena. The output of the Ti:sapphire ring laser was recorded with a 1 GHz detector [17].

Interestingly, the strongest pulses were found when the tuning birefringent filter with 20 GHz linewidth was optimized for the 760 nm output. Fig. 9.8 shows unstable subnanosecond laser pulses at 760 nm and 800 nm laser output. The amplitudes of the 800 nm pulses are about 3 times lower than these for the 760 nm pulses [17].

In order to figure out if the enhanced damaging effect is of biological origin or a result of the enhanced mode-beating effects in the multimode cw laser, modifications on the laser resonator were performed.

By introducing a special etalon into the laser resonator, our multimode laser was transformed into a "perfect" cw single frequency laser. Now mode-beating effects could be avoided and the power output became stable without any occurrence of transient laser pulses. As expected, trapped human sperm now survived longer in the trap. However, photo-killing effects could not be avoided. In fact, trap-induced cell death was now observed after (406 ± 160) s in single-frequency cw traps instead of (65 ± 20) s in multimode cw traps [17].

9.10 Conclusion

Laser traps have become important optical micromanipulation tools for single cell handling and pN force transducers [29]. Typically, tightly focused Nd:YAG laser at 1064 nm, frequency doubled erbium:YAG laser fibers at 760 nm, tunable cw Ti:sapphire ring lasers, and laser diodes are employed as cw NIR trapping sources. Interestingly, these intense NIR laser beams can induce two-photon effects such as two-photon excited fluorescence and two-photon induced "UV-like" photochemical destructive effects. Therefore, NIR traps can induce cell damage including cell death. Trapping wavelengths higher than 800 nm are safer than short-wavelength NIR. The use of Nd:YAG lasers at 1064 nm is recommended. Also, single-frequency cw lasers are safer than multimode cw lasers due to possible mode-beating effects. This should be considered in particular for clinical applications, such as laser-assisted IVF. Meanwhile, the first "laser babies" have been born. To my knowledge, the first reported laser-assisted IVF with human gametes and successful birth occurred in Germany [30]. A first larger LAZ study on 179 patients was performed in Italy by Antinori et al. [31]. Today, laser-assisted hatching is used in many IVF clinics worldwide [32]. So far, laser tweezers for sperm transportation as well as to test the quality of sperm, in particular of the individual sperm to be used for IVF procedures, are not yet used clinically.

Trap-induced fluorescence can be employed as a novel tool for nonlinear cell diagnostics. In general, two-photon microscopy including two-photon lithography can also be performed with low-cost continuous wave (cw) laser sources [33, 34]. However, the use of femtosecond laser sources is recommended in two-photon microscopes due to the far more efficient process when using transient kW laser peak powers and low sub-mW and mW mean femtosecond laser powers.

References

[1] Ashkin A. Forces of a single-beam gradient laser trap on a dielectric sphere in the ray optics regime. Biophys J. 1992;61:569–582.
[2] Ashkin A. Acceleration and trapping of particles by radiation pressure. Phys Rev Lett. 1970;24:156.
[3] Ashkin A, Dziedzic JM, Bjorkholm JE, Chu S. Observation of a single-beam gradient force optical trap for dielectric particles. Opt Lett. 1986;11:288.
[4] Chu S, Bjorkholm JE, Ashkin A, Cable A. Experimental observation of optically trapped atoms. Phys Rev Lett. 1986;57:314.
[5] Chu S. Laser manipulation of atoms and particles. Science. 1991;253:861–866.
[6] Ashkin A, Dziedzic JM. Optical tapping and manipulation of viruses and bacteria. Science. 1987;235:1517–1520.
[7] Ashkin A, Dziedzic JM, Yamane T. Optical trapping and manipulation of single cells using infrared laser beams. Nature. 1987;330:769–71.
[8] Liu Y, Cheng DK, Sonek GJ, Berns MW, Chapman CF, Tromberg BJ. Evidence for localized cell heating induced by near infrared optical tweezers. Biophys J. 1995;68:2137–2144.
[9] Ashkin A, Schütze K, Dziedzic JM, Eutenauer U, Schliwa M. Force generation of organelle transport measured in vivo by an infrared laser trap. Nature. 1990;348:346–348.
[10] Block SM, Goldstin LRS, Schnapp BJ. Bead movement by single kinesin molecules studied with optical tweezers. Nature. 1990;348:348–352.
[11] Wright WH, Sonek GJ, Tadir Y, Berns MW. Laser trapping in cell biology. IEEE J. Quant. Electronics. 1990;26:2148–2157.
[12] Block SM. Making light work with optical tweezers. Nature. 1992;360:493–495.

[13] Perkins TT, Quake SR, Smith DE, Chu S. Relaxation of a single DNA molecule observed by optical microscopy. Science. 1994;264:822–826.

[14] König K, Liang H, Berns MW, Tromberg BJ. Cell damage by near-IR microbeams. Nature. 1995;377:20–21.

[15] König K, Liu Y, Sonek GJ, Berns MW, Tromberg BJ. Autofluorescence spectroscopy of optically-trapped cells during light stress. Photochem Photobiol. 1995;62:830–835.

[16] Liu Y, Sonek GJ, Berns MW, König K, Tromberg BJ: Two-Photon fluorescence excitation in continuous-wave infrared optical tweezers. Optics Letters. 1995;20:2246–2248.

[17] König K, Liang H, Berns MW, Tromberg BJ. Cell damage in near infrared multimode optical traps due to multi-photon absorption. Optics Letters. 1996;21:1090–1092.

[18] König K, Svaasand L, Liu Y, Sonek GJ, Patrizio P, Tadir J, Berns MW, Tromberg BJ. Determination of motility forces of human spermatozoa using an 800 nm optical trap. Cell Mol Biol. 1996;42:501–509.

[19] König K, Tadir Y, Patrizio P, Berns MW, Tromberg BJ. Effects of Ultraviolet Exposure and Near Infrared Optical Traps on Spermatozoa. Human Reproduction. 1996;11:2162–2164.

[20] Strick TR, Allemand JF, Bensimon D, Bensimon A, Croquette V. The elasticity of single supercoiled DNA molecule. Science. 1996;271:1835–1837.

[21] Quake SR, Babcock H, Chu S. The dynamics of partially extended single moleucle of DNA. Nature. 1997;388:151–154.

[22] Schnitzer MJ, Block SM. Kinesin hydrolysis one ATP per 8 nm step. Nature. 1997;389:387–390.

[23] König K, Boehme S, Leclerc N, Ahuja R. Time-gated autofluorescence microscopy of motile green microalga in an optical trap. Cell Mol Biol. 1998;44, 763–770.

[24] König K. Laser tweezers are sources of two-photon excitation. Cell Mol Biol. 1998;44:721–734.

[25] König K. Robert Feulgen Prize Lecture 2000. Laser Tweezers and Multiphoton Microscopes in Life Sciences. Histochem Cell Biol. 2000;114:79–92.

[26] Wang X, Chen S, Kong M, Costa KD, Li R, Sun D. Enhanced cell sorting and maniulation with combined optical tweezer and microfluidic chip technologies. Lab on a Chip. 2011;11:3656–62.

[27] Tadir Y, Wright WH, Vafa O, Ord T, Asch RH, Berns MW. Micromanipulation of sperm by a laser-generated optical trap. Fertil Steril. 1989;52:870–873.

[28] Tadir Y, Neev J, Ho P, Berns MW. Lasers for gamete micromanipulation. Basic concepts. J Assist Reprod Genet. 1993;10:121–125.

[29] Sheetz MP, editor. Laser tweezers in cell biology. Methods in cell biology vol. 55. Academic Press; 1998.

[30] Wiedemann R, Montag M. Laser-assisted oocyte microinsemination and direct transuterine tubal transfer in male infertility. Poster. International symposium on male factor in human infertility; 21–22 April 1994; Paris.

[31] Antinori S, Selman HA, Caffa B, Panci C, Dani GL, Versaci C. Zona opening of human embryos using a non-contact UV laser for assited hatching in patients with poor prognosis of pregnancy. Human Reprod. 1996;11:2488–2492.

[32] Hammadeh ME, Fischer-Hammadeh C, Ali KR. Assisted hatching in assisted reproduction: a state of the art. J Assist Reprod Genet. 2011;28:119–128.

[33] Hänninen PE, Soini E, Hell SW. Continuous wave excitation in two-photon fluorescence microscopy. J Microsc. 1994;176:222–225.

[34] Thiel M, Fischer J, von Freymann G, Wegener M. Direct laser writing of three-dimensional submicron structures using a continuous-wave laser at 532 nm. Appl Phys Lett. 2010;97.

[35] König K. Biomedical applications by optical micromanipulation and two-photon excitation of living cells using near infrared laser microscopy [habilitation]. Jena: Shaker Verlag; 1999. ISBN: 3-8265-6788-9.

Marina Shirmanova, Tatiana Sergeeva, Irina Druzhkova,
Aleksandra Meleshina, Maria Lukina, Varvara Dudenkova,
Vladislav Shcheslavskiy, Wolfgang Becker, Vsevolod Belousov,
Nataliya Mishina, and Elena Zagaynova

10 Metabolic shifts in cell proliferation and differentiation

Abstract: Alteration of cellular energy metabolism is a principal feature of tumor and stem cells. Here we analyze the metabolic interactions between cancer cells and fibroblasts in a co-culture model and the metabolic heterogeneity of tumors and metabolic changes in mesenchymal stem cells during adipogenic differentiation based on the fluorescence of the metabolic cofactors NADH, NADPH, and FAD. We registered a metabolic switch from oxidative phosphorylation to glycolysis with slight acidification of the cytosol in cancer cells in a co-culture model. In the tumor tissue we detected metabolic heterogeneity with more glycolytic metabolism of cancer cells in the stroma-rich zones. The shift of cellular energy metabolism from glycolysis to oxidative phosphorylation and the activation of lipogenesis were observed in adipocytes. Data about metabolic alterations in cancer and stem cells are important for monitoring the progression of cancers, the development of anticancer drugs and stem cell therapy.

10.1 Introduction

Cell metabolism is defined as the sum of the chemical reactions taking place within each cell of a living organism and providing energy for vital processes. It is known that the main way to generate adenosine triphosphate (ATP) for providing living cells with energy is by oxidative phosphorylation (OXPHOS). The metabolic cofactors oxidized flavin adenine dinucleotide (FAD) and reduced nicotinamide adenine dinucleotide (NAD(P)H) are the primary electron acceptor and donor, respectively, in this process. Since these cofactors have a capability for fluorescence, and the fluorescence lifetimes differ for different states, the free or protein-bound, fluorescence intensity and lifetime measurements can be used to monitor the metabolic activity of the cells [1–5]. Many enzymes bind to NAD(P)H and FAD in the different metabolic pathways [6]. However, in cell proliferation and differentiation an active role in metabolism belongs to glycolysis. Rapidly proliferating cancer cells as well as stem cells tend toward glycolysis because it provides intermediates for the biosynthesis of macromolecules and a fast production of ATP to support their high growth rate [7–9]. Partial breakdown of glucose through glycolysis and the pentose phosphate pathway provides a compromise between the catabolic generation of ATP and reducing cofactors and production of biosynthetic substrates to meet the cells' anabolic requirements [10].

https://doi.org/10.1515/9783110429985-012

Multiphoton fluorescence microscopy and fluorescence lifetime imaging (FLIM) are powerful techniques for the non-invasive characterization and long-term monitoring of the functional changes that underlie cellular metabolism. The possibilities for investigating cell metabolism in normal and pathological conditions in vitro and in vivo using these methods have been demonstrated widely [11–14].

Here we analyze the metabolic interaction between cancer cells and fibroblasts in a co-culture model, the metabolic heterogeneity of tumors in vivo and metabolic changes in mesenchymal stem cells (MSCs) during adipogenic differentiation, based on the fluorescence of the metabolic cofactors NADH, NADPH, and FAD. Cellular metabolism was examined by monitoring the optical redox ratio (FAD/NAD(P)H), the fluorescence lifetime contributions of the free and bound forms of NADH and the bound form of NADPH. Two-photon fluorescence microscopy combined with FLIM was used to analyze this fluorescence in living cells.

10.2 Materials and methods

10.2.1 Cancer cells and tumor model

HeLa Kyoto, human cervical carcinoma, cells and human skin fibroblasts (huFb) were used in the study. For intracellular pH (pHi) measurements a HeLa cell line stably expressing the cytoplasmic pHi-sensor SypHer2 (HeLa-SypHer2) was used. The huFb were obtained from the Koltzov Institute of Developmental Biology Russian Academy of Science (Moscow, Russia). Genetically transfected cell line was generated and characterized in the Institute of Bioorganic Chemistry RAS (Moscow, Russia).

The cells were cultured in DMEM containing 100 mg/ml penicillin, 100 mg/ml streptomycin sulfate and 10 % fetal bovine serum (FBS) at 37 °C in a humidified atmosphere with 5 % CO_2.

The protocol for the co-culturing of cancer cells and fibroblasts was modified from that of [15]. For co-culturing, the huFb were plated in glass-bottom FluoroDishes in complete DMEM media without phenol red (Life Technologies) and then HeLa Kyoto cells were seeded within 4 hours. The total number of cells per dish in the co-culture was 1×10^5 with a 1 : 5 fibroblast-to-cancer cell ratio. In parallel, monotypic cultures were plated at the same quantity as in the corresponding co-culture. The day of plating was defined as a day 0. The day after plating (day 1), the medium was changed to DMEM with 5 % FBS, and afterwards the medium was changed every other day. The cells were analyzed over a period of 5 days.

All animal protocols were approved by the Ethics Committee of Nizhny Novgorod State Medical Academy. Female athymic nude mice of 20–22 g body weight were inoculated subcutaneously in the left flank with HeLa cells (2×10^6 in 200 μl PBS). Imaging started 21 days after the cell injection, when the tumors had reached ≈ 10 mm in diameter. Before fluorescence imaging the mice were anesthetized intramuscularly with a mixture of Zoletil (40 mg/kg, Virbac SA, Carros, France) and 2 % Rometar (10 mg/kg, Spofa, Czech Republic), a skin flap over the tumor was surgically opened, and the objective was placed directly on the tumor surface. After image acquisition, the animals were sacrificed by cervical dislocation and the tumors were excised for histopathology.

10.2.2 Stem cells and adipogenic differentiation

All procedures were conducted according to N. K. Koltzov Institute of Developmental Biology Ethic Committee approval. Human bone marrow mesenchymal stem cells (MSCs) were isolated from bone marrow of healthy donors with informed consent according to the Institutional Guidelines under the approved protocol. The MSCs were isolated and characterized as described previously [16].

The MSCs were cultured in MesenCult™ MSC Basal Medium (Human) (STEMCELL TECHNOLOGIES, Canada) supplemented with 10 % fetal bovine serum (FBS) (Hyclone), 0.58 mg/ml L-glutamine (PanEco) and 40 U/ml gentamicin. The cell culture was maintained at 37 °C in a 5 % CO_2, humidified atmosphere.

Differentiation was induced by incubating the MSCs in MesenCult™ Adipogenic Differentiation Medium (Human) (STEMCELL TECHNOLOGIES, Canada). The medium was replaced every 3–4 days over the experimental period of up to 4 weeks.

For microscopic imaging, 4×10^5 cells were transferred into a sterile dish with a cover-glass bottom (0.17 mm thick) and incubated for one day until they attached to the glass surface. The cells were imaged before the induction of differentiation (day 0) and on days 5, 12, 19 and 26 subsequently. The cells were washed twice using phosphate-buffered saline, and then placed in FluoroBrite™ DMEM (Gibco) with 10 % FBS and 0.58 mg/ml L-glutamine (PanEco) and 40 U/ml gentamicin.

10.2.3 Two-photon fluorescence microscopy and FLIM

The two-photon exited fluorescence intensity and FLIM images of NAD(P)H and FAD of the cultured cells were obtained on an LSM 710 (Carl Zeiss, Germany) inverted laser scanning confocal microscope. For two-photon fluorescence microscopy and FLIM of NAD(P)H in tumors in vivo a multiphoton tomograph MPTflex (JenLab, Germany) was used. Both systems are equipped with time-correlated single-photon counting (TCSPC) modules (Becker & Hickl GmbH, Germany).

The images of cells in vitro were acquired through a 40×, 1.2 NA water immersion objective. During image acquisition, the cells were maintained at 37 °C and 5 % CO_2. NAD(P)H and FAD fluorescence was excited with a Chameleon Vision II (Coherent, USA) Ti:Sa femtosecond laser, using an 80 MHz repetition rate and a pulse duration of 140 fs at wavelengths of 750 nm and 900 nm, respectively. Emission was detected in the ranges 455–500 nm for NAD(P)H, and 500–550 nm for FAD.

The images in vivo were acquired through a 40×, 1.3 NA oil immersion objective. NAD(P)H fluorescence was excited at the wavelength of 740 nm using a tunable 80 MHz, 200 fs Ti:Sa laser MaiTai (Spectra Physics, USA) and detected in the range 410–660 nm. Second-harmonic generation (SHG) was exited at 750 nm and detected in the range 373–387 nm to identify collagen in the tumor tissue.

The optical redox ratio FAD/NAD(P)H was calculated from corresponding two-photon fluorescence images of FAD and NAD(P)H after subtracting the background on a pixel-by-pixel basis using ImageJ 1.39p software (NIH, USA).

The analysis of the FLIM data was performed using SPCImage software (Becker & Hickl GmbH, Germany). The lifetime decay curve was fitted to a double- or triple-exponential decay model in the case of NAD(P)H, and to a double-exponential decay model in the case of FAD. The goodness of the fit, the χ^2 value, was 1.2–0.1. An average of 5000–8000 photons were assessed per decay curve for the region of interest (cell cytoplasm).

10.3 Metabolic shifts in cancer

10.3.1 Metabolic interaction of cancer cells and fibroblasts

An important role in modifying cancer cell metabolism belongs to the tumor stroma, and, especially, to cancer-associated fibroblasts (CAFs). Multiple studies report on the CAF's ability to promote tumorigenesis and to regulate cancer cell motility and stemness through the secretion of a number of growth factors, hormones and cytokines [17]. Metabolic coupling between cancer cells and CAFs occupies a special role in these processes because metabolic reorganization underlies further changes leading to adaptation of the cancer cells. Interestingly, in some tumors CAFs undergo metabolic alterations and switch their metabolism to aerobic glycolysis when they interact with cancer cells, while the cancer cells themselves remain oxidative, the phenomenon being known as the "reverse Warburg effect" [18].

To estimate the overall cellular metabolic activity, measurements of the fluorescence intensity of FAD to NAD(P)H were performed in the co-culture of cancer cells and fibroblasts and in the corresponding monocultures. Cancer cells and fibroblasts in the co-culture were identified by their cellular morphology. Cancer cells are much smaller than fibroblasts, have a round shape and large nucleus with a thick cytoplasmic layer. The fibroblasts have a typically spindle-shaped morphology, much greater size and a larger volume of cytoplasm compared to the cancer cells.

Monitoring of the redox ratio FAD/NAD(P)H in the co-culture showed that in the cancer cells the redox ratio significantly decreased on day 5 of co-culturing, indicating an increase in metabolic activity (Fig. 10.1 (a)). Meanwhile, the fibroblasts showed a gradual increase of their redox ratio from day 1 to day 5, indicating a probable metabolic shift toward OXPHOS [19].

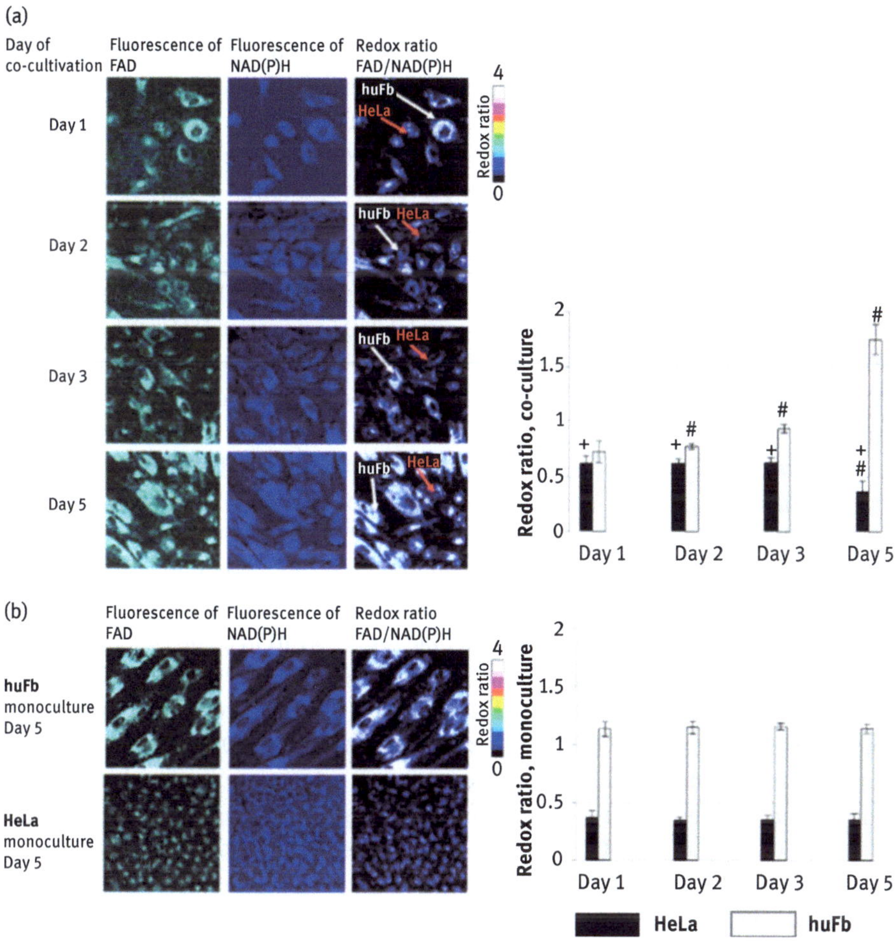

Fig. 10.1: Fluorescence and optical redox ratio images (FAD/NAD(P)H) of cancer cells and fibroblasts in co-culture. For NAD(P)H – excitation: 750 nm, detection: 455–500 nm. For FAD – excitation: 900 nm, detection: 500–550 nm. Image size is 213 μm × 213 μm (1024 × 1024 pixels). *M* ± *SD*. # statistically significant difference from day 1, $p \leq 0.005$. + statistically significant difference from fibroblasts on the same day of cultivation, $p \leq 0.05$.

In the monocultures the redox ratio in the fibroblasts was higher than in the cancer cells, which testifies to higher oxidative metabolism in the stromal cells, but did not change with time in either cell type (1.4 vs. 0.35, $p = 0.000000$) (Fig. 10.1 (b)) [19].

Our results are consistent with the data by Ostrander et al., who showed that the MCF-10A and HMEC cell lines of normal mammary epithelial cells had an increased redox ratio when compared to nine different breast cancer cell lines [20]. Martinez-Outschoorn et al. also detected increased mitochondrial activity in fibroblasts when compared with the MCF-7 cell line in monoculture [15]. Similar results were reported by de Andrade Natal et al. for biopsies of breast cancer patients [21]. Cancer cells, where aerobic glycolysis predominates, display a decreased redox ratio FAD/NAD(P)H, as has been demonstrated for different cancer cell lines and tumors [14, 20].

As expected for FLIM, the fluorescence decay curves for NAD(P)H and FAD were best fitted to a double-exponential decay model, indicating the presence of two distinctly different lifetimes for the free and protein-bound forms of the cofactors. The fluorescence lifetimes of the free (τ_1) and protein-bound (τ_2) NAD(P)H were measured to be ≈ 0.4 and ≈ 2.6 ns, respectively. For the free (τ_2) and protein-bound (τ_1) FAD fluorescence lifetimes were ≈ 2.7 and ≈ 0.4 ns, respectively. Fluorescence lifetimes documented in the literature lie in the range of 0.2–0.5 ns for free NAD(P)H, 1.8–2.7 ns for protein-bound NAD(P)H, 0.15–0.4 ns for protein-bound FAD and 2–2.8 ns for free FAD [5, 14, 22]. Therefore, the values recorded in our study agree well with singular data.

We have found no difference in the fluorescence lifetimes between cancer cells and fibroblasts of one cell type in the process of co-cultivation in vitro. However in work by Skala, increased fluorescence lifetimes of free and protein-bound NAD(P)H and free FAD, and decreased fluorescence lifetimes for protein-bound FAD in normal tissue compared to oral precancerous tissue were detected in vivo [23]. McGintly et al. found an increased mean fluorescence lifetime for NAD(P)H in colon cancer but a decreased lifetime for precancerous colon lesions compared to the normal colon [24].

The relative contribution of free NAD(P)H (a_1) in cancer cells co-cultured with fibroblasts increased from 76.9 % on day 1 to 79.45 % ($p = 0.000001$) on day 2, and remained at that increased level during whole period of co-culturing. The relative contribution of free FAD (a_2) in these cells increased on day 3 from 27.9 to 32.4 % ($p = 0.000000$), and then did not change. The observed changes in the relative contributions of free NAD(P)H and FAD testify to an increased bias toward a glycolytic metabolism. By contrast, for the fibroblasts in the co-culture, the relative contributions of free NAD(P)H and FAD gradually decreased starting from day 2, indicating a shift toward oxidative metabolism (Fig. 10.2). All cells in the population displayed the described changes.

In monocultures of cancer cells and fibroblasts the relative contributions of the cofactors were fairly stable throughout the 5 days of cultivation (≈ 76 % for free NAD(P)H and ≈ 30 % for free FAD) without any statistical difference for the NAD(P)H and only a slight difference for FAD ($p = 0.000011$).

Previously, in vivo, a decreased contribution of protein-bound NAD(P)H and FAD had been demonstrated for precancer development [24]. A decreased amount of NAD(P)H and a decreased free/bound NAD(P)H ratio were also shown in oral squamous carcinoma cells when compared with nonmalignant cells [13]. A glycolytic phenotype of cervical cancer had been demonstrated for cell cultures and human tumors. For example, Rossignol et al. reported that HeLa cells generate energy predominantly by glycolysis, but can change ATP generation to exclusively OXPHOS when the availability of glucose is limited [25]. It was found by Herst et al. that HeLa

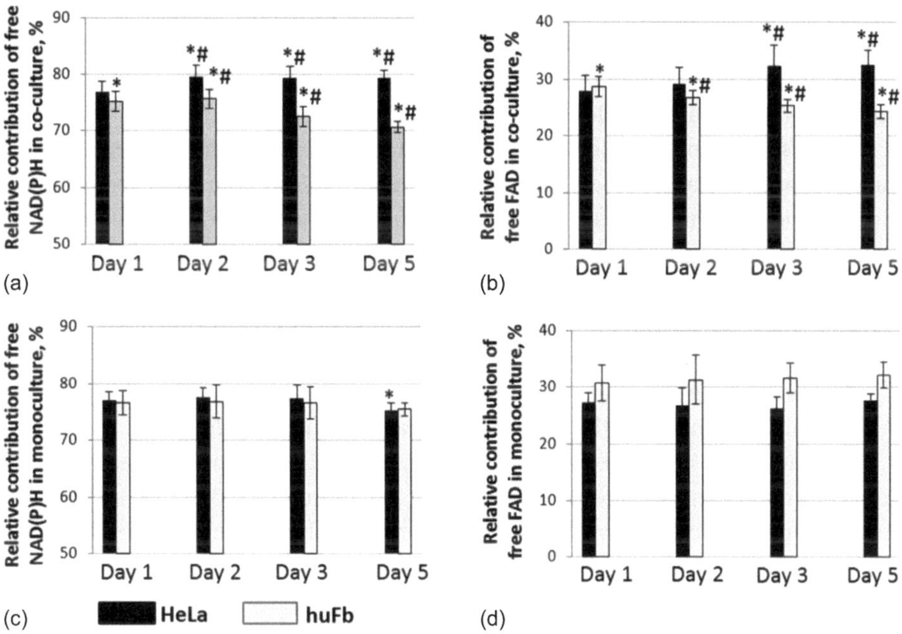

Fig. 10.2: Relative contributions of metabolic cofactors in HeLa cells and fibroblasts. Free NAD(P)H (a_1) in co-culture (a), free FAD (a_2) in co-culture (b), free NAD(P)H (a_1) in monoculture (c), free FAD (a_2) in monoculture (d). * statistically significant difference from the same day in monoculture, $p \leq 0.05$. # statistically significant difference from day 1, $p \leq 0.005$.

cells that rely preferentially on OXPHOS in highly oxygenated conditions became more glycolytic upon hypoxia [26]. Glycolytic metabolism of HeLa cells was also shown for tumor spheroids [27]. At the same time, fibroblasts retain a significant level of OXPHOS even if the glucose level is high [28, 29]. Nevertheless, there have been very few studies in this field performed on the co-culture model. For example, a higher glucose uptake and increased lactate production in co-cultures of MCF-7 and CAFs compared with monocultures were demonstrated by Brauer et al., but in that work the cancerous and the normal cells were not separated [30]. Martinez-Outschoorn et al. demonstrated the reversed Warburg effect (OXPHOS in cancer cells and glycolysis in fibroblasts) in co-cultures of MCF7 cancer cells and fibroblasts [15].

Therefore, the FLIM measurements of the relative contributions of protein-bound and free NAD(P)H and FAD in cancer cells and fibroblasts showed a switch of the HeLa cells toward a glycolytic phenotype and a switch of the huFb toward OXPHOS as a result of co-cultivation.

Although there are some papers demonstrating the dependence of the metabolic state on the stage of cell culture growth, in particular a decrease in the fluorescence lifetime of both free and protein-bound NADH and the contribution of protein bound

NADH at higher cell densities [31, 32], the observed metabolic changes in our study with co-culture were most probably not related with this. Control monocultures were grown under similar conditions and had the same density as co-culture, but did not show any fluctuations of redox ratios and fluorescence lifetimes.

Analysis pHi in cancer cells in a co-culture model was performed using a new ratiometric genetically-encoded pH sensor SypHer2. SypHer2 is an improved pH sensitive ratiometric indicator based on the cpYFP fluorophore. SypHer2 has two excitation peaks, at 420 nm and 500 nm, and one emission peak at 516 nm [33, 34]. The excitation peak at 420 nm decreases with pH proportionally to the increase in the peak at 500 nm, allowing ratiometric (dual excitation) imaging of intracellular pH in living cells.

In monoculture of cancer cells stably expressing the sensor, the fluorescence ratio I_{500}/I_{420} did not change with time, indicating a stable pHi (Fig. 10.3).

On day 1 of co-cultivation with fibroblasts, the SypHer2 ratio had already started to show a reduction compared to that in the monoculture ($p = 0.028$) because the pHi became more acidic. The lower SypHer2 ratio remained throughout further cultivation.

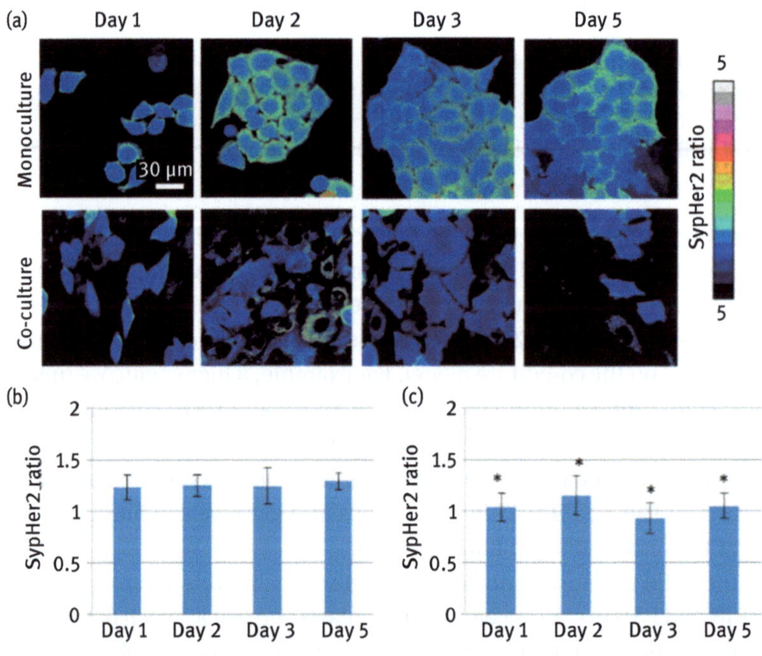

Fig. 10.3: Analysis of pHi in cancer cells in monoculture and in co-culture with fibroblasts using genetically encoded sensor SypHer2. Ratiometric images I_{500}/I_{420} (a), SypHer2 ratio in monoculture (b) and in co-culture (c) from days 1 to 5 of culturing. * statistically significant difference from monoculture on the same day, $p \leq 0.05$.

A minor acidification of the cytosol of cancer cells co-cultured with fibroblasts corresponds well with metabolic shift to glycolysis shown by NAD(P)H and FAD fluorescence imaging: decreased pHi is associated with increased lactic acid production in conditions of increased glycolysis. The combination of lactic acidosis and decreased cytosolic pH is a specific feature of cancer cells with high rates of glycolysis [35]. It is known that the accumulation of lactate during tumor development may affect the pHi and drive metabolic changes [36]. At the same time, for successful realization of glycolysis, an elevated pH level is required [37].

It should be noted that there is a great diversity of metabolic phenotypes among tumors, and besides the above mentioned "Warburg type" and "reverse Warburg type" there are tumors where both cancer cells and stroma are glycolytic or nonglycolytic [38]. Our HeLa-huFb system demonstrated metabolic behavior similar to Warburg type tumors, where cancer cells use increased glycolytic and stromal cells tend toward increased nonglycolytic metabolism.

10.3.2 Metabolic heterogeneity of tumors

The great degree of metabolic heterogeneity, both inter- and intratumoral, and the high plasticity of cancer cells significantly complicate the development of effective therapeutic strategies and result in nonuniform response to therapy. In general, the metabolism of cancer cells is more flexible compare to normal cells, which gives them the opportunity to survive in unfavorable conditions. Such factors as hypoxia, glucose deprivation, lactate accumulation and extracellular acidosis contribute a great deal to tumor progression through forming a specific microenvironment and thus they determine, to a great extent, the metabolic phenotype [39]. It has been demonstrated that some cancer cells can reversibly switch between fermentation and oxidative metabolism, depending on the absence or the presence of glucose and oxygen and other environmental conditions [40]. Therefore, cancer metabolism is neither as homogenous nor as reproducible as initially suspected. Rather, the metabolic activity of cancer cells is a complex, heterogeneous, and nuanced process that may be a key to successful treatment.

Fluorescence lifetimes of free and protein-bound forms of NAD(P)H and their contributions were measured in vivo in the HeLa tumors in nude mice [41]. Fluorescence intensity of FAD was very low in the tumor tissue, which did not allow fluorescence lifetime measuring of this cofactor.

Macroscopically, the tumors had a multinodular, fleshy appearance with blood vessels, plus yellowish and red-colored areas.

The fluorescence lifetimes of the free (τ_1) and protein-bound (τ_2) NAD(P)H measured in vivo in cancer cells were 0.47 ± 0.8 ns and 2.3 ± 0.3 ns, respectively.

The relative contributions of free (a_1) and protein-bound (a_2) NAD(P)H in cancer cells were different in different tumor sites (Fig. 10.4). In some areas the relative

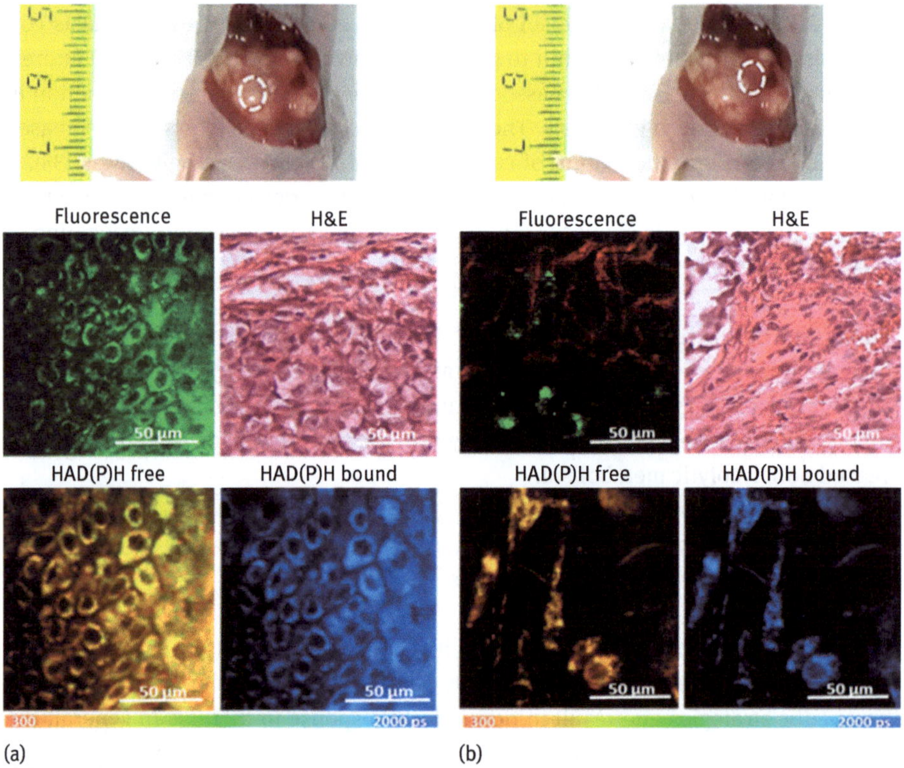

Fig. 10.4: Metabolic heterogeneity of HeLa tumor in vivo. Two-photon fluorescence microscopy (green: autofluorescence, red: SHG from collagen), FLIM of NAD(P)H and histopathology of two representative areas of the tumor (a: cell-rich area, b: stroma-rich area). For NAD(P)H – excitation: 740 nm, detection: 410–660 nm. For SHG – excitation: 750 nm, detection: 373–387 nm. Field of view is 213 × 213 µm.

contribution of free NAD(P)H was 75.5 ± 2.4 %, while in other zones it increased to 80.8 ± 2.7 % ($p = 0.000001$), indicating heterogeneity of cellular metabolism within a tumor node. Intravital fluorescence intensity and SHG microscopy and subsequent histological examination showed that the areas with greater contribution of free NAD(P)H, and consequently more glycolytic metabolism, were enriched with connective tissue fibers and included not numerous cancer cells. Whereas more oxidative areas, with greater contribution of the protein-bound NAD(P)H, had more dense cellular structure with low content of fibrotic stroma.

Therefore, we demonstrated for the first time the metabolic heterogeneity in the HeLa tumor in vivo using FLIM. Cancer cells in cell-rich and stroma-rich areas differed in their energy metabolism. In the literature, metabolic heterogeneity has been shown for various tumor types [42]. It is considered that all cancerous cells that make up a tumor do not behave in a uniform fashion since genes can be regulated at the single cell level [43].

It is known that cancer-associated extracellular matrix (ECM) actively contributes to histopathology and behavior of tumors. Meanwhile, the interplay between ECM mechanics and energy metabolism of cancer cells is not fully understood. Emerging evidence consistently indicates that ECM stiffness and increased ECM deposition may influence cellular metabolism, thereby promoting both cell proliferation and survival [44], as well as facilitating oncogenic transformation and tumor metastasis [45]. It is supposed that the stiffened ECM, often enriched for type I collagen, promotes glucose uptake and aerobic glycolysis through the mechanical activation of the protumorigenic signaling pathways (PI3K/Akt) [46].

10.4 Metabolic shifts in stem cells

Energy metabolism is known to be an important regulator of the ability of stem cells to undergo both differentiation and self-renewal. We investigated the metabolic changes in living MSCs during adipogenic differentiation [47].

MSC differentiation was assessed by noting morphological changes and the development of lipid vacuoles. On days 5 and 12 of differentiation the MSCs had a spindle-shaped morphology; the cell population was homogeneous. By days 19 and 26, 14 % and 36 %, respectively, of the total number of cells had become polygonal in morphology and contained lipid vacuoles.

To estimate the general level of metabolic activity of the cells during adipogenic differentiation, the fluorescence intensities of NAD(P)H and FAD were measured and represented as their redox ratio (FAD/NAD(P)H). No significant changes in the redox ratio values were detected until 19 days after the induction of differentiation. On days 19 and 26 a decrease in the redox ratio could be observed (Fig. 10.5).

The changes in the cofactor fluorescence and, consequently, in the redox ratio for differentiated adipocytes may be associated with both energy metabolism and the biosynthesis of lipids.

The optical redox ratio is sensitive to the balance between the rates of ATP consumption and glucose catabolism in a cell. Undifferentiated MSCs produce ATP primarily through glycolysis, and this is accompanied by NADH production, while those undergoing differentiation display a shift towards OXPHOS for their energy needs. It is generally accepted that glycolytic cells show decreased redox ratios, while in oxidative cells the ratio increases [48].

Since a metabolic shift to OXPHOS is expected during the differentiation of stem cells and that would lead to an increase in the redox ratio, we speculated that the observed decrease of the ratio resulted from the accumulation of NADPH in the process of lipogenesis.

To estimate the contribution of energy metabolism and lipogenesis in the metabolic profile, a separate analysis of NADH and NADPH is required. To separate NADH and NADPH in differentiating stem cells we used FLIM with a three-exponential fitting

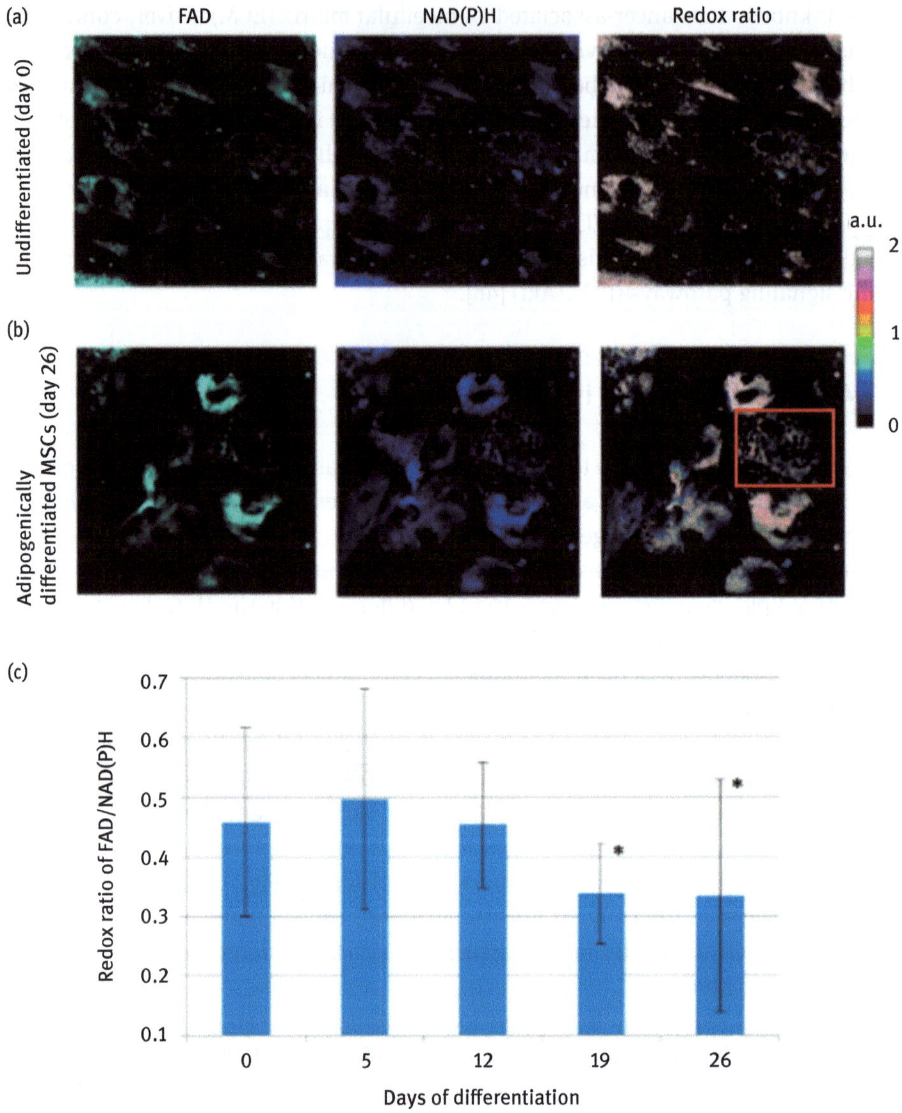

Fig. 10.5: Fluorescence and optical redox images [FAD/NAD(P)H] of undifferentiated MSCs (a) and adipogenically differentiated MSCs (b) and the dynamic of the redox ratio at all stages of differentiation (c). For NAD(P)H – excitation: 750 nm, detection: 455–500 nm; for FAD – excitation: 900 nm, detection: 500–550 nm. The image size is 213 μm × 213 μm (1024 × 1024 pixels). Adipogenically differentiated MSCs are shown by red square. $M \pm SD$. * statistically significant difference with “0 day differentiation”, $p \leq 0.05$.

of the fluorescence decay curves, using the method developed be Blacker et al. [49]. In 2014, Blacker et al. were the first to use FLIM for the separation of NADH and NADPH fluorescence in live cells (HEK293) and tissues (mammalian cochlea). Using numerical methods in combination with biochemical manipulations on the cells, the intracellular fluorescence lifetimes of bound NADH and bound NADPH were evaluated to be 1.5 ± 0.2 and 4.4 ± 0.2 ns, respectively [49].

Since the differentiation of the MSCs may result in changes in the fluorescence lifetimes and the amplitudes of the free and the two bound forms of NAD(P)H, it would be natural to use the three fluorescence lifetime components in a fitting model of the fluorescence decay.

It was found that the contribution of free NADH did not change up to day 19. A statistically significant decrease in the short fluorescence lifetime contribution was shown in differentiated cells (days 19 and 26) in the regions of cytoplasm at the cell periphery as compared with undifferentiated MSCs (day 0) (Fig. 10.6 (a)).

The reverse trend was observed in the contribution of bound NADH on days 19 and 26. It is interesting that the fluorescence lifetime contribution of bound NADH increased only in the regions of cytoplasm at the periphery of the cells. The statistically significant decrease in the contribution of the bound form of NADH on day 12 of differentiation is probably associated with the redistribution of bound NADH to bound NADPH (Fig. 10.6 (a)).

The contribution of bound NADPH was higher at all time points of adipogenic differentiation when compared with undifferentiated MSCs. The maximum elevation of bound NADPH was detected in differentiated cells (days 19 and 26) without, however, any differences between the zones of the cytoplasm at the cell periphery and around the lipid vacuoles (Fig. 10.6 (a)).

It can be seen from Fig. 10.6 (b) that the distribution of the $a_{\text{boundNADH}}/a_{\text{boundNADPH}}$ ratio in adipocytes is inhomogeneous, varying from 0 in the areas of the cytoplasm next to lipid vacuoles (which indicates a contribution to the ratio only from NADPH), to 2 in the other parts of the cytoplasm.

Therefore, the FLIM data, processed with the three-exponential fitting model, on the increase in the contribution of protein-bound NADH in the cytoplasm at the cell periphery testify to the metabolic switch from glycolysis to OXPHOS during the differentiation of stem cells. The rise in the contribution of bound NADPH is probably associated with activation of fatty acid biosynthesis, where NADPH is involved.

Similar to our work, Quinn et al. showed a decrease in the optical redox ratio associated with lipogenesis during adipogenic differentiation [50]. They suggested that adipogenic differentiation, at least in vitro, is associated with an increase in flux through the metabolic pathways using enzymes expressed in the mitochondria. Citrate can then be shuttled out of the mitochondria and cleaved in the cytosol so that acetyl-CoA can be used as a carbon supply for fatty acid synthesis. This process would require the conversion of NAD^+ to NADH by the LipDH-containing pyruvate dehydrogenase complex (PDHC). Carbon storage in lipid droplets through de novo fatty

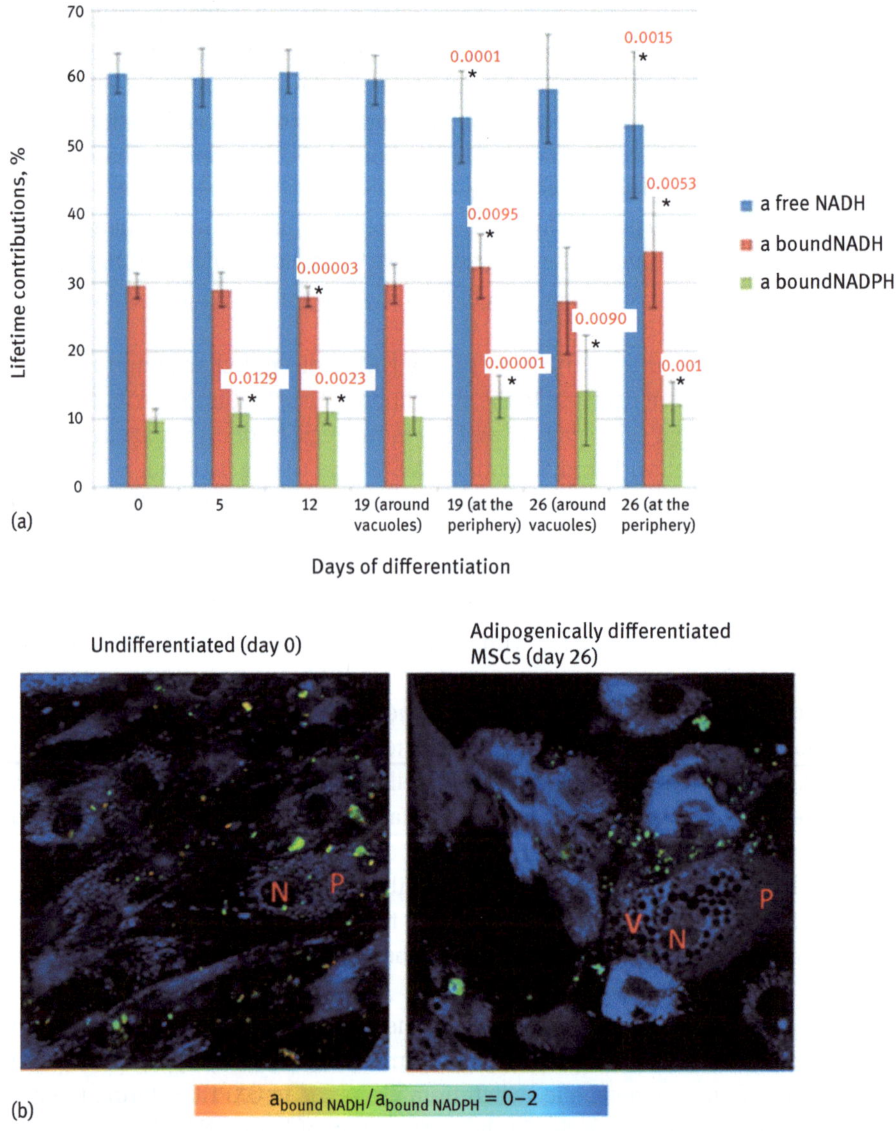

Fig. 10.6: The dynamic of fluorescence lifetimes contributions of free NADH, and protein-bound forms of NADH and NADPH in undifferentiated MSCs and cells at all stages of differentiation (a). $M \pm SD$. * statistically significant difference with day 0. p values are shown. The distribution of NADH and NADPH in undifferentiated (day 0) and adipogenically differentiated MSCs (day 26) (b). False color-coded images of the $a_{\text{bound NADH}}/a_{\text{bound NADPH}}$ ratio. Field of view is 213 µm × 213 µm (512 × 512 pixels). V: vacuoles, N: nucleus, P: periphery of cell.

acid synthesis would probably not occur when substantial ATP production is necessary, and consequently the reduced form of NADH may accumulate. The remaining cytosolic oxaloacetate can be converted to pyruvate by malic enzyme (NADPH is also produced during its conversion) for use in long chain fatty acid synthesis [51]. A decreased redox ratio in metabolically active, adipogenically differentiated cells was also detected in [12].

König et al. showed an increase in the bound forms of NAD(P)H during the differentiation of stem cells into adipocytes [52], Stringari et al. demonstrated a higher free/bound NADH ratio in undifferentiated neural progenitor and stem cells than in differentiated neurons, and that stem cells followed a metabolic trajectory from a glycolytic phenotype to an OXPHOS phenotype through the different stages of differentiation [53].

Our results about lower contribution of NADPH compared to NADH to the total fluorescent signal are consistent with other studies. Blacker et al. showed that the concentration of bound NADPH decreased ≈ 3-fold in HEK293 cell lines [49]. Quinn et al. demonstrated on pancreatic islet and porcine heart that the intracellular concentration of NADH was 33.7 ± 13.1-fold higher than NADPH [50]. Klaidman et al. showed that the concentration of the fluorescent reduced NADH was 5 times greater than that of the fluorescent NADPH in mouse hippocampus [54]. Moreover, the enhancement of mitochondrial NADH quantum yield due to environmental effects has been estimated to be a factor of 1.25–2.5 greater than that of NADPH [55]. Protein binding can also affect the fluorescence quantum yield of NADH, with up to a 10-fold increase in the fluorescence intensity of mitochondrial protein-bound NADH relative to free NADH [56].

To the best of our knowledge, the changes in the contribution of bound NADPH in MSCs during adipogenic differentiation have never previously been measured using FLIM with three-exponential fitting of the fluorescence decay curves. However, using the contribution of bound NADPH and NADH for the study of phenotypic changes requires further investigation and appropriate metabolic measurements.

10.5 Conclusions

Effective control of cell proliferation and differentiation is a great challenge because of the complex relationships between the different signaling pathways, the extracellular microenvironment, and the metabolic requirements of the cell. There are many methods for assessing the metabolic status of cells, including immunocytochemistry or immunohistochemistry for metabolic markers (for example, the monocarboxylate transporters MCT1 and MCT4); measuring glucose uptake; lactate production and oxygen consumption; evaluation of the key glycolytic enzymes (for example, hexokinase-II, phosphofructokinase and pyruvate kinase) and the microarray analysis of glycolytic gene expression. However, these methods are invasive, require the introduction of exogenous labels into the cells and do not provide complete information about the physiological changes in the living cells during their growth, proliferation and differentiation. In contrast, multiphoton fluorescence microscopy combined with FLIM is an emerging modality that permits repeated nondestructive measurements of the components within living cells. To date, multiphoton microscopy of NAD(P)H and FAD provides an established technique for metabolic imaging in vitro and in vivo. The data about metabolic changes in tumor-stroma co-evolution are important for the development of anticancer drugs targeted on metabolic pathways in both cancer and adjacent stromal cells. Understanding the metabolic peculiarities of stem cells and the changes accompanying the differentiation process is extremely important for the development of new treatment strategies in regenerative medicine and stem cell therapy.

Acknowledgment: The investigation of metabolism and pH was carried out using grant #14-15-00646 of the Russian Science Foundation. The study of metabolic shifts in stem cells was performed as a part of grant #17-75-20178 of the Russian Science Foundation.

References

[1] Chance B, Schoener B, Oshino R, Itshak F, Nakase Y. Oxidation-reduction ratio studies of mitochondria in freeze-trapped samples. NADH and flavoprotein fluorescence signals. J Biol Chem. 1979;254:4764–4771.

[2] Lakowicz JR, Szmacinski H, Nowaczyk K, Johnson ML. Fluorescence lifetime imaging of free and protein-bound NADH. Proc Natl Acad Sci U S A. 1992;89:1271–1275.

[3] Walsh AJ, Shah AT, Sharick JT, Skala MC. Fluorescence lifetime measurements of NADH in live cells and tissue in advanced time-correlated single photon counting applications. Becker W, editor. Advanced time-correlated single photon counting applications. Springer; 2015. p. 435–456.

[4] Sanchez WY, Pastore M, Haridass I, König K, Becker W, Roberts MS. Fluorescence lifetime imaging of the skin in advanced time-correlated single photon counting applications. Becker W, editor. Advanced time-correlated single photon counting applications. Springer; 2015. p. 457–508.

[5] Blacker TS, Mann ZF, Gale JE, Ziegler M, Bain MJ, Szabadkai G, Duchen MR. Separating NADH and NADPH fluorescence in live cells and tissues using FLIM. Nature Communications. 2014;5:3936.

[6] Zheng J. Energy metabolism of cancer: Glycolysis versus oxidative phosphorylation [Review]. Oncology Letters. 2012;4:1151–1157.

[7] Pfeiffer T, Schuster S, Bonhoeffer S. Cooperation and competition in the evolution of ATP-producing pathways. Science. 2001;292:504–507.

[8] Heiden Vander MG, Cantley LC, Thompson CB. Understanding the Warburg effect: the metabolic requirements of cell proliferation. Science. 2009;324:1029–1033.

[9] Zhang J, Nuebel E, Daley GQ, Koehler CM, Teitell MA. Metabolism in pluripotent stem cell self-renewal, differentiation, and reprogramming. Cell Stem Cell. 2012;11:589–595.

[10] Kondoh H, Lleonart ME, Gil J, et al. Glycolytic enzymes can modulate cellular life span. Cancer Res. 2005;65:177–185.

[11] Uchugonova A, König K. Two-photon autofluorescence and second-harmonic imaging of adult stem cells. Journal of Biomedical Optics. 2008;13:1–8.
[12] Rice WL, Kaplan DL, Georgakoudi I. Two-photon microscopy for noninvasive, quantitative monitoring of stem cell differentiation. PLoS One. 2010;5:e10075.
[13] Rück A, Hauser C, Mosch S, and Kalinina S. Spectrally resolved fluorescence lifetime imaging to investigate cell metabolism in malignant and nonmalignant oral mucosa cells. Journal of Biomedical Optics. 2014;19:096005.
[14] Skala M, Ramanujam N. Multiphoton redox ratio imaging for metabolic monitoring in vivo. Methods Mol Biol. 2010;594:155–162.
[15] Martinez-Outschoorn UE, Lin Z, Trimmer C, et al. Cancer cells metabolically "fertilize" the tumor microenvironment with hydrogen peroxide, driving the Warburg effect. Implications for PET imaging of human tumors. Cell Cycle. 2011;10:2504–2520.
[16] Meleshina AV, Cherkasova EI, Shirmanova MV. et al. Influence of mesenchymal stem cells on the metastases development in mice in vivo. Stem Cell Research &Therapy. 2015;6:15.
[17] Cirri P, Chiarugi P. Cancer associated fibroblasts: the dark side of the coin. Am J Cancer Res. 2011;1:482–497.
[18] Pavlides S, Whitaker-Menezes D, Castello-Cros R, et al. The reverse Warburg effect aerobic glycolysis in cancer associated fibroblasts and the tumor stroma. Cell Cycle. 2009;8:3984–4001.
[19] Druzhkova IN, Shirmanova MV, Lukina MM, Dudenkova VV, Mishina NM, Zagaynova EV. The metabolic interaction of cancer cells and fibroblasts – coupling between NAD(P)H and FAD, intracellular pH and hydrogen peroxide. Cell Cycle. 2016;15:1257–1266.
[20] Ostrander JH, McMahon CM, Lem S, et al. Optical redox ratio differentiates breast cancer cell lines based on estrogen receptor status. Cancer Res. 2010;70:4759–4766.
[21] de Andrade Natal R, Pelegati VB, Bondarik C, et al. Increased metabolic activity detected by FLIM in human breast cancer cells with desmoplastic reaction: a pilot study. Proc SPIE 9536; 2015. p. 95360L.
[22] Guo HW, Chen CT, Wei YH, et al. Reduced nicotinamide adenine dinucleotide (NADH) fluorescence lifetime separates human mesenchymal stem cells from differentiated progenies. Journal of Biomedical Optics. 2008;13:050505.
[23] Skala MC, Riching KM, Gendron-Fitzpatrick A, et al. In vivo multiphoton microscopy of NADH and FAD redox states, fluorescence lifetimes, and cellular morphology in precancerous epithelia. PNAS. 2007;104:19494–19499.
[24] McGinty J, Galletly NP, Dunsby C, et al. Wide-field fluorescence lifetime imaging of cancer. Biomedical Optics Express. 2010;1:627–640.
[25] Rossignol R, Gilkerson R, Aggeler R, Yamagata K, Remington SJ, and Capaldi RA. Energy substrate modulates mitochondrial structure and oxidative capacity in cancer cells. Cancer Research. 2004;64:985–993.
[26] Herst PM, Berridge MV. Cell surface oxygen consumption: A major contributor to cellular oxygen consumption in glycolytic cancer cell lines. Biochim Biophys Acta. 2007;1767:170–177.
[27] Rodríguez-Enríquez S, Gallardo-Pérez JC, Avilés-Salas A, et al. Energy metabolism transition in multi-cellular human tumor spheroids. J Cell Physiol. 2008;216:189–197.
[28] Reitzer L, Wice B, Kennel D. Evidence that glutamine, not sugar, is the major energy source for cultured Hela cells. J Biol Chem. 1979;254:2669–2676.
[29] Donnely M, Scheffler I. Energy metabolism in respiration deficient and wildtype Chinese hamster fibroblasts in culture. J Cell Physiol. 1976;89:9–52.
[30] Brauer HA, Makowski L, Hoadley KA, et al. Impact of tumor microenvironment and epithelial phenotypes on metabolism in breast Cancer. Clin Cancer Res. 2013;19:571–585.

[31] Bird DK, Yan L, Vrotsos KM, et al. Metabolic mapping of MCF10A human breast cells via multiphoton fluorescence lifetime imaging of the coenzyme NADH. Cancer Res. 2005;65: 8766–8773.
[32] Ghukasyan V, Buryakina T, Kao FJ. Metabolic mapping of cell culture growth by NADH fluorescence lifetime imaging. Proc SPIE 7183; 2009. p. 718309.
[33] Matlashov ME, Bogdanova YA, Ermakova GV, et al. Fluorescent ratiometric pH indicator SypHer2: Applications in neuroscience and regenerative biology. Biochim Biophys Acta. 2015;1850:2318–2328.
[34] Shirmanova MV, Druzhkova IN, Lukina MM, et al. Intracellular pH imaging in cancer cells and tumors in vivo using new genetically encoded sensor SypHer 2. Biochim Biophys Acta. 2015;1850:1905–1911.
[35] Brauer HA, Makowski L, Hoadley KA, et al. Impact of tumor microenvironment and epithelial phenotypes on metabolism in breast cancer. Clin Cancer Res. 2013;19:571–585.
[36] Aredia F, Scovassi AI. Multiple effects of intracellular pH modulation in cancer cells. Cancer Cell & Microenvironment. 2014;1:e136.
[37] Webb BA, Chimenti M, Jacobson MP, Barber DL. Dysregulated pH: a perfect storm for cancer progression. Nat Rev Cancer. 2011;11:671–677.
[38] Öhlund D, Elyada E, Tuveson D. Fibroblast heterogeneity in the cancer wound. J Exp Med. 2014;211:1503–1523.
[39] Vaupel P. Metabolic microenvironment of tumor cells: A key factor in malignant progression. Exp Oncol. 2010;32:125–127.
[40] Diaz-Ruiz R, Rigoulet M, Devin A. The Warburg and Crabtree effects: On the origin of cancer cell energy metabolism and of yeast glucose repression. Biochim Biophys Acta. 2011;1807:568–576.
[41] Lukina M, Shirmanova M, Dudenkova V, Druzhkova I, Shumilova A, Zagaynova E. Analysis of energy metabolism of HeLa cancer cells in vitro and in vivo using fluorescence lifetime microscopy. Proc SPIE 9887; 2016. p. 98872S.
[42] Sengupta D, Pratx G. Imaging metabolic heterogeneity in cancer. Molecular Cancer. 2016;15:1–12.
[43] Marusyk A, Polyak K. Tumor heterogeneity: Causes and consequences. Biochimica et Biophysica Acta – Reviews on Cancer. 2009;1:105–117.
[44] Paszek MJ, Zahir N, Johnson KR, et al. Tensional homeostasis and the malignant phenotype. Cancer Cell. 2005;8:241–254.
[45] Levental KR, Yu H, Kass L, et al. Matrix crosslinking forces tumor progression by enhancing integrin signaling. Cell. 2009;139:891–906.
[46] Tung JC, Barnes JM, Desai SR, et al. Tumor mechanics and metabolic dysfunction. Free Radic Biol Med. 2015;79:269–280.
[47] Meleshina AV, Dudenkova VV, Shirmanova MV, et al. Probing metabolic states of differentiating stem cells using two-photon FLIM. Scientific Reports. 2016;6:21853.
[48] Folmes CDL, Dzeja PP, Nelson TJ, Terzic A. Metabolic plasticity in stem cell homeostasis and differentiation. Cell Stem Cell. 2012;11:596–606.
[49] Blacker TS, Mann ZF, Gale JE, et al. Separating NADH and NADPH fluorescence in live cells and tissues using FLIM. Nature Communications. 2014;5:3936.
[50] Quinn KP, Sridharan GV, Hayden RS, Kaplan DL, Lee K, Georgakoudi I. Quantitative metabolic imaging using endogenous fluorescence to detect stem cell differentiation. Scientific Reports. 2013;3:3432.
[51] Quinn KP, Bellas E, Fourligas N, Lee K, Kaplan DL, Georgakoudi I. Characterization of metabolic changes associated with the functional development of 3D engineered tissues by non-invasive, dynamic measurement of individual cell redox ratios. Biomaterials. 2012;33:5341.

[52] König K, Uchugonova A, Gorjup E. Multiphoton fluorescence lifetime imaging of 3D-stem cell spheroids during differentiation. Microsc Res Tech. 2011;74:9.

[53] Stringari C, Sierra R, Donovan PJ, Gratton E. Label-free separation of human embryonic stem cells and their differentiating progenies by phasor fluorescence lifetime microscopy. J Biomed Opt. 2012;17:046012.

[54] Klaidman LK, Leung AC, Adams JD. High-performance liquid chromatography analysis of oxidized and reduced pyridine dinucleotides in specific brain regions. Anal Biochem. 1995;228: 312–317.

[55] Avi-Dor Y, Olson JM, Doherty MD, Kaplan NO. Fluorescence of pyridine nucleotides in mitochondria. J Biol Chem. 1962;237:2377–2383.

[56] Vishwasrao HD, Heikal AA. Kasischke KA, Webb WW. Conformational dependence of intracellular NADH on metabolic state revealed by associated fluorescence anisotropy. J Biol Chem. 2005;280:25119–25126.

Karsten König

11 Femtosecond laser nanoprocessing

Abstract: The integration of an ultrafast laser into an optical microscope was a milestone as it has allowed for nonlinear imaging and precise nano- and microprocessing. Highly focused femtosecond laser pulses can be used to drill nanoholes, to perform nano- and microsurgery, and to generate sub-100 nm structures even with low-energy near-infrared (NIR) radiation. NIR nanoprocessing far below Abbe's diffraction limit is based on multiphoton processes within the central part of the submicron focal spot, such as multiphoton ionization and plasma formation. With two-beam laser systems, structures with a feature size even in the sub-10 nm range can be generated. Nanoprocessing can be performed at the surface as well as inside transparent materials, such as glass, polymers, biological cells and inside tissues, e.g. cornea. Transient TW/cm^2 intensities are required. Interestingly, these enormous light intensities can be achieved with mean laser powers of a few milliwatt when using sub-20 fs laser pulses. Applications of fs NIR laser microscopes (nanoscopes) include 3D nanolithography, nanomachining such as nanowire production, optical cleaning, targeted transfection, optical reprogramming, and low power nanosurgery.

11.1 Laser microscopes for material processing and analysis

Laser microscopes have existed for more than 50 years. In 1962, just two years after the invention of the laser, Bessis et al. reported on the use of a ruby laser microscope to expose cell organelles with a red 680 nm focused pulsed laser microbeam [1] and Brech and Cross [2] introduced laser-induced breakdown spectroscopy (LIBS). The LIBS laser system was commercialized by the US company *Jarrell-Ash* ("Laser Microprobe") and the East German company *VEB Carl Zeiss* ("laser micro analyzer LMA 1").

With the availability of powerful short UV laser pulses from nanosecond excimer lasers, microscopes became widely employed as medical microprocessing tools for eye surgeries in which the cornea is reshaped for the treatment of shortsightedness, a technique called LASIK. In fact, excimer laser surgery microscopes have been used for LASIK surgery on millions of short-sighted people to replace eye glasses and contact lenses [3, 4].

However, a major disadvantage of the UV laser microscope with its low light penetration depth is the limitation to 2D micromachining. "In-bulk" processing is not possible with UV laser.

This is a huge problem for refractive eye surgery where the inner part of the cornea (the stroma) should be laser-removed, but not the outermost epithelium layer. In order to preserve the epithelium, mechanical cutting with so-called microkeratomes has

https://doi.org/10.1515/9783110429985-013

been performed for decades to generate a tiny tissue flap that is put back after laser ablation of the stroma.

Nowadays, near-infrared (NIR) femtosecond lasers are more and more replacing the microkeratome ("Femtosecond LASIK"). Fig. 11.1 demonstrates a LASIK system in a typical eye clinic consisting of a nanosecond UV excimer laser microsurgery microscope based on photoablation and an additional femtosecond NIR laser refractive microsurgery microscope based on photodisruption. The fs laser creates a 20 mm flap by side cuts and a single horizontal cut based on multiphoton induced plasma-filled microbubbles. The high photon energy of the UV laser photons breaks the bindings of the macromolecules of the stroma resulting in the precise tissue ablation.

In the near future, the fs NIR laser will also replace the UV excimer ablation laser. This novel all-in-one femtosecond laser refractive surgery approach is called "SMILE" ("small incision lenticule extraction") [5, 6]. In that case, a < 4 mm incision (opening) in the epithelium will be performed to remove an inner tissue area with a diameter of 7.5 mm (lenticule). The incision as well as the lenticule production is performed with the same NIR femtosecond microsurgery laser. A flap is no longer required. Fig. 11.1 shows also the principle of different LASIK procedures.

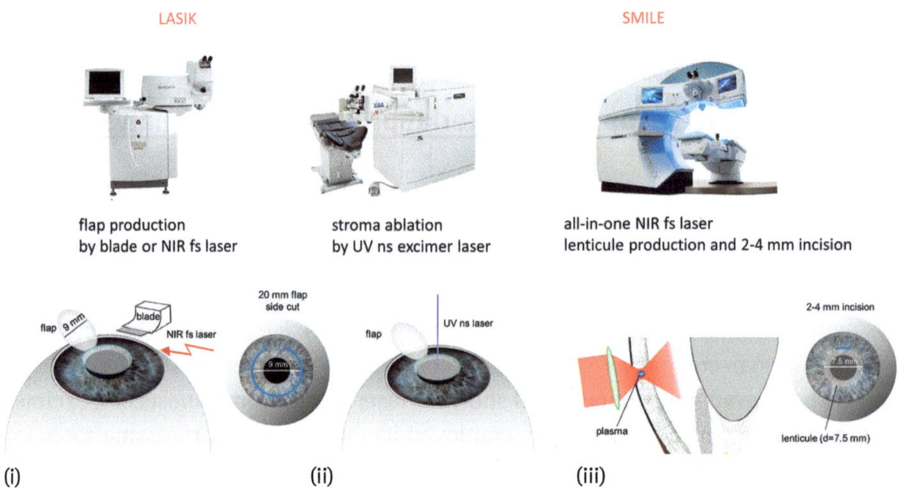

Fig. 11.1: Current refractive surgery microprocessing procedures include: (i) microkeratome for flap production + UV ns laser ablation, (ii) NIR fs laser for flap production + UV ns laser ablation, and (iii) NIR fs laser for lenticel production/extraction as well as 4 mm incision (ReLEx-SMILE).

These examples demonstrate the million-fold medical applications of pulsed laser microscopes for microsurgery. But how can we realize even nanosurgery or other forms of nanoprocessing with hole sizes and cut sizes of less than 100 nanometers using laser microscopes/nanoscopes? And how can we realize in-bulk nanoprocessing?

According to Ernst Abbe's famous diffraction formula:

$$d = \frac{\lambda}{2\text{NA}}$$

the spot size d of a perfectly focused laser beam depends on the wavelength λ and the numerical aperture (NA) of the focusing optics with NA being the product of the refractive index of the environment times the sinus of half the incident beam angle after transmission through the optics. NA values can be 1.3 or higher, when using immersion oil.

That means that the typical minimum spot size d for non-UV radiation ($\lambda >$ 400 nm) is about 200 nm. According to Abbe's formula, nanospots ($d <$ 100 nm) can be achieved only when using ultraviolet light < 200 nm. Indeed, the best way optical nanoprocessing can be performed is using excimer lasers such as the ArF laser with a light wavelength of 193 nm. In fact, 70 nm feature sizes have already been realized at the beginning of this century ("Pentium technology"). The latest generation of nanoprocessors employs 22 nm structures on 300 mm wafers using short wavelength UV. But there is a major drawback when using UV light sources. Unfortunately, "in wafer bulk" nanomachining is not possible due to the strong UV absorption of most materials.

However, multiphoton effects in combination with laser light of high penetration depth can overcome the "UV problem" and realize "in bulk" 3D nanoprocessing.

Tightly focused NIR femtosecond laser radiation induces highly localized multiphoton effects in the focal volume. Many materials including biological cells and tissues appear nearly transparent in the spectral range of 700–1100 nm; i.e., in the "optical window". Nonlinear absorption starts only at high transient light intensities of at least 0.1 GW/cm^2.

For comparison: In spite of high light power, the sun's light intensity on earth is only 0.1 W/cm^2. In contrast to sunlight, laser light can be focused down to submicron spots. Therefore, high laser intensities in the range of GW/cm^2 and TW/cm^2 can be achieved even with low mean milliwatt laser powers (kW peak powers, respectively) by tight focusing using high NA objectives. For example, two-photon effects occur in a 0.1 femtoliter volume corresponding to a focal spot of 0.3 µm (radial) and 0.7 µm (axial) in diameter when employing an NA 1.3 objective. Higher-order multiphoton effects such as direct ionization, where 4 or more photons are involved, are confined to an even smaller volume within the central part of the focal volume. Therefore, the spatially confined multiphoton effects provide the possibility for sub-100 nm nanoprocessing even when using NIR light. Surface processing and, more interesting, in-bulk 3D laser nanoprocessing can be performed. Fig. 11.2 illustrates the two- and multiphoton excitation scheme with various deactivation pathways exploitable for imaging and material processing.

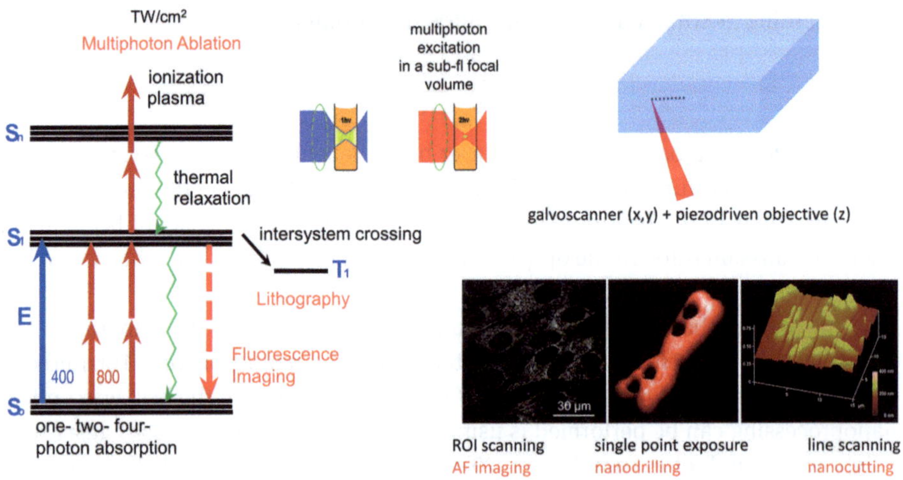

Fig. 11.2: Multiphoton processes for nanoprocessing and imaging. Nonresonant two-photon absorption is used for autofluorescence (AF) imaging and nonlinear 3D photolithography. Ablation, drilling, and cutting can be realized by multiphoton ionization and plasma formation when the galvanometer scanner operates (i) in the region-of-interest (ROI) scanning mode, (ii) in the single point laser exposure mode, and (iii) in the line scanning mode.

During the exposure of cells and tissues to very intense femtosecond laser beams, a significant amount of free electrons is produced in dependence on intensity due to multiphoton ionization even below the optical breakdown threshold. Dissection and hole drilling is already possible with multiple pulses at a high MHz repetition rate in this low-density plasma regime, likely due to extremely confined free-electron-induced chemical decomposition (bond breaking). At slightly higher intensities, optical breakdown and the formation of a nano- or microplasmas may occur resulting in thermomechanical effects such as long-lasting bubbles and shock waves termed photodisruptive effects. When operating close to the threshold of optical breakdown the size of the plasma filled cavitation bubbles can be limited to the submicron range. This also enables nanoprocessing [7]. However, when larger bubbles are produced, cells will not survive. Large plasma-filled bubbles and shock waves, however, can be employed for tissue microsurgery, such as LASIK and SMILE refractive procedures.

The combination of ultrafast laser pulses with a microscope was a milestone. The very first ultrashort laser scanning microscope was realized by our group in Jena in the 1980s. A picosecond dye laser microscope was used for fluorescence lifetime imaging (FLIM) [8, 9]. Shortly later, Denk et al. used a sub-picosecond dye laser to realize two-photon imaging with submicron resolution as well as two-photon induced highly localized chemical reactions within cells, such as photolysis to release biologically active “caged” chemicals such as calcium ions [10].

Two-photon effects can be also employed to realize "in bulk" chemical reactions such as in silicon and glass wafers with "thick layers" of photoresists. Such two-photon 3D microlithography was first demonstrated by Maruo et al. in 1997 [11].

Finally, in 1999 our group reported on femtosecond laser nanoprocessing based on multiphoton processes. A feature size of less than 100 nm was achieved by employing an 80 MHz, 100 fs titanium:sapphire NIR scanning microscope [12, 13]. Sub-100 nm structures such as nanoholes and nanocuts could be realized with low nanojoule pulse energy in biological samples such as intratissue plant samples [14, 15], human chromosomes [16, 17], and living animal cells [12, 13]. A single chromosome within a live PTK cell could be nanodissected without any collateral effects. The cell survived the "in bulk" laser nanomachining and started cell division. Laser nanosurgery was born [12, 13, 17] (Fig. 11.3).

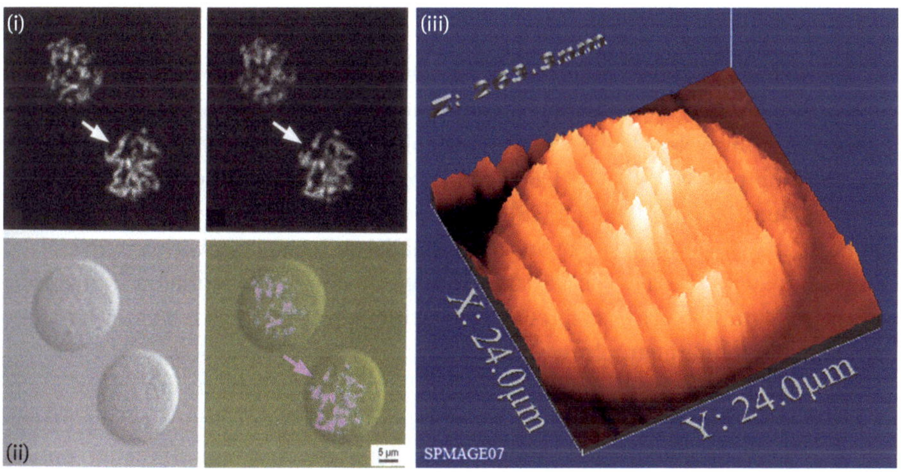

Fig. 11.3: Intracellular femtosecond laser nanosurgery was introduced in the 1990s. The figure left demonstrates the first intracellular chromosome nanodissection in a living cell. The cell survived and started cell division afterwards [12]. Sub-100 nm laser cuts can per performed by low-power nanosurgery when using ultrashort femtosecond laser pulses. The figure right demonstrates nanocuts in a cell nucleus (< 10 mW at 12 fs, 85 MHz).

In 2002, laser-induced nanoholes in the cell's membrane were used to introduce nanomaterial that normally would not enter the cell. In particular, foreign DNA could be introduced in live cells via this optoporation. We called this procedure targeted transfection by femtosecond laser [18, 19]. Also in 2002, intratissue cutting within the cornea without destroying the epithelium was realized with this nanojoule NIR fs laser nanoprocessing microscope [20–22].

Interestingly, sub-100 nm laser-induced periodic surface structures (LIPSS) were discovered by König's group in 2005 when exposing silicon wafers with the nanojoule femtosecond NIR laser nanoprocessing microscope. We observed periodicities as low as 70 nm on crystalline silicon using a 90 MHz Ti:sapphire resonator at 800 nm [23]. Under oil confinement using high NA 1.3 objectives (with a higher refractive index than water) it was possible to perform nanoprocessing likely due to regular self-assembly and to obtain reproducible structures within large area arrays.

These structures with a very low sub-wavelength lateral period were termed 'high spatial frequency LIPSS' (HSFL) or more simply "nanoripples".

LIPSS themselves ("ripples") were already discovered by Birnbaum in 1965 on semiconductor surfaces using millisecond-pulsed ruby laser light [24]. However, such structures have always been periodic within the range of the laser wavelength. They are generally explained by interference between a surface scattered wave with the incident laser beam [25]. Typically, LIPPS are produced with nanosecond UV and VIS lasers at a fluence close to the ablation threshold. These ripples with periodicities in the micron and submicron spatial range are termed low spatial frequency LIPSS (LSFL).

Surprisingly, only the application of femtosecond laser pulses results in the generation of the sub-wavelength LIPSS ("nanoripples"). Meanwhile, both parallel and perpendicular subwavelength LIPSS have been observed in a variety of materials, such as on lithium niobate surfaces [26, 27]. Nanoripple crests feature typical widths of 20–40 nm and heights of 70 nm [28]. Since a central fs laser wavelength of 800 nm was used, these feature sizes are one order smaller than the laser wavelength [23].

Also in 2005, the first sub-100 nm features by non-UV two-photon lithography were manufactured by Juodkazis et al. who fashioned 30 nm thick rod structures [29]. Nonlinear 3D nanolithography was born.

Nowadays, even sub-10 nm features sizes have been realized by nonlinear lithography. In that case, two laser beams are employed, one for two-photon polymerization and the other one for single-photon inhibition. This deep sub-diffraction optical 3D two-photon beam lithography enabled 9 nm features sizes [30]. Stimulated Emission Depletion (STED) nanoscopy, invented by the 2014 Nobel laureate Stefan Hell, is also based on the use of two laser beams rather than one. In STED nanolithography, the stimulated emission is used to deplete the polymerization starters in the outer rim of the point spread function. Applications of STED nanolithography include the generation of nanoanchors for antibodies, where one anchor carries exactly one antibody [31].

Also nanoparticles in combination with fs NIR laser nanoprocessing microscopes have been employed for nanoprocessing. Using a functionalized gold-silver nanoparticle, König's group realized molecular surgery. In particular, DNA nanosurgery as precise as 40 nm was performed using nanoparticles specifically bound to genomic regions of interest as destructive "light antenna". In particular, a certain genomic region within a human chromosome was optically knocked out by the generation of a 40 nm hole at the original nanoparticle position [32, 33].

Besides a variety of lab microscopes, a first commercial NIR femtosecond laser nanoprocessing microscope *FemtOgene* (JenLab GmbH, Germany) was manufactured [34]. Interestingly, that nanoprocessing tool employs a sub-20 femtosecond laser resonator at high 80 MHz repetition rate [34].

11.2 The sub-20 femtosecond laser scanning microscope for nanoprocessing and two photon imaging

Nanoprocessing efficiency can be enhanced by a shorter pulse width, as the two-photon excitation probability follows a P^2/τ function and the three-photon effect a P^3/τ^2 relation. Therefore, the threshold for material ablation can be reached with low-power laser systems when using extremely short laser pulses. In that case, two-photon 3D nanolithography as well as two-photon imaging can be performed even with sub-mW mean laser powers when using a MHz femtosecond laser oscillator. Interestingly, hole drilling, cutting and laser-assisted etching can be realized in a variety of materials with mean powers as low as a few milliwatts [35, 36]. This opens the way to novel nanomachining tools using low-power ultracompact ultrashort femtosecond laser sources in the near future. In order to compensate pulse broadening in the optical system due to group velocity dispersion effects, chirped mirror technology has been employed to maintain the ultrashort laser pulse width also within the focal plane of the microscope.

Fig. 11.4 demonstrates the nanoprocessing microscope employing a 10 fs laser, a step-motor driven stage, an *x*-/*y*-galvanometer scanning unit, and a piezodriven *z*-focusing optics. Normally, the in-situ pulse width would be in the picosecond range after transmission through the microscope including the 40× objective (NA 1.3). However, when employing a dispersion pre-compensation chirped mirror unit, an excellent ultrashort pulse width of 12 fs only can be achieved on the target. At a central wavelength of 800 nm, a pulse width of 12 fs, and a laser spot diameter of $0.61\lambda/\text{NA} = 375$ nm with an exposure area of 0.11 μm^2, respectively, 1 mW mean power is equivalent to 1 kW transient power, 11.8 pJ pulse energy, 0.9 TW/cm^2 peak intensity, and 10.8 mJ/cm^2 fluence per pulse [34]. The maximum scanning area of this particular nanoprocessing and two-photon imaging microscope is 512 × 512 pixels covering 450 × 450 μm^2.

Using galvanometer scanners, nanoprocessing can be performed in three exposure modes:

(i) scanning of a region of interest (ROI) for ablation and photoresist exposure,
(ii) single point illumination for hole drilling, and
(iii) cutting with the galvanometer scanners performing a line scan.

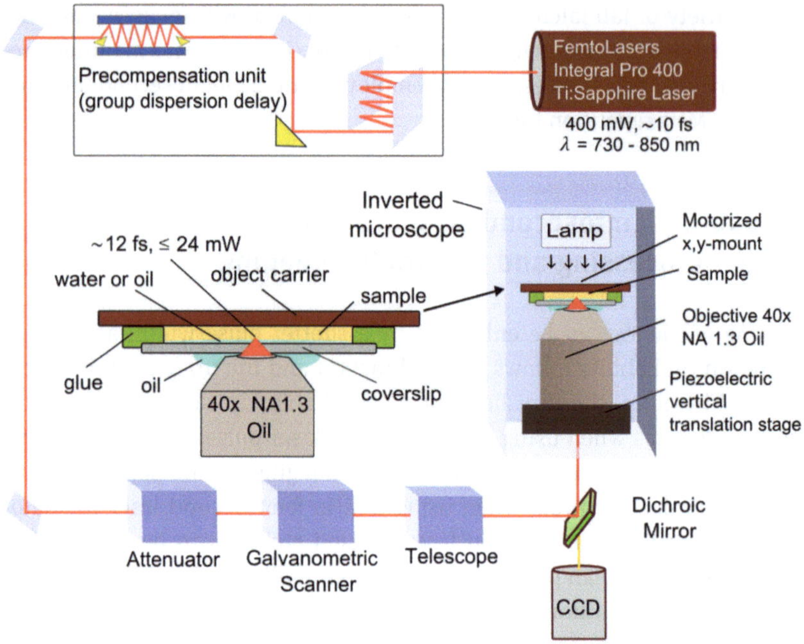

Fig. 11.4: Sub-20 fs laser nanoprocessing scanning microscope. The laser source is a 10 fs titanium:sapphire laser at 85 MHz repetition frequency (spectral bandwidth 120 nm). Chirped mirror technology is employed to maintain an ultrashort in situ pulse width of 12 fs at the target. Nanoprocessing of a variety of materials can be performed with a few milliwatt mean power [34, 35].

A CCD camera attached to the side port enables on-site monitoring of the target during nanoprocessing and a photomultiplier (PMT) is employed to detect two-photon fluorescence, SHG, and plasma luminescence.

11.3 Two-photon lithography with broadband pulses

When using the 10 fs NIR laser source, two-photon 3D nanolithography can be performed with a variety of photoresists due to the 120 nm broad excitation spectrum that "covers" the UV one-photon absorption bands of the photoresists.

Fig. 11.5 shows an example of two-photon nanolithography with the well-established negative photoresist SU-8 (MicroChem Corp.) [37]. The photoresist was spin-coated on a 170 µm thick glass cover slip, baked, and laser-exposed. The post-exposure bake and wet chemistry was performed in a class 100 cleanroom. Interestingly, in some of the generated structures the aspect ratio exceeded 50 : 1. This may create instability. Stable nanostructures in an upright orientation as vertical walls were generated, e.g., with a width of 95 nm and a wall size of $8.0 \times 6.5\,\mu m^2$, within 6 s (0.7 mm/s line speed using 0.3 mW mean power).

Fig. 11.5 presents a 3D cell container of 20 μm in length featuring nanotopographical features on walls and bottom. Such cell containers for microscopic live cell investigations mimic the “real life” of intratissue cells in their nano- and microenvironment much better than the current “gold standard” of a cell monolayer on a flat piece of glass. In real nature, the intratissue cells with a typical diameter of 20 μm are often embedded in the extracellular matrix (ECM) with the major components elastin and collagen. Elastin and collagen fibrils are nanostructures. They build up a 3D network hosting. Researchers nowadays use 3D cell containers and try to understand the interaction of cancer cells with the ECM matrix and, e.g., to figure out how the differentiation of stem cells is influenced by the 3D nano- and microenvironment.

The bottom wall and side wall pattern of the fs-laser generated cell container in Fig. 11.5 consists of nanostructures that were built up by laser scans with laser beam increments of 20 nm. It was demonstrated that the special artificially generated nanotopography in the 3D cell chamber exerted an influence on the alignment of PC-3 cells compared to adjacent planar glass surface regions [37].

Transient changes of optical properties [38], the influence of SHG and self-organization effects [39] have been proposed as mechanisms of nanoripple formation as well as interference between bulk plasma waves and the incident laser light for periodic in-bulk modifications and transiently in-bulk plasma nanobubbles [40]. As shown recently by a Drude-like surface plasmon polariton model, standing surface plasma waves of the electron-hole plasma generated transiently in the tight focus of a high numerical aperture objective result in the generation of nanoripples [41].

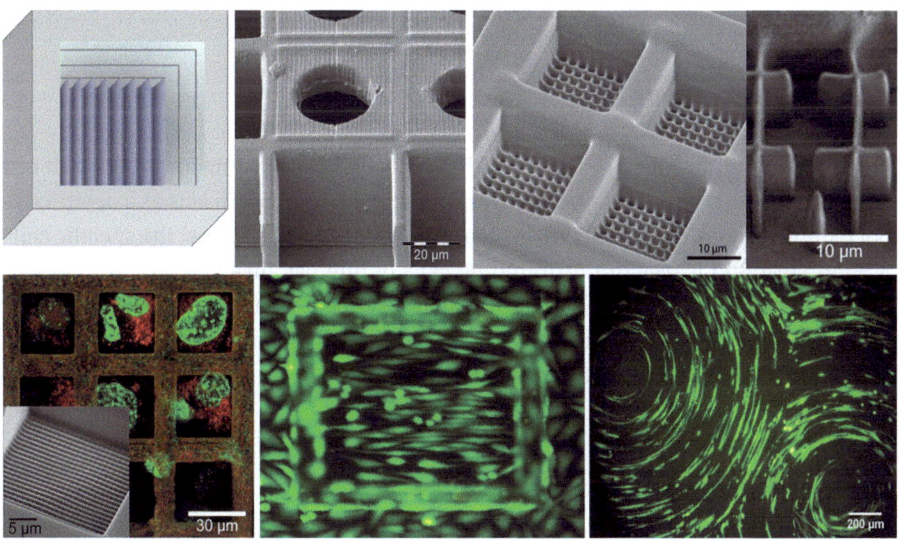

Fig. 11.5: Example of 3D micro- and nano-photolithography. Cell cages with nanotopography on bottom and side walls have been generated within the photoresist SU-8. These cages mimic the natural 3D extracellular micro- and nanoenvironment of intratissue cells. Cancer cells and pulp stem cells elongate along nanostructures.

11.4 Nanowire production by laser-assisted etching

Nanofabrication of a single indium-tin-oxide (ITO) nanowire (NW) from a sputtered film can be performed by direct laser writing and a subsequent etching process. The transparent conducting oxide ITO is widely used, e.g., in organic light emitting diodes (OLEDs), in photovoltaics, and in gas sensors to detect NO_2 and ozone.

Nanowires down to a diameter of 60 nm have been achieved when using 15 mW (12 fs, 85 MHz) and line scans at a speed of 300 µm/s on sputtered films to modify the degree of crystallization and subsequent etching. The exposed lines were found to be more hydrochloride (HCl) resistant than nonexposed areas, which disappeared after a few minutes of 10 % HCl exposure. The combination of fs laser exposure and wet chemistry results in NWs [35, 36, 42].

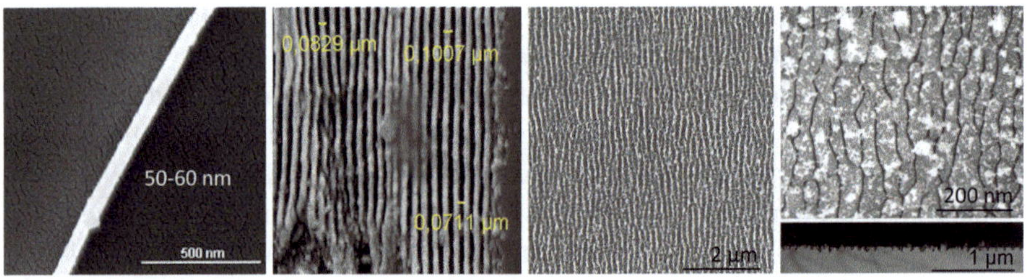

Fig. 11.6: Left: ITO nanowire production by 12 femtosecond NIR laser exposure and subsequent HCl etching. Right: Nanoripples with typical widths of 20–40 nm and a height as well as interspace of 70 nm have been produced in Si(100) silicon with NA 1.3 oil immersion objectives.

11.5 Optical cleaning

The "contamination" of cell cultures with nondesired cells is a major problem in cell culturing. A new concept is optical cleaning where the laser beam acts as a destructive tool without any collateral effect. The beam optically knocks out just the specific cells of no interest and leaves the surrounding cells of interest alive [43]. Fig. 11.7 demonstrates an example where an NR femtosecond laser beam was employed to destroy a nondesired cell by single point exposure with sufficient mean power to induce a large plasma cavitation bubble that destroyed the cell.

11.6 Targeted transfection

Low-power nanosurgery can be employed to generate transient nanoholes in the cell's membrane in order to overcome the barrier function. If the hole is small enough, the cell will survive by a self-repairing process. Typically, the membrane of laser-

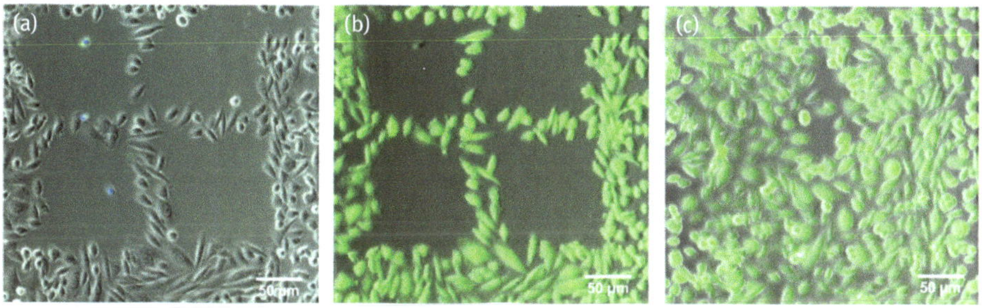

Fig. 11.7: Optical cleaning with NIR femtosecond laser pulses. Cells within the ROI have been destroyed by multiphoton-induced plasma formation. The laser-exposed cells disappear. Nonexposed neighbor cells are not affected. They stay alive, reproduce, and migrate into the cleaned field the day after laser exposure [47].

perforated CHO cells can be repaired within 5 seconds when drilling a hole with a diameter of less than 10 nm. The transient opening can be used to introduce foreign molecules such as DNA and RNA [18, 19, 44, 45] as well as nanoparticles from the outside medium by diffusion.

Interestingly, the drilling of a second hole in the nuclear membrane is not required to introduce foreign DNA into the target cell genome (transfection). It is sufficient to wait until the cell division process when the nuclear membrane is dissolved. At that moment, the foreign DNA that is localized inside the cytoplasm can be integrated into the cell's genome. In the case of CHO cells, the integration of, e.g., a GFP plasmid and the subsequent green fluorescent protein expression will take about 24 hours. CHO cells divide typically twice a day. The procedure of femtosecond laser transfection by transient opening of the cell's membrane through multiphoton plasma formation was invented and published by König and Tirlapur in 1995 [18, 19]. Meanwhile, high-throughput flow-cytometers for laser-assisted transfection employing Bessel beams have been realized [46].

11.7 Optical reprogramming

Induced pluripotent stem cells (iPS) closely resemble embryonic stem cells in many aspects such as differentiation potency, protein and gene expressions, proliferation, morphology, and embryonic body formation. IPS cells derived from a patient's own cells have a tremendous potential for transplantation therapies by circumventing problems associated with immune compatibility.

Typically, iPS cells are generated through the reprogramming of adult somatic cells by forced expression of a combination of appropriate transcription factors. IPS cells were first established from mouse fibroblasts by Yamanaka's group in Japan in 2006 (Nobel Prize in 2012). It was shown that the introduction of the transcription fac-

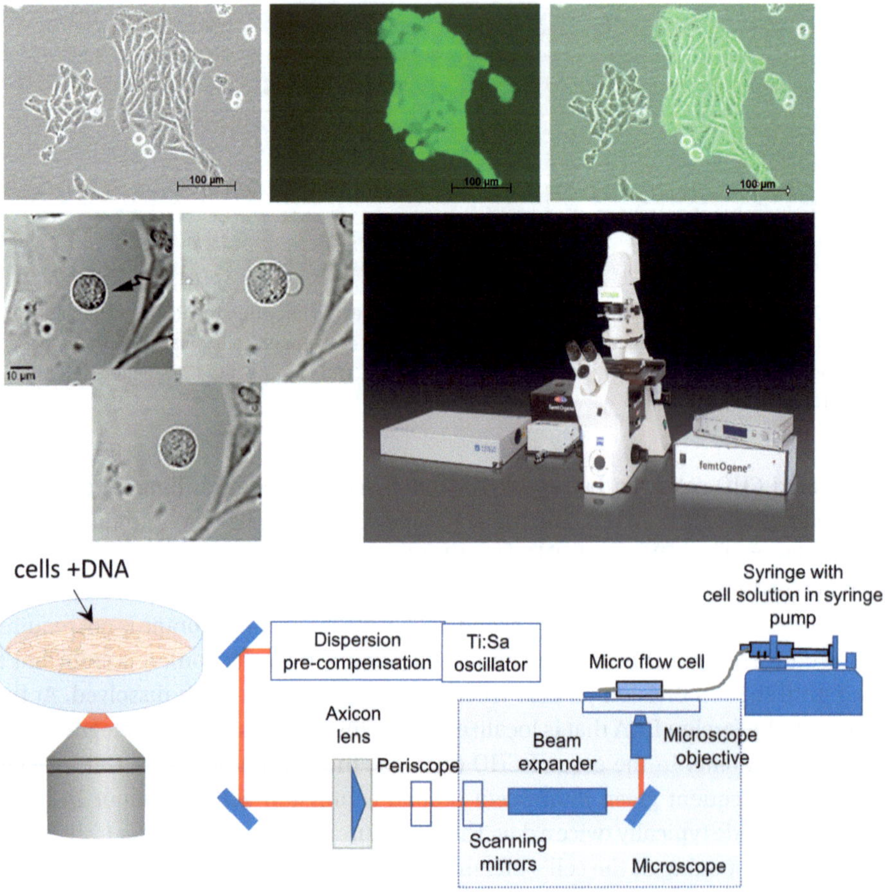

Fig. 11.8: Femtosecond laser transfection can be realized by drilling a transient nanohole in the cell's membrane and subsequent introduction of foreign DNA molecules. The membrane is sealed within seconds by self-repairing processes. Above: Laser-transfected cells as well as daughter cells express green fluorescence when the GFP plasmid is introduced via optoporation. Micro- and submicron plasma-filled bubbles are seen after exposure with the femtosecond laser transfection microscope *FemtOgene*. The fs laser transfection can also be performed on nonattached cells when using a special flow cytometer with femtosecond laser Bessel beams.

tors Oct4, Klf4, Sox2, and c-Myc transferred fibroblasts into iPS cells. Within the last few decades, iPS cells have also been generated from peripheral blood cells, hepatocytes, keratinocytes, oral mucosa, umbilical cord blood, as well as dental pulp cells.

The reprogramming of somatic cells into induced pluripotent stem (iPS) cells can be evoked through retro/lenti-viruses that deliver genes/transcription factors as well as by integrating the transcription factors into that of the host genome. However, the use of viruses may limit the use for clinical applications due to the risk of cancer formation.

We proposed a new approach for virus-free cell reprogramming by low-power nanosurgery with ultrashort laser pulses. Picojoule near-infrared 12 fs laser pulses at a low mean power of 10 mW (85 MHz) can be employed to introduce four or more required transcription factors into human adult cells, such as dermal fibroblasts, by the transient generation of a nanohole [47–49].

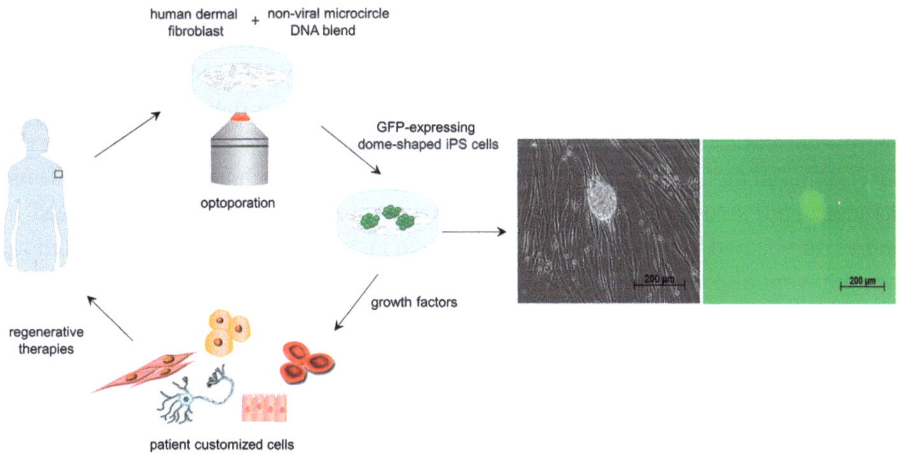

Fig. 11.9: Optically induced reprogramming of human skin cells into pluripotent iPS cells by transient nanohole laser fabrication and subsequent introduction of up to 5 plasmids. Within a few days, the laser-exposed successfully reprogrammed cells exhibit a dome-shaped morphology, green protein fluorescence and express pluripotency markers such as SSEA4.

11.8 Outlook

Femtosecond laser nanoprocessing was introduced by König et al. in the 1990s [12, 50]. Since then, multiphoton effects induced by VIS or NIR femtosecond laser pulses have been employed for nanomachining including 3D nanolithography, nanoripple production, laser-assisted etching, nanodrilling, nanocutting, and in-bulk material nanoablation. The mechanisms behind these are
(i) nonresonant two-photon absorption,
(ii) multiphoton ionization and low density plasma formation, and
(iii) multiphoton plasma formation with submicron cavitation bubble generation [50].

Femtosecond laser nanoprocessing of biological cells and tissues is employed for low-power nanosurgery, targeted transfection, optical cleaning, and optical reprogramming [18, 19, 43, 47–53]. Femtosecond laser transfection and optical nanoinjection is used in a variety of research groups due to the high viability, its selective targeting capability, and safety [54–68].

Optically generated pluripotent iPS cells that are capable of developing into all types of cells of the body may become of high interest due the avoidance of viruses or chemicals to realize gene transfer into the cell and would advance the clinical translation of iPS cells. In the future, femtosecond laser nano- and microprocessing will influence ophthalmic and other tissue surgery significantly.

Interestingly, low mean powers of less than 10 mW are sufficient to perform 3D nanoprocessing in a variety of materials when employing ultrashort femtosecond laser systems, such as a 10 fs NIR 85 MHz high NA laser microscope. Sub-100 nm features without collateral thermal or mechanical effects can be generated at picojoule pulse energies. A very interesting feature of this technology is the capability to induce morphological crystallographic changes in some materials (e.g. ITO), rendering the laser-exposed areas resistant against a subsequent attack of chemical etchants. Thus, nanostructures such as nanowires can be generated directly, without the need of more complicated lithographic processes which include the deposition, illumination, development and the removal of a photoresist layer on top of the material as well as the subsequent highly anisotropic etching step [34, 36, 42].

The unique 3D capability of fs NIR lithography cannot be reached even with the most advanced, extremely expensive UV lithography systems reaching down to feature sizes below 40 nm in planar structural arrangements. 3D nanolithography opens up the way to generate complex nonplanar 3D nanostructures and nano-electromechanical systems (e.g., conical jets, lenses, resonators) with many applications, such as optical data communication, fluidic systems for microthrusters in space technology and trace detection of hazardous materials. The laser-induced nanowire fabrication is attractive for biological and chemical analysis and for gaining new insight into the quantum behavior of such structures. All these capabilities may impact the field of optical storage, rapid prototyping in particular for tissue implants (cardiovascular stents, intraocular lenses, catheters, tooth material, prosthetics), optical communication, and nanomachining [51, 69].

These fs NIR nanoprocessing results may influence the future production of small-sized, low cost femtosecond lasers for biophotonics and material processing. Ultracompact chiller-free femtosecond laser oscillators may be directly attached to a nanoprocessing and imaging microscope. When using microscopes with a sub-20 fs source, the broad laser emission spectrum may facilitate the excitation of multiple fluorophores and photoresists without the need for wavelength tuning [70]. It is also an ideal source for optical coherence microscopy (OCM), optical coherence tomography (OCT), and multiphoton tomography (MPT). Of high interest is the development of ultrafast fiber lasers that may become the most important ultrashort laser sources due to their low cost, versatility, liability, and less complex architecture.

With novel commercial low-cost femtosecond lasers, laser nanoprocessing of non-biological and biological materials as well as live cell microscopy and medical diagnostics will be revolutionized in the next ten years.

References

[1] Bessis M, Gires F, Mayer G, Nomarski G. Irradiation des organites cellulaires a l'aide d'un laser a rubis. Compte Rendus de l'Academie des Sciences. 1962;255:1010–1012

[2] Brech F, Cross L. Optical microemission stimulated by a ruby MASER. Applied Spectroscopy. 1962;16:59.

[3] Stern D, Schoenlein RW, Puliafito CA, Dobi ET, Birngruber R, Fujimoto JG. Corneal ablation by nanosecond picosecond and femtosecond lasers at 532 and 625 nm. Arch Ophthalmol. 1989;107:587–592.

[4] Ratkay-Traub I, Ferincz IE, Juhasz T, Kurtz RM, Krueger RR. First clinical results with the femtosecond neodynium-glass laser in refractive surgery. J Refract Surg. 2003;19(2):94–103.

[5] Sekundo W, Kunert K, Russmann C,Gille A, Bissmann W, Stobrawa G, Sticker M, Bischoff M, Blum M. First efficacy and safety study of femtosecond lenticule extraction for the correction of myopia: six-month results. J Cataract Refract Surg. 2008;34:1819.

[6] Reinstein DZ, Archer TJ, Gobbe M. Small incision lenticule extraction (SMILE) history, fundamentals of a new refractive surgery technique and clinical outcomes. Eye and Vision. 2014;1:3.

[7] Vogel A, Noack J, Hüttman G, Paltauf G. Mechanisms of femtosecond laser nanosurgery of cells and tissues. Appl Phys B. 2005;81:1015.

[8] Bugiel I, König K, Wabnitz H. Investigation of cells by fluorescence laser scanning microscopy with subnanosecond time resolution. Lasers Life Sciences. 1989;3:1–7.

[9] König K. PhD thesis. Beiträge zur Photochemotherapie und optischen Diagnostik von Tumoren. Contributions to photochemotherapy and optical diagnostics of tumors. Archive University Jena, Germany; 1989.

[10] Denk W, Strickler J, Webb W. Two-photon laser scanning fluorescence microscopy. Science. 1990;248:73–76.

[11] Maruo S, Nakamura O, Kawata S. Three-dimensional microfabrication with two-photon absorbed photopolymerization. Optics Letters. 1997;22:132–134.

[12] König K, Riemann I, Fischer P, Halbhuber KJ. Intracellular nanosurgery with near infrared femtosecond laser pulses. Cell Mol Biol. 1999;45:195–201.

[13] König K, Fischer P, Riemann I, Halbhuber KJ. Intracellular nanosurgery with compact femtosecond laser. In: 18th Congress of the International Commission for Optics: Optics for the next Millennium. Proc SPIE 3749; 1999. p. 390.

[14] Tirlapur U, König K. Near-infrared femtosecond laser pulses as a novel non-invasive means for dye permeation and 3D imaging of localised dye-coupling in the Arabidopsis root meristem. The Plant Journal. 1999;20:363–370.

[15] Tirlapur U, König K. Femtosecond near-infrared laser pulses as a versatile non-invasive tool for intratissue nanoprocessing in plants without compromising viability. The Plant Journal. 2002;31:365–374.

[16] König K, Riemann I, Fritzsche W. Nanodissection of human chromosomes with near infrared femtosecond laser pulses. Optics Letters. 2001;26:819–821.

[17] König K, Riemann I, Krauss O, Fritzsche W. Invited paper: Nanodissection of human chromosomes and ultraprecise eye surgery with nanojoule near infrared femtoseond laser pulses. Proc SPIE 4633; 2002. p. 11–22.

[18] Tirlapur U, König K. Targeted transfection of cells by femtosecond near-infrared laser pulses. Nature. 2002;418:290–291.

[19] König K, Tirlapur U. Patents: König: EP 1509611, AU2003243902, CN ZL 03810559.4.

[20] König K, Krauss O, Riemann I. Intratissue surgery with 80 MHz nanojoule femtosecond laser pulses in the near infrared. Optics Express. 2002;10:171–176.

[21] König K, Riemann I, Stracke F, Le Harzic R. Nanoprocessing with nanojoule near infrared femtosecond laser pulses. Med Laser Appl. 2005;20:169–184.
[22] König K, Schuck H, Sauer D, Bauerfeld F, Stracke F, Velten T, Tchernook A, Martin S, Le Harzic R. Invited paper: Femtosecond laser nanoprocessing using near infrared nanojoule pulses at MHZ repetition frequency. Proc SPIE 6400; 2006.
[23] Le Harzic R, Schuck H, Sauer T, Anhut T, Riemann I, Velten T, König K. Sub-100 nm nanostructering of silicon by ultrashort laser pulses. Optics Express. 2005;13:6651–6656.
[24] Birnbaum M. Semiconductor surface damage produced by ruby lasers. Journal of Applied Physics. 1965;36:3688–3689.
[25] Fauchet PM, Siegman AE. Surface ripples on silicon and gallium arsenide under picosecond laser illumination. Applied Physics Letters. 1982;40:824–826.
[26] Daminelli G, Krüger J, Kautek W. Femtosecond laser interaction with silicon under water. Thin Solid Films. 2004;467:334–341.
[27] Straub M, Weigand B, Afshar M, Feili D, Seidel H, König K. Periodic subwavelength ripples and random nanocone patterns on lithium niobate surfaces generated by tightly focused sub-15 fs sub-nanojoule pulsed near-infrared laser light. Journal of Optics. 2013;15:055601.
[28] Straub M, Afshar, M, Seidel H, König K. Surface plasmon polariton model of high- spatial frequency laser-induced periodic surface structure generation in silicon. Journal of Applied Optics. 2012;111:124315-1–124315-6.
[29] Juodkaziz S, Mizeikis V, Seet KK, Miwa M, Misawa H. Two-photon lithography of nanorods in SU-8 photoresist. Nanotechnology. 2005;16:846–849.
[30] Gan Z, Cao Y, Evans RA, Gu M. Three-dimensional deep sub-diffraction optical beam lithography with 9 nm feature size. Nature Communications. 2013;4:2061.
[31] Klar TA. STED lithography and protein nanoanchors. In: K. König, A. Ostendorf, editors. Optically induced nanostructures. Biomedical and technical applications. De Gruyter; 2015. p. 303–323. ISBN 9783110337181.
[32] König K, Garwe F, Czaki A, Maubach G, Riemann I, Fritzsche W. Nanoprocessing of DNA with femtosecond laser. Proc SPIE 5462; 2004.
[33] Czaki A, Garwe F, Steinbrück A, Maubach G, Festag G, Weise A, Riemann I, König K, Fritzsche W. A parallel approach for sub-wavelength molecular surgery using gene-specific positioned metal nanoparticles as laser light antennas. Nanoletters. 2007;2:247–253.
[34] www.jenlab.de
[35] König K, Uchugonova A, Straub M, Zhang H, Licht M, Afshar M, Feili D, Seidel H. Sub-100 nm material processing and imaging with a sub-15 femtosecond laser scanning microscope. Journal of Laser Application. 2012 Jul;24:042009-1–042009-9.
[36] Afshar M, Straub M, Voellm H, Feili D, König K, Seidel H. Sub-100 nm structuring of indium-tin-oxide thin films by sub-15 femtosecond pulsed near-infrared laser light. Optics Letters. 2012;37:563–565.
[37] Licht M, Uchugonova A, König K, Straub M. Sub-15 fs multiphoton lithography of three-dimensional structures for live cell applications. J Opt. 2012;14:065601–065607.
[38] Dufft A, Rosenfeld A, Das SK, Grunwald R, Bonse J. Femtosecond laser-induced periodic surface structures revisited: A comparative study on ZnO. Journal of Applied Physics. 2009;105:034908.
[39] Wu Q, Ma Y, Fang R, Liao Y, Yu Q, Chen X, Wang K. Femtosecond laser-induced periodic surface structure on diamond film. Applied Physics Letters. 2003;82:1703–1705.
[40] Shimotsuma Y, Kazansky PG, Qiu J, Hirao K. Self-organized nanogratings in glass irradiated by ultrashort light pulses. Physical Review Letters. 2003;91:247404.

[41] Straub M, Afshar M, Seidel H, König K. Surface plasmon polariton model of high-spatial frequency laser-induced periodic surface structure generation in silicon. Journal of Applied Optics. 2012;111:124315-1–124315-6.
[42] Afshar M, Preiss EM, Sauerwald T, Rodner M, Feili D, Straub M, König K, Schütze A, Seidel H. Indium-tin-oxide single-nanowire gas sensor fabricated via laser writing and subsequent etching. Sensors and Actuators B. 2015;215:525–535.
[43] Uchugonova A, Isemann A, Gorjup E, Tempea G, Bückle R, Watanabe W, König K. Optical knock out of stem cells extremely ultrashort femtosecond laser pulses. J Biophotonics. 2008; 1:463–469.
[44] Földes-Papp Z, König K, Studier H, Bückle R, Breunig HG, Uchugonova A, Kostner GM. Trafficking of mature miRNA-122 into the nucleus of live liver cells. Current Pharmaceutical Biotechnology. 2009;10:569–578.
[45] Uchugonova A, König K, Bueckle R, Isemann A, Tempea G. Targeted transfection of stem cells with sub-20 femtosecond laser pulses. Opt Express. 2008;16:9357–9364.
[46] Breunig HG, Uchugonova A, Batista A, König K. High-throughput continuous flow femtosecond laser-assisted cell optoporation and transfection. Mic Res Tech. 2014;77:974–979.
[47] Uchugonova A. Optical reprogramming of human cells in an ultrashort femtosecond laser microfluidic transfection platform. J Biophoton. 2015:1–6.
[48] Uchugonova A, Breunig HG, Augsburger C, Monaghan M, Schenke-Layland K, König K. Optical reprogramming and optical characterization of cells using femtosecond lasers. In: König K, Ostendorf A, editors. Optically induced nanostructures. Biomedical and technical applications. De Gruyter; 2015. p. 159–178. ISBN 9783110337181.
[49] König K, Uchugonova A. Patent on optical reprogramming of living cells. European Patent EP16154825.0. Filed 2017, Mar 27.
[50] König K. Patent on optical processing of cellular structures and biomolecules. Patent 19719345.5. Filed 1997, May 7.
[51] König K, Ostendorf A, editors. Optically induced nanostructures. Biomedical and technical applications. De Gruyter; 2015. ISBN 9783110337181.
[52] Uchugonova A, Lessel M, Nietzsche S, Zeitz C, Jacobs K, Lemke C, König K. Nanosurgery of cells and chromosomes using near infrared twelve femtosecond laser pulses. J Biomed Optics. 2012;17:101502.
[53] Baxter J. Low-power nanosurgery. Nature Photonics. 2012;6:573.
[54] Zeira E, et al. Femtosecond infrared laser – an efficient and safe in vivo gene delivery system for prolonged expression. Mol Ther. 2003;8:342–350.
[55] Mohanty SK, Sharma M, Gupta PK. Laser-assisted microinjection into targeted animal cells. Biotech Lett. 2003;25:895–899.
[56] Yanik MF, Cinar H, Cinar HN, Chisholm AD, Jin Y, Ben-Yakar B. Neurosurgery: Functional regeneration after laser axotomy. Nature. 2004;432:822.
[57] Stracke F, Riemann I, König K. Optical nanoinjection of macromolecules into vital cells. J Photochem Photobiol B. 2005;81:136–142.
[58] Barrett LE. et al. Region-directed phototransfection reveals the functional significance of a dendritically synthesized transcription factor. Nat Methods. 2006;3:455–460.
[59] Gu L, Mohanty S. Targeted microinjection into cells and retina using optoporation. J Biomed Opt. 2011;16:128003–128006.
[60] Soman P, Zhang WD, Umeda A, Zhang ZJ, Chen SC. Femtosecond Laser-Assisted optoporation for drug and gene delivery into single mammalian cells. J Biomed Nanotechnol. 2011;7:334–341.
[61] Peng C, Palazzo RE, Wilke I. Laser intensity dependence of femtosecond near-infrared optoinjection. Phys Rev E. 2007;75:041903

[62] Tsampoula X, et al. Fibre based cellular transfection. Opt Express. 2008;16:17007–17013.
[63] Lei M, Xu HP, Yang H, Yao BL. Femtosecond-laser assisted microinjection into living neurons. J Neurosci Meth. 2008;174:215–218.
[64] Kohli V, Elezzabi AY. Laser surgery of zebrafish (Danio rerio) embryos using fs laser pulses. BMC Biotechn. 2008;8:7.
[65] Mthunzi P, Dholakia K, Gunn-Moore F. Phototransfection of mammalian cells using femtosecond laser pulses: optimization and applicability to stem cell differentiation. J Biomed Opt. 2010;15:041507.
[66] Wu T, et al. Neuronal growth cones respond to laser-induced axonal damage. J Royal Soc Interface. 2012;9:535–547.
[67] Dhakal K, Black B, Mohanty S. Introduction of impermeable actin-staining molecules to mammalian cells by optoporation. Sci Rep. 2014;4:6553.
[68] Dhakal K, Batabyal S, Wright W, Kim YT, Mohanty S. Optical delivery of multiple opsin-encoding genes leads to targeted expression and white-light activation. Light: Science and Applications. 2015;4:e352.
[69] Straub M. Habilitation thesis. Archive Saarland University; 2016.
[70] König K, Liu T, Anderson P, Breunig HG. Multiphoton imaging with a novel compact diode-pumped Ti:sapphire oscillator. Microsc Res Technol. 2015;78:1154–1158.

Hans Georg Breunig and Karsten König

12 Cryomultiphoton imaging

Abstract: We describe multiphoton imaging with sample-temperature control to monitor animal cells and cells of intact plants during freezing, thawing and heating processes based on autofluorescence intensity and lifetime. The sample temperature can be set with a heating and freezing stage to any value in the range between liquid nitrogen temperature (–196 °C; 77 K) and +600 °C (873 K) and changed with adjustable heating/freezing rates between 0.01 K/min and 150 K/min. Multiphoton imaging is realized with near-infrared femtosecond-laser excitation with different setups employing different laser sources. To illustrate the capabilities, imaging of animal cell samples with and without a cryoprotectant during freezing at cooling rates is presented. Lowering the temperature led to a significant increase of the intracellular fluorescence intensity and modifications. Fluorescence lifetime imaging indicated an increase of the mean lifetime with decreasing temperature. Furthermore, to illustrate imaging of plant samples, *Arabidopsis thaliana* leaves were employed. The measurements revealed thermally-induced changes of fluorescence lifetime and intensity as well as morphological alterations in the distribution of chloroplasts. The measurements illustrate the general usefulness of multiphoton imaging to investigate freezing and thawing effects on animal and plant cells even at temperatures commonly used for cryopreservation.

12.1 Introduction

The basis for cryopreservation, nowadays a routine technique for long-term preservation, is the fact that basically all enzymatic or chemical activities inside biological material stop at low enough temperatures [1]. A variety of human and animal cells, tissues such as cornea, skin, pancreatic tissue, liver slices, and heart valves as well as plant cells, sprouts, seeds, and even whole plants can be stored in this way for long time periods [2] with the possibility to recover the original function of the material provided that hazards like ice formation, which may damage cellular structures, are limited. Most freezing methods aim to prevent ice formation inside cells by applying suitable freezing conditions through the control of the freezing rate and/or the addition of cryoprotectants, i.e., substances that protect biological tissue from freezing damage, or by dehydration of the material prior to freezing [3]. In fact, cryopreservation methods range from very slow cooling to instant vitrification of a sample by placing it directly in liquid nitrogen [4, 5]. The success of the different methods is related to microscopic processes which are induced by the freezing of extra- and intracellular liquids [4]. To be able to monitor a sample during the freezing process on-line on a microscopic scale could be helpful to optimize freezing protocols and to better

https://doi.org/10.1515/9783110429985-014

understand fundamental temperature-induced processes [6]. Both are scientifically as well as economically significant due to the general importance of cryopreservation. With this background in mind, we have combined multiphoton imaging systems with precise sample temperature control. This combination provides multiphoton imaging capability based on fluorescence intensity, lifetime and spectral characteristics of externally applied or intrinsic fluorophores during freezing/thawing/heating processes with subcellular resolution. The temperature control in the range from liquid nitrogen temperatures up to +600 °C is realized with a heating/freezing microscope stage. To illustrate the imaging capabilities, mammal cells were imaged at different temperatures with and without the addition of a cryoprotectant as well as parts of plant leaves.

Microscopic imaging of cooled or frozen samples can in principle also be realized with a conventional optical microscope combined with a cryostage. However, this limits the application to the observation of surfaces or very thin samples. In contrast, multiphoton imaging, which relies on near-infrared excitation, focusing with high NA microscope objectives, and a laser-scanning setup provides intrinsic 3D-imaging capability with high penetration depth, i.e., also below the surface inside a sample, due to low near-infrared absorption and scattering rates in biological material as well as further well-known advantages like low out-of-focus phototoxicity [7–9]. The method is an established microscopy technique which has already been extensively used to imagine human/animal/plant materials with a variety of objectives ranging from basic research to clinical diagnosis of diseases by in vivo imaging [8–14] and can be adapted for precise material processing as well [15].

12.2 Materials and methods

12.2.1 Heating and cooling stage

The heating and cooling stage (MDS 600, Linkham, "cryostage") controls the sample temperature in the range between −196 °C and +600 °C, i.e., between 77 K and 873 K, respectively. A photograph of the stage is shown in Fig. 12.1 [16]. The sample is placed onto a metal block (labelled "11" in Fig. 12.1) whose temperature is continuously measured and precisely controlled. The user can set desired values for temperature and cooling/heating rate in a control software interface. The rate can be set to zero, i.e., the temperature is kept fixed, or to any value in the range between ±0.1 K/min and ±150 K/min. Based on this input, control electronics dynamically regulate the amount of electrical heating and/or the flow rate of liquid-nitrogen through the stage.

For imaging, different kinds of microscope objectives can be combined with the cryostage. Dry objectives, with large working distances, provide the possibility to close the interior of the stage with a sealed glass window above the sample which improves the thermal isolation of the sample. In Fig. 12.1, the stage is shown with this cover in place. Nevertheless, also high NA oil-immersion objectives can be used. High NA ob-

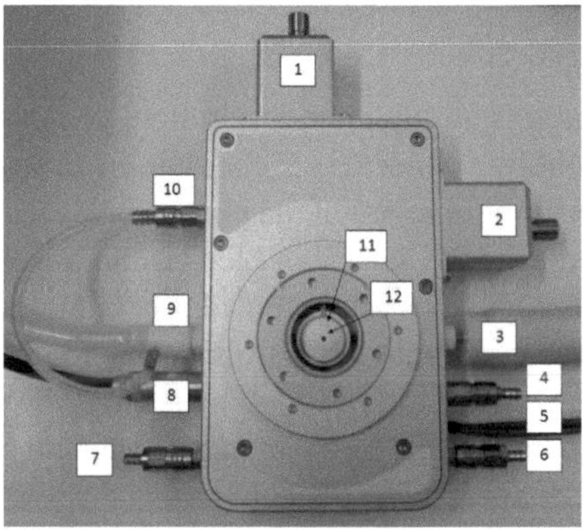

Fig. 12.1: Cooling and heating stage Linkham MDS 600 viewed from the top. 1, 2: micro-step motors for *x-y* positioning, 3: tube for liquid nitrogen in-flow, 4: nitrogen exhaust gas for housing flushing (outlet), 5: *x-y* motor control cable, 6: cooling water connection for housing (inlet valve), 7: cooling water connection for housing (outlet valve), 8: temperature control cable, 9: nitrogen gas (outlet), 10: nitrogen exhaust for flushing (inlet) of stage housing, 11: metal block where sample is placed, 12: illumination aperture. The interior of the stage is covered with a removable lid with a glass window for thermal shielding. The outer dimensions of the stage housing are 94 mm × 149 mm. From [16].

jectives are advantageous for two-photon imaging because with the same mean laser power a higher in-focus intensity can be realized which results in higher two-photon signals. When using oil-immersion objectives, the interior of the cryostage with the sample holder cannot be sealed but still be somewhat closed with a cooling jacket which fits around the objective to provide some thermal isolation. Nevertheless, in this case during cooling/heating, the oil still provides thermal contact between the temperature-controlled sample and the objective which is at ambient temperature. This may lead to a temperature gradient between sample surface and bottom, and some uncertainty of the precise sample temperature in particular for very low temperatures. Throughout this chapter all temperatures values given relate to the values measured inside the metal block onto which the sample was placed. We found that oil-immersion objectives can still be used down to −80 °C, although the immersion oil (Zeiss immersion oil 518F) solidifies at temperatures below −50 °C. In addition to the possibility to regulate the temperature, the stage provides the option to place the sample in a small dish which can be moved in *x* and *y* directions by microstep motors with a resolution of 0.05 µm and a maximum travel range of 15 mm in either direction. This option in principle allows centering a specific region of the sample in the focus of the objective or mapping a larger region of interest with mosaic imaging by stitching together individual images from neighboring regions.

12.2.2 Multiphoton imaging

The basis for multiphoton fluorescence imaging is multiphoton absorption, i.e., the simultaneous absorption of two or more photons by a molecule which results in an energetic excitation that corresponds to the sum of the energies of the photons involved [7, 17]. Molecules with one-photon-absorption bands in the UV or blue-visible wavelength region can, in this way, be excited by simultaneous absorption of two (or more) low energy photons in the near-infrared wavelength region. However, such a multiphoton absorption occurs at a significant rate only at high enough light intensities (MW/cm^2 to GW/cm^2) [7]. Those can be realized inside the focal volume (in the range of a femtoliter [11]) when ultrashort laser pulses are focused with high NA objectives. Outside the focal volume, the intensity is then not sufficient for multiphoton absorption to occur. This provides intrinsic 3D imaging capability with high resolution. For image acquisition, the laser focus is raster scanned across the sample. Fluorescence is collected (in case of reflection geometry) by the same focusing optics, separated from the incoming laser light by a dichroic mirror and a short-pass filter, which further discriminates residual laser light, and then detected for each pixel. Besides fluorescence, also other nonlinear signals like second-harmonic generation (SHG) can arise from the interaction between the intense laser light in the focal region and the sample depending on the specific sample properties as well as excitation conditions and can be used for imaging as well [17].

In order to elucidate the possibility to non-invasively image a section from the inside of a frozen sample, i.e., below the surface with the two-photon cryomicroscope, horizontal sections of an ice block containing fluorescent microspheres are shown in Fig. 12.2 [16]. The microspheres can be clearly identified against the nonfluorescent ice. Imaging up to a depth of 1.5 mm limited only by the working distance of the objective was possible, however, a higher laser power was necessary for imaging inside ice compared to imaging inside liquid water [16].

12.2.3 Imaging systems

The cryostage has been combined with several multiphoton imaging systems (*2PM Cryo* [16], *DermaInspect* [18], *MPTflex* [14]; all JenLab GmbH) that are all based on the well-known laser-scanning microscope setup with samples kept fixed, scanning of the laser focus, and signal detection in reflection geometry (Fig. 12.3 (a)) [8]. The platform *2PM Cryo* (Fig. 12.3 (b)) which was used for the animal cell imaging comprises of the cryostage (Fig. 12.1) in combination with a modified optical microscope (Axioskop, Zeiss). For cell imaging, the *2PM Cryo* was outfitted with a high NA objective (Zeiss, 40×, NA 1.3) which provided a maximum field of view of 250 μm × 250 μm and image resolutions of 0.5 μm in lateral and 1–2 μm in axial directions, respectively. Other objectives (oil immersion or dry) could in principle also be used for imaging.

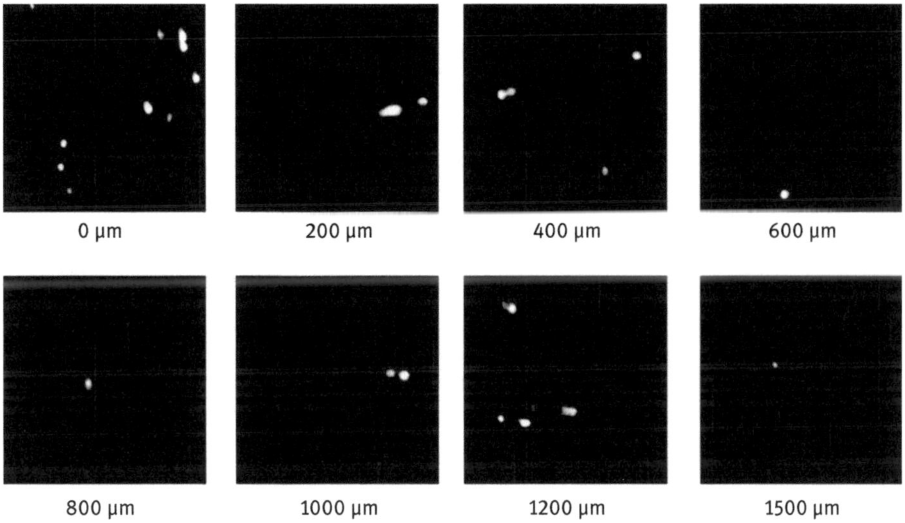

Fig. 12.2: Optical horizontal two-photon sections through an ice block which contains fluorescent microspheres. Indicated is the imaging depth inside the ice below the surface. The images show sections of 60 × 60 μm^2. From [16].

The multiphoton cryomicroscope *2PM Cryo* can be combined with different excitation lasers. For the measurements presented here, either a Ti:sapphire laser (Vitesse 800, Coherent; 800 nm center wavelength, 100 fs pulse width, 80 MHz repetition rate) or the frequency-doubled output of a fiber laser (mfiber, MenloSystems, 780 nm center wavelength, 250 fs pulse width, 250 MHz repetition rate) was used [19].

Alternatively, the cryostage can be combined with the clinical multiphoton tomographs *DermaInspect* and *MPTflex* (JenLab GmbH), respectively [11, 18]. These systems typically incorporate an fs laser (MaiTai XF, Newport Spectra Physics) which provides 100 fs pulses at a repetition rate of 80 MHz. The center wavelength of the pulses is tunable in the near-infrared wavelength range from 710 nm to 920 nm. The *DermaInspect* was used for imaging of the *Arabidosis thaliania* samples (Section 12.3.2) with the laser wavelength set at 800 nm and a maximum mean laser power of a few tens of mW [20].

In a further experimental setup, the cryostage was combined with a mobile multiphoton-imaging system (*MPTflex*, JenLab GmbH; Fig. 12.3 (c)). The *MPTflex* includes an articulated mirror arm for beam delivery and a mechanical arm that allows to freely position the scan-detector head such that otherwise hard-to-reach sample areas become accessible [21]. Depending on the configuration, the *MPTflex* also features coherent anti-Stokes Raman scattering (CARS) [22, 23] and fluorescence lifetime imaging (FLIM) capability [24]. To realize the FLIM capability, time-correlated single-photon counting (TCSPC) with a temporal resolution of about 170 ps (Becker & Hickl, SPC card 830) was employed in the system used here. FLIM images were obtained by pseudo-color coding the mean fluorescence decay times which were obtained from

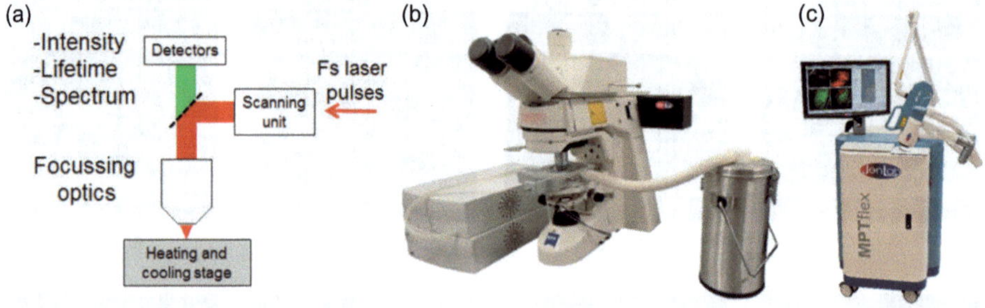

Fig. 12.3: (a) Scheme of multiphoton imaging systems based on laser scanning setup and signal detection in reflection geometry. (b) Imaging system *2PM Cryo* (without laser). The cryostage is combined with a laser scanning microscope. Next to the microscope are the control electronics for the stage and a Dewar with liquid nitrogen. (c) The cryostage can also be combined with the mobile multiphoton imaging system *MPTflex*. The *MPTflex* is a multipurpose multiphoton imaging system which incorporates fluorescence intensity, lifetime, SHG, and CARS imaging capability and features a scan-detector head attached to an articulated optical arm.

pixelwise fitting a two-exponential decay to the fluorescence data [24, 25]. The mean fluorescence decay time $\tau_m = a_1\tau_1 + a_2\tau_2$ was calculated from the decay time fitting parameters τ_1, τ_2 and the associated normalized amplitudes $a_{1,2}$. Intensity and FLIM images were typically obtained without further averaging at imaging speeds of 13.4 s per frame with 512 × 512 pixels and 256 × 256 pixels, respectively. Spectra were measured with a fiber-coupled thermoelectrically-cooled CCD-array spectrometer (B & W Tek, BTC112). The spectrometer had a wavelength-dependent resolution of a few nm. During the spectral measurements, the laser focus was continuously scanned over the sample and the signals summed for 30 s. To increase the signal-to-noise ratio, all spectra were smoothed by 30-point adjacent averaging. The spectra were not corrected for the spectral transmission of the system and detector sensitivity.

12.2.4 Sample preparation

12.2.4.1 CHO cells

As a model for animal cells, Chinese hamster ovary K1 cells (CHO-K1 cells ECACC, Sigma Aldrich #07K011) grown in a monolayer were imaged. Before starting the imaging procedure, the fluorescent growth medium was rinsed off the samples with phosphate buffered saline (PBS). The cells were placed on microscope cover glasses which were positioned on the temperature-controlled metal block of the cryostage (Fig. 12.1). The cryoprotectant dimethyl sulphoxide (DMSO) [3] was added to some of the cell samples up to a concentration of 10 vol.% to investigate its effect on the freezing process.

12.2.4.2 Plant samples

For imaging plant cells, small parts (≈ 2 × 2 mm^2) of leaves of *Arabidopsis thaliana* (ecotype Colombia – 0) were cut and fixed between two microscope coverslips which were then placed inside the cryostage on top of the temperature-controlled metal block (Fig. 12.1). No additional cleaning or treatments of the samples were performed. Cooling/heating rates of 10 K/min were used in between the measurements.

Furthermore, a leaf of Dieffenbachia, a popular house plant purchased at a local store, was imaged without further preparation or treatment before the measurement. A part of one of the leaves (few cm^2) was fixed on the temperature-controlled metal block inside the cryostage and cooled while the rest of the plant was at ambient temperature. During the imaging procedure, the leaf was not detached from the rest of the plant. The images were recorded a few minutes after the metal block reached the set temperature to allow the imaged region to reach a temperature equilibrium. Only a small part (≈ 200 µm × 200 µm) of the cooled region was imaged.

12.3 Results and discussion

12.3.1 Cell monolayer imaging

Cooling a cell that is inside a solution or in tissue below the freezing point leads initially to ice formation inside the extracellular environment because the cell membrane prevents (at first) intracellular ice formation. In this case, the supercooled intracellular water remains at a higher chemical potential than the partially frozen extracellular solution. This leads to a thermodynamic nonequilibrium which causes, depending on the cooling rate, subsequent cellular dehydration or intracellular ice formation [26–28]. With a slow cooling rate, the intracellular liquid can flow out of the cell through the semipermeable membrane and join the extracellular matrix before solidifying. This results in changes of the pH value as well as the ion concentrations inside the extracellular region and may cause the tertiary structure of proteins to fold, effectively eliminating the original cell properties [29, 30]. Furthermore, severe damage to cells may result from mechanical interactions with extracellular ice crystals [1]. Intracellular ice formation threatens intracellular structures and the cell membrane [31]. The balance between the water permeability of the membrane and the rate of intracellular ice formation leads to different optimal cooling rates for different cell types. For cryopreservation, typically, a slow freezing with cooling rates between 0.3 K/min and 10 K/min with 5–10 % cryoprotective agent in the medium is beneficial. An alternative approach is to immediately place the sample with a high cryoprotective agent concentration between 40 % and 70 % into liquid nitrogen (vitrification) [6]. This however may kill many cells.

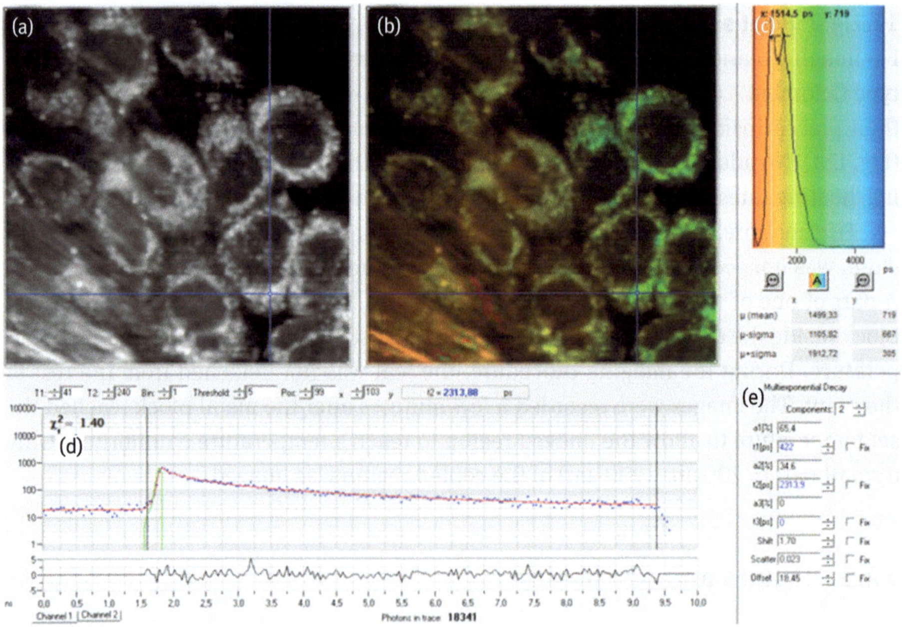

Fig. 12.4: FLIM measurement of CHO cells at a temperature of 0 °C. (a) Two-photon fluorescence image, (b) FLIM image (fluorescence decay time τ_2 color-coded according to color distribution shown in (c), (c) decay time distribution of the FLIM image shown in (b) and color code for τ_2 decay time values, (d) fluorescence decay curve at the position at the crossing of the blue lines in (a) and (b). The blue dots are measurement data, the red line a two exponential fit, (e) fit results of the data shown in (d). Laser parameters: wavelength: 780 nm; laser power: 13 mW. From [52].

An example for FLIM imaging of animal cells (CHO-K1 cells) is shown in Fig. 12.4. Starting at room temperature, the temperature was set to 0 °C with a freezing rate of 10 K/min. In the fluorescence intensity (Fig. 12.4 (a)) and FLIM (Fig. 12.4 (b)) images, the nonfluorescent cell nuclei appear as dark round regions inside the cells. With a laser wavelength of 780 nm, which was employed here, NAD(P)H molecules are efficiently excited by two-photon absorption molecules and constitute the main source of the fluorescence signals [18]. NAD(P)H is abundant inside the cell cytoplasm as well as inside mitochondria. The intracellular space outside the cells is presumably filled with water and/or ice, does not contain fluorescent material, and appears black. The fluorescent decay data were fitted with two-exponential decay functions. Fig. 12.4 (d) and 12.4 (e) exemplary show decay time data with the fitting curve and the corresponding fit parameters. The distribution of the values of the longer fluorescence decay time τ_2, shown in Fig. 12.4 (c), exhibits a maximum around 1000 ps. Fig. 12.5 shows the situation after further cooling to a set temperature of −100 °C at a rate of 10 K/min. Again, fluorescence intensity and FLIM images are shown in the figure. Comparing the FLIM images (Fig. 12.4 and 12.5) reveals an increase of bluish colored pixels which indicates a general increase of the τ_2 values with decreasing temperature. This can also be seen in the change of the distribution of the τ_2-decay times (compare Fig. 12.4 (c) and 12.5 (c)), which is shifted to higher values. The characteristics of the representative decay curve in Fig. 12.5 (d) changed to a single-exponential decay. These temperature-induced changes reflect the temperature-dependent characteristics of the NAD(P)H molecules which can also be observed when freezing pure NADH in vitro as described in [32].

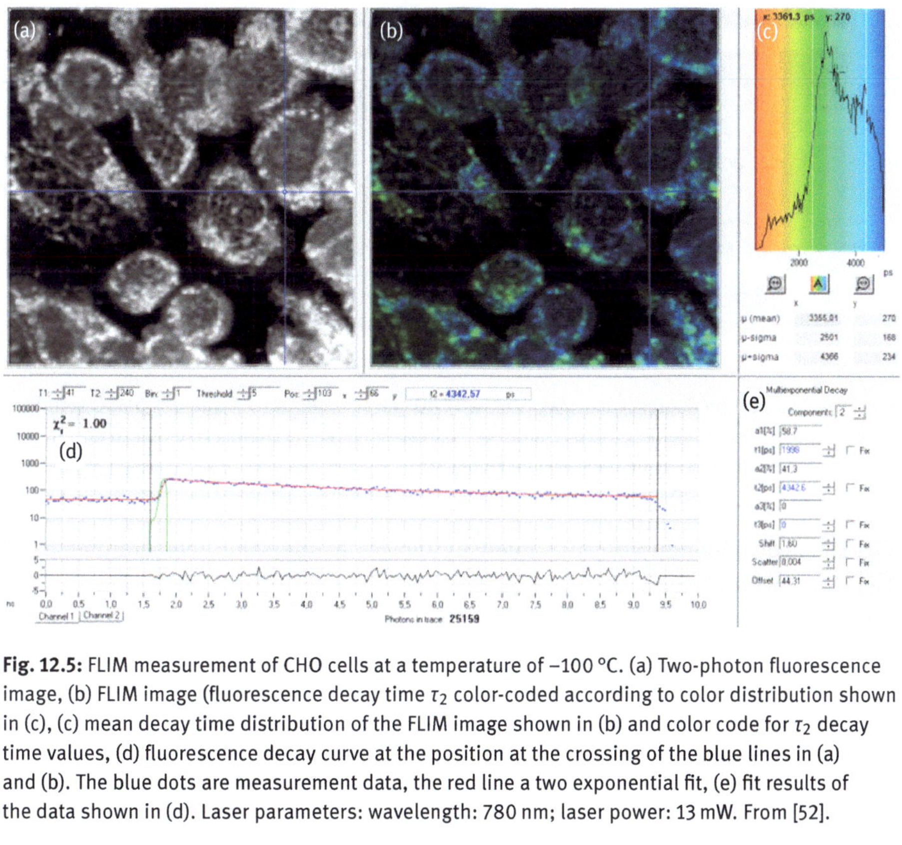

Fig. 12.5: FLIM measurement of CHO cells at a temperature of −100 °C. (a) Two-photon fluorescence image, (b) FLIM image (fluorescence decay time τ_2 color-coded according to color distribution shown in (c), (c) mean decay time distribution of the FLIM image shown in (b) and color code for τ_2 decay time values, (d) fluorescence decay curve at the position at the crossing of the blue lines in (a) and (b). The blue dots are measurement data, the red line a two exponential fit, (e) fit results of the data shown in (d). Laser parameters: wavelength: 780 nm; laser power: 13 mW. From [52].

The effect of the cryoprotectant DMSO and the influence of different cooling rates on cell samples are illustrated in Fig. 12.6 and 12.7. The sample was cooled with a rate of 10 K/min starting at room temperature and imaged at 25 °C, −40 °C and −80 °C. During the time taken to record an imaging, the temperature was kept constant. As can be seen (Fig. 12.6 and 12.7), the fluorescence intensity increases with deceasing temperature and in addition the cell morphology as well as the fluorescence pattern change significantly. The most intense fluorescence stems from the cell membranes. In the sample with 10 % DMSO the cell shape morphology is also affected by the temperature.

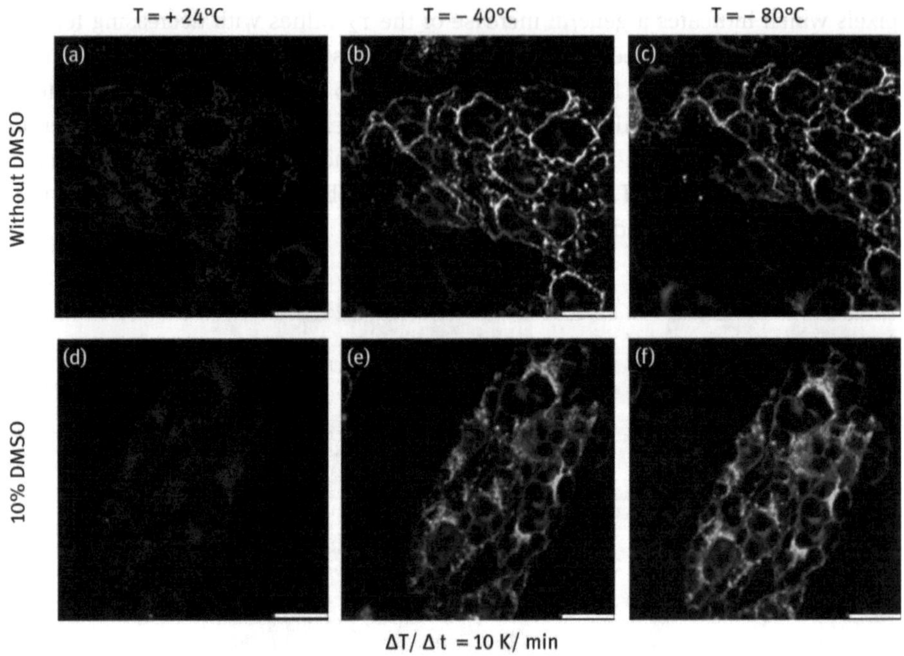

Fig. 12.6: Two-photon fluorescence images of two CHO cell samples in 10 % and 0 % DMSO solution at sample temperatures of 0 °C, −20 °C and −80 °C, respectively. The cooling rate was 10 K/min between the recordings. Scale bars: 20 μm. From [32].

Fig. 12.7 shows autofluorescence intensity images at the same temperatures as in Fig. 12.6 of samples cooled with a higher cooling rate of 100 K/min. In Fig. 12.7, the cell morphology appears to be less affected than the morphology of the cells that underwent the slower freezing procedure. The same strong increase of fluorescence intensity with decreasing temperature is nevertheless observable.

12.3.2 In situ imaging of plants

In general, plants contain quite a few intrinsic fluorophores that, in combination, generate fluorescence in the range from the UV to the far-red spectral region. These fluorophores are molecules and substances like chlorophyll a and b, ferulic acid, NADPH, FAD and rubisco [33]. Chlorophyll, which is present inside the chloroplasts of green mesophyll cells of leaves, generates fluorescence in the red (maximum around 680 nm) and far-red (maximum around 740 nm) spectral range [34, 35] whereas fluorescence emitted by ferulic acids, which are bound to cell walls, is maximum near 440 nm and 520 nm [33, 34, 36, 37]. Fluorescence imaging is an important tool to study plant conditions. The technique has been applied to investigate spatial and tempo-

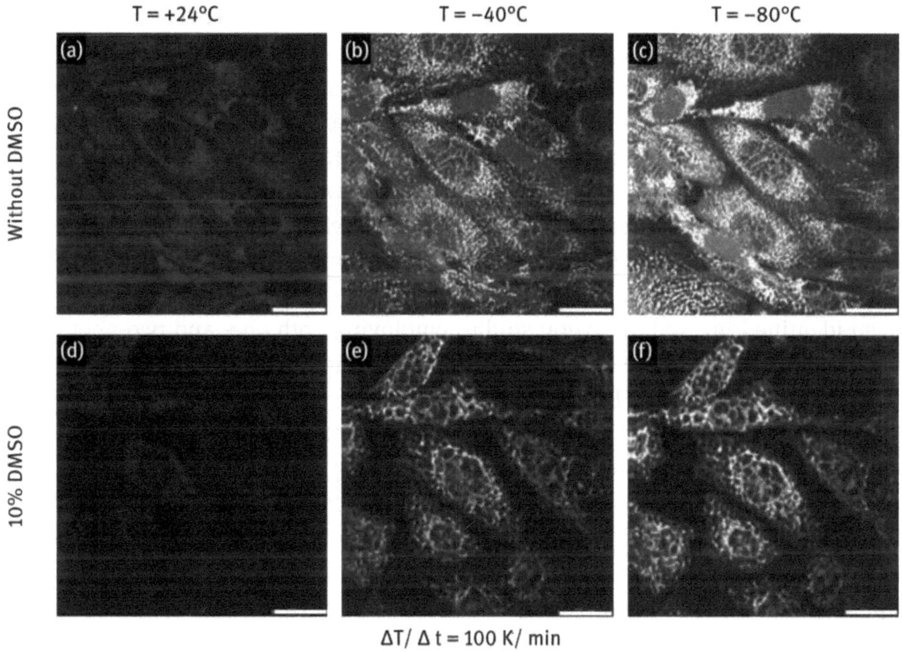

Fig. 12.7: Fluorescence images of two CHO cell samples in 10 % and 0 % DMSO solution at sample temperatures of 0 °C, −20 °C and −80 °C, respectively. The cooling rate was 100 K/min between the recordings. Scale bars: 20 μm. From [32].

ral changes due to stress put on plants, environmental effects, herbicide resistance, and pathogen toxin responses on a microscopic level [37–44] as well as to monitor photosynthetic chlorophyll activity [35, 45, 46].

The cryostage in combination with a multiphoton imaging system can be used to monitor temperature-induced effects in plants. As an example, fluorescence from an *Arabidopsis thaliana* leaf was recorded at temperatures of 0 °C and −20 °C (Fig. 12.8 (a) and (b)). The sample was cooled, starting from room temperature at a rate of 10 K/min. The fluorescence was excited by two-photon absorption with a laser power of 8 mW at 800 nm. The images show horizontal "optical cuts" of the same region of the mesophyll cell layer. Chloroplasts inside a cell can be seen in high detail (Fig. 12.8 (a)); a second cell is partly visible at the bottom part of the figure. The chloroplasts appear due to the strong chlorophyll fluorescence as bright oval objects with sizes of a few μm and are aligned along the cell walls. Further structures and shapes between the chloroplasts appear less bright and more diffuse. Besides fluorescence intensity, fluorescence lifetime can also be recorded and depicted in pseudo-color coded FLIM images. The corresponding FLIM image (Fig. 12.8 (c)) based on pseudo-color coding of the mean decay time τ_m indicates two different main sources of fluorescence. Pixels with mean decay times below and above 0.8 ns are colored in red and green, respectively. By this coloring, it becomes clear that the regions with short decay times contain chloroplasts which appear in red and the regions with longer decay times are the surrounding more diffuse structures which appear in green. The chloroplasts contain chlorophyll that emits strong and short-lived fluorescence signals. Therefore, mean decay times below 0.8 ns can be attributed to chlorophyll, decay times around and above 1 ns to fluorescence of molecules bound to cell walls, in particular ferulic acids, whose fluorescence is typically emitted at shorter wavelengths and decays slower than chlorophyll fluorescence [20, 34, 35, 37]. The FLIM images were obtained from two-exponential fitting. Two- and even three-exponential decays of the fluorescence have been identified [47, 48] in several studies employing both one- and two-photon excitation [44, 49, 50]. Fig. 12.8 (b) and (d) show the situation after further cooling of the sample at a rate of 10 K/min to a temperature of –20 °C. Freezing of the sample by further cooling leads to a compression of the plant cell and a confinement of the chloroplasts inside a smaller region. Such a shrinking occurred in several samples. It could be a result of purely temperature-induced effects, i.e., a sudden freezing of extracellular liquids which presses the cell and the contained chloroplasts together, or a response to the laser. The latter has been reported by Hasegawa et al. during multiphoton fluorescence imaging at room temperature to cause chloroplast relocation movements, however, mainly in z-direction [39]. Whereas in Fig. 12.8 (c) the individual chloroplasts are clearly distinguishable, in Fig. 12.8 (d) they appear as a continuous

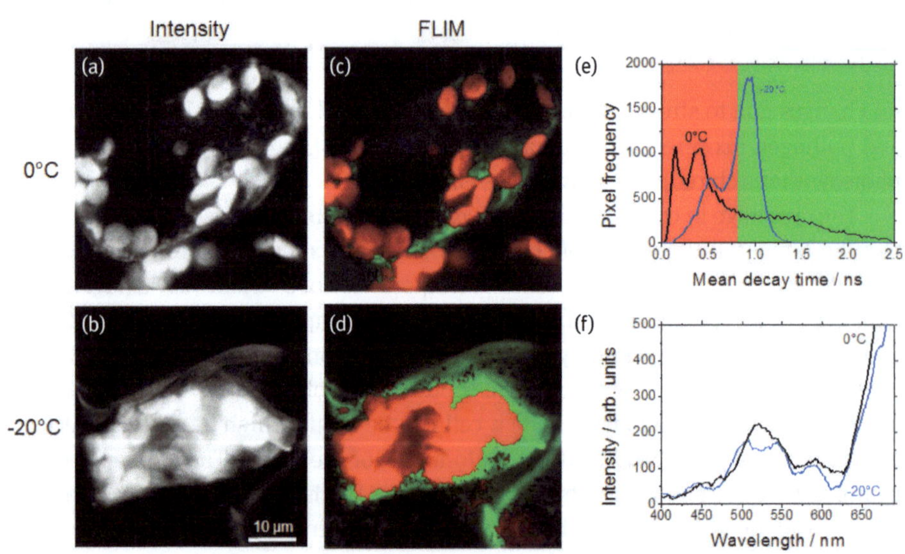

Fig. 12.8: (a)–(d) Multiphoton imaging of a plant leaf (*Arabidopsis thaliana*). (a) and (b) fluorescence intensity images, (c) and (d) color-coded mean fluorescence lifetime (FLIM) images, red: pixels with decay time below 0.8 ns, green: pixels with decay time above 0.8 ns. The color code is illustrated in (e) which shows the decay time distributions. (f) Micro-spectra collected from the imaged regions. Excitation: 800 nm pulses from Ti:sapphire laser. From [53].

red region in the image with fast fluorescence decays. The histograms of the mean decay time (Fig. 12.8 (e)) show a striking change from 0 °C (peaks below 0.5 ns, and around 1.0 ns) to −20 °C (maximum around 0.5 ns and strong maximum around 1 ns). Spectra of the fluorescence from the imaged region (Fig. 12.8 (f)) indicate signals in the range between 400 and 600 nm stemming from cell wall fluorescence, and additional signals above 600 nm as a result of the chlorophyll fluoresce [34, 36]. Further temperature effects as well as additional local maxima are present in the spectra which reflect the temperature influence on chlorophyll, and cell wall fluorescence and indicate an even more complex situation. A more detailed discussion of these temperature effects and the participating fluorescent molecules as well as their temperature-dependent dynamics is, however, beyond the scope of this chapter.

The cryostage in combination with multiphoton imaging can also be used to image parts of leaves which are still attached to the rest of the plant. To illustrate the possibility to investigate temperature-induced effects on fluorescence intensity and lifetime on a plant leaf, the cryostage was combined with the multiphoton tomograph *MPTflex*, which due to its articulated arm and freely positionable scan-detector head offers easier accessibility to large samples than the microscope platform. The leaf was pressed onto the cooling block of the cryostage and the scan-detector head placed onto the sample from above. Fig. 12.9 shows three FLIM images of the same leaf region after cooling of the cryostage to set temperatures of −25 °C, −50 °C and −80 °C, respectively. The chloroplasts containing chlorophyll as the main fluorophore exhibit shorter decay times than the other intracellular regions inside the cells which appear in green and blue pseudo-colors. The intracellular fluorescence results from other fluorophores like ferulic acids [36] which exhibit longer fluorescence lifetimes than chlorophyll [51]. The chloroplasts appear mainly red in Fig. 12.9 (a). With decreasing temperature (Fig. 12.9 (b) and (c)), more green and blue false-colored pixels appear indicating longer decay times. This is similar to the observations described before with *Arabidopsis thaliana* leafs (Fig. 12.8) [51] including probably temperature-induced structural changes in the imaged region, i.e., chloroplast relocations which are indicated by the arrows in Fig. 12.9 (a) and (b). Also, a movement of the whole sample resulting from the cooling is recognizable.

The images of animal cell and plant samples discussed above were based on detecting fluorescence light. Therefore, water and ice, which are nonfluorescent, were not directly visible. However, the actual forming of ice and its distribution inside a sample is of particular interest because ice formation is a major cause of cell damage. In principle, water and ice as well as different ice structures can be detected by their specific Raman transitions [27]. While Raman imaging is very time consuming, nonlinear Raman processes like CARS [29] could be a worthwhile additional imaging modality for multiphoton cryoimaging as already realized in a recent version of the *MPTflex* which incorporates additional CARS imaging capability [29].

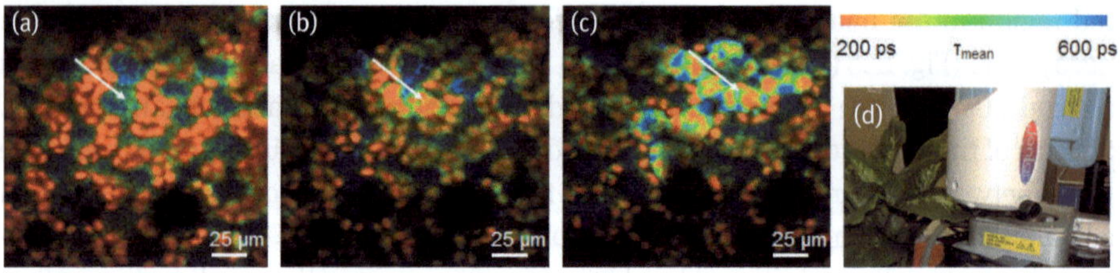

Fig. 12.9: In vivo FLIM images of a plant leaf. The plant with the imaged region was cooled by the cooling block of the cryostage while the rest of the plant was still at room temperature. Set temperatures: (a) −25 °C, (b) −50 °C, (c) −80 °C. The pseudo colors encode mean fluorescence decay times according to the color legend. The arrows point in each image to the same region where temperature-induced structural changes are visible. Laser parameters – wavelength: 820 nm; laser power: 13 mW. (d) During the measurement the leaf was not detached from the living plant. Imaging is possible in regions inside the cooled as well as uncooled parts. Adapted from [52].

12.4 Conclusion

We have described temperature-dependent multiphoton imaging of fluorescence intensity and lifetime of animal and plant cell samples. The sample temperature could be controlled by a stage in the range between −196 °C and +600 °C with cooling/heating rates from 0.1 K/min to 150 K/min by balancing the amounts of liquid nitrogen cooling and electrical heating. Multiphoton imaging was realized by near-infrared femtosecond-laser pulse excitation and laser imaging scanning systems (*2PM Cryo*, *MPTflex*, *DermaInspect*) in reflection geometry. For the animal cells, an increase of fluorescence intensity and lifetime with decreasing temperature was observed. Influences of a cryoprotectant (DMSO) on the freezing process could be clearly seen to affect the cell morphology. In an in vitro measurement, parts of plant leaves were imaged; "in vivo" results were presented from a leaf that was still attached to the plant during the imaging procedure. For the plant samples, temperature-induced changes like chloroplast relocation and changes in the fluorescence decay characteristics were observed as well as a trend to larger decay values with decreasing temperature in line with properties of chlorophyll reported in the literature.

References

[1] Meryman HT. Cryopreservation of living cells: principles and practice. Transfusion. 2007; 47(5):935–945.

[2] Day J, Stacey G. Biobanking. Mol Biotechnol. 2008;40(2):202–213.

[3] Fuller BJ. Cryoprotectants: the essential antifreezes to protect life in the frozen state. Cryo-Letters. 2004;25(6):375–388.

[4] Day JG, McLellan MR, editors. Cryopreservation and freeze-drying protocols. 1995 ed. Totowa, NJ: Humana Press; 1995.

[5] Sakai A, Engelmann F. Vitrification, encapsulation-vitrification and droplet-vitrification: a review. Cryo Letters. 2007;28(3):151–172.

[6] Mullen SF, Critser JK. The science of cryobiology. Oncofertility fertility preservation for cancer survivors. Springer; 2007. p. 83–109.

[7] Zipfel WR, Williams RM, Webb WW. Nonlinear magic: multiphoton microscopy in the biosciences. Nature Biotechnology. 2003;21(11):1369–1377.

[8] Denk W, Strickler JH, Webb WW. Two-photon laser scanning fluorescence microscopy. Science. 1990;248(4951):73–76.

[9] Zoumi A, Yeh A, Tromberg BJ. Imaging cells and extracellular matrix in vivo by using second-harmonic generation and two-photon excited fluorescence. Proceedings of the National Academy of Sciences of the United States of America. 2002;99(17):11014–11019.

[10] So PT, Dong CY, Masters BR, Berland KM. Two-photon excitation fluorescence microscopy. Annu Rev Biomed Eng. 2000;2:399–429.

[11] König K. Clinical multiphoton tomography. J Biophotonics. 2008;1(1):13–23.

[12] Zipfel WR, Williams RM, Christie R, Nikitin AY, Hyman BT, Webb WW. Live tissue intrinsic emission microscopy using multiphoton-excited native fluorescence and second harmonic generation. PNAS. 2003;100(12):7075–7080.

[13] Breunig H, Weinigel M, König K. In vivo imaging of ZnO nanoparticles from sunscreen on human skin with a mobile multiphoton tomograph. BioNanoSci. 2015;5(1):42–47.

[14] Weinigel M, Breunig HG, Uchugonova A, König K. Multipurpose nonlinear optical imaging system for in vivo and ex vivo multimodal histology. Journal of Medical Imaging. 2015;2(1): 016003-1–016003-10.

[15] König K, Ostendorf A, editors. Optically induced nanostructures, biomedical and technical applications. Berlin: De Gruyter; 2015.

[16] Breunig HG, Köhler C, König K. Two-photon cryomicroscope. Proc SPIE 8225; 2012. p. 822516-1-7.

[17] Boyd RW. Nonlinear optics. San Diego: Academic Press; 2003.

[18] Breunig HG, Studier H, König K. Multiphoton excitation characteristics of cellular fluorophores of human skin in vivo. Optics Express. 2010;18(8):7857–7871.

[19] König K, Uchugonova A, Breunig HG. High-resolution multiphoton cryomicroscopy. Methods. 2014;66(2):230–236.

[20] Breunig HG, Tumer F, König K. Multiphoton imaging of freezing and heating effects in plant leaves. J Biophotonics. 2012;6(8):622–630.

[21] Weinigel M, Breunig HG, Gregory A, et al. A novel flexible clinical multiphoton tomograph for early melanoma detection, skin analysis, testing of anti-age products, and in situ nanoparticle tracking. Proc SPIE 7589; 2010. p. 758908-1-7.

[22] Weinigel M, Breunig HG, Kellner-Höfer M, et al. New developments in clinical CARS. Proc SPIE 8588; 2013. p. 85881E-1-9.

[23] Breunig HG, Bückle R, Kellner-Höfer M, et al. Combined in vivo multiphoton and CARS imaging of healthy and disease-affected human skin. Microscopy Research and Technique. 2012;75(4):492–498.

[24] Suhling K, Hirvonen LM, Levitt J A, et al. Fluorescence lifetime imaging (FLIM): Basic concepts and some recent developments. Medical Photonics. 2015;27:3–40.

[25] Becker W, Bergmann A, Haustein E, et al. Fluorescence lifetime images and correlation spectra obtained by multidimensional time-correlated single photon counting. Microscopy Research and Technique. 2006;69(3):186–195.

[26] Devireddy RV, Swanlund DJ, Roberts KP, Pryor JL, Bischof JC. The effect of extracellular ice and cryoprotective agents on the water permeability parameters of human sperm plasma membrane during freezing. Hum Reprod. 2000;15(5):1125–1135.

[27] Mazur P. Cryobiology: the freezing of biological systems. Science. 1970;168(3934):939–949.
[28] Choi J, Bischof JC. Review of biomaterial thermal property measurements in the cryogenic regime and their use for prediction of equilibrium and non-equilibrium freezing applications in cryobiology. Cryobiology. 2010;60(1):52–70.
[29] Meryman HT. Cryopreservation of living cells: principles and practice. Transfusion. 2007;47(5): 935–945.
[30] Pegg DE. Principles of cryopreservation. Methods in Molecular Biology. 2007;368:39–57.
[31] Pegg DE. The relevance of ice crystal formation for the cryopreservation of tissues and organs. Cryobiology. 2010;60(Suppl 3):S36–S44.
[32] König K, Uchugonova A, Breunig HG. High-resolution multiphoton cryomicroscopy. Methods. 2014;66(2):230–236.
[33] Cerovic ZG, Samson G, Morales F, Tremblay N, Moya I. Ultraviolet-induced fluorescence for plant monitoring: present state and prospects. Agronomie. 1999;19(7):543–578.
[34] Buschmann C, Langsdorf G, Lichtenthaler HK. Imaging of the blue, green, and red fluorescence emission of plants: An overview. Photosynthetica. 2000;38(4):8.
[35] Krause GH, Weis E. Chlorophyll fluorescence and photosynthesis: the basics. Annual Review of Plant Physiology and Plant Molecular Biology. 1991;42(1):313–349.
[36] Lichtenthaler HK, Schweiger J. Cell wall bound ferulic acid, the major substance of the blue-green fluorescence emission of plants. Journal of Plant Physiology. 1998;152:272–282.
[37] Meyer S, Cartelat A, Moya I, Cerovic ZG. UV-induced blue-green and far-red fluorescence along wheat leaves: a potential signature of leaf ageing. J Exp Bot. 2003;54(383):757–769.
[38] Gilroy S. Fluorescence Microscopy of Living Plant Cells. Annu Rev Plant Physiol Plant Mol Biol. 1997;48:165–190.
[39] Hasegawa M, Shiina T, Terazima M, Kumazaki S. Selective excitation of photosystems in chloroplasts inside plant leaves observed by near-infrared laser-based fluorescence spectral microscopy. Plant Cell Physiol. 2009;51(2):225–238.
[40] Chaerle L, Leinonen I, Jones HG, Van Der Straeten D. Monitoring and screening plant populations with combined thermal and chlorophyll fluorescence imaging. J Exp Bot. 2007;58(4): 773–784.
[41] Barbagallo RP, Oxborough K, Pallett KE, Baker NR. Rapid, noninvasive screening for perturbations of metabolism and plant growth using chlorophyll fluorescence imaging. Plant Physiol. 2003;132(2):485–493.
[42] Wild E, Dent J, Barber JL, Thomas GO, Jones KC. A novel analytical approach for visualizing and tracking organic chemicals in plants. Environmental Science & Technology. 2004;38(15): 4195–4199.
[43] Frolec J, Krchňák P, Sušila P, Nauš J. Irreversible changes in barley leaf chlorophyll fluorescence detected by the fluorescence temperature curve in a linear heating/cooling regime. Photosynthetica. 2008;46(4):10.
[44] Feijo JA, Moreno N. Imaging plant cells by two-photon excitation. Protoplasma. 2004;223(1): 1–32.
[45] Baker NR, Oxborough K, Lawson T, Morison JI. High resolution imaging of photosynthetic activities of tissues, cells and chloroplasts in leaves. J Exp Bot. 2001;52(356):615–621.
[46] Maxwell K, Johnson GN. Chlorophyll fluorescence – a practical guide. J Exp Bot. 2000;51(345): 659–668.
[47] Moya I, Silvestri M, Vallon O, Cinque G, Bassi R. Time-resolved fluorescence analysis of the photosystem II antenna proteins in detergent micelles and liposomes. Biochemistry. 2001; 40(42):12552–12561.

[48] Tirlapur UK, König K. Two-photon near-infrared femtosecond laser scanning microscopy in plant biology. In: Diaspro A, editor. Confocal and two-photon microscopy: foundations, applications and advances. New York: Wiley-Liss; 2002.
[49] van Grondelle R, Dekker JP, Gillbro T, Sundstrom V. Energy transfer and trapping in photosynthesis. Biochimica et Biophysica Acta (BBA) – Bioenergetics. 1994;1187(1):1–65.
[50] Broess K, Borst JW, van Amerongen H. Applying two-photon excitation fluorescence lifetime imaging microscopy to study photosynthesis in plant leaves. Photosynth Res. 2009; 100(2):89–96.
[51] Breunig HG, Tumer F, König K. Multiphoton imaging of freezing and heating effects in plant leaves. J Biophotonics. 2013;6(8):622–630.
[52] Breunig HG, Uchugonova A, König K. Multiphoton imaging of biological samples during freezing and heating. Proc SPIE 8948; 2014. p. 894819-1–894819-7.
[53] Breunig H, Uchugonova A, König K. Multiphoton cryo microscope with sample temperature control. Proc SPIE 8588; 2013. p. 858819-1–858819-6.

Part III: **Nonlinear tissue imaging**

Karsten König

13 Multiphoton Tomography (MPT)

A novel imaging tool for high-resolution multimodal in-vivo histology

Abstract: Multiphoton tomography is a novel tissue imaging method with superior subcellular spatial resolution and picosecond temporal resolution. It provides non-invasively and label-free optical biopsies within seconds based on autofluorescence (AF) and second-harmonic generation (SHG). Besides 3D morphology, MPT enables optical metabolic imaging (OMI) by fluorescence lifetime imaging (FLIM) of the coenzymes NAD(P)H and flavins, the sensitive biosensors of cellular metabolism and oxidative stress. The add-on modules "CARS" and "microendoscope" provide further chemical information and deep tissue endomicroscopy.

The multiphoton tomographs *DermaInspect*, *MPTflex*, and *MPTflex-CARS* have become CE certified medical devices. They are used in clinics around the world as well as in research centers of cosmetic and pharmaceutic companies. Major applications include early skin cancer detection, evaluation of skin modifications after long-term space flights, testing of anti-ageing substances, and intratissue nanoparticle tracking. Recently, the first intraoperative MPT-imaging during brain tumor surgery as well as MPT of human cornea have been reported.

Besides clinical applications, the novel multiphoton tomographs are used as upright and inverted two-photon/SHG/FLIM microscopes to watch stem cells at work in transgenic mice, to perform in situ pharmacokinetics in the liver of small animals, to screen ex vivo human biopsies directly in the operation theater, and to image cell monolayers and 3D cell clusters.

13.1 Introduction

The word tomography is derived from the Greek words "tomos" ("slice") and "graphein" ("write") and refers to imaging by sectioning. The most common medical tomography is based on X-rays which were discovered in Würzburg/Germany on November 8, 1895 by *Wilhelm Conrad Röntgen*. Röntgen became the first Nobel Prize winner. X-ray tomography was finally introduced in the 1930s by the Italian radiologist *Vallebona* and further developed as computer tomography (CT) by the electrical UK engineer *Sir Hounsfield* and the South African physicist *Cormack*. In 1971, the first CT scan was performed in man on a cerebral cyst patient in London. Both, the electrical engineer and the physicist, were honored with the Nobel Prize in Physiology or Medicine in 1979 [1].

https://doi.org/10.1515/9783110429985-015

Further medical tomographs using radioactive tracers are positron-electron tomographs (PET) based on electron-positron annihilation and SPECT based on gamma rays. PET was invented by the US American physicists *Brownell* and *Sweet* at the *Massachussetts General Hospital* (MGH) in Boston in the 1950s and further developed by the German-Armenian *Ter-Pogossian* and the US mathematician and chemists *Phelps* [2, 3]. Single photon emission computed tomography (SPECT) was invented by the US medical doctor *Kuhl* and the engineer *Edwards* in 1963. SPECT became a clinical method in the 1980s [4].

Magnetic resonance tomography (MRT) is based on nonionizing radiation, in particular on magnetic fields and radio frequency waves. It was invented by the US American chemist *Lauterbur* and the English physicist *Sir Mansfield* (Nobel Prize in Physiology or Medicine in 2003). Also the Soviet Union scientist *Ivanov* claims to be the inventor of MRT based on a patent filed in 1960 [5]. The first MRI human body scan was performed on July 3, 1977 [6].

Also today's major clinical diagnostic imaging technology, medical ultrasound imaging, can be considered as a tomographic tool when performing 3D scans such as in breast imaging (USCT, [7]). The Austrian brothers (neurologist and physicist) *Dussik* performed the first ultrasound imaging in the human body (brain) in 1941. Also the US American and MD *Ludwig* wrote a paper on diagnostic ultrasound but it was not released to the public by the *Department of Defense* until 1949. In the 1950s, *Wild et al.* used ultrasound to detect cancer [8, 9].

The highest spatial resolution can be achieved when using optical radiation. Medical optical tomography based on reflected/scattered photons are known as confocal reflectance laser scanning microscopy (reflectance CLSM) and optical coherence tomography (OCT).

The slit lamp used in ophthalmology can be considered as some kind of first confocal device. It was developed by the Swiss ophthalmologist *Goldman* in 1939 [10]. However, the US scientist *Minsky* is considered as inventor of the confocal microscope based on a single point scanner. He built a first prototype for brain imaging already in 1955 ("pre-laser time") and filed a patent in 1957 [11, 12]. In the 1960s, the Czech medical doctor *Petran* developed the tandem scanning microscope based on multiple excitation and detection pinholes produced by a rotating Nipkow disk [13]. In 1969, *Egger and Davidovits* published on the first confocal *laser* scanning microscope based on a fixed point scanner and stage scanning [14].

Brakenhoff et al. demonstrated the elimination of out-of-focus blur in a biological specimen. Finally, the British scientists Amos, Durbin, Fordham and White built in Cambridge the first confocal beam scanning microscope with a variable pinhole and licensed their technology to the US company *Bio-Rad*. At the end of 1987, the first commercial confocal beam scanning microscopes were sold [15–17]. Around that time and before the unification of Germany, the East German ZEISS company in Jena, together with our lab at the University Jena, built the first confocal ultrashort laser scanning microscope with picosecond time resolution [18–21].

The first papers using confocal microscopy for imaging human skin in vivo appeared in 1993 and 1994 [22–25]. *Radjadhyaksha* reported on the use of a CLSM for in vivo skin imaging in 1995 [26]. Today's commercialized skin imaging system *Vivascope* is based on a CLSM in the reflection mode using a continuous wave (cw) NIR laser diode. It is mainly used to detect nonmelanoma skin cancer [27].

Confocal microscopy is also applied to image the human eye in vivo [28]. In this case, confocal CLSM is used in both modes, the reflection mode as well as in the fluorescence mode. The fluorescence mode requires the application of exogenous fluorescent dyes for angiography. (The use of confocal fluorescence CLSM for skin imaging did not succeed due to the required use of an exogenous fluorophore.) Early work on scanning laser ophthalmoscopy dates back to the 1980s [29, 30]. In the 1990s, confocal microscopy of the living eye was performed [31, 32].

Also in the 1990s, OCT was pioneered by *Fujimoto's* group at MIT [33–35]. The first in vivo OCT images showing human retinal structures were taken in 1993 [34] and the first certified medical OCT product was released in 1996. OCT is now the most well-established optical tomography method in medicine, in particular for retina imaging. The wide-spectrum NIR light sources (about 100 nm wavelength range) include superluminescent diodes, ultrashort pulsed lasers, and supercontinuum lasers. Frequency domain OCT has the advantages of high speed and signal-to-noise ratio over spectral OCT [36].

The first laser scanning microscope that employed an ultrashort laser was built in Jena at the end of the 1980s [18, 19, 21]. It was based on a picosecond argon-ion/dye laser and was used for fluorescence lifetime imaging (FLIM). Shortly after, two-photon microscopy based on a sub-picosecond dye laser was invented by *Denk, Strickler, and Webb* in the US [37]. A patent was filed and licensed to *Bio-Rad*. Later on, ZEISS purchased *Bio-Rad*. Nowadays, two-photon microscopes are the major tool for live cell imaging [38].

Imaging of the human skin in vivo was first performed by Masters et al. at the University of Illinois at Urbana at the end of the 1990s on the right forearm of a lab group member positioned at the stage of a custom-made 80 MHz femtosecond laser scanning ZEISS Axiovert 35 microscope with FLIM capability in the frequency domain shortly after the investigation of two-photon autofluorescence of photostressed cultured cells with this lab microscope [39, 40]. Intratissue skin cells could be monitored based on two-photon and three-photon excited NAD(P)H and flavoprotein autofluorescence at 730 nm and 960 nm excitation wavelength down to 80 mm skin depth. The large difference between the phase and modulation fluorescence lifetimes indicated multiple species in the excitation volume with mean values between 0.2 ns and 3.1 ns. Point emission spectra revealed two broad maxima at 440 nm (FWHM: 75 nm) and 480 nm (FWHM: 75 nm) when excited at 730 nm and a narrow unknown peak at 420 nm (FWHM: 30 nm) as well as a broad band at 520 nm (FWHM: 100 nm) when excited at 960 nm.

A major step to the medical use of two-photon imaging was made by the company *JenLab GmbH*. Within the first years of the new century, clinical MPT was introduced by JenLab based on the CE certified medical product called *DermaInspect* [41–43]. Since then, multiphoton tomography has emerged as a powerful non-invasive label-free multimodal imaging technique for in vivo skin characterization. MPT has superior submicron spatial resolution, single-photon sensitivity, and optical metabolic imaging (OMI) capability. MPT provides images of collagen by SHG and chemical information by coherent anti-Stokes Raman spectroscopy (CARS). Recently, the application field was enhanced to cornea imaging and to intraoperative human brain imaging.

Medical imaging based on photoacoustic tomography (PAT) is relatively new. Typically, nanosecond visible laser radiation leads to a thermoelastic expansion resulting in wideband ultrasound waves that can be detected by ultrasonic transducers. First medical devices focus on imaging of blood vessels and their oxygen concentration based on light absorption by hemoglobin [44, 45].

Fig. 13.1 and Tab. 13.1 show an overview on different tomograph methods and in particular the various resolution values and signal depths.

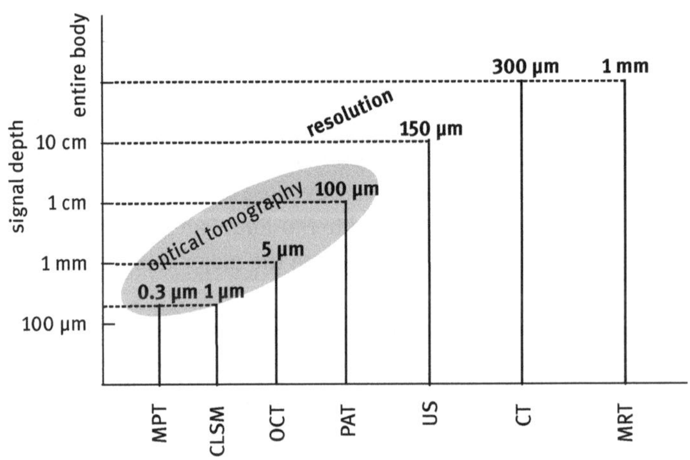

Fig. 13.1: Resolution and signal depths in medical tomography.

13.2 Principle of Multiphoton Tomography (MPT)

MPT is based on nonlinear effects of near-infrared (NIR) femtosecond laser pulses. So far, the major source is the mode-locked titanium:sapphire laser with its tuning range of 670 nm to 1000 nm, which corresponds to the first optical window of tissues with low absorption coefficients (e.g., 0.1 cm^{-1} at 1000 nm for water as the major biomolecule) and low scattering coefficients. Meanwhile, also higher NIR wavelengths

Tab. 13.1: Comparison of medical tomography methods.

Method	Introduction	Key element	Resolution	Signal depth	Price
US	1941	ultrasound	> 100 μm	> 10 cm	€
PET	1950s	positron-electron	> 1 mm	> 10 cm	€€€
CT	1971	X-rays	> 1 mm	> 10 cm	€€
MRT	1977	magnetic fields radio waves	> 1 mm	> 10 cm	€€€
SPECT	1980s	γ-rays	> 1 mm	> 10 cm	€€€
OCT	1991	near infrared	> 5 μm	> 500 μm	€
CLSM	1992	near infrared	> 1 μm	< 200 μm	€
MPT	2002	near infrared	< 1 μm	< 1000 μm	€€
PAT	2010	near infrared	> 500 μm	< 3 cm	€€

are employed, such as at 1700 nm, where even higher signal depths of more than 1 mm in brain tissue may be achieved.

The inherent optical sectioning originates in the nonlinear absorption process within the sub-femtoliter laser excitation focal volume when using high NA objectives. NIR light can penetrate deeper in highly scattering tissues such as skin than confocal microscopes operating in the UV and visible spectral range. Also the light collection efficiency is higher than in CLSM because spatial filters (pinholes) and "descanning" optics are not required. All fluorescence photons obtained by multiphoton excitation can be detected even when traveling large distances by multiple scattering in the turbid medium with large area detectors close to the focusing optics. Furthermore, MPT is less phototoxic than CLSM as demonstrated on long photoexposed hamster embryos [46] and the evaluation of DNA damage of UV- and NIR fs laser exposed human skin [47]. Therefore, MPT is the perfect tool to perform long-term studies on living intratissue cells.

The major signal source for MPT is fluorescence. It is relatively easy to detect fluorophores with high fluorescence quantum yield such as the DNA marker Hoechst, fluorescein or fluorescent proteins such as the green fluorescent protein GFP or dsRed. However, a major advantage of MPT is the possibility to detect the weak tissue autofluorescence (AF) based on a variety of biomolecules with low fluorescence quantum yield. AF can act as source of natural contrast for clinical applications providing therefore a label-free high-resolution imaging method.

Skin AF originates from endogenous intratissue biomolecules such as reduced pyridine nucleotides (NADH and NADPH) and flavins/flavoproteins. The major absorption band of NAD(P)H has a maximum around 340 nm. Ultraviolet (UV) skin excitation at 365 nm could be used as excitation source for cell monolayers. However, UV would not allow to detect NAD(P)H in the deep skin layers stratum spinosum and stratum basale due to the low UV penetration depth on the order of 20 μm. In contrast, with two-photon NIR excitation, these important fluorescent biomolecules with broad two-photon absorption spectra can easily be detected for laser excitation wavelength of 700–800 nm.

Only the reduced forms of the pyridine nucleotides are fluorescent. NADH and NADPH possess the same emission spectrum but different fluorescence lifetimes. Both reduced coenzymes as well as the flavin coenzymes in its free form and bound form (NAD(P)H-protein, flavoproteins) play an important role in cellular metabolism, namely the respiratory chain, and act as sensitive biosensors. Detection of their fluorescence in the mitochondria as well as the cytoplasm allows optical metabolic imaging (OMI) [48–51], in particular when using FLIM. Interestingly, the measurement of the fluorescence lifetimes allows to distinguish between free and bound coenzymes. NAD(P)H has a picosecond lifetime in its free form but a long 2–4 ns fluorescence lifetime when bound to proteins. In contrast, flavoproteins possess a shorter lifetime than free flavins. Also tryptophan, keratin, melanin, elastin, and porphyrins are AF sources.

MPT relies not only on fluorescence signals. Another major advantage compared to CLSM is the generation of SHG, providing therefore a signal at exactly half the laser wavelength. A major SHG source is collagen. Although SHG is generated in forward direction, it can be easily detected in all types of tissue because of backscattering effects.

A further signal is inelastic Raman scattering, where the signal wavelength is slightly shifted from the incident laser wavelength due to interactions with molecular vibrations. Detection of the wavelength shift would provide a chemical fingerprint to identify the molecule. However, spontaneous Raman scattering is extremely weak and superimposed by the intense scattered laser light. Imaging of spontaneous Raman scattering is not an option for live tissue imaging where beam dwell times per pixel of some microseconds are required.

Raman signals are not detected during one-beam MPT. However, when providing a second laser beam at a wavelength that matches the vibration band of interest, CARS can be realized. CARS provides the same information on vibration bands as Raman but allows a high imaging speed in the range of some seconds per optical section because the CARS signal is several orders stronger. CARS is based on multiphoton excitation (third-order nonlinear process) that produces a coherent signal in forward direction [52]. As in the case of SHG, the CARS signal in turbid media such as skin can easily be detected by backscattering.

The two-beam tomograph *MPTflex-CARS* possesses an add-on CARS module based on white light generation with a photonic crystal fiber (PCF) that allows the detection of nonfluorescent and non-SHG-active biomolecules and tissue structures, such as lipids and intratissue water. Typically, the CARS signal is measured in the red spectral range. It is composed of a resonant coherent signal and a nonresonant background [52, 53]. Fig. 13.2 shows multiphoton processes used in commercial MPT devices.

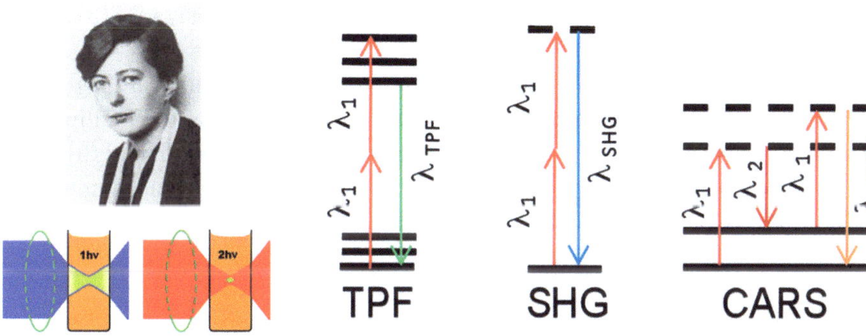

Fig. 13.2: Multiphoton processes employed in medical multiphoton tomography. The PhD student Maria Göppert predicted the two-photon absorption process in Göttingen in 1932. The two-photon excited fluorescence (TPF), second-harmonic generation (SHG), and Coherent Anti-Stokes Raman Spectroscopy (CARS) signals are detected in a sub-femtoliter nonlinear excitation volume for multimodal MPT imaging.

13.3 Multiphoton tomographs

Fig. 13.3 demonstrates the CE certified medical devices *DermaInspect*, *MPTflex*, and *MPTflex-CARS* that are currently available. These systems are the only certified femtosecond laser systems worldwide for medical imaging. They possess a tunable 80 MHz mode-locked titanium:sapphire laser with a typical tuning range of 730–960 nm and an output pulse width of about 100 fs. The typical laser excitation wavelength for NAD(P)H/collagen imaging is 780 nm. Flavins/flavoproteins can be excited without "NAD(P)H background fluorescence" at 840 nm, for melanin the best is in the range around 900 nm. The in situ mean laser power at the tissue is maximal 50 mW as required by the *Notified Bodies*, even when using the two-beam CARS system. Typically, MPT on skin is performed with 20 mW mean power or less. The laser light is focused with NA 1.3 optics and a working distance of 200 μm when used on humans but can be changed for ex-vivo samples and live animal studies.

The scanning is achieved by a galvanometer-driven x-y scanner that allows also single-point illumination, fast line and region-of-interest (ROI) scans. A typical optical section consists of 512×512 pixels covering a field-of-view (FOV) of 0.3×0.3 mm^2 and takes 1–6 s when imaging skin. A 3D stack of horizontal images ("en face") can be obtained by controlling the tomograph's focusing optics with either a piezomotor or a stepper motor. The typical depth increment is 5 μm or 10 μm but can be reduced to the submicron range.

Also vertical images (x, z sections) can be obtained that fit more to the typical images of the pathologists who use conventional physically taken histological sections.

The signals are obtained by multiple photomultipliers in the analog mode (gray levels) as well as in the single photon counting (SPC) mode. For FLIM, time-correlated SPC (TCSPC) is employed with a typical temporal resolution of 170 ps. For multimodal imaging, a four-detector head is employed to measure simultaneously AF, SHG, FLIM, and CARS within one scan.

Furthermore, a two-photon microendoscope with a high numerical aperture of 0.8 can be magnetically attached to the scan-detection module of the multiphoton tomographs. This special high NA endoscope is based on a combination of Grin lens and a semisphere. A typical outer diameter is 1.2 mm. The two-photon microendoscope is in clinical use in German hospitals (dermatology) to evaluate wound healing effects.

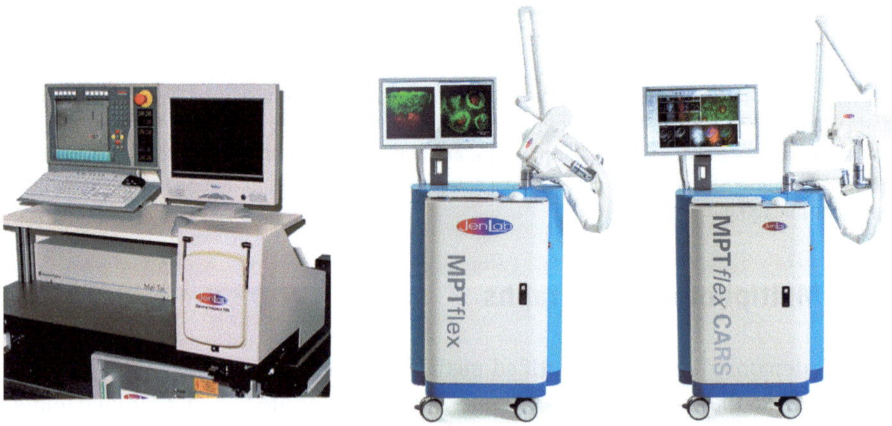

Fig. 13.3: Photographs of the certified medical multiphoton tomographs *DermaInspect* (first system), *MPTflex*, and *MPTflex-CARS*.

13.4 In vivo histology based on MPT

The gold standard for histology is the piece of tissue taken invasively (e.g., by a 3 mm punch biopsy) that is fixed (e.g., with formalin), sliced into hundreds of thin sections (e.g., 7 µm thick) and stained with the dye mixture hematoxylin-eosin (HE). HE was introduced in the 19th century but is still the most widely used stain in medical diagnosis [52]. It involves a basic complex of aluminum ions and hematein that colors the nuclei blue as well as acidic eosin Y that colors various cell structures in red, pink, and orange. Some features, such as elastin, cannot be imaged with the HE stain. Elastin imaging requires further staining procedures. A typical pathologist will examine these stained histological sections with a conventional white light microscope in the trans-illumination mode with objectives of typical numerical apertures (NA) from 0.2–1.0 (air) and a resolution between 0.5 µm and 2 µm, respectively. In most cases, vertical sections will be analyzed. The typical feedback time to the dermatologist/surgeon with the diagnosis of the pathologist is on the order of 3–7 days after the biopsy was taken.

MPT provides non-invasively label-free optical biopsies within seconds with a superior resolution of 0.3 µm lateral and 1–2 µm axial. This excellent resolution fits therefore perfectly the requirement of the conventional pathologist. The nuclei can be easily identified due to the lack of fluorescence/SHG. They appear as dark features. The cytoplasm and in particular the mitochondria can be easily imaged due to NAD(P)H fluorescence. Furthermore, single elastin fibers can be visualized by autofluorescence without any further staining. Also collagen can be easily imaged by AF and SHG. Fig. 13.4 shows non-invasively taken label-free MPT sections imaged with the femtosecond laser tomograph in comparison with conventional HE stained skin sections obtained from an invasively taken punch biopsy.

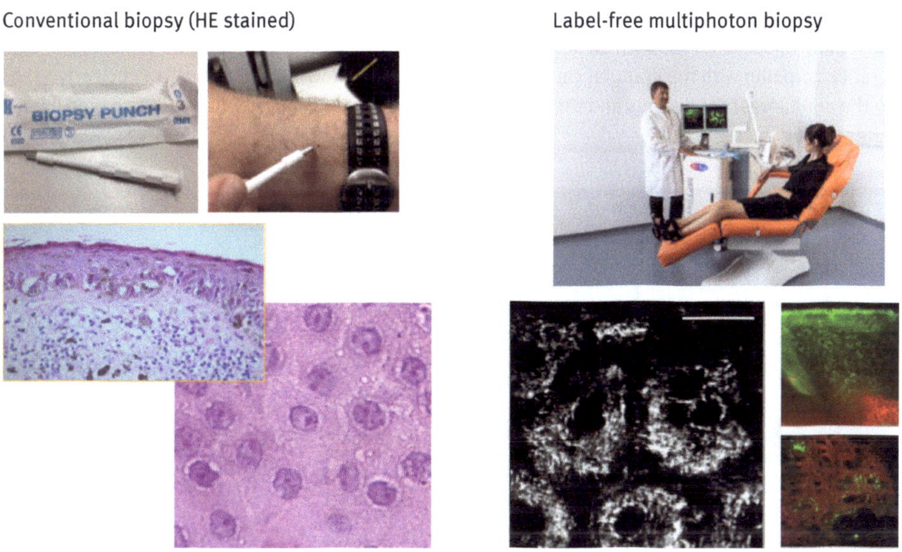

Fig. 13.4: HE stained sections of a punch biopsy in comparison with label-free multiphoton sections. Green: AF, red: SHG (collagen).

MPT provides axial, diagonal, and horizontal sections due to galvoscanning regimes. It allows also digital slicing out of the cube/stack of optical sections (Fig. 13.5). Furthermore, larger tissue areas can be imaged in the mosaic mode, where consecutive scans are taken tiled-wise (Fig. 13.6). Examples of a "vertical" skin section and a horizontal section ("en face") near the epidermis-dermis junction as well as a large-field 4×4 mm^2 horizontal MPT skin section are shown.

The superior high spatial resolution of MPT enables the in vivo imaging of single intratissue mitochondria and single elastin fibers as well as collagen bundles in human skin.

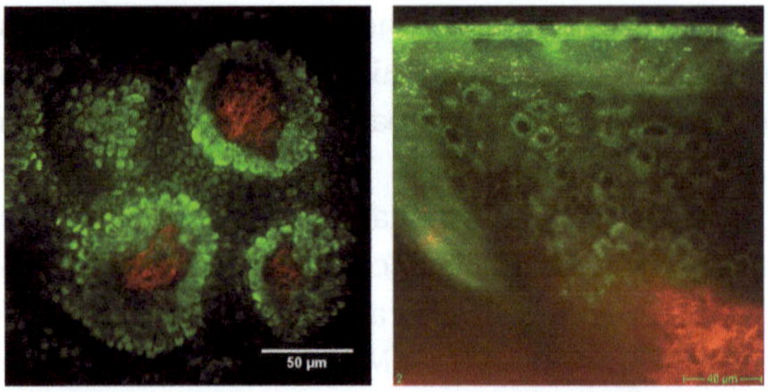

Fig. 13.5: Label-free horizontal "en face" multiphoton section and vertical section of in vivo human skin. Green: AF (NADH, flavins, melanin, keratin, single elastin fibers), red: SHG (collagen). The "en face" section near the epidermis-dermis junction exhibits a multiphoton cut through 3 dermal papillae with collagen (red) in the center, surrounded by highly fluorescent melanin-containing basal cells. The vertical section demonstrates the strong fluorescent outermost layer stratum corneum with the granular cell layer beneath. Further down, the living cells are clearly seen by the nonfluorescent nucleus and the fluorescent mitochondria. Finally, SHG-active collagen dominates the signal.

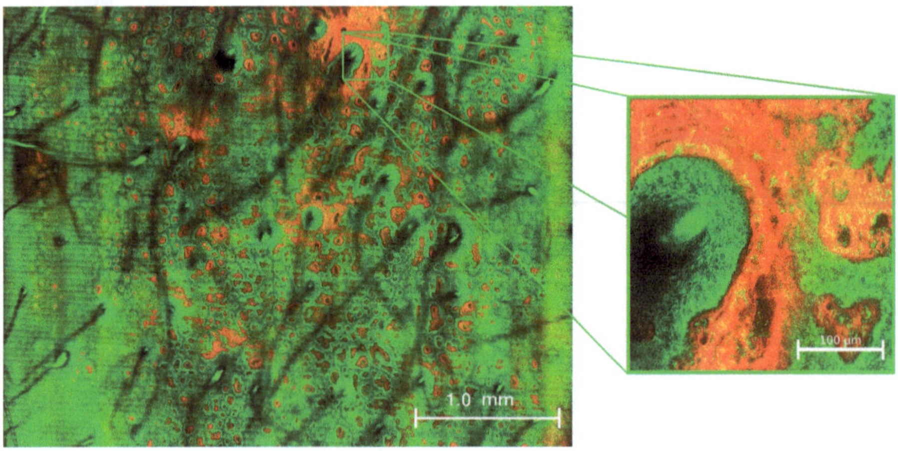

Fig. 13.6: Large-scale optical sectioning (> 10 mm^2) can be obtained by mosaics (stitching). The high spatial resolution (NA 1.3: 300 nm lateral resolution, 1–2 µm axial resolution) is preserved.

13.5 Optical Metabolic Imaging (OMI) based on two-photon FLIM

The fluorescence lifetime is characteristic for the biomolecule and its microenvironment and can be used as a further contrast mechanism in imaging. In particular, it can be employed to distinguish between fluorophores with similar spectral characteristics such as NADH and NADPH as well as to probe the binding status (nanosecond lifetime in NADH-protein complexes vs. picosecond lifetime for free NADH). NAD(P)H provides the major contribution to cellular autofluorescence (AF). It is also a well-known biosensor of the cell's metabolism [48–51].

Clinical MPT studies have been performed to evaluate the potential of clinical use of FLIM data. Interestingly, significant changes of mean fluorescence lifetime data were found in tumors [55], such as brain tumors and skin tumors, as well as in inflammation sites, such as dermatitis [56], compared with normal healthy surrounding tissue.

Fig. 13.7 demonstrates MPT-FLIM based on two-photon AF detected by TCSPC. The fluorescence decay curve depicted is based on the measurement of the arrival times of about 100 fluorescence photons after the femtosecond laser pulse excited the molecules in the tiny sub-femtoliter excitation volume within one single mitochon-

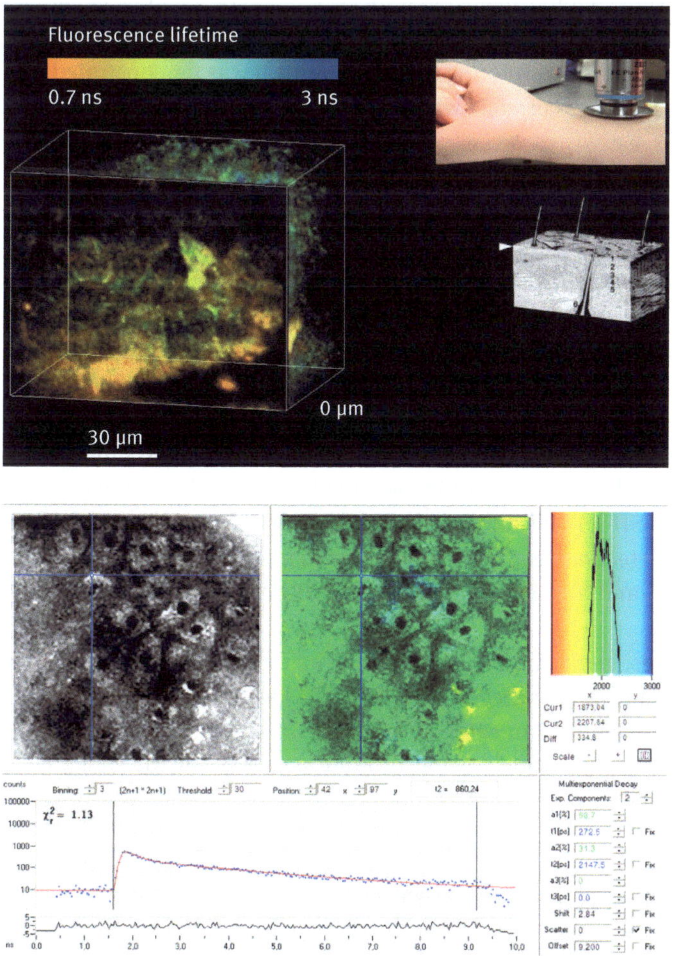

Fig. 13.7: Four-dimensional multiphoton imaging with the color-coded fluorescence lifetime as the 4th dimension. FLIM is performed in the TCSPC mode with a typical frame rate (512 × 512 pixels) of 0.1 s^{-1} when imaging skin AF. The bi-exponential fit allows the differentiation between free (non-bound) coenzyme molecules and protein-bound molecules.

drium, marked with a crosshair, of a particular intratissue skin cell. The arrival times are collected in a histogram consisting of 128 time channels, each 128 ps broad. Fitting of this histogram reveals a fluorescence decay curve. This curve does not simply follow a mono-exponential decay indicating that at least two major fluorescent components are present in the excitation volume. The two-exponential fit provides a short-lived fluorophore with a fluorescence lifetime of $\tau_1 = 270$ ps and an amplitude of 69 % as well as a long-lived component with $\tau_2 = 2.1$ ns ($a_2 = 31$ %). Although the long-lived component has only a percentage of 31 % of the two-fluorophore blend, it contributes most to the fluorescence intensity ($a_2\tau_2/a_1\tau_1$). The values τ_1 or τ_2 or the mean value τ_m can be depicted as a false-color-coded FLIM image, where the color represents the fluorescence lifetime. Thus, a four dimensional AF (x, y, z, τ) stack can be generated (Fig. 13.7).

In a large study with patients suffering from dermatitis it was demonstrated that the mean fluorescence lifetimes correlate with the stage of the disease [56].

13.6 Multimodal imaging (AF, SHG, FLIM, CARS)

The medical tomographs *DermaInspect* and *MPTflex* allow the simultaneous detection of autofluorescence and SHG. In most devices, AF is registered by two photomultipliers, one working in the analog mode, the second one operating in the TCSPC mode.

When adding the CARS module, the tomographs *DermaInspect-CARS* and *MPTflex-CARS* can also register CARS signals such as from lipids or intratissue water.

Fig. 13.8 demonstrates multimodal multiphoton imaging of in vivo human skin with two NIR femtosecond laser beams within the articulated optical arm and multiple detectors to measure AF, SHG, FLIM, and CARS simultaneously during beam scanning. The flexible 360° scan/detection head consists of beam shaping optics, galvoscanners, a piezodriven objective, and also four miniaturized PMT-detectors with appropriate beam splitters.

In feasibility studies, multiphoton tomography was also combined with dermoscopy, ultrasound, and swept-source OCT [57–59]. The idea behind this is to detect suspicious lesions with wide-field imaging and further on to perform in vivo histology by MPT in that particular lesion (region of interest). However, so far no such commercial hybrid system exist.

13.7 Skin cancer detection with MPT

A major MPT application is in situ skin cancer detection based on AF and SHG. Interestingly, melanocytes can be easily detected and distinguished from Langerhans cells due to their melanin autofluorescence (Fig. 13.9). In 2009, our group demonstrated in the University Hospital Jena the detection of malignant melanoma with a high sensi-

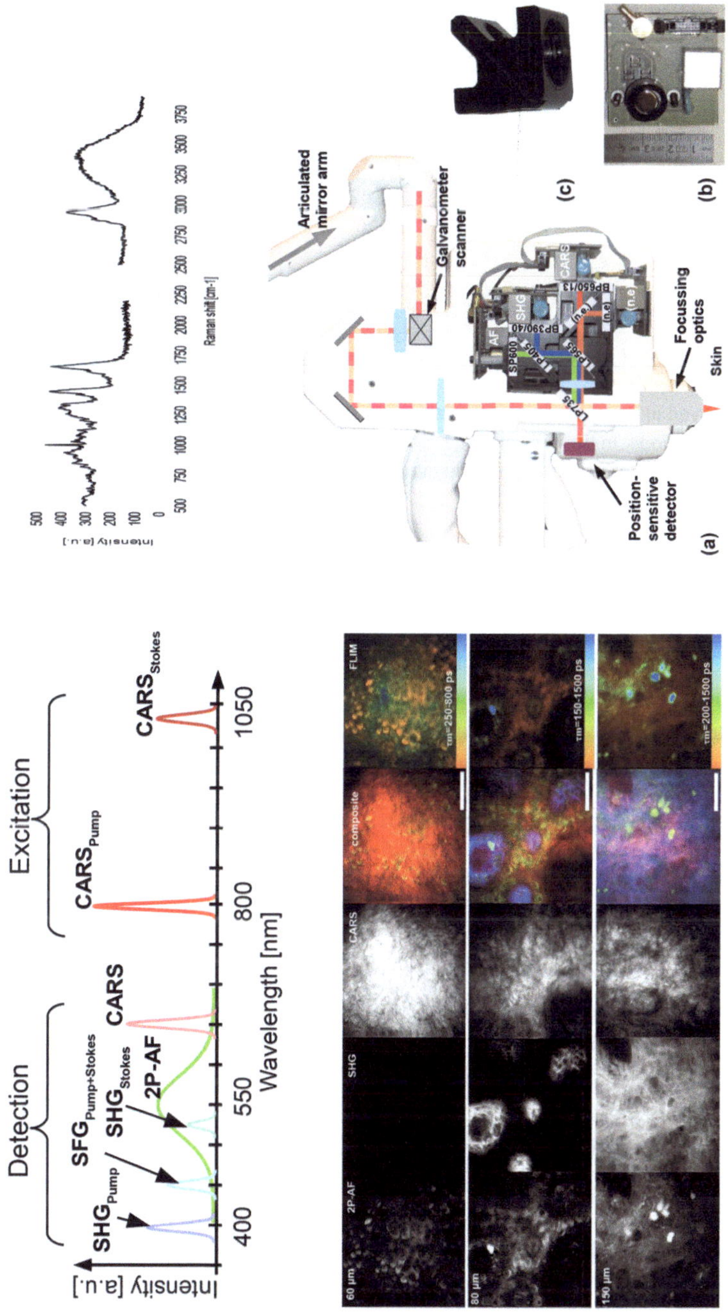

Fig. 13.8: Label-free multimodal multiphoton imaging of in vivo human skin with two NIR femtosecond laser beams and four detectors to obtain simultaneously signals from AF, SHG, FLIM, and CARS. AF originates from NAD(P)H, flavins, melanin, keratin, and elastin; SHG from collagen, CARS from lipids.

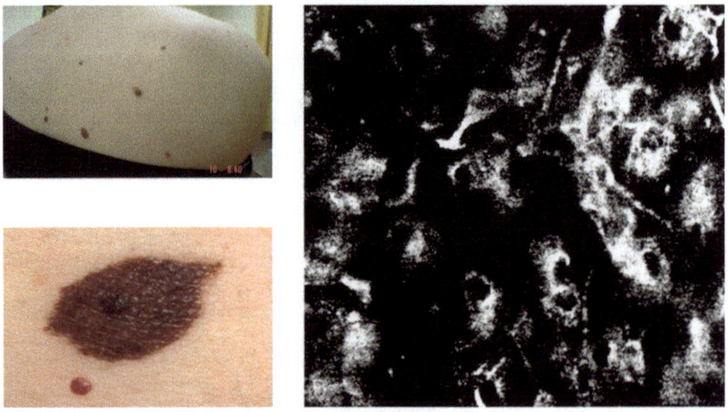

Fig. 13.9: MPT is employed in hospitals for early detection of black skin cancer (malignant melanoma MM) in pigmented skin lesions. In the case of MM, single dendritic cancer cells (melanocytes with spines) can be detected in the upper epidermis and dermis due to the strong melanin fluorescence.

tivity and specificity of up to 97 % and 95 %, respectively, using the tomograph *DermaInspect*. In this very first clinical feasibility MPT study on skin cancer, five dermatologists evaluated optical MPT sections and defined parameters such as the occurrence of melanocytes, cell-cell distances, and atypical morphology [60]. Further MPT parameters for the detection of malignant melanoma were defined recently by the dermatologists at the Beckman Laser Institute and Medical Center in Irvine/California [61]. Also nonmelanoma skin cancer can be detected with multiphoton tomography [62–64].

13.8 The skin ageing index SAAID

The ratio elastin/collagen changes with age and modifications of the environment such as UV exposure and lifestyle, such as smoking. An index SAAID based on the measurement of the AF of elastin and the SHG signal of collagen was introduced by the Taiwanese group using skin biopsies [65]. Our in vivo MPT studies on volunteers confirmed the dependence of the SAIID index on biological age [66]. Currently, major global players in the cosmetics industry employ multiphoton tomographs to evaluate anti-ageing effects of their future products [67, 68].

Besides the SAAID index, also the NAD(P)H fluorescence in the skin layer stratum granulosum was found to show a dependence on age. Researches at P & G developed new cosmetics with antioxidants based on MT studies conducted in Germany and Japan [69].

13.9 MPT in space medicine

MPT was employed to evaluate skin modifications of three astronauts after long-term space flights [65, 66]. The investigated European astronauts worked at least 5 months on board the international space lab ISS. MPT measurements were taken before and after their flights in the ESA center in Cologne (Köln). Six sites of the left arm were investigated within 45 minutes using AF, SHG, and FLIM stack imaging at 780 nm excitation [70].

Interestingly, a significant shrinking of the epidermis of up to 20 % was measured in all three astronauts as well as an increase in collagen in the upper dermis. Fig. 13.10 shows a photograph of astronaut Alexander Gerst during the MPT session.

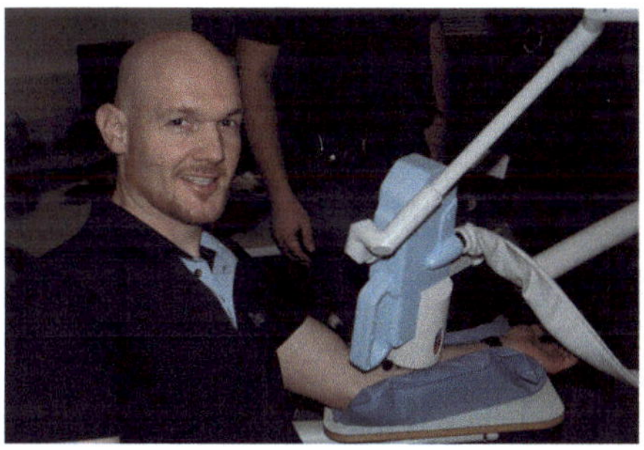

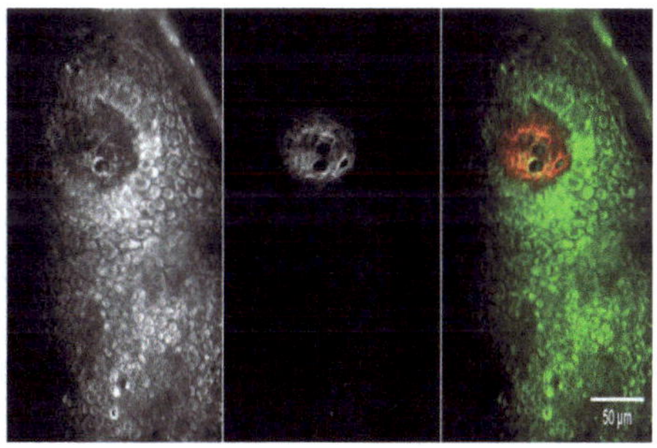

Fig. 13.10: The skin of astronauts is tested with the certified multiphoton tomograph *MPTflex* prior the mission and after long-term space flights in order to monitor spaceflight-related skin modifications and to answer the question if accelerated skin ageing and impaired wound healing may occur in space.

13.10 Multiphoton tomography of in vivo human brain

The first MPT in human brain was conducted in 2014 at the hospital in Mainz during brain tumor surgery [55]. The purpose of this first intrasurgical MPT was to discriminate glioma tissue and adjacent brain tissue intraoperatively by the detection of the two-photon excited time-resolved AF.

Malignant glioma belongs to the most dreaded diseases, in particular the most aggressive histopathological subtype glioblastoma multiforme (GBM). Unfortunately, most gliomas are surrounded by functional brain tissue such as speech areas or motor cortical areas. The exact determination of the tumor margins is therefore of high interest.

In a first step, a glioma mouse model as well as human brain tumor biopsies have been investigated by TCSPC-MPT. It was found that two-photon FLIM is able to define the tumor borders without any further labeling. Based on these promising preliminary data sets, a first feasibility MPT study on a GBM patient was performed in Mainz.

This first intrasurgical MPT was successful and provided optical sections of high contrast. However, there is still a long way to establish routine intraoperative MPT application. One problem to be solved is the noise by ambient operation light. Furthermore, the current field of view and the signal depth are rather small. Nevertheless, multiphoton tomography may have a great potential to become an important imaging tool for intraoperative resection guidance in neurosurgery.

13.11 Two-photon imaging of cornea transplants

MPT has been employed to characterize the quality of cornea transplants [71]. Corneal transplantation is the most common transplantation procedure in the world. However, there are not enough donors to satisfy the demand and a large number of donated samples are not suitable due to poor tissue quality. Current quality checks on donated corneas are performed by slit-lamps. MPT can provide a better spatial resolution and additional information on the metabolism and therefore may optimize the storage procedure and the evaluation of the right time for transplantation. In a first clinical MPT study on human cornea at the Saarland Medical Center together with the Lion's Cornea Bank, the morphology of the corneal layers, the intratissue cells and the collagen network as well as the metabolism by optical metabolic imaging was investigated. Interestingly, the metabolism of many intratissue cells increased with storage time after donation, whereas the metabolism of the important endothelial cells decreased.

13.12 Watching stem cells at work in transgenic mice

Pluripotent stem cells can be found in the hair bulge and the dermal papillae of the hair. Interestingly, these stem cells can be easily isolated and used to generate nerve cells and other cell types. They are likely involved in skin wound repair mechanisms. These stem cells can be tagged with a fluorescent protein such as GFP (green fluorescent protein) based on nestin expression. The tomograph was employed at the company AntiCancer Inc. in San Diego to probe if stem cells can be monitored in the hair follicles of live transgenic mice completely non-invasively and over long time periods.

Mice were coupled to the tomograph via the magnetic interface with the 160 μm thick glass window without any surgical opening of the skin. Interestingly, the magnetic force of the coupler was strong enough to avoid major animal movements. The mice could be investigated over time periods up to 4 hours even without anesthesia.

Nestin-expressing green-fluorescent stem cells could be also clearly identified by their unique morphology. They consist of a 7 μm long cell body with one or two 10 μm long extrusions. We were able to detect stem cell migration within the hair bulge and

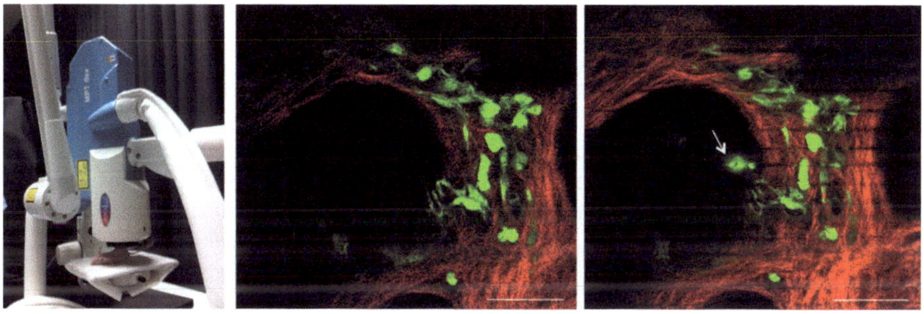

Fig. 13.11: Multiphoton tomographs with the flexible optomechanical arm and the 360° measurement head are employed to on-line monitor the movement of GFP expressing stem cells in the hair bulge of live transgenic mice over time periods of several hours [72, 73].

hypothesize that hair bulge and DP stem cells have the same origin. Based on the unique morphological shape we were also able to identify stem cells in the hair bulge of normal, nontransgenic mice even without any labeling or the use of fluorescent proteins. The Fig. 13.11 shows intrahair stem cells in a live mouse taken with the tomograph MPTflex [72, 73].

13.13 Conclusion

MPT is a novel label-free optical medical tomography method with superior subcellular spatial resolution and picosecond temporal resolution [74]. Furthermore, metabolic optical imaging based on coenzyme autofluorescence can be performed and chemical information can be extracted.

Certified multiphoton tomographs are in clinical use in Europe, US, China, Japan, Russia, and Australia. The current major application field is in vivo skin imaging. The multiphoton tomographs with the flexible mechano-optical arm and the 360° measurement head can also be employed as upright and inverted two-photon microscopes for cell and animal studies as well as for rapid investigation of biopsies.

The major disadvantage of current MPT devices is the large size and the high price of the required femtosecond laser source as key element of the tomograph. However, ultracompact and chiller-free NIR MHz femtosecond laser systems at an affordable price will be available within the next few years [75]. This development will likely push MPT as an interesting imaging tool for non-invasive and rapid in vivo histology and may help to reduce the number of physically taken invasive biopsies. MPT will also help to test the efficacy of anti-ageing and pharmaceutical products and to reduce the number of animal studies [76].

References

[1] Raju TN. The Nobel Chronicles. The Lancet. 1999;354:1653–1656. doi: 10.1016/S0140-6736(05)77147-6.

[2] Brownell GL. A history of positron imaging; 1999.

[3] Ter-Pogossian MM, Phelps ME, Hoffman EJ, Mullani NA. A Positron-Emission Transaxial Tomograph for Nuclear Imaging (PETT). Radiology. 1975;114(1):89–98. doi: 10.1148/114.1.89.

[4] Hutton BF. The origins of SPECT and SPECT/CT. Eur J Nucl Med Mol Imaging. 2014;41:3–16.

[5] MacWilliams B. News & Views: Russian claims first in magnetic imaging. Nature. 2003;426(6965):375. doi: 10.1038/426375a. Cited in: PubMed; PMID 14647349.

[6] Hinshaw WS, Bottomley PA, Holland GN. Radiographic thin-section image of the human wrist by nuclear magnetic resonance. Nature. 1977;270:722–723.

[7] Ruiter NV, Zapf M, Hopp T, Dapp R, Kretzek E, Birk M, Kohout B, Gemmeke H. 3D ultrasound computer tomography of the breast: A new era? European Journal of Radiology. 2012:133–134.

[8] Wild JJ, Neal D. Use of high-frequency ultrasonic waves for detecting changes of texture in living tissues. The Lancet. 1951;257:655–657. doi: 10.1016/S0140-6736(51)92403-8. Cited in: PubMed; PMID 14814827.

[9] Wild JJ, Reid JM. Application of echo-ranging techniques to the determination of structure of biological tissues. Science. 1952;115:226–230. doi:10.1126/science.115.2983.226. Cited in: PubMed; PMID 14913217.

[10] Goldmann H. Spaltlampenphotographie und -photometrie. Ophthalmologica. 1939;98:257–270. doi: 10.1159/000299716.

[11] Boyde A, Minsky M. Memoir on inventing the confocal scanning microscope. Scanning. 1988;10:128–138.

[12] Minsky M, inventor. Microscopy Apparatus. United States Patent US 3,013,467. 1961 December 19.

[13] Egger MD, Petrăn M. New reflected-light microscope for viewing unstained brain and ganglion cells. Science. 1967;157:305–307. doi: 10.1126/science.157.3786.305. Cited in: PubMed; PMID 6030094.

[14] Davidovits P. Scanning laser microscope. Nature. 1969;223:831. doi: 10.1038/223831a0. Cited in: PubMed; PMID 5799022.

[15] Amos WB, White JG. How the confocal laser scanning microscope entered biological research. Biol Cell. 2003;95:335–342.

[16] Masters BR. Confocal microscopy and multiphoton excitation microscopy. the genesis of live cell imaging. Bellingham, WA: SPIE Press; 2006. p. 120–121.

[17] Sheppard CJR. Confocal microscopy. The development of a modern microscopy [Internet]. Imaging & Microscopy; 2009. www.imaging-git.com/science/light-microscopy/confocal-microscopy.

[18] Gärtner W, Gröbler B, Schubert D, Wabnitz H, Wilhelmi B. Fluorescence scanning microscopy combined with subnanosecond time resolution. Experimental Techniques in Physics. 1988;36:443–451.

[19] König K. PhD thesis. Beiträge zur Photochemotherapie und optischen Diagnostik von Tumoren. Contributions to photochemotherapy and optical diagnostics of tumors. Archive University Jena, Germany; 1989.

[20] Bugiel I, König K, Wabnitz H. Investigation of cells by fluorescence laser scanning microscopy with subnanosecond time resolution. Lasers Life Sciences. 1989;3:1–7.

[21] König K, Wabnitz H. Fluoreszenzuntersuchungen mit hoher zeitlicher, spektraler und räumlicher Auflösung. Labortechnik. 1990;23:26–31.

[22] Corcuff P, Bertrand C, Leveque JL. Morphometry of human epidermis in vivo by real-time confocal microscopy. In: Archives of Dermatological Research. Springer. 1993;285:475–481.
[23] Corcuff P, Leveque JL. In vivo vision of the human skin with the tandem scanning microscope. Dermatology. 1993;186:50–54.
[24] Pierard GE. In vivo confocal microscopy: A new paradigm in dermatology. Dermatology. 1993;186:4–5.
[25] Bertrand C, Corcuff P. In vivo spatio-temporal visualization of the human skin by real-time confocal microscopy. Wiley Online Library. 1994;16:150–154.
[26] Rajadhyaksha M, Gonzalez S, Zavislan JM, Anderson RR, Webb RH. In vivo confocal scanning laser microscopy of human skin II: Advances in instrumentation and comparison with histology. J Invest Dermatol. 1999;113:293–303.
[27] Hofmann-Wellenhof R, Pellacani G, Malvehy J, Soyer HP, editors. Reflectance confocal microscopy for skin diseases. Berlin, Heidelberg: Springer Verlag; 2012. doi: 10.1007/978-3-642-21997–21999.
[28] Guthoff RF, Baudouin C, Stave J. Atlas of confocal laser scanning in-vivo microscopy in ophthalmology. Berlin, Heidelberg, New York: Springer; 2007.
[29] Webb RH, Hughes GW. Scanning laser ophthalmoscope. IEEE Transactions on Biomedical Engineering. 1981;28:488–92.
[30] Dilly PN. Tandem scanning reflected light microscopy of the cornea. Scanning. 1988;10:153–156.
[31] Cavanagh HD, Jester JV, Essepian J, Shields W, Lemp MA. Confocal microscopy of the living eye. CLAO J. 1990;16:65–73.
[32] Masters BR. Scanning slit confocal microscopy of the in vivo cornea. Optical Engineering. 1995;34:684-692.
[33] Huang D, Swanson EA, Lin CP, Schuman JS, Stinson WG, Chang W, Hee MR, Flotte T, Gregory K, Puliafito CA, Fujimoto JG. Optical coherence tomography. Science. 1991;254:1178–1181.
[34] Swanson EA, Izatt JA, Hee MR, Huang D, Lin CP, Schumann JS, Puliafito CA, Fujimoto JG. In vivo retinal imaging by optical coherence tomography. Opt Lett. 1993;18:1864–1866.
[35] Fercher AF, Hitzenberger CK, Drexler W, Kamp G, Sattmann H. In vivo optical coherence tomography. Am J Ophthalmol. 1993;116:113–114.
[36] Drexler W, Fujimoto JG. Retinal optical coherence tomography. Heidelberg: Springer; 2008.
[37] Denk W, Strickler JH, Webb WW. Two-photon laser scanning fluorescence microscopy. Science. 1990;248:73–76.
[38] https://www.gov.uk/cma-cases
[39] König K, So PTC, Mantulin WW, Tromberg BJ, Gratton E. Two-photon excited lifetime imaging of autofluorescence in cells during UVA and NIR photostress. J Microsc. 1996;183:197–204.
[40] Masters BR, So PTC, Gratton E. Multiphoton excitation microscopy of in vivo human skin. Functional and morphological optical biopsy based on three-dimensional imaging, lifetime measurements and fluorescence spectroscopy. Ann NY Acad Sci. 1998;838:58–67.
[41] König K. Verfahren und Anordnung zur nicht-invasiven dreidimensionalen optischen Untersuchung und Therapie der Haut. Patent DE 10065146 A1. 2000 Dec 22.
[42] König K, Wollina U, Riemann I, Peuckert C, Halbhuber KJ, Fünfstück V, Fischer TW, Elsner P. Optical tomography of human skin with subcellular spatial and picosecond time resolution. SPIE-Proceedings. 2002;4620:191–201.
[43] König K, Riemann I. High-resolution multiphoton tomography of human skin with subcellular spatial resolution and picosecond time resolution. Journal Biomedical Optics. 2003;8:432–439.
[44] Wang LV, Hu S. Photoacoustic tomography: in vivo imaging from organelles to organs. Science. 2012;335:1485–1462 .

[45] Valluru KS, Willmann JK. Clinical photoacoustic imaging of cancer. Ultrasonography. 2016;35:267–280.
[46] Squirrell JM, Wokosin DL, White JG, Bavister BD. Long-term two-photon fluorescence imaging of mammalian embryos without compromising viability. Nat Biotechnol. 1999;17:763–767.
[47] Fischer F, Volkmer B, Puschmann S, Greinert R, Breitbart E, Kiefer J, Wepf R. Assessing the risk of skin damage due to femtosecond laser irradiation. J Biophotonics. 2008; 1:470–477.
[48] Chance B, Cohen P, Jobsis F, Schoener B. Intracellular oxidation-reduction states in vivo. Science. 1962;137:499–508.
[49] Chance B, Jobsis F. Changes in fluorescence in a frog sartorius muscle following a twitch. Nature. 1959;184:195–196.
[50] Chance B, Leigh JS, Miyake H, Smith DS, Nioka S, Greenfeld R, Finander M, Kaufmann K, Levy W, Young M. Comparison of time-resolved and –unresolved measurements of deoxy-hemoglobin in brain. Proc Natl Acad Sci USA. 1988;85:4971–4975.
[51] Walsh AJ, Cook RS, Manning HC, Hicks DJ, Lafontant A, Arteaga CL, Skala MC. Optical metabolic imaging identifies glycolytic levels, subtypes, and early-treatment response in breast cancer. Cancer Res. 2013;73:6164–6174.
[52] Evans CL, Xie XS. Coherent anti-Stokes Raman scattering microscopy: chemical imaging for biology and medicine. Annu Rev Anal Chem. 2008;1:883–909.
[53] König K, Breunig HG, Bückle R, Kellner-Höfer M, Weinigel M, Büttner E, Sterry W, Lademann J. Optical skin biopsies by clinical CARS and multiphoton fluorescence/SHG tomography. Laser Physics Letters. 2011;8:1–4.
[54] Mayer P. Ueber das Färben mit haematoxylin. Mitt Zool Stat Neapel. 1891;10:170-186.
[55] Kantelhardt S, Kalasauskas D, König K, Kim E, Weinigel M, Uchugonova A, Giese A. In vivo multiphoton tomography and fluorescence lifetime imaging of human brain tumor tissue. J Neurooncol. 2016;127(3):473–482. doi: 10.1007/s11060-016-2062-8.
[56] Huck V, Gorzelanny C, Thomas K, Getova V, Niemeyer V, Zens K, Unnerstall TR, Feger JS, Fallah MA, Metze D, Ständer S, Luger TA, König K, Mess C, Schneider SW. From morphology to biochemical state-intravital multiphoton fluorescence lifetime imaging of inflamed human skin. Sci Rep. 2016;6:22789. doi: 10.1038/srep22789.
[57] König K, Speicher M, Köhler MJ, Scharenberg R, Kaatz M. Clinical application of multiphoton tomography in combination with high-frequency ultrasound for evaluation of skin diseases. J Biophotonics. 2010;3:759–773. doi: 10.1002/jbio.201000074.
[58] König K, Speicher M, Bückle R, Reckfort J, McKenzie G, Welzel J, Koehler MJ, Elsner P, Kaatz M. Clinical optical coherence tomography combined with multiphoton tomography of patients with skin diseases. Journal of Biophotonics. 2009;2:389–397. doi: 10.1002/jbio.200910013.
[59] König K. Hybrid multiphoton multimodal tomography of in vivo human skin. IntraVital. 2012;1:8.
[60] Dimitrow E, Ziemer M, Koehler MJ, Norgauer J, König K, Elsner P, Kaatz M. Sensitivity and specificity of multiphoton laser tomography for in vivo and ex vivo diagnosis of malignant melanoma. Journal of Investigative Dermatology 2009;129:1752–1758. doi: 10.1038/jid.2008.439.
[61] Balu M, Kelly KM, Zachary CB, Harris RM, Krasieva TB, König K, Durkin AJ, Tromberg BJ. Distinguishing between benign and malignant melanocytic nevi by in vivo multiphoton microscopy. Cancer Res. 2014;74:2688–2697. doi: 10.1158/0008-5472.
[62] Seidenari S, Arginelli F, Bassoli S, Cautela J, Cesinaro AM, Guanti M, Guardoli D, Magnoni C, Manfredini M, König K. Diagnosis of BCC by multiphoton laser tomography. Skin Res Tech. 2013;19:e297–304. doi: 10.1111/j.1600-0846.2012.00643.

[63] Klemp M, Meinke M, Weinigel M, Roewert-Huber HJ, König K, Ulrich M, Lademann J, Darvin M. Comparison of morphologic criteria for actinic keratosis and squamous cell carcinoma using in vivo multiphoton tomography. Exp Dermatol. 2016;25:218–222. doi: 10.1111/exd.12912.
[64] Balu M, Zachary CB, Harris RM, Krasieva TB, König K, Tromberg BJ, Kelly KM. In vivo multiphoton microscopy of basal cell carcinoma. JAMA Dermatol. 2015;151(10):1068–1074. doi: 10.1001/jamadermatol.2015.0453.
[65] Lin SJ, Wu R Jr, Tan HY, Lo W, Lin WC, Young TH, Hsu CJ, Chen JS, Jee SH, Dong CY. Evaluating cutaneous photoaging by use of multiphoton fluorescence and second-harmonic generation microscopy. Opt Lett. 2005;30:2275–2277.
[66] Koehler MJ, König K, Elsner P, Bückle R, Kaatz M. In vivo assessment of human skin aging by multiphoton laser scanning tomography. Opt Lett. 2006;31:879–881.
[67] Puschmann S, Rahn CD, Wenck H, Gallinat S, Fischer F. Approach to quantify human dermal skin aging using multiphoton laser scanning microscopy. J Biomed Opt. 2012;17:036005.
[68] Decencière E, Tancrède-Bohin E, Dokládal P, Koudoro S, Pena AM, Baldeweck T. Automatic 3D segmentation of multiphoton images: a key step for the quantification of human skin. Skin Res Technol. 2013;19:115–124.
[69] Miyamaoto K, Kudoh H. Quantification and visualization of cellular NAD(P)H in young and aged female facial skin with in vivo two-photon tomography. Br J Dermatol. 2013;169:25–31.
[70] König K, Weinigel M, Pietruszka A, Bückle R, Gerlach N, Heinrich U. Multiphoton tomography of astronauts. Proc SPIE 9329; 2015.
[71] König K. Multiphotonen-Tomographie der humanen Kornea. Ophthalmologische Nachrichten. Ophthalmologic News. 2014;Sept.
[72] Uchugonova A, Hoffman RM, Weinigel M, Koenig K. Watching stem cells in the skin of living mice non-invasively. Cell Cycle. 2011;10:2017–2020.
[73] Uchugonova A, Cao W, Hoffman RM, König K. Comparison of label-free and GFP multiphoton imaging of hair-follicle-associated pluripotent (HAP) stem cells in mouse whiskers. Cell Cycle. 2015;14:3430–3433. doi: 10.1080/15384101.2015.1090064.
[74] König K. Clinical Multiphoton Tomography. J Biophotonics. 2008;1:13–23.
[75] König K, Liu T, Anderson P, Breunig G. Multiphoton imaging with a novel compact diode-pumped Ti:sapphire oscillator. Microsc Res Technol. 2015; 78:1154–1158.
[76] König K. Multiphoton tomography. In: Humbert P, et al., editors. Agache's Measuring the Skin. New York: Springer International Publishing; 2017. p. 1177–1189. doi: 10.1007/978-3-319-32383-1_46.

Martin Weinigel, Hans Georg Breunig, and Karsten König

14 Clinical multimodal CARS imaging

Abstract: Nonlinear contrast methods such as two-photon excited autofluorescence (in combination with fluorescence lifetime imaging) and second-harmonic generation have been combined with a further multiphoton contrast mechanism called coherent anti-Stokes Raman scattering. We describe the principle and the instrumentation for its implementation in tomographs for dermatological applications and present multimodal optical skin biopsies.

14.1 Introduction

14.1.1 CARS

Raman scattering, named after C. V. Raman, is an inelastic scattering process that originates from light's interaction with molecular vibrations. The interaction leads to a characteristic energy shift of the photon energy which contains information about the vibrational modes of a system. A shift to higher energies leads to so-called anti-Stokes Raman scattering whereas a shift to lower energy results in the Stokes signal (Fig. 14.1). Raman spectroscopy was realized already in the late 1920s [1] to identify molecules and became, in particular with the advent of the laser, a fundamental method commonly used in chemistry, solid state physics, and biomedicine [2–4]. A typical Raman spectrum of complex tissue like human skin in vivo is illustrated in Fig. 14.1 [5]. It represents a unique signature of the molecular tissue composition. Linear Raman scattering however, is a very inefficient "low-signal process". Therefore, several Raman spectroscopy methods have been developed to increase the sensitivity [6], among them nonlinear Raman techniques like coherent anti-Stokes Raman scattering (CARS). In contrast to spontaneous Raman, CARS is based on a coherently driven transition which involves three photons with different frequencies: pump, probe, and Stokes photons. Technically, these photons can be provided by

(i) one (spectrally broadband or broadened) laser,
(ii) two lasers, or
(iii) three lasers with different frequencies (ω_{pump}, ω_{probe}, ω_{Stokes}).

The experimental setup must ensure the spatial and temporal overlapping of the pulses and focal volumes, respectively, in the Raman medium. To reduce the complexity of the setup, pump and probe photons from one excitation source are often used ($\omega_{\text{pump}} = \omega_{\text{probe}}$), and a second source is employed to generate the Stokes photons. Resonant CARS signals are generated in the case where the difference frequency $\Omega = \omega_{\text{pump}} - \omega_{\text{Stokes}}$ of pump and Stokes photons with frequencies ω_{pump}

https://doi.org/10.1515/9783110429985-016

and ω_{Stokes}, respectively, matches the frequency of a Raman-active molecular vibration [7]. The CARS signal itself is generated at the anti-Stokes Raman frequency $\omega_{\text{CARS}} = \omega_{\text{pump}} + \omega_{\text{probe}} - \omega_{\text{Stokes}}$ (Fig. 14.1) [8].

In the case of high molecular concentrations, the CARS signal is stronger than the spontaneous Raman signal because the CARS signal arises from the coherent addition of all signals from multiple molecules. In contrast, the total spontaneous Raman signal is the sum of signals from individual molecules without coherent addition. The coherent addition of the individual CARS signals is additive when phase matching is fulfilled, i.e., when the condition $\Delta k = k_{\text{pump}} + k_{\text{probe}} - k_{\text{Stokes}} - k_{\text{CARS}} = 0$ is satisfied (Fig. 14.1). The CARS signal propagates in a beam along the direction k_{CARS} while the total spontaneous Raman signal is emitted in all directions. In practice, this leads to the possibility to collect CARS signals up to 10^6 times faster than spontaneous Raman scattering. This allows faster (µs beam dwell time per pixel instead of seconds per pixel) and more detailed imaging (more image pixels) [9]. The total CARS signal, however, also contains an inherent nonresonant background [10]. This nonresonant background arises from off-resonance transitions which also generate signals that coherently add up, but are molecule unspecific (Fig. 14.1). Thus, the sensitivity is reduced by the ratio of resonant to nonresonant background (R/NRB). The nonresonant back-

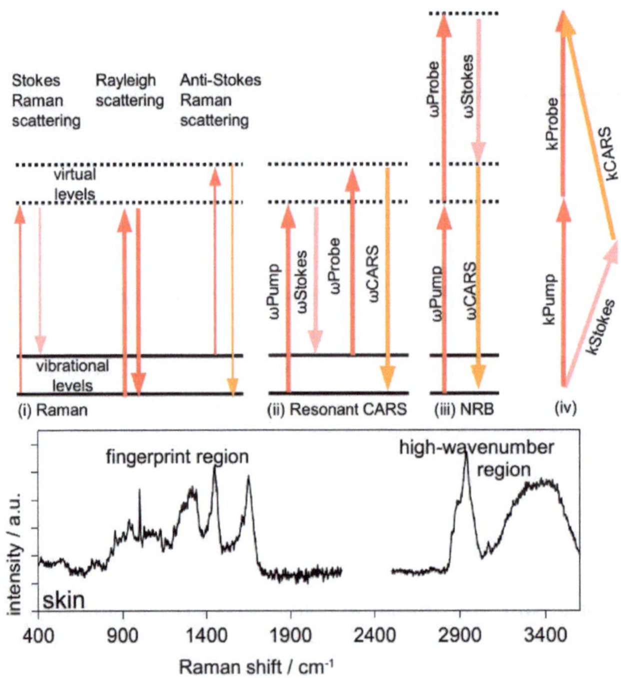

Fig. 14.1: Top: energy schemes of (i) spontaneous Raman, (ii) resonant CARS and (iii) nonresonant CARS background, (iv) phase-matching condition for the CARS process; bottom: Raman spectrum of human skin (with kind permission adapted from [5]).

ground causes the CARS spectrum to differ from spontaneous Raman spectra. This is often regarded as a problem and a drawback for the application of CARS. Therefore, several methods have been developed to minimize and suppress the nonresonant background in the CARS signal or correct the recorded signals for nonresonant contributions [11–15].

14.1.2 CARS microscopy

Historically, it was assumed the above mentioned phase-matching condition $\Delta k = 0$ can only be fulfilled in a noncollinear beam geometry. In such a beam geometry with the excitation beams entering the back-focal plane of the focusing optics at different positions and angles (which required a nonfilled back-focal plane) the effective numerical aperture (NA) is lowered. Such a CARS microscope setup was first demonstrated by Duncan et al. in 1982 [16]. However, this first CARS microscope was complicated and the tuning of the excitation wavelengths was challenging. Additionally, the configuration-immanent low NA limited the spatial resolution [16].

Interestingly, almost two decades later, Sunney Xie's group at Harvard showed that the phase-matching condition can also be fulfilled in a collinear excitation geometry using high-NA focusing optics [17]. The collinear excitation geometry not just simplifies the setup but also allows for diffraction limited focusing. The limited interaction length of the ($\mathrm{CARS_{pump}}$ and $\mathrm{CARS_{Stokes}}$) excitation radiation and the nonlinear intensity dependence of the CARS signal ensure that signals are generated only inside the focal volume. This intrinsic three-dimensional imaging capability (optical sectioning capability) is the basis for high-resolution CARS microscopy [17, 18]. By scanning the focal volume across the sample with a laser scanning setup (or moving the sample across the fixed focal region) and collecting CARS signals spatially resolved in forward and/or backward direction, images based on the CARS-signal intensity per pixel can be generated.

Different CARS microscope setups have been developed using mostly

(i) ps pulses (ps-CARS) or
(ii) fs pulses (fs-CARS),
(iii) a combination of both or
(iv) single broadband pulses which contain $\mathrm{CARS_{pump}}$ and $\mathrm{CARS_{Stokes}}$ photons [7, 18, 19].

While ps-CARS provides in general a higher spectral selectivity than fs-CARS, the use of fs pulses is favorable for a multimodal approach where CARS and additional nonlinear optical image modalities like two-photon excited autofluorescence (2P-AF) and second-harmonic generation (SHG) are intended to be implemented in one common system [20]. Furthermore, the low spectral resolution of fs-CARS has been countered by the development of methods to increase and even tune the spectral resolution with fs pulses in the range of 10–100 cm^{-1} [21]. This increased spectral resolution is possible because the spectrum of the vibrational excitation is not determined by the whole width of the fs pulse spectra but only by the effective spectrum of their temporal overlap. This temporal overlap can be altered by individually chirping the pump and Stokes pulses. By inducing equal linear chirps on pump and Stokes pulses a constant instantaneous frequency difference exists between them. This "spectral focusing" results in a vibrational excitation centered at the instantaneous frequency difference with an instantaneous bandwidth of the chirped pulses which is only determined by the stretched pulse duration but not by the whole bandwidth of the broadband pulses [22].

Multimodal optical imaging enables simultaneous recording of fluorescent and nonfluorescent (but Raman- or SHG-active) molecules to provide corresponding and complementing information on a sample. In biological samples, the CH_2 transition in particular has been used as a contrast mechanism for CARS which allows detecting nonfluorescent lipids [7, 23]. CARS microscopy was recently introduced to clinical imaging and has been evaluated as a new in vivo diagnostic tool for skin diseases [24–28].

14.2 Setup of CARS tomographs

The general setup of a CARS imaging system consists of two parts: the excitation source(s) and the imaging module (Fig. 14.2). The excitation source(s) for CARS can be realized by "single beam CARS" [19] using a single broadband fs oscillator which provides the $CARS_{pump}$ and $CARS_{Stokes}$ photons. More common is a two-beam setup. Technically, this can be a combination of an fs oscillator (e.g., an 80 MHz fs Ti:sapphire laser) with
(i) an optical parametric oscillator (OPO) [17],
(ii) a photonic crystal fiber (PCF) [29],
(iii) an additional fs laser or by a single dual-output laser [30].

In vivo CARS imaging of human skin has been realized by two-beam setups with OPO and PCF technology [24–27] as described below.

OPO technology: The OPO contains a periodically poled potassium titanyl phosphate (KTP) crystal which is used to generate so-called signal and idler pulses at longer wavelengths. Two temporally synchronized outputs are provided by the OPO, the residual fundamental oscillator pulses (which can be employed as $CARS_{pump}$ pulses) and the newly generated signal pulses which can be employed as $CARS_{Stokes}$ pulses. The advantages of the OPO are:
(i) high output power, in particular a high spectral power density,
(ii) the possibility for computer-controlled tunability and
(iii) a high pulse-to-pulse stability.

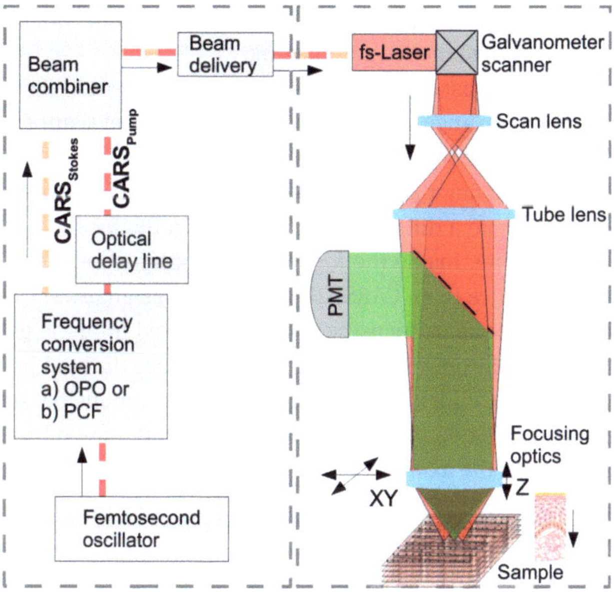

Fig. 14.2: Schematic diagram of a CARS Tomograph; left: CARS excitation sources; right: scanning unit; PMT: photomultiplier tube.

However, the large footprint of the OPO is nonoptimal for a compact device.

PCF technology: $CARS_{Stokes}$ photons can also be provided from PCFs. In particular, coupling the pulse energy of an fs oscillator into the PCF core can result in a strong spectral broadening of the transmitted pulses. Due to the small core diameter of about 1–2 µm and the small group velocity dispersion of these fibers close to their zero dispersion wavelength(s), the high optical intensities are mostly maintained over the length of the fiber [31]. Over this long interaction length nonlinear optical effects such as self-phase modulation, stimulated Raman scattering and four-wave mixing [32] are induced. This leads to strong broadening of the output spectrum – also called white-light continuum or supercontinuum, which is temporally synchronized with the fs oscillator. Advantageously, already moderate mean powers of the fs oscillator are sufficient for a strong spectral broadening which, in combination with the original pulses, can be used for CARS excitation in the spectral range from the so-called fingerprint (< 1500 cm^{-1}) up to the high wavenumber region around 3000 cm^{-1}. The employed PCF fibers are commercially available and integrated in easy to handle and cost-efficient modules (e.g., FemtoWHITE CARS, NKT Photonics, Denmark) with small footprint. However, their integration and stable long-term operation is challenging due to the high requirements regarding the beam pointing stability of the incoming fs laser beam. Even in a stable breadboard design, minor temperature differences can affect this beam pointing stability (thermal drift) and thus the PCF coupling efficiency. Additionally, limited output powers (limited by material damage threshold) and a low spectral power density – superimposed by a reduced spectral amplitude stability (spectral amplitude noise) – challenge the use for imaging.

The imaging module in Fig. 14.2 is based on beam scanning. Briefly described, the incoming laser beam experiences an angular deflection by galvanometer scanners which is translated through the scan- and tube-lens system into an angular deflection inside the back-focal plane of the focusing optics. The focusing optics itself focuses the incoming collimated laser beam into a diffraction limited spot in the focal plane or more specifically in the object plane. Inside the focal spot the interaction of the radiation with the molecules can induce (e.g., CARS, 2P-AF and SHG) signals. These can be collected in epi direction by the same focusing optics, separated from the excitation radiation by dichroic beam splitters and detected by PMTs. Since the position of the focal spot depends on the angle of incidence of the excitation beam inside the back-focal plane, raster-scanning can be performed point-by-point (and line-by-line) for two-dimensional imaging. Moving the axial position of the focusing optics allows for a movement of the object plane inside the tissue.

14.2.1 Requirements for CARS imaging

For CARS imaging the interaction of $CARS_{pump}$ and $CARS_{Stokes}$ photons with the molecules in focus is necessary. When $CARS_{pump}$ and $CARS_{Stokes}$ photons are generated by OPO or PCF technology they are temporally synchronized but not overlapped in time and space. The temporal overlapping can be realized by an optical delay line in one of the beams which compensates the optical path difference. The required spatial overlap of $CARS_{pump}$ and $CARS_{Stokes}$ pulses in the focal volume inside the tissue can be realized by the collinear superposition of both beam paths by a dichroic beam splitter (in this context called beam combiner, see Fig. 14.3).

14.2.2 First clinical CARS tomographs

The first CARS tomographs for in vivo imaging of human skin were realized by extending the imaging capabilities of the clinical tomograph *DermaInspect* (JenLab, Germany). Two different implementations with this rigid but stable imaging platform are put into practice by the introduction of the *DermaInspect CARS-MPT* – one based on OPO and one on PCF technology (see Fig. 14.4).

The OPO setup employs a *Chameleon OPO* (APE GmbH, Germany) which is pumped by a high power fs oscillator (*MaiTai HP*, repetition rate 80 MHz, pulse duration about 100 fs, Newport Spectra Physics, USA). The signal photons of the OPO are used as $CARS_{Stokes}$ photons which are combined (collinear alignment) with the residual oscillator pulses ($CARS_{pump}$ photons) by a beam combiner (Fig. 14.3). Temporal overlap of $CARS_{pump}$ and $CARS_{Stokes}$ photons inside the focal volume is

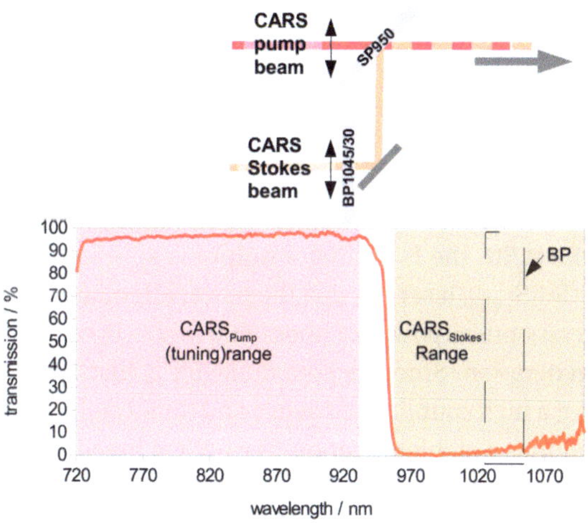

Fig. 14.3: Top: Collinear superposition of parallel polarized $CARS_{pump}$ and $CARS_{Stokes}$ excitation beams by a 45° (angle of incidence) dichroic beam splitter; bottom: typical spectral characteristic of a dichroic beam splitter (SP950) and indicated Stokes band-pass filter (BP1045/30).

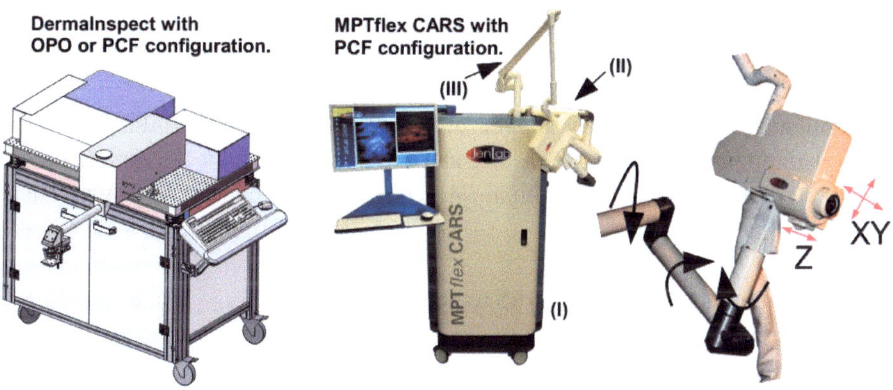

Fig. 14.4: CARS tomographs for in vivo imaging of human skin. Left: DermaInspect CARS-MPT with (i) OPO and (ii) PCF technology; middle: MPTflex CARS with PCF technology; right: demonstration of the flexibility of the scan head.

ensured by tuning the length of a temporal delay line. By tuning both the center wavelength of the OPO signal pulses and the pump pulses, CARS excitation of molecular vibrations in the range between about 2500 cm^{-1} and 4000 cm^{-1} can be achieved. For CARS skin imaging, the symmetric CH_2 stretch vibration (at 2845 cm^{-1}) and the OH-water resonance (at about 3250 cm^{-1}) are probed. The maximum output power of the OPO is about 120 mW, which is attenuated by the internal optical power attenuator of the scan detector unit of the *DermaInspect* to about 17 mW. The maximum optical power applied to the skin is limited to 50 mW ($P_{pump} \pm P_{Stokes} \leq 50$ mW), which is below harmful power levels [33].

The excitation sources of the second *DermaInspect*-based CARS tomograph are a "wavelength extension unit" (*WEU*, Newport-Spectra Physics, USA) containing a PCF module [34] and an external fs oscillator (*MaiTai XF-1*). In addition to the PCF module, the WEU includes further manually adjustable modules, e.g., an optical delay line and power attenuators. After entering the WEU, the fs oscillator output is split into two parts. Part (I) is used directly as the $CARS_{pump}$ laser passing the power attenuator and the delay line. Part (II) – with a mean optical power of about 500 mW – is coupled into the PCF for supercontinuum generation. Since its core diameter is just 1.4 µm, beam pointing stability is essential for a high coupling efficiency and thus a stable output power level. Technically, this requires a stable vibration-free optical breadboard, which is provided by the design of the tomograph *DermaInspect* and a temperature-controlled environment. The latter is of importance to minimize temperature-induced drifts of the optical components. In accordance with the specifications of the PCF (FemtoWHITE CARS, NKT Photonics, Denmark) the PCF-pump wavelength of the fs oscillator is set to 800 nm. For CARS excitation, only a small part of the supercontinuum is needed. This required part is isolated by
(i) a long-pass filter (LP780) and
(ii) a band-pass filter (BP1045/30).

The collimated and spectrally shaped output then provides the $CARS_{Stokes}$ beam with maximum spectral power densities at a wavelength of about 1035 nm. Collinear aligned with the $CARS_{pump}$ beam, lipid-CARS contrast at 2845 cm^{-1} can be generated.

14.2.3 The flexible multimodal CARS tomograph

The requirement for more flexibility, compactness and mobility of CARS tomographs is fulfilled by the second generation imaging device [27] – the *MPTflex CARS* (see Fig. 14.4). The design is based on the established flexible platform of the multimodal (2P-AF, SHG and FLIM) tomograph *MPTflex* [35]. Briefly described, it consists of a mobile and compact housing (I) and a scan head (II) which is connected to the housing (I) by an articulated mirror arm (III). The mirror arm provides free-space beam delivery and significant flexibility that are necessary for a high accessibility and time-efficient imaging (Fig. 14.4). The CARS excitation sources of the *MPTflex CARS* are similar to those described above of the *DermaInspect CARS-MPT* with WEU extension. However, the desired flexibility and compactness of the *MPTflex CARS* required a stacked optical breadboard design and active beam stabilization modules. More details on the setup are given in Fig. 14.5.

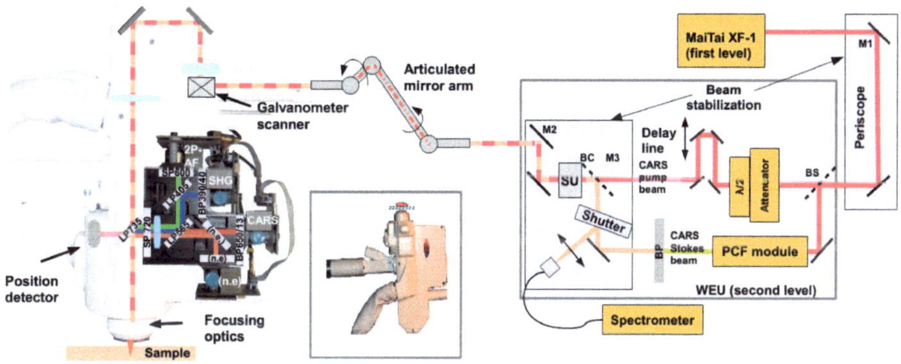

Fig. 14.5: Left: Scan head with four-channel detector; middle: inverted scan head position; right: beam path of CARS excitation sources. BC: beam combiner; M1-3: beam stabilization modules; WEU: wavelength extension unit; SU: safety unit; BS: beam splitter (with kind permission adapted from [27, 28]).

The optical system setup inside the opto-electronic housing includes a lightweight carbon-fiber breadboard with a two-level design. The first level holds the Ti:sapphire fs oscillator. Its output (1100 mW) is guided by a periscope to the second level. A polarizing beam splitter inside the WEU splits the pulse train into two parts where the first part (about 500 mW) is fed into the core of the PCF module. Its collimated spectrally broadened output provides the $\text{CARS}_{\text{Stokes}}$ photons. As mentioned before, coupling the laser into the small diameter of the PCF core is technically challenging. Minor changes to the beam pointing induced by thermal drift strongly influence the coupling efficiency and thus the stability of the output spectrum. Stable operation of the system is ensured by the integration of an active beam stabilization. Based on a feedback algorithm, the transmission of the PCF is monitored and a motorized kinematic mirror mount (M1), directly behind the fs oscillator, is electronically adjusted to optimize transmission. The PCF spectrum can be additionally monitored by a fiber-coupled spectrometer.

The $\text{CARS}_{\text{pump}}$ pulses are provided by the residual part of the fundamental beam. These pulses pass a polarization-maintaining optical power attenuator and an optical delay line, which are both computer controlled. The polarization orientation parallel to the $\text{CARS}_{\text{Stokes}}$ beam is set by a half-wave plate. Before entering the articulated mirror arm the $\text{CARS}_{\text{pump}}$ beam passes a second beam stabilization (M2), which centers the centroid of the beam profile at the center of the back-focal plane of the focusing optics. This beam stabilization is necessary to compensate displacements due to internal mechanical tolerances, tensions inside the mechanical support structure and alignment tolerances of the articulated mirror arm. The feedback signal is provided by a detector located inside the scan head, close to the back-focal plane of the focusing optics (see Fig. 14.5) [36]. The collinear superposition of $\text{CARS}_{\text{pump}}$ and $\text{CARS}_{\text{Stokes}}$

beams is realized by the superposition of the centroids of both beams on the surface of a beam combiner (SP850 or alternatively SP950 from Fig. 14.3). It is important to note that angular deviations of the $CARS_{Stokes}$ beam direction can be induced by thermal drift as well. Since the long distance of the following beam path in the articulated mirror arm translates these deviations into a spatial displacement in the back-focal plane of the focusing optics, it results in a noncentered illumination of the image area. Furthermore, due to the induced deviation between the angle of incidence of $CARS_{pump}$ and $CARS_{Stokes}$ beams at the back-focal plane, the required spatial overlap of $CARS_{pump}$ and $CARS_{Stokes}$ in the focal volume at the sample position is disturbed. Thus, a third active beam stabilization is required, which is incorporated into the mechanical mount of the beam combiner (motorized kinematic mirror mount M3). The same sensor as for M2 is used for M3 in a time-multiplex manner. This is the basis for the spatial overlap of the focal volumes in the two-beam setup of the *MPTflex CARS*. The required temporal overlap for the interaction of $CARS_{pump}$ and $CARS_{Stokes}$ photons in the object position is achieved by adjusting the motorized delay line. In case sole 2P-AF and SHG imaging is required, CARS imaging can be suppressed by blocking the $CARS_{Stokes}$ beam with an internal computer-controlled mechanical shutter.

14.2.4 Multichannel-detection

The detection is typically realized with a two-channel PMT-based detection module, where channel (1) is used for CARS and channel (2) for 2P-AF and SHG signal acquisition [27]. The demand for even more detection channels triggered the development of a four-channel PMT-based detection module for simultaneous detection in individual channels. Fig. 14.5 presents the individual detection channels for 2P-AF, SHG_{pump}, and CARS signals. The included PMTs have a small footprint and provide a fast temporal response that is necessary for fluorescence lifetime imaging (FLIM). The signal light inside the detection beam path is discriminated from the residual backscattered excitation light by a short pass filter (SP720). A long-pass dichroic beam splitter (LP565) separates the CARS signal from the 2P-AF/SHG signal. The 2P-AF/SHG beam path is further divided by a long-pass dichroic beam splitter (LP405) into a 2P-AF-detection channel (1) with a short-pass (SP600) filter to block the residual CARS signals. The SHG-detection channel (2) is narrowed by a 40 nm (FWHM) band-pass filter centered at 390 nm to further reduce the weak wing of the superimposed AF spectrum. To suppress non-CARS signals a 13 nm (FWHM) band-pass filter centered at 650 nm is used to narrow the CARS-detection channel (3). (Channel (4) is not employed (*n.e.*) but offers further detection possibilities.) The detector electronics were developed for simultaneous single photon counting (SPC) as well as time-correlated single photon counting (TCSPC) to enable FLIM. Images with 512 × 512 pixels are recorded at a mean dwell time of about 25 µs/pixel. The corresponding lifetime measurements can be accomplished in parallel to SPC operation by internal TCSPC hardware with a standard image resolution of 128 × 128 pixels (alternative resolution settings are possible). In this chapter, lifetime imaging is presented for detection channel (1) only. However, signals from all four channels can be recorded using the integrated TCSPC router electronics. The analysis of the FLIM data shown below is done with the following settings: double exponential best-fit, 3-fold binning.

Ex vivo imaging is demonstrated using the inverted scan head position (Fig. 14.5). The animal studies were carried out according to ethical standards. Murine tissue samples were retrieved about 5 h post mortem. Brain, liver, spinal cord, abdominal fat, intestine and muscle tissue from a hairy mouse were placed on a microscope coverslip which was fixed by the sample adapter ring. XY and Z positioning of the tissue relative to the optical axis of the system was realized by the integrated motorized stage with maximum vertical and axial travel ranges of 5 × 5 mm² and 5 mm, and a resolution (step size) of 12.7 μm and 1.25 μm, respectively. Images at sample depths of 5–30 μm were acquired with optical powers ($P_{pump} + P_{stokes}$) in the range of 10–40 mW without averaging. For the benefit of optimal print visualization, individual images are contrast enhanced using ImageJ software [37].

Human skin imaging is realized by placing an adapter ring with a fixed coverslip in between skin and focusing optics. Immersion oil (type 518 F; Zeiss, Germany) and water are placed at the interface focusing optics/coverslip and coverslip/skin, respectively. Imaging is performed on healthy human skin (volar side of male forearm) at a skin depth of about 70 μm. The maximum output power ($P_{pump} + P_{Stokes}$) is limited by the safety unit (SU) to 50 mW at the skin.

14.3 Results and Discussion

14.3.1 Demonstration of the spatio-temporal overlap

In the case where the spatio-temporal overlap is fulfilled, the signal spectrum may include CARS signal(s), 2P-AF signals, SHG signals (of the $CARS_{pump}$ und $CARS_{Stokes}$ fundamentals), and the sum frequency (SFG) from both excitation beams (see Fig. 14.6 [27]). As a test sample for the quality of the spatio-temporal pulse overlap at the sample position, urea crystals (grown from urea solution) can be used. If there is only spatial but no temporal overlap, the urea crystals produce the SHG signals of the $CARS_{pump}$ (from 800 nm) and $CARS_{Stokes}$ (from, e.g., 1035 nm) pulses at 400 nm and 517 nm, respectively. In the case of a good spatio-temporal overlap, an additional SFG signal is produced at 451 nm. The strength of this signal can be used to optimize the spatio-temporal overlap by fine adjustment of the spatial overlap on the beam combiner and subsequent readjustment of the spatial overlap at the focal position. CARS contrast at 2845 cm^{-1} can then be tested using a sample with a high concentration of lipids such as grease. In this case, the CARS signal is generated at a wavelength of about 651 nm. It is worth noting that during skin imaging the SHG signal of the $CARS_{Stokes}$ as well as

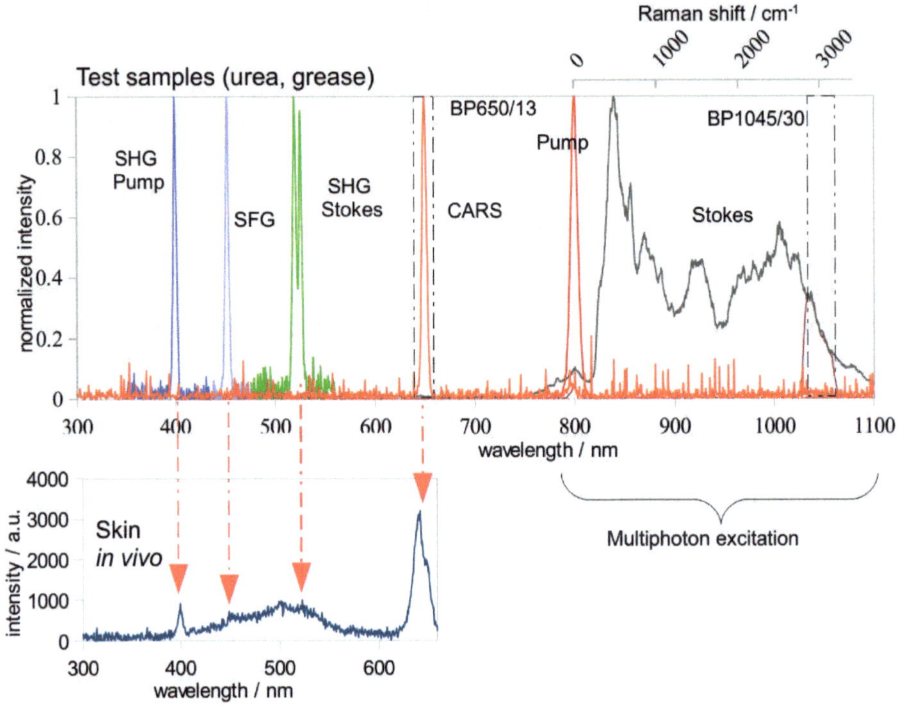

Fig. 14.6: Top: excitation and emission spectra of in vitro test samples. All signals are normalized to their maximum. The band-pass filters (BP) to isolate the $CARS_{Stokes}$ excitation spectrum and the CARS emission spectrum are schematically indicated by dashed lines (adapted from [27]); bottom: in vivo spectrum of human skin at the epidermal-dermal junction (adapted from [26]).

the $SFG_{pump+Stokes}$ signal can be present in the spectral range of 400–600 nm, which is mostly used for 2P-AF detection of the skin. However, in practice, the SFG signal is not visible inside the epidermis (absence of collagen) and just slightly visible inside the spectrum of epidermal-dermal junction of human skin [26], as can be seen in Fig. 14.6. In particular, with animal studies (reduced thickness of skin, e.g., for mice) these collagen fibers and/or muscles can be abundant and can create a significant structure "cross talk" between the detector channels. In this case, an effective suppression of the SHG_{Stokes} and $SFG_{pump+Stokes}$ signals inside the 2P-AF image can be accomplished by enhanced optical filter design such as multiband rejection filters or by time gating [38]. The latter can be realized by the inherent TCSPC hardware, which allows the temporal separation of the instantaneous SHG and SFG signals (fast components) from the decayed fluorescence lifetime components.

14.3.2 Ex vivo imaging

Multimodal imaging of fresh murine tissue is shown in Fig. 14.7. The abdominal fat tissue reveals a strong CARS signal due to a high CH_2 concentration inside the monovacuolar adipocytes. It seems that the adipocytes are connected to each other by reticular collagen fibers which generate SHG signals. 2P-AF contrast is visible only from the periphery of the cells. Interestingly, the lifetime image significantly enhances contrast and a capillary-like structure appears (arrowhead) which is hardly visible in the 2P-AF or SHG channel. Even blood cells are visible inside this capillary. A high lipid content

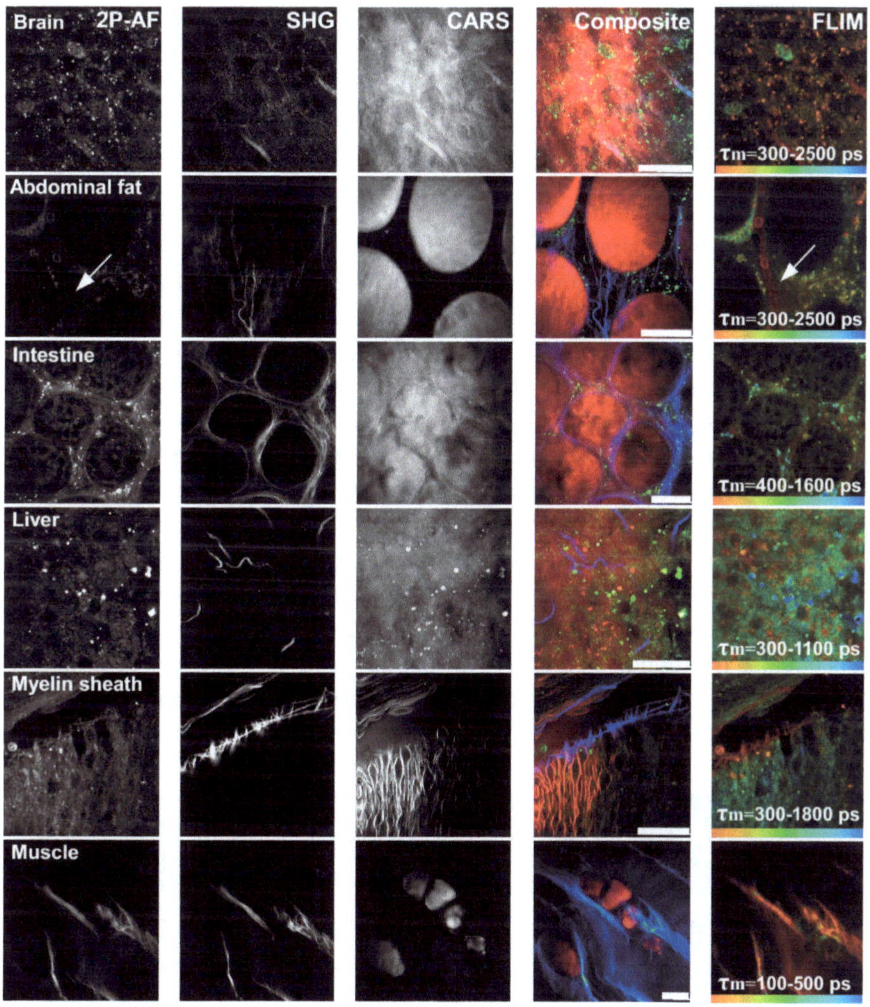

Fig. 14.7: Ex vivo images of murine tissues ($CARS_{pump}$ laser: 800 nm; $CARS_{Stokes}$ laser: 1035 nm). Scale bar: 30 µm (adapted from [28]).

and thus strong CARS signals are additionally present inside the myelin sheaths of the spinal cord, intracellular lipid droplets/vacuoles in liver tissue, and in adipocytes between muscle tissue. Furthermore, the images from muscle tissue show highly periodic structures (possibly from myofilaments) which are almost identical for the SHG- and 2P-AF-images. This congruence is most likely induced by nonlinear interaction of both $CARS_{pump}$ and $CARS_{Stokes}$ excitation light at the muscle fibrils producing an SHG_{Stokes} signal (at a wavelength of 517 nm) and possibly also an $SFG_{pump+Stokes}$ signal (at a wavelength of 451 nm).

These experiments present the detection capabilities of the system. Of course these ex vivo images can also be generated in vivo, as has already been demonstrated with living mice with the 2P-AF and SHG imaging capability of the tomograph *MPTflex* [39].

14.3.3 In vivo human skin imaging

Multimodal imaging of healthy human skin at the dermal-epidermal junction at a depth of 70 µm is demonstrated in Fig. 14.8. Inside the 2P-AF channel, basal cells are arranged in a circular manner surrounding elastin fibers. Further contrast is presented by the associated lifetime image (FLIM), which superimposes the intensity contrast of the 2P-AF image by visualization of a heterogeneity of lifetimes in pseudo-colors. The spatially resolved representation of these lifetimes helps to distinguish between cell types or structures in skin. For instance, the basal cells with a high content of melanin reveal short lifetimes (orange color coded) and thus can be clearly isolated from the elastin fibers (blue-green color coded), which reveal longer lifetimes [40] compared to melanin. The image from the SHG-detection channel visualizes corresponding endogenous collagen fibers by detecting SHG from the nonlinear light-collagen interaction. The image obtained in the CARS-detection channel presents additional morphological information about the skin. Fine structured details of separated cellular details but also the roundish pattern of the dermal papillae are visible. Especially in between the fibers of the dermal-epidermal structures further information can be extracted from the sample.

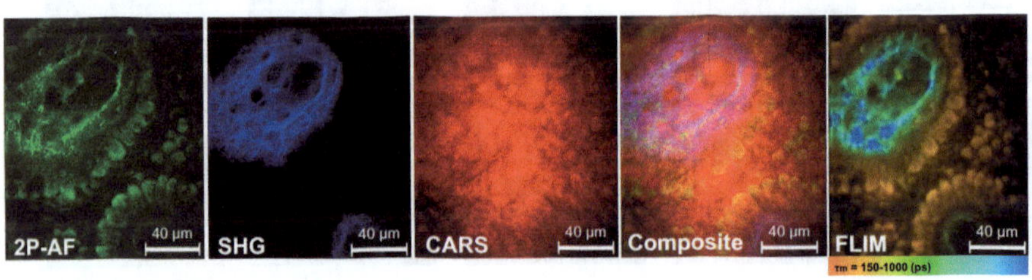

Fig. 14.8: Optical biopsy: multimodal imaging of human skin in vivo at an imaging depth of 70 µm (dermal papillae) (adapted from [28]).

The images in Fig. 14.8 show no significant 2P-AF detector channel cross talk. However, deeper skin regions provide the possibility to produce SHG_{Stokes} and $SFG_{pump+Stokes}$ signals by interaction with collagen structures of the dermis. From the emission spectrum in Fig. 14.6 (b) it can be expected that their signal strengths are small compared to the 2P-AF signal. Still, if required they can be eliminated by the same methods discussed above.

14.3.4 Chemical contrast

The spectral resolution of the fs-CARS system is determined by the spectral width of the $CARS_{pump}$ laser. A typical pulse width of about 100 fs (transform limited) at 800 nm presents a spectral width of about 150 cm^{-1} which is significantly broader than the line width of many Raman resonances in the fingerprint region [41]. Spectral parts of the excitation spectrum which do not overlap with the Raman-line width excite only the nonresonant background and reduce the overall chemical contrast. This challenges the application of fs pulses for imaging in the fingerprint region [7]. However, in the high wavenumber region around 3000 cm^{-1} the situation is different. For instance, the broad Raman resonances of lipids (2845 cm^{-1}) and proteins (2950 cm^{-1}) [42] have a width which is similar to the spectral width of the fs pulses. Thus, the sensitivity and the ratio R/NRB is improved for these resonances, which has been verified by Chen et al. who compared the lipid contrast for ps- and fs-CARS systems [20].

These results are also confirmed by our experiments. Ex vivo fresh murine myelin sheaths (see Fig. 14.7) reveal a contrast ratio of $> 7:1$, similar to contrast ratios determined by other groups [20]. For imaging on human skin the ratio R/NRB is significantly smaller, apparently due to lower concentrations of lipids. However, in some skin regions such as the exit of sebaceous glands a higher lipid concentration is present (sebum) leading to R/NRB ratios in the range between 10 to 20 [28].

14.4 Conclusion

This chapter described multiple multimodal imaging systems capable of acquiring optical sections of human skin in vivo as well as animal tissue samples. In a single scan of a few seconds a multitude of modalities (2P-AF/FLIM, SHG, and CARS) generate a multitude of images with subcellular resolution. A variety of applications can benefit from these systems, such as rapid ex vivo imaging of tumor biopsies [43] and the in vivo investigation of skin barrier problems [44].

Acknowledgment: The work is supported by the German Federal Ministry of Education and Research (BMBF; Chemoprävent 13N10507 and Biofeedback 13N12594) as well as the European Union's Horizon 2020 research and innovation program under grant agreement No 726666 (LASER-HISTO).

References

[1] Raman CV, Krishnan KS. A new type of secondary radiation. Nature. 1928;121:501–502.
[2] Ellis DI, Cowcher DP, Ashton L, O'Hagan S, Goodacre R. Illuminating disease and enlightening biomedicine: Raman spectroscopy as a diagnostic tool. Analyst. 2013;138:3871–3884.
[3] Kong K, Kendall C, Stone N, Notingher I. Raman spectroscopy for medical diagnostics – From in-vitro biofluid assays to in-vivo cancer detection. Advanced Drug Delivery Reviews. 2015;89:121–134.
[4] Movasaghi Z, Rehman S, Rehman IU. Raman spectroscopy of biological tissues. Applied Spectroscopy Reviews. 2007;42:493–541.
[5] Weinigel M, Breunig HG, Darvin M E, Klemp M, Rowert-Huber J, Lademann J, König K. Impact of refractive index mismatches on coherent anti-Stokes Raman scattering and multiphoton autofluorescence tomography of human skin in vivo. Phys Med Biol. 2015;60:6881–6899.
[6] Kneipp K, Kneipp H, Itzkan I, Dasari RR, Feld MS. Ultrasensitive chemical analysis by Raman spectroscopy. Chemical Reviews. 1999;99:2957–2976.
[7] Evans CL, Xie XS. Coherent anti-stokes Raman scattering microscopy: chemical imaging for biology and medicine. Annual review of analytical chemistry. 2008;1:883–909.
[8] Bloembergen N. A quarter century of stimulated Raman scattering. Pure and Applied Chemistry. 1987;59:1229–1236.
[9] Downes A, Elfick A. Raman spectroscopy and related techniques in biomedicine. Sensors. 2010;10:1871–1889.
[10] Müller M, Zumbusch A. Coherent anti-Stokes Raman scattering microscopy. Chem Phys Chem. 2007;8:2156–2170.
[11] Ganikhanov F, Evans CL, Saar BG, Xie XS. High-sensitivity vibrational imaging with frequency modulation coherent anti-Stokes Raman scattering (FM CARS) microscopy. Opt Lett. 2006;31:1872–1874.
[12] Rocha-Mendoza I, Langbein W, Watson P, Borri P. Differential coherent anti-Stokes Raman scattering microscopy with linearly chirped femtosecond laser pulses. Opt Lett. 2009;34:2258–2260.
[13] Jurna M, Korterik JP, Otto C, Herek JL, Offerhaus HL. Background free CARS imaging by phase sensitive heterodyne CARS. Optics Express. 2008;16:15863–15869.
[14] Volkmer A, Book LD, Xie XS. Time-resolved coherent anti-Stokes Raman scattering microscopy:Imaging based on Raman free induction decay. Applied Physics Letters. 2002;80:1505–1507.
[15] Cheng J-X, Volkmer A, Book LD, Xie XS. An epi-detected coherent anti-Stokes Raman scattering (E-CARS) microscope with high spectral resolution and high sensitivity. J Phys Chem B. 2001;105:1277–1280.
[16] Duncan MD, Reintjes J, Manuccia TJ. Scanning coherent anti-Stokes Raman microscope. Opt Lett. 1982;7:350–352.
[17] Zumbusch A, Holtom G, Xie S. Three-dimensional vibrational imaging by coherent anti-Stokes Raman scattering. Physical Review Letters. 1999;82:4142–4145.

[18] Cheng J-X, Xie XS. Coherent anti-Stokes Raman Scattering microscopy: Instrumentation, theory, and applications. J Phys Chem B. 2003;108:827–840.
[19] Dudovich N, Oron D, Silberberg Y. Single-pulse coherently controlled nonlinear Raman spectroscopy and microscopy. Nature. 2002;418:512–514.
[20] Chen H, Wang H, Slipchenko MN, Jung Y, Shi Y, Zhu J, Buhman KK, Cheng JX. A multimodal platform for nonlinear optical microscopy and microspectroscopy. Optics Express. 2009;17:1282–1290.
[21] Langbein W, Rocha-Mendoza I, Borri P. Single source coherent anti-Stokes Raman microspectroscopy using spectral focusing. Applied Physics Letters. 2009;95:081109.
[22] Hellerer T, Enejder AMK, Zumbusch A. Spectral focusing: High spectral resolution spectroscopy with broad-bandwidth laser pulses. Applied Physics Letters. 2004;85:25–27.
[23] Le TT, Yue S, Cheng J X. Shedding new light on lipid biology with coherent anti-Stokes Raman scattering microscopy. J Lipid Res. 2010;51:3091–3102.
[24] König K, Breunig HG, Bückle R, Kellner-Höfer M, Weinigel M, Büttner E, Sterry W, Lademann J. Optical skin biopsies by clinical CARS and multiphoton fluorescence/SHG tomography. Laser Physics Letters. 2011;8:465–8.
[25] Breunig HG, Bückle R, Kellner-Höfer M, Weinigel M, Lademann J, Sterry W, König K. Combined in vivo multiphoton and CARS imaging of healthy and disease-affected human skin. Microscopy Research and Technique. 2012;75:492–498.
[26] Breunig HG, Weinigel M, Bückle R, Kellner-Höfer M, Lademann J, Darvin ME, Sterry W, König K. Clinical coherent anti-Stokes Raman scattering and multiphoton tomography of human skin with a femtosecond laser and photonic crystal fiber. Laser Physics Letters. 2013;10:025604–025609.
[27] Weinigel M, Breunig HG, Kellner-Höfer M, Bückle R, Darvin ME, Klemp M, Lademann J, König K. In vivo histology: optical biopsies with chemical contrast using clinical multiphoton/CARS tomography. Laser Physics Letters. 2014;11:055601–055609.
[28] Weinigel M, Breunig HG, Uchugonova A, König K. Multipurpose nonlinear optical imaging system for in vivo and ex vivo multimodal histology. Journal of Medical Imaging. 2015;2:016003–016010.
[29] Murugkar S, Brideau C, Ridsdale A, Naji M, Stys PK, Anis H. Coherent anti-Stokes Raman scattering microscopy using photonic crystal fiber with two closely lying zero dispersion wavelengths. Opt Express. 2007;15:14028–14037.
[30] Baldacchini T, Zadoyan R. Multimodal microscopy with high resolution spectral focusing CARS. Proc SPIE 8948; 2014. p. 894812–894816.
[31] Kano H. Molecular Spectroscopic Imaging Using a White-Light Laser Source. Bulletin of the Chemical Society of Japan. 2010;83:735–743.
[32] Klarskov P, Isomaki A, Hansen KP, Andersen PE. Supercontinuum generation for coherent anti-Stokes Raman scattering microscopy with photonic crystal fibers. Opt Express. 2011;19:26672–26683.
[33] Fischer F, Volkmer B, Puschmann S, Greinert R, Breitbart W, Kiefer J, Wepf R. Risk estimation of skin damage due to ultrashort pulsed, focused near-infrared laser irradiation at 800 nm. J Biomed Opt. 2008;13:041320.
[34] Zadoyan R, Baldacchini T, Carter J, Kuo C-H, Ocepek D. CARS module for multimodal microscopy. Proc SPIE (San Francisco) 7903; 2011. p. 79031–79039.
[35] Weinigel M, Breunig H G, Gregory A, Fischer P, Kellner-Höfer M, Bückle R, König K. A novel flexible clinical multiphoton tomograph for early melanoma detection, skin analysis, testing of anti-age products, and in situ nanoparticle tracking. Proc SPIE (San Francisco) 7589; 2010. p. 758908–758915.

[36] König K, Weinigel M. Flexibles nichtlineares Laserscanning-Mikroskop zur nicht-invasiven dreidimensionalen Detektion. German patent DE102011115944 B4. 2011.
[37] Rasband WS. ImageJ. Available from: http://imagej.nih.gov/ij/ 1997–2014 [cited 2014 November 12,2014].
[38] Gerritsen HC, Draaijer A, van den Heuvel DJ, Agronskaia AV. In: Pawley J, editor. Fluorescence lifetime imaging in scanning microscopy in handbook of biological confocal microscopy. 3rd edition. New York: Springer; 2006. p. 516–532.
[39] Uchugonova A, Hoffman RM, Weinigel M, König K. Watching stem cells in the skin of living mice noninvasively. Cell Cycle. 2011;10:2017–2020.
[40] König K. Clinical multiphoton tomography. Journal of Biophotonics. 2008;1:13–23.
[41] Caspers PJ, Lucassen GW, Puppels GJ. Combined in vivo confocal Raman spectroscopy and confocal microscopy of human skin. Biophysical Journal. 2003;85:572–580.
[42] Caspers PJ, Lucassen GW, Carter EA, Bruining HA, Puppels GJ. In vivo confocal Raman microspectroscopy of the skin: noninvasive determination of molecular concentration profiles Journal of Investigative. Dermatology. 2001;116:434–442.
[43] Kantelhardt SR, Leppert J, Krajewski J, Petkus N, Reusche E, Tronnier VM, Hüttmann G, Giese A. Imaging of brain and brain tumor specimens by time-resolved multiphoton excitation microscopy ex vivo. Neuro-Oncology. 2007;9:103–112.
[44] Proksch E, Brandner JM, Jensen JM. The skin: an indispensable barrier. Experimental Dermatology. 2008;17:1063–1072.

Mihaela Balu, Kristen M. Kelly, Ronald M. Harris, Karsten König, Christopher B. Zachary, and Bruce J. Tromberg

15 In vivo multiphoton microscopy of human skin

Abstract: The development of clinical multiphoton technologies has led to new, label-free approaches for non-invasive, in vivo imaging of human skin. Recent studies have shown that multiphoton imaging can be used to assess a wide range of biological processes, including cancer, cellular metabolism, and the effects of skin treatments. The imaging devices *MPTflex* and its earlier version, *DermaInspect*, have been employed in a broad range of clinical applications spanning from characterizing and understanding keratinocyte metabolism to malignant melanoma detection and diagnosis, pigment biology, cosmetic treatments, and skin aging. The promising results indicate that in the near future, real-time non-invasive nonlinear "optical biopsies" can be performed at the bedside.

15.1 Introduction

Microscopic evaluation of skin is required in many areas of dermatology such as diagnosing skin diseases, assessing the effects of cosmetic treatments, and understanding skin function. However, optical imaging of skin is challenging due to the fact that it is relatively thick, scatters light strongly, and has very little contrast. As a result, the standard approach for skin imaging with conventional optical microscopy is to slice it upon fixation, mount and stain the slices on microscope slides, and analyze each slide one at a time. This technique has two main limitations:

(1) It is time consuming and it involves many steps in the process that can be prone to errors.
(2) The tissue needs to be removed from its natural environment; while the overall tissue morphology is preserved during the slicing and staining process, the main features related to tissue functionality such as metabolism, are lost.

Over the past 20 years, optical technologies based on in vivo laser scanning microscopy have been developed to address these limitations. This method utilizes focused illumination of a laser beam to provide access to the submicron scale and raster-scanning across the sample to create high-resolution images. Each image plane (about 2 µm thick) forms an optical section and can be used in reconstructing a 3D-view of the skin by scanning at multiple depths. How laser scanning microscopy (LSM) can overcome high scattering and low contrast of unstained tissue is an area of current research. The scattering limitation has been solved only partially. Using a laser beam as a light source allows for wavelength selection that affects the scattering process. Longer wavelengths are scattered less than shorter wavelengths by a turbid

https://doi.org/10.1515/9783110429985-017

medium such as skin making near-infrared light (NIR) particularly attractive for LSM techniques. Nevertheless, limited depth penetration in skin due to light scattering remains a major limitation of these techniques and a permanent challenge that is being addressed as the technology advances.

Image contrast of unstained skin tissue, on the other hand, has been improved substantially and is based on different mechanisms corresponding to particular LSM techniques. Among these, reflectance confocal microscopy (RCM) and multiphoton microscopy (MPM) have been the most widely used in research studies related to applications in dermatology. RCM contrast is based on variations in tissue refractive index, which provide gray scale images with submicron resolution. MPM contrast in skin is derived from second-harmonic generation (SHG) of collagen and two-photon excited fluorescence (TPEF) of tissue components such as the cofactors NADH and FAD, elastin, keratin, and melanin.

Due to its dual contrast mechanism, MPM provides multicolor images that distinguish cellular features from the extracellular matrix. MPM imaging is unique among other optical imaging technologies in that it provides 3D, near real-time submicron resolved label-free images of living tissues in their native environment with contrast that closely resembles the histological sections dermatopathologists use for diagnosis. MPM technology has been translated into clinical settings through the development of the MPM-based clinical tomographs, DermaInspect [1] and MPTflex by JenLab GmbH (Jena, Germany). This chapter summarizes the results of the most recent clinical studies in the dermatology field, performed by employing either the DermaInspect or the MPTflex tomographs.

15.2 MPM technology and translation into the clinic

The first experiments using TPEF and SHG laser scanning microscopy were performed in 1990 [2] and 1993 [3], respectively. The first experiments on in vivo TPEF imaging of human skin were performed in the late 1990s [4–6]. TPEF and SHG imaging techniques were combined in a single MPM-based tomograph (DermaInspect, Fig. 15.1 (a)) for in vivo skin imaging in 2003 [1]. This system was developed by JenLab, GmbH (Jena, Germany) and it is CE-marked for clinical use along with its most advanced version MPTflex (Fig. 15.1 (b)), a compact, portable device that features an articulated arm to allow imaging almost any region of the body.

The system consists of a compact Ti:sapphire femtosecond laser, an articulated arm with near-infrared optics, and beam scanning module. The system has two photomultiplier tube detectors for parallel acquisition of two-photon excited fluorescence (TPEF) and second-harmonic generation (SHG) signals. A customized metallic ring taped on the subject's skin attaches magnetically to the objective holder in the articulated arm, minimizing motion artifacts. The images acquired by MPTflex and DermaInspect have a lateral spatial resolution of $\approx 0.5\,\mu m$ and an axial resolution $< 2\,\mu m$. The

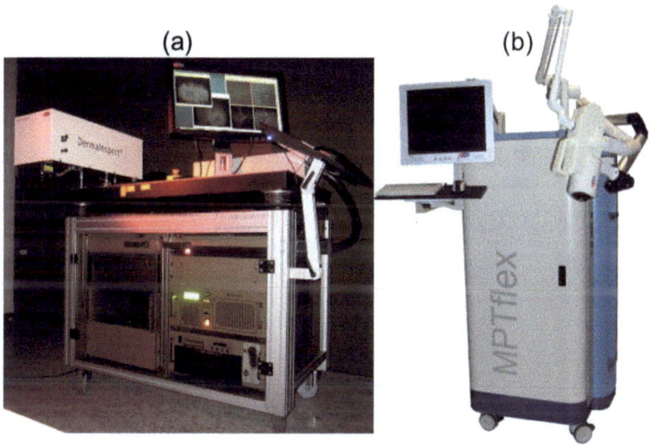

Fig. 15.1: MPM-based clinical tomographs, (a) DermaInspect and (b) MPTflex.

imaging depth in skin is about 200 µm, depending on the skin or lesion types imaged. The area imaged (field of view) is $250 \times 250\,\mu m^2$. The field of view can be increased to a few mm^2 by implementing a mosaic “tiling” feature (acquisition of adjacent fields of view).

MPTflex and its earlier version, DermaInspect, have been employed in a broad range of clinical applications spanning from characterizing and understanding keratinocyte metabolism [7] to skin cancer detection and diagnosis [8–10], pigment biology [11], cosmetic treatments [12], and skin aging [13–16].

15.3 Applications of MPM-based clinical tomographs in dermatology

15.3.1 In vivo MPM imaging of normal skin

Commonly, MPM imaging is used in en-face (horizontal section) mode, but cross-sectional (vertical) sections can also be acquired in real time. Fig. 15.2 shows a representative example of MPM images from normal skin acquired with MPTflex. The en-face images were acquired as a z-stack of horizontal images every 5 µm. En-face images corresponding to different skin layers were selected and shown in Fig. 15.2. The stratum corneum is visualized through TPEF fluorescence from keratin, a thin bright layer on the top surface of the skin in the cross-sectional image and an acellular bright fluorescent layer in the corresponding en-face image. The epidermis shows normally distributed keratinocytes imaged by the TPEF fluorescence from NADH/FAD, keratin and melanin (in the case of pigmented skin). The dermal-epidermal junction (DEJ) is clearly delineated in the cross-sectional view as it separates the basal cell layer from the dermis. Pigmented keratinocytes in the basal

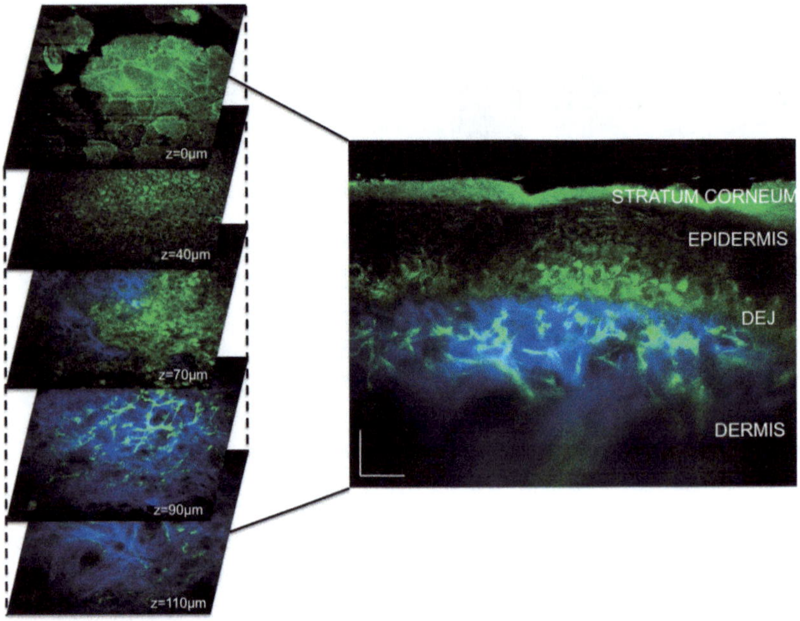

Fig. 15.2: In vivo MPM imaging of normal skin (volar forearm). (Left) MPM horizontal sections (XY scans) at different depths showing images of the stratum corneum ($z = 0$ µm), keratinocytes normally distributed in the stratum spinosum ($z = 40$ µm), the basal cells (green) surrounding dermal papilla (blue) ($z = 70$ µm), collagen (blue) and elastin (green) in the dermis ($z = 90$ µm and 110 µm). (Right) Cross-sectional view (XZ scan) corresponding to a vertical plane through the horizontal sections on the left. The image shows the well-delineated layers of the stratum corneum, epidermis, dermalepidermal junction (DEJ) and the superficial dermis.

layer appear as bright fluorescent cells along the DEJ due to their melanin content. In the en-face images of the DEJ, the basal cells are imaged as surrounding the tips of the dermal papillae. Dermal papillae and the dermis are visualized through the SHG signal from collagen and TPEF signal from elastin fibers. Occasionally, blood vessels and capillaries are imaged in the dermis as shown in Fig. 15.2.

15.3.2 In vivo MPM imaging can determine the depth dependent sensitivity of human epidermis to vascular oxygen supply

The reduced nicotinamide adenine dinucleotide (NADH) concentration within mitochondria is closely related to the cellular oxygen supply [7, 18, 19]. Monitoring and imaging the mitochondrial NADH fluorescence, as a way of evaluating the supply of oxygen within cells, has been an area of interest in research over the past 50 years. While most of the pioneering work in this field was based on UV excitation sources, two-photon induced fluorescence employing NIR excitation light has the advantage of reduced cellular photodamage and improved cell viability, allowing in vivo probing of mitochondrial NADH fluorescence over an extended period of time. MPM allows in vivo imaging of human keratinocytes through the TPEF signal from NADH. In a recent study, we used this device to monitor in vivo and non-invasively the changes in NADH fluorescence of human epidermal cells induced by ischemia through blood-supplied oxygen deprivation [7]. Ischemia-induced changes in NADH fluorescence of keratinocytes in layers close to the stratum basale were compared to NADH fluorescence of keratinocytes from epidermal layers closer to the skin surface. The intensity results recorded as a time series before, during, and after arterial occlusion for epidermal cells in layers close to the stratum basale and skin surface, are shown in Fig. 15.3, along with representative TPEF images at different time points. The results show that ischemia induced by blood-supplied oxygen deprivation is associated with a strong increase in NADH fluorescence of keratinocytes in layers close to the stratum basale, whereas keratinocytes from epidermal layers closer to the skin surface are not affected. This implies that the metabolic processes of keratinocytes in deeper layers such as stratum spinosum and stratum basale are regulated by capillary oxygen supply, whereas the keratinocytes in upper layers such as stratum granulosum are either supplied by atmospheric oxygen [20] or are functionally anaerobic as suggested in previous studies [21].

NADH fluorescence changes were monitored concurrently with changes in tissue oxy- and deoxyhemoglobin concentration during oxygen deprivation using the functional imaging method, spatial frequency domain imaging (SFDI [22]) (Fig. 15.3 (c)). After blood flow was occluded at $t = 3$ min, the tissue deoxyhemoglobin concentration (ctHHb) increased at the expense of oxyhemoglobin (ctO_2Hb), because oxygen was extracted from blood by the epidermal cells and surrounding tissue. The occlusion-induced ischemia reduces the rate of oxidative phosphorylation, resulting in a higher concentration of NADH and higher fluorescence from the epidermal cells that are sensitive to capillary oxygen supply. After the occlusion release at $t = 6$ min, the tissue was rapidly reperfused resulting in an increase in ctO_2Hb and decrease in ctHHb. There was an initial overshoot of the oxy- and deoxyhemoglobin values followed by a return to baseline. These results were similar to previous in vivo measurements of human tissue hemodynamics during arterial occlusion using SFDI [23] and other techniques [24].

The combined MPM/SFDI measurements allowed us to estimate an oxygen consumption rate by basal layer keratinocytes. The rate of decrease in oxyhemoglobin concentration (ctO_2Hb) is related to the rate of oxygen consumption (OC) concentration by:

$$\mathrm{OC} = -4\frac{\partial}{\partial t}ctO_2\mathrm{Hb}, \tag{15.1}$$

where the 4 accounts for the hemoglobin to oxygen molar ratio [25]. From the average decrease in oxyhemoglobin concentration of 0.025 µmol/l/s during the arterial occlusion measured by SFDI (Fig. 15.3 (c)), the oxygen consumption rate can be estimated as 0.1 µmol/l/s by using equation (15.1). By calculating the TPEF imaged volume of the keratinocytes close to the basal layer, the oxygen consumption rate of these cells can be estimated. Therefore, in the TPEF imaged volume of about 200 picoliters containing around 200 cells, the oxygen consumption rate can be estimated to be about 0.035 µmol/10^6 cells/hour. This is the first attempt to measure the oxygen consumption rate by human keratinocytes in vivo.

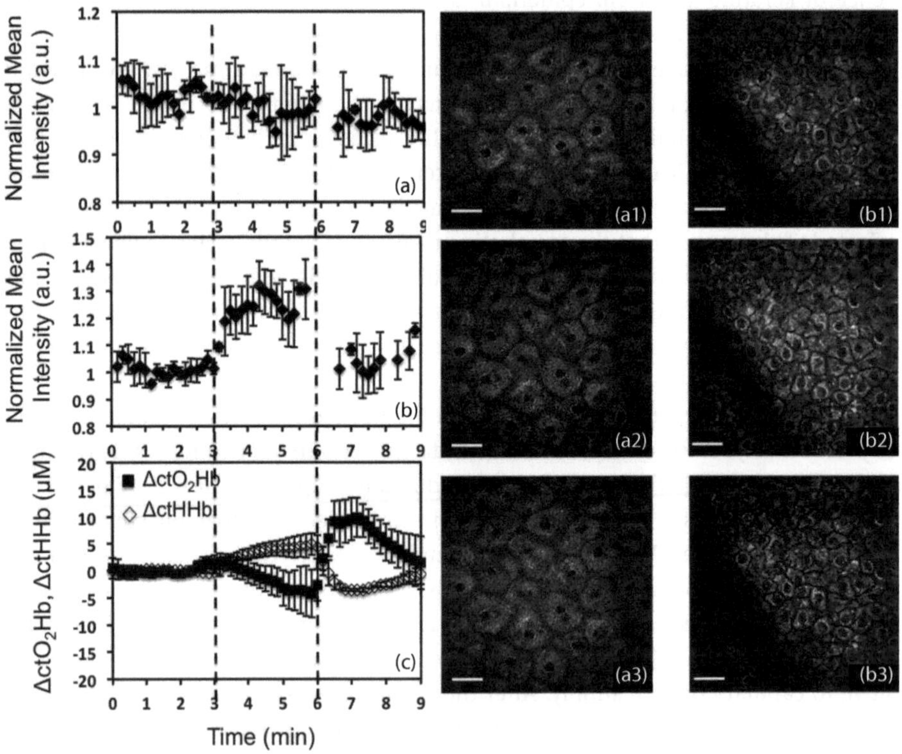

Fig. 15.3: Normalized mean intensity of TPEF images as a function of time before, during and after arterial occlusion for human keratinocytes in stratum granulosum (a) and in a layer close to stratum basale (b). Representative TPEF in vivo images recorded before (a1, b1), during (a2, b2) and after (a3, b3) occlusion corresponding to human keratinocytes in stratum granulosum (a1, a2, a3) and in a layer close to stratum basale (b1, b2, b3). Scale bar is 20 µm. (c) Hemodynamic response to arterial occlusion; oxyhemoglobin concentration ctO_2Hb (square), deoxyhemoglobin concentration ctHHb (diamond). The dashed lines represent the start and the release points of the occlusion. Error bars represent standard deviation of five measurements. Reproduced with permission from [7].

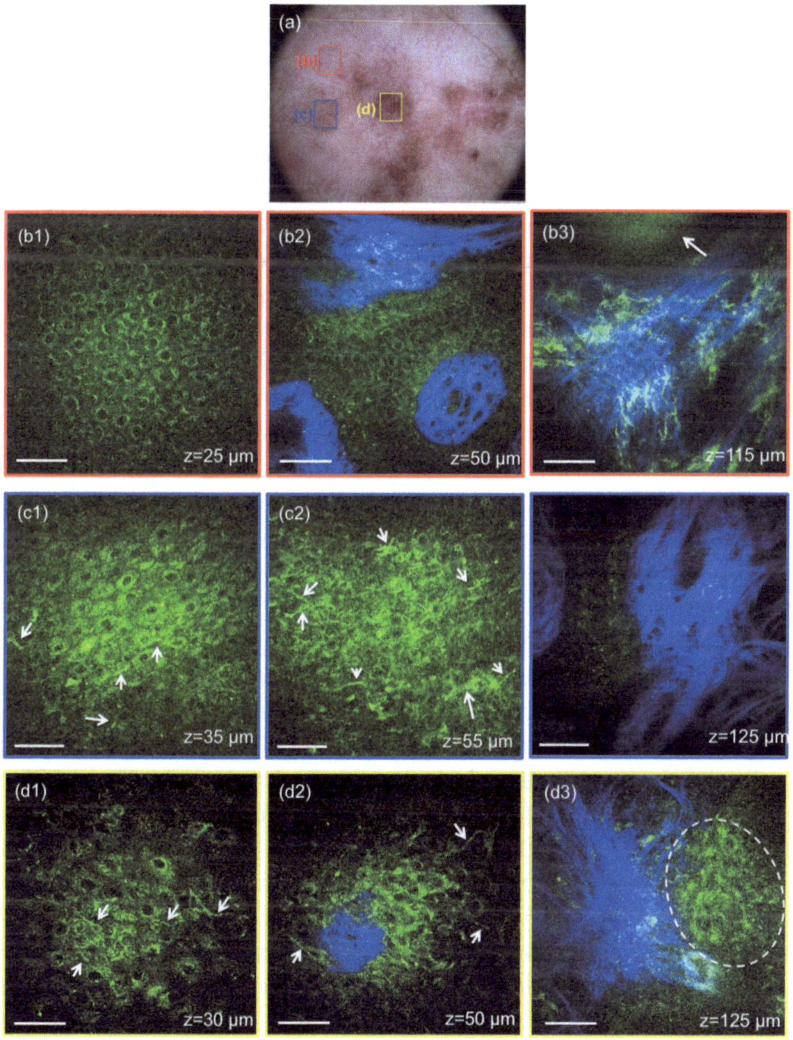

Fig. 15.4: MPM images of a micro-invasive melanoma lesion on a patient's forehead (a) Clinical image (DermLite FOTO, Dermlite Inc.). The markers represent the locations where the images shown in (b1–b3), (c1–c3) and (d1–d3) were acquired. (b1–b3) MPM images corresponding to normal skin adjacent to lesion (red square in (a)) showing normal distribution of keratinocytes (b1), the basal cells (green) surrounding dermal papilla (blue) at 50 µm depth (b2) and collagen (blue) and elastin (green) fibers surrounding a hair follicle (arrow) in the dermis at a depth of 115 µm (b3). (c1–c2) MPM images of the epidermal layers corresponding to different depths of the lesion area indicated by the blue square in (a). The images show ascending melanocytes (white arrows) (c1–c2) and no obvious features of malignancy in the papillary dermis (c3). (d1–d3) MPM images of the lesion area indicated by the yellow square in (a). The images show ascending melanocytes (white arrows) in upper epidermal layers (d1), proliferation of melanocytes (white arrows) at DEJ (d2) and of pigmented cells with different shapes and enlarged nuclei (cellular atypia) in the dermis at a depth of 125 µm. Scale bar is 40 µm. Reproduced with permission from [17].

15.3.3 In vivo MPM imaging of melanoma

Melanoma is the most severe form of skin cancer. It arises from melanocytes, the cells responsible for producing pigment. There are several subtypes of melanoma. Their features have differences and similarities, but generally the following are suggestive of malignancy: presence of melanocytes within the upper portion of the epidermis singly or in groups (Pagetoid spread); irregular junctional activity (atypical melanocytes, architectural disorder); and invasion of tumor cells into the dermis [9, 26, 27]. These features have been identified in melanoma lesions in a study published in 2009, where a DermaInspect MPM tomograph was used to establish sensitivity and specificity criteria for melanoma diagnosis [8]. The assessment by different observers of these features led to overall sensitivity and specificity values for melanoma diagnosis of 75 % and 80 %, respectively. In that study as well as in most of the RCM studies on in vivo imaging of melanoma, particular features of atypical/dysplastic nevi are not extensively discussed.

In a more recent pilot study, our group employed MPM (MPTflex, JenLab, Germany) to identify characteristic features of 15 melanocytic nevi at three different stages: common nevi without dysplastic changes, dysplastic nevi with structural and architectural atypia, and melanoma (5 in each group) [9]. We proposed and developed a quantitative approach by translating the qualitative features used by dermatopathologists in histopathology into quantitative parameters that can be uniquely extracted from 3D in vivo MPM images. We defined a numerical “multiphoton melanoma index (MMI)” based on 3D in vivo image analysis that scores signals derived from TPEF, SHG, and melanocyte morphology features on a continuous 9-point scale. Indices corresponding to common nevi (0–1), dysplastic nevi (1–4) and melanoma (5–8) were significantly different ($p < 0.05$), suggesting the potential of the method to distinguish between melanocytic nevi in vivo [9]. Although promising, a more comprehensive study of a larger number of patients is necessary in order to validate the proposed scoring algorithm and evaluate how well MPM technology can distinguish dysplastic nevi from common nevi and melanoma.

Fig. 15.4 illustrates representative MPM images of a micro-invasive melanoma lesion along with the MPM images corresponding to normal skin adjacent to lesion. The lesion was imaged prior to biopsy. The intraepidermal proliferation of melanocytes and the upward intraepithelial spread shown in the MPM images of Fig. 15.4 were confirmed by histopathology.

15.3.4 In vivo MPM imaging of basal cell carcinoma (BCC)

BCC is a form of skin cancer that originates from the basal cells of the epidermis and associated follicular structures. Nonmelanoma skin lesions, such as BCC have been imaged in vivo by MPM only recently in two pilot research clinical studies [10, 28].

In a first pilot study evaluating the in vivo MPM imaging of BCC lesions, the identified features were compared to the ones provided by RCM imaging [28]. Besides the two main features: elongated, polarized nuclei and tumor nests showing peripheral palisading identified by both techniques, MPM had the ability to evaluate changes in the nucleus/cytoplasm ratio and in cell density across the epidermal layers of BCC lesions. In a more recent study, the MPM (MPTflex, JenLab, Germany) was used in order to evaluate whether histopathologic criteria can be identified in MPM images [10]. In this study, the MPTflex articulating arm allowed imaging of lesions on different parts of the body rather than being limited to lesions on the extremities, as in the previous work with the DermaInspect [28]. This demonstrated the ability of MPM to identify the main histopathologic criterion for BCC diagnosis in all the lesions imaged: tumor nests of basaloid cells at the DEJ and/or in the dermis that often showed palisading. The ability of MPM to resolve cellular structures inside tumor nests is critical for distinguishing BCC from other types of neoplasms that involve the presence of nests in the dermis. Fig. 15.5 shows representative MPM images of a BCC lesion imaged by MPM.

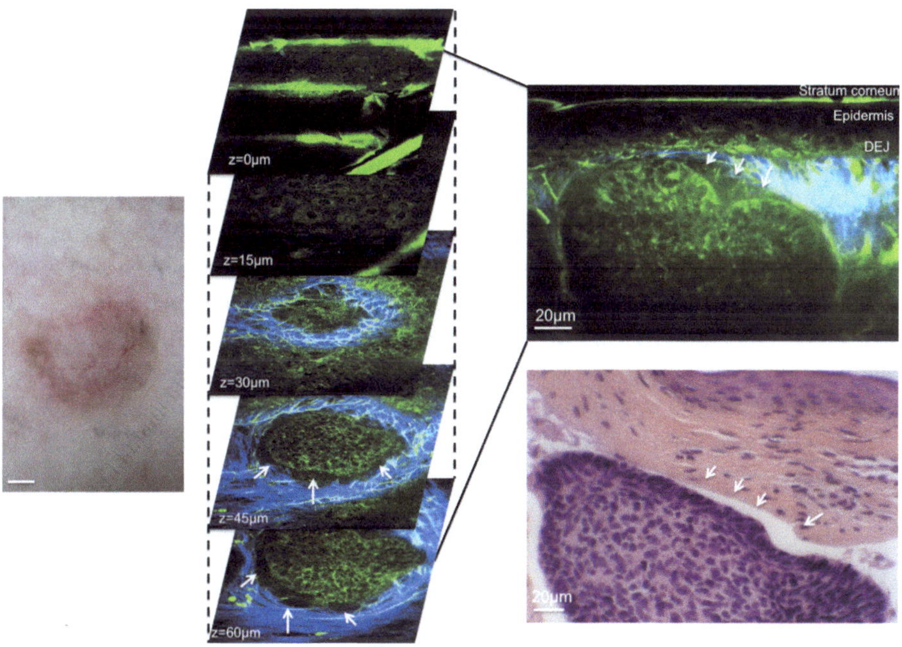

Fig. 15.5: MPM images of a superficial BCC lesion on the shoulder of a patient. (a) Clinical image (DermLite FOTO, Dermlite Inc.), scale bar is 2 mm. (b1–b5) MPM en-face images (XY scans) of the stratum corneum at $z = 0$ µm (b1), keratinocytes in the stratum spinosum at $z = 15$ µm (b2), a nest of basaloid cells (green) surrounded by collagen (blue) and elastin fibers (green) imaged at different depths: $z = 30$ µm, 45 µm, 60 µm (b3–b5). (c) Cross-sectional view (XZ scan) corresponding to a vertical plane through the same interrogating volume shown on the left. (d) H & E stained histologic section of the lesion. Both the MPM and the histologic images show a mucinous stroma adjacent to the tumor mass (arrows). Reproduced with permission from [10].

15.3.5 Other applications of MPM in dermatology

Aside from melanoma and nonmelanoma skin cancers, MPM clinical tomographs have also been employed to image other skin conditions such as actinic keratosis [29, 30], atopic dermatitis [31] and pemphigus vulgaris [29]. The morphological features identified in actinic keratosis lesions were associated with wide intercellular spaces between keratinocytes, an increase in the ratio of nuclear to cellular size of keratinocytes and an increase in epidermal thickness compared to adjacent normal skin [30]. In atopic dermatitis lesions, changes in the cellular metabolism of the lesions have been assessed in vivo through protein-bound and free NADH fluorescence lifetime measurements [31].

Monitoring the effects of skin treatments, a key area of focus in the cosmetic industry, is another field of interest for MPM imaging. In a study published in 2010, Bazin et al. used a DermaInspect MPM tomograph to evaluate the effects of a cosmetic product (topical cream) on the collagen and elastin fibers of the forearms of 24 subjects [12]. The ratios between the signals from collagen and elastin were measured at different depths in the dermis for each subject. After 12 weeks of treatment, an increase in the collagen/elastin ratio was measured for the cosmetic product containing active components (soy and jasmine) known to increase collagen synthesis. The treatment with a placebo topical cream (no active ingredients) resulted in no significant variation in the collagen/elastin ratio over the same treatment period of time.

In order to evaluate cosmetic treatments targeting skin anti-aging, approaches to quantify skin aging non-invasively have been investigated by using MPM imaging. Lin et al. proposed a dermis index that estimates the skin age by the ratio of the SHG signal from collagen and the TPEF signal from elastin [32]. The correlation of this index with age was demonstrated by in vivo MPM measurements of sites on subjects' forearms and faces using a DermaInspect tomograph [13, 15]. Kaatz performed depth-resolved measurements of the proposed dermis index and its correlation with age [14]. MPM imaging has also been used by Koehler et al. to evaluate in vivo and non-invasively dermal elastosis, a characteristic of skin photoaging [16].

15.4 Discussion

The development of clinical MPM technologies has led to new, label-free approaches for non-invasive, in vivo imaging of human skin. Recent studies have shown that MPM can be used to assess a wide range of biological processes, including cancer, cellular metabolism, and the effects of skin treatments. While these are promising results indicating that real-time non-invasive "optical biopsies" can be performed at the bedside, translation of this technology into clinical practice is limited by several technical and practical challenges.

In this early stage of MPM technology clinical translation, the limited field of view (about $250 \times 250\,\mu m^2$) and penetration depth (about 200–300 µm) are the main technical hurdles. The field of view can be increased by implementing a mosaic "tiling" feature (acquisition of adjacent fields of view) or by re-design of optical components such as the tube lens, the scan lens, and the objective. This technical limitation is being addressed by combining tiling with improved lens design in an MPM platform we have recently developed [33]. Penetration depth can be improved by employing dispersion compensation to decrease the laser pulse duration, but the gain would be limited. Generally, this technology is intended as an aid for dermatologists to improve their clinical diagnosis of early stage skin diseases when the uncertainty of their decision is likely to be higher than in the case of advanced disease.

Practical challenges are also related to the need to establish a correlation between MPM and histological images. This is due to the fact that MPM provides high-resolution images of horizontal (x–y) optical sections within a tissue. In histopathology, histological cross-sections (x–z) of the biopsied tissue are used for diagnosis. In addition, while MPM provides images of cells and extracellular matrix in 3D intact tissue, histological sections contain artifacts due to tissue sectioning, processing, and staining. A relevant example is the different appearance of melanocytes in vivo vs sections. In non-invasive MPM images they have easily detectable dendritic processes that can extend from the cell body into the upper epidermis while in hematoxylin and eosin (H & E) stained histological sections they appear in well-defined lacunae as highly pigmented cells due to a fixation retraction artifact [34]. Finally, the different colors of MPM and histology images might also be a barrier to clinical acceptance of MPM. In MPM, images are usually color-coded using red, green and blue to represent TPEF (red and green) and SHG (blue) signals. Histology is based on stains and images that have real color. For instance, H & E histology is based on two stains (hematoxylin for nuclei, eosin for cellular cytoplasm and extracellular matrix of dermis) and the images appear purple and pink. This potential barrier can be addressed by "digitally staining" the MPM images to mimic the appearance of histology [35]. Generally, overcoming these practical challenges related to translation of MPM in clinical practice requires training by both microscopy specialists and dermatopathologists in order to generate a common language that eases communication.

References

[1] König K, Riemann I. High-resolution multiphoton tomography of human skin with subcellular spatial resolution and picosecond time resolution. Journal of Biomedical Optics. 2003;8(3):432–439.

[2] Denk W, Strickler JH, Webb WW. Two-photon laser scanning fluorescence microscopy. Science. 1990;248(4951):73–76.

[3] Campagnola PJ, et al. High resolution non-linear optical microscopy of living cells by second harmonic generation. Biophys J. 1999;77:3341–3349.

[4] Masters BR, So PT, Gratton E. Multiphoton excitation fluorescence microscopy and spectroscopy of in vivo human skin. Biophys J. 1997;72(6):2405–2412.
[5] Masters BR, So PT, Gratton E. Multiphoton excitation microscopy of in vivo human skin. Functional and morphological optical biopsy based on three-dimensional imaging, lifetime measurements and fluorescence spectroscopy. Ann N Y Acad Sci. 1998;838:58–67.
[6] Masters BR, So PT, Gratton E. Optical biopsy of in vivo human skin: Multi-photon excitation microscopy. Lasers in Medical Science. 1998;13(3):196–203.
[7] Balu M, et al. In vivo multiphoton NADH fluorescence reveals depth-dependent keratinocyte metabolism in human skin. Biophysical J. 2013;104(1):258–267.
[8] Dimitrow E, et al. Sensitivity and specificity of multiphoton laser tomography for in vivo and ex vivo diagnosis of malignant melanoma. Journal of Investigative Dermatology. 2009;129(7):1752–1758.
[9] Balu M, et al. Distinguishing between benign and malignant melanocytic nevi by in vivo multiphoton microscopy. Cancer Res. 2014;74(10):2688–2697.
[10] Balu M, et al. In vivo multiphoton microscopy of basal cell carcinoma. JAMA Dermatol. 2015;151(10):1068–1074.
[11] Dancik Y, et al. Use of multiphoton tomography and fluorescence lifetime imaging to investigate skin pigmentation in vivo. Journal of Biomedical Optics. 2013;18(2).
[12] Bazin R, et al. Clinical study on the effects of a cosmetic product on dermal extracellular matrix components using a high-resolution multiphoton tomograph. Skin Res Technol. 2010;16(3):305–310.
[13] Koehler MJ, et al. In vivo assessment of human skin aging by multiphoton laser scanning tomography. Opt Lett. 2006;31(19):2879–2881.
[14] Kaatz M, et al. Depth-resolved measurement of the dermal matrix composition by multiphoton laser tomography. Skin Res Technol. 2010;16(2):131–136.
[15] Sugata K, et al. Evaluation of photoaging in facial skin by multiphoton laser scanning microscopy. Skin Res Technol. 2011;17(1):1–3.
[16] Koehler MJ, et al. Non-invasive evaluation of dermal elastosis by in vivo multiphoton tomography with autofluorescence lifetime measurements. Exp Dermatol. 2012;21(1):48–51.
[17] Balu M, Tromberg BJ. Multiphoton microscopy for non-invasive optical biopsy of human skin. Romanian Journal of Clinical and Experimental Dermatology. 2015;2(3):160–166.
[18] Chance B, et al. Basic principles of tissue oxygen determination from mitochondrial signals. Advances in Experimental Medicine and Biology. 1973;37A:277–292.
[19] Lehninger AL, Nelson DL, Cox MM. Lehninger principles of biochemistry. 5th edition. New York: W.H. Freeman; 2008.
[20] Stucker M, et al. The cutaneous uptake of atmospheric oxygen contributes significantly to the oxygen supply of human dermis and epidermis. Journal of Physiology. 2002;538(3):985–994.
[21] Ronquist G, et al. Human epidermal energy metabolism is functionally anaerobic. Experimental Dermatology. 2003;12(5):572–579.
[22] Cuccia DJ. Spatial frequency domain imaging (SFDI): a technology overview and validation of an LED-based clinic friendly device. Proc SPIE 8254; 2012. p. 825405–825406.
[23] Mazhar A, et al. Wavelength optimization for rapid chromophore mapping using spatial frequency domain imaging. Journal of Biomedical Optics. 2010;15(6):061716.
[24] Nishidate I, et al. Noninvasive imaging of human skin hemodynamics using a digital red-green-blue camera. Journal of Biomedical Optics. 2011;16(8):086012.
[25] Boas DA, Franceschini MA. Haemoglobin oxygen saturation as a biomarker: The problem and a solution. Phil Trans Royal Soc. A Math Phys Eng Sci. 2011;369(1955):4407–4424.
[26] Lever WF, Schaumburg-Lever G. Histopathology of the skin. Sixth edition: Melanocytic nevi and malignant melanoma. Lippincott Williams and Wilkins; 1990.

[27] McKee PH, Calonje E, Granter SR. Pathology of the skin, vol. 2: Melanocytic nevi. 3rd edition. Philadelphia: Mosby; 2005. p. 1250–1258.

[28] Ulrich M, et al. In vivo detection of basal cell carcinoma: comparison of a reflectance confocal microscope and a multiphoton tomograph. J Biomed Opt. 2013;18(6):61229.

[29] Koehler MJ, et al. Clinical application of multiphoton tomography in combination with confocal laser scanning microscopy for in vivo evaluation of skin diseases. Exp Dermatol. 2011;20(7):589–594.

[30] Koehler MJ, et al. Keratinocyte morphology of human skin evaluated by in vivo multiphoton laser tomography. Skin Res Technol. 2011;17(4):479–486.

[31] Huck V, et al. Intravital multiphoton tomography as an appropriate tool for non-invasive in vivo analysis of human skin affected with Atopic Dermatitis. Photonic Therapeutics and Diagnostics Vii. 2011;7883.

[32] Lin SJ, et al. Evaluating cutaneous photoaging by use of multiphoton fluorescence and second-harmonic generation microscopy. Optics Letters. 2005;30(17):2275–2277.

[33] Balu M, et al. Large field of view multiphoton microscopy of human skin. Proceedings SPIE; 2016. Multiphoton microscopy in the biomedical sciences XVI.

[34] McKee PH, Calonje E, Granter SR. Pathology of the skin, Vol. 2: Melanocytic nevi. 3rd edition. Philadelphia: Mosby; 2005. p. 1288.

[35] Bini J, et al. Confocal mosaicing microscopy of human skin ex vivo: spectral analysis for digital staining to simulate histology-like appearance. Journal of Biomedical Optics. 2011;16(7):076008.

Ana Batista, Hans Georg Breunig, Christoph Donitzky, and Karsten König

16 Two-photon microscopy and fluorescence lifetime imaging of the cornea

Abstract: Dysfunctions and dystrophies severely affect the cornea's function. In point of fact, cornea diseases are the second major cause of blindness worldwide. Corneal diagnosis in clinical practice heavily relies on imaging techniques such as slit lamp microscopy, confocal microscopy, or optical coherence tomography. However, these fail to provide information on the cell's metabolic state or the structural organization of the corneal stroma. With two-photon microscopy and fluorescence lifetime imaging this information can be obtained. Therefore, corneal pathology diagnosis may be improved. The feasibility of corneal characterization by two-photon imaging has been demonstrated in ex vivo samples and in vivo animal models.

In this chapter, we report on the use of two multiphoton microscopy instruments for imaging the human cornea: a 5D multiphoton laser scanning microscope and the multiphoton tomograph *MPTflex*. Human corneas unsuitable for transplantation but otherwise normal and pathological samples obtained after surgery were imaged and characterized based on their autofluorescence and second-harmonic generation signals. Two possible clinical applications of two-photon microscopy are discussed: (i) the assessment of tissue viability before corneal transplantation and (ii) the differential diagnosis of corneal pathologies, further demonstrating the advantages of this imaging modality for corneal diagnosis.

16.1 Cornea anatomy, histology, and physiology

The cornea is the transparent tissue in the anterior part of the eye. Its primary function is the refraction and transmission of light that enters the eye (it accounts for two-thirds of the eye's refractive power) [1–3]. However, it also plays an important role as a biodefense system [1]. Structurally, it is approximately 0.5 mm thick in the center and its thickness increases gradually towards the periphery where it can reach 0.7 mm [1, 2]. Histologically, it is composed of five layers (Fig. 16.1): the epithelium, Bowman's layer, stroma, Descemet's membrane, and endothelium [1, 2]. In this section the characteristics of each corneal layer are discussed.

https://doi.org/10.1515/9783110429985-018

16.1.1 Epithelium

The epithelium is the outermost layer of the cornea and acts as a protective barrier, preventing the entrance of foreign particles [1]. It is approximately 50 μm thick and is composed of 5 to 7 layers of stratified, squamous cells with quick regeneration: superficial cells, wing cells, and basal cells [1, 2, 4]. Basal cells form the innermost layer of the epithelium. They are the only cells with mitotic activity and differentiate progressively into wing and superficial cells [1, 2]. Basal cell renewal is accomplished by the limbus, a band in the cornea periphery composed of stem cells [1, 2]. Wing cells, located in the middle of the epithelium, are wing-like shaped cells under differentiation [1, 2]. Superficial cells are fully differentiated and do not proliferate [1, 2]. They are continuously sloughed off from the tissue and replaced by new cells [2]. The epithelium is attached posteriorly to the basement membrane, a membrane secreted by the basal cells [1, 2].

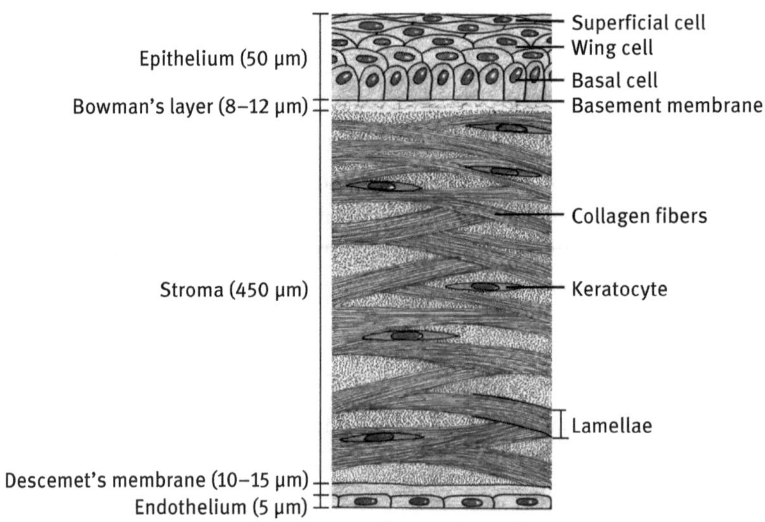

Fig. 16.1: Cross-sectional representation of the human cornea.

16.1.2 Bowman's layer

Bowman's layer is an acellular membrane, 8 to 12 μm thick, that lies between the epithelium and the stroma. It is composed of interwoven, randomly organized collagen fibers, with diameters between 20 to 30 nm [1, 2]. The posterior boundary of this layer is not well defined, as the organization of collagen fibers gradually changes [2].

16.1.3 Stroma

The stroma accounts for approximately 90 % of the total cornea thickness [1, 2] and is the layer that contributes most to the cornea's shape, strength and transparency [1]. This layer is composed of highly organized collagen fibers, keratocytes, and extracellular matrix components [1, 2].

The collagen fibers, with uniform diameters (between 25 to 35 nm) and lengths are arranged in parallel to each other forming bundles called lamellae [1, 2]. The stroma is composed of 200 to 300 lamellae that lie parallel to each other and run through the entire surface of the cornea [1, 2]. Keratocytes are cellular components that lie between the collagen lamellae [1, 2]. They account for only 3 to 5 % of the total volume of the stroma [4]. These metabolically active cells have two major roles:

(i) stromal preservation through the synthesis of collagen and extracellular matrix components [2];
(ii) stromal wound healing [1, 4].

The spaces between the collagen fibers, lamellae, and cells are filled with extracellular matrix components [1, 2]. The most abundant matrix components are proteoglycans, composed of a carbohydrate glycosaminoglycan chain and a core protein. These molecules play an important role in stroma hydration, and collagen fibers and lamellae regular spacing, through the attraction and binding of water molecules [1, 2].

16.1.4 Descemet's membrane

Descemet's membrane, also considered the basement membrane of the endothelium, is composed of collagen molecules and glycoproteins such as laminin and fibronectin [1, 2]. The thickness of this layer increases throughout life due to the constant deposition of collagen synthetized by the endothelial cells. At birth, it is approximately 3 to 5 µm thick and it can reach 10 to 15 µm in the elderly [1, 2].

16.1.5 Endothelium

The endothelium is the innermost layer of the cornea. It is formed by a single layer of polyhedral cells with regular shape (approximately 70 % to 80 % are hexagonal), a thickness of about 5 µm, and a width of about 20 µm [1, 2]. The endothelial cells are rich in cellular organelles, particularly mitochondria, which reflects their high metabolic activity [2]. They play a pivot role in corneal hydration. Throughout their membranes water channels (aquaporins) [2] and sodium-potassium pumps can be found. These enable the removal of excess water molecules, thus regulating the water content [1, 2].

16.1.6 Corneal nourishment and cells metabolism

The cornea is an avascular tissue. The absence of blood vessels is essential to maintain its transparency [2]. However, being composed of metabolically active cells, the supply of oxygen and nutrients (usually delivered via blood supply) is essential [1, 2]. The nourishment of the cornea is performed via diffusion from the aqueous humor and from the capillary networks in the limbic region [2]. Corneal oxygenation occurs by diffusion from the tear film [1].

In normal conditions, corneal cells produce energy in the form of adenosine triphosphate (ATP) via aerobic respiration. However, in specific situations, such as when the eye lids are closed during sleep, the oxygen supply is interrupted and ATP is produced via anaerobic mechanisms [1]. Pathological conditions may also affect the metabolic pathways of the corneal cells.

16.2 Current clinical corneal imaging methodologies

The human cornea is commonly imaged in clinical practice using slit lamp microscopy, confocal microscopy, and optical coherence tomography (OCT).

Slit lamp microscopy is considered the most important technique available to ophthalmologists for the examination of the eye's anterior segment. It was first developed in 1911 by the Noble laureate Allvar Gullstrand [5]. It consists of an adjustable binocular microscope and an illumination source in which width, height, and incidence angle can be adapted [5]. This system provides an overall picture of the healthy and diseased cornea [5]. However, it does not provide information at a cellular level.

Marvin Minsky is considered the pioneer of confocal microscopy [6]. The basic idea behind it is the use of pinholes in both illumination and detection paths. In this way, a single point of tissue is illuminated and simultaneously imaged, while out-of-focus light is blocked [7, 8]. This gives confocal microscopy enhanced lateral and axial resolutions, and enables optical sectioning of thick samples [8]. This imaging modality was first used in the ophthalmological field in vivo in 1986 [9]. When applied to the cornea, confocal microscopy allows the visualization of cell anatomy and density as well as nerve density. Therefore, it has important clinical applications such as differential diagnosis of corneal dystrophies or tissue monitoring after surgery [10]. However, confocal microscopy fails to provide information on the cell's metabolism. Furthermore, the major layer of the cornea is disregarded, since it cannot be used to assess stroma collagen fiber organization.

Optical coherence tomography (OCT), a noncontact imaging modality demonstrated in 1991 by David Huang and co-workers at James G. Fujimoto's laboratory, is based on interferometry to measure the delay of reflected light [11, 12]. OCT was first used to image the human retina in vivo in 1993 [13] and was later introduced for imaging the eye's anterior segment by Joseph Izatt [14]. However, the system specifications were optimized for retinal imaging. The first OCT specifically designed for imaging of the anterior segment (AS-OCT) was proposed only in 2001 [15]. This system was characterized by faster scanning times, scanning configuration adapted to the cornea shape, and laser sources with a longer wavelength (1310 nm). These optimizations strongly reduced image distortion. Furthermore, by using 1310 nm laser sources, the scattering in opaque tissues was reduced, enhancing light depth penetration for sclera and limbus imaging [16, 17]. AS-OCT offers clinicians cross-sectional views of the cornea, thus it provides information on corneal thickness and allows corneal layer discrimination [18]. This is helpful for patient follow-up after corneal surgery [17]. AS-OCT is also useful for the diagnosis of closed angle glaucoma [16, 17]. A major drawback of this imaging modality is its low resolution [19]. Furthermore, AS-OCT fails to provide information on the corneal cells' metabolic state.

16.3 Two-photon corneal imaging

Novel clinical imaging methods based on two-photon microscopy may be used to improve the diagnosis of corneal pathologies. With two-photon microscopy, 3D images of the cornea may be acquired with subcellular resolution. The use of near-infrared laser sources allows high imaging penetration depths. Furthermore, photobleaching and photodamage are reduced due to the confinement of the excitation to the focal volume [20]. Besides morphological information, already accessible by the current methodologies, two-photon microscopy can also provide two new levels of information: cell metabolism and stroma structural organization.

The fluorescence of endogenous fluorophores such as NAD(P)H (nicotinamide adenine dinucleotide and nicotinamide adenine dinucleotide phosphate) and flavins (flavin mononucleotide/flavin adenine dinucleotide) can be used to obtain information on the cell's metabolic state – optical metabolic imaging (OMI) [21, 22]. NAD(P)H and flavins are metabolic cofactors involved in the fundamental aerobic and anaerobic metabolic pathways [23]. Therefore, their autofluorescence intensity and lifetime are correlated with the cell's metabolism and are commonly used for its assessment. The structural organization of corneal stroma can be assessed by second-harmonic generation imaging (SHG) of the collagen fibers. The characteristics of SHG and autofluorescence allow the spectral separation of the emitted signals. Therefore, they are often combined for imaging of biological tissues.

To this date, two-photon excitation-based clinical instruments in ophthalmology are nonexistent. However, a clinical instrument based on one-photon excited autofluorescence lifetime for human retina imaging [24], capable of differential diagnosis of retinal pathologies [25, 26], was developed by Schweitzer and co-workers. The first two-photon excited autofluorescence lifetime certified medical imaging device was developed by König and co-workers [27–30].

In this section, the autofluorescence and SHG properties of corneal endogenous fluorophores as well as their applications for corneal imaging and/or diagnosis are discussed.

16.3.1 Autofluorescence intensity imaging

The use of endogenous fluorophores to evaluate the metabolism of the cells is only possible due to the unique fluorescence characteristics of each fluorophore. The fluorescence can be characterized by its spectra, quantum yield, intensity, and lifetime. The individual excitation and emission spectra of NAD(P)H and flavins allow their selective excitation and/or spectral separation of the emitted signals. NAD(P)H can be selectively excited by one-photon absorption using ultraviolet (UV) light sources (maximum at 340 nm) and has maximum emission at approximately 460 nm, while FAD can be selectively excited using visible light sources (maximum at 450 nm) and has emission maximum at 525 nm [21, 31]. The use of pulsed near-infrared (NIR) laser sources allows simultaneous excitation of both metabolic cofactors by two-photon absorption.

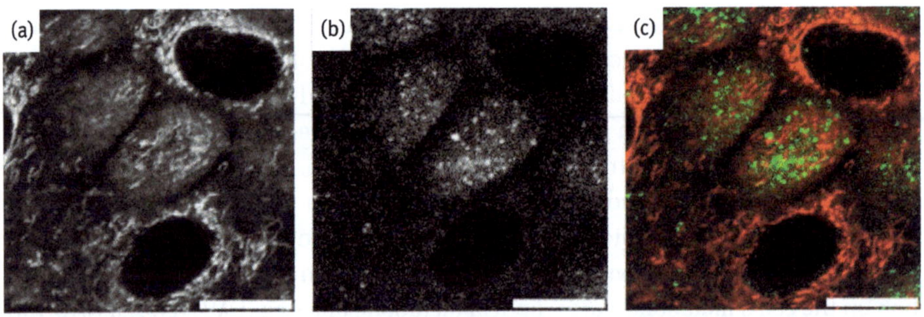

Fig. 16.2: Two-photon autofluorescence intensity images of the human corneal epithelial cells. (a) NAD(P)H autofluorescence arising in the cell's cytoplasm and mitochondria. An excitation wavelength of 760 nm was used for image acquisition. (b) Flavins autofluorescence arising from the cell's mitochondria. Selective excitation of flavins was performed using a wavelength of 850 nm. (c) Overlap image of NAD(P)H (red) and flavins (green) autofluorescence intensity. The lack of NAD(P)H and flavins in the cell's membranes and nuclei, hence the lack of autofluorescence, allows single cell visualization. Images were acquired with the multiphoton tomograph *MPTflex*. Scale bar: 10 µm.

The relation between the metabolic cofactors' quantum yield, hence the autofluorescence intensities of their oxidized and reduced forms, provides information on the metabolic state. NAD(P)H exhibits autofluorescence in its reduced forms, whereas for flavins only the oxidized forms are autofluorescent [21, 22, 31]. Oxidation reduction reactions, in which the metabolic cofactors are reduced (NAD(P)H/$FADH_2$) and oxidized ($NAD(P)^+$/FAD), occur during energy production. Therefore, their autofluorescence intensity provides an indirect evaluation of the oxidation reduction reactions from which the cell's metabolic state can be inferred. The autofluorescence of each metabolic cofactor is a good indicator of cellular metabolism [22]. However, the most commonly used metabolic indicator is the ratio between the flavins and NAD(P)H autofluorescence intensity [21].

Two-photon autofluorescence intensity was first used to image the cornea ex vivo in 1995 by Piston and co-workers [32]. Since then it has already been applied to image and characterize ex vivo mouse [33], porcine [34–36] and human corneas [37], as well as rabbit's cornea in vivo [38]. Examples of two-photon autofluorescence intensity images of the human corneal epithelium acquired by our group are shown in Fig. 16.2. As visible in the figure, NAD(P)H autofluorescence signals arise from the cell's cytoplasm and mitochondria (Fig. 16.2 (a)), while flavins autofluorescence is limited to the mitochondria (Fig. 16.2 (b)). Cell membrane and nuclei appear black due to the lack of NAD(P)H and flavins (Fig. 16.2 (c)). Therefore, autofluorescence intensity is a good parameter for individual cell visualization and characterization, hence corneal cell layer discrimination. The feasibility of two-photon excitation autofluorescence intensity to discriminate between the corneal epithelium at multiple depths, the stroma, and the endothelium has already been demonstrated [32–38].

Possible clinical applications of two-photon excited autofluorescence intensity have also been evaluated. It is known that pathological conditions affect the cell's metabolism and that those changes influence the metabolic cofactors' autofluorescence. An increase in autofluorescence intensity in epithelial and endothelial cells, as well as in stromal keratocytes, is observed following cornea incubation with cyanide (inhibitor of the electron transport chain) [32]. Therefore, autofluorescence intensity may be used for diagnostic purposes. Combined with SHG, two-photon autofluorescence intensity can be used to image ex vivo the pathogens responsible for corneal infectious keratitis as well as disruptions in stromal collagen fiber organization [39]. It may also be used for monitoring the cornea after keratoplasty and wound healing [40, 41]. Furthermore, by using the same laser sources for imaging and intratissue processing, cornea monitoring may be accomplished not only before and after but also during surgery [34, 35]. Corneal neovascularization may be monitored using this imaging modality and has been demonstrated in animal models of the pathology in vivo [42, 43].

16.3.2 Autofluorescence lifetime imaging

Fluorescence lifetime is influenced by the fluorophore microenvironment. Therefore, the same fluorophore has different lifetimes in its free configuration and when bound to proteins [21, 44]. Furthermore, binding to different types of proteins is also reflected in its lifetime [31]. Since the oxidation reduction reactions of NAD(P)H and flavins are catalyzed by enzymes (proteins) [23], the ratio between the free and protein-bound components of the metabolic cofactors is also an indirect indicator of the oxidation reduction ratio.

Autofluorescence intensity could also be used for discriminating between free and protein-bound components of the same fluorophore, since they can be separated spectrally. Protein-bound NAD(P)H has an emission maximum at 445 nm while for the free form the maximum occurs at 460 nm [45]. However, the proximity between the spectral peaks and their large spectral overlap makes the discrimination a difficult task [45]. Using autofluorescence lifetime their discrimination is straightforward. A shift of NADH autofluorescence lifetime to longer values occurs upon protein binding, while the opposite occurs for flavins [31, 44]. These alterations were first described in 1992 by Lakowicz et al. [44]. Unlike the intensity, autofluorescence lifetime is also independent of the molecular concentration, which reduces the influence of artifacts such as photobleaching in cell's metabolic assessment [20].

Two-photon autofluorescence lifetime has already been applied to image and characterize the corneal metabolism of freshly enucleated mouse, rabbit and porcine, as well as nonpathological human corneas [46–49]. Using this imaging modality the metabolic status of different cell layers as well as different cell types of the same layer can be determined [49]. A correlation between epithelium depth, hence cell type, and the metabolic state based on the metabolic cofactors' autofluorescence lifetime was demonstrated for porcine cornea [49]. The potential of the autofluorescence lifetime for corneal clinical applications was also addressed. This imaging modality proved to be an effective tool to characterize cornea [50] and eye [51] pathologies, as well as to monitor the alterations occurring during corneal epithelium wound healing both in human epithelial cell cultures and in erosion models of organ culture explants from murine eyes [52]. Combined with SHG, it can be also used for corneal evaluation after medical procedures such as corneal collagen crosslinking (CXL). An increase in both the autofluorescence intensity and lifetime was observed in rabbit corneal stroma following CXL [53].

16.3.3 Second-harmonic generation imaging (SHG)

Along with the assessment of corneal cells' metabolic state, the characterization of the stroma's structural organization is of major importance. Alterations in the collagen fiber organization are one of the first signs of pathology. Therefore, their evaluation by SHG may be used to improve corneal pathology diagnosis.

SHG was first introduced for imaging biological tissues in 1971 [54]. Ever since, this imaging modality has surfaced as an important and valid non-invasive imaging method, due to its high resolution, high contrast, and absence of photobleaching and photodamage [55, 56]. SHG is generated by structures with non-centrosymmetric molecular organization, such as collagen [55]. Therefore, it is ideal for the assessment of stroma structural organization. Being a coherent process, SHG photons propagate mainly in forward direction. However, in the cornea, backward-direction SHG can also be detected [55], thus enabling the in vivo application of this imaging modality.

As aforementioned, SHG is often used in combination with two-photon excited autofluorescence. As such, some of the clinical applications of this imaging modality have already been mentioned. SHG is of particular interest for the evaluation of pathologies in which the stroma organization is disrupted and it can be used to evaluate the alterations due to pathologies such as keratoconus [57] or medical procedures such as CXL [58, 59]. The assessment of corneal alterations due to induced hyperglycemia (diabetes model) in rats is also possible [60].

16.3.4 Two-photon instrumentation

The experimental systems typically used for two-photon imaging consist of (modified) microscopes with a near-infrared (NIR) Ti:sapphire laser for sample excitation, an *x*-*y*-*z* scanning module, a beam expander, an attenuator, a microscope objective, and a photomultiplier tube (PMT) for detection of autofluorescence and SHG generated photons.

In this section, we focus on the use of two multiphoton laser systems for imaging the human cornea:
(i) a 5D laser scanning microscope (Fig. 16.3 (a)), and
(ii) a multiphoton tomograph *MPTflex* (JenLab GmbH, Jena, Germany) (Fig. 16.3 (c)).

The setups of both systems have already been described elsewhere [30, 61]. Briefly, in the 5D laser scanning microscope sample excitation is accomplished using a mode-locked NIR Ti:sapphire laser Integral Pro 400 (Femtolasers Produktions GmbH, Austria) generating ≈10 fs pulses at a frequency of 85 MHz. This laser source has

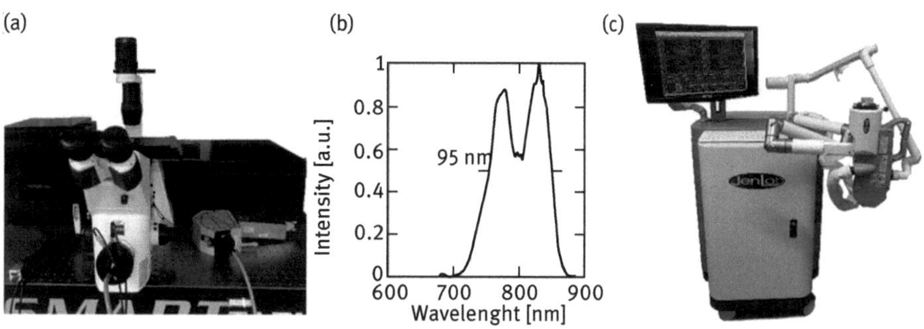

Fig. 16.3: Multiphoton laser scanning systems used for two-photon imaging of the human cornea. (a) Custom built 5D laser scanning microscope with (b) a broad bandwidth laser for sample excitation and 16-channel PMT detector for signal collection, and (c) multiphoton tomograph *MPTflex* (JenLab GmbH, Jena, Germany).

a fixed wavelength of 800 nm and an M-shaped spectral profile with approximately 95 nm bandwidth (Fig. 16.3 (b)). Chirped mirrors are used to guarantee ultrashort laser pulses (12 fs) behind the objective. Autofluorescence and SHG signals are detected using a spectral photomultiplier, consisting of a grating polychromator and a multi-anode PMT for simultaneous detection in 16 channels, coupled to a time-correlated single-photon counting SPC-150 module (Becker & Hickl GmbH, Berlin, Germany) for fluorescence lifetime imaging.

Using the 5D laser scanning microscope, metabolic information on the corneal epithelium and endothelium based on NAD(P)H and flavins autofluorescence can be obtained simultaneously. The broad bandwidth of this laser source enables the simultaneous excitation of both metabolic cofactors, while the 16-channel PMT detector allows the spectral separation of the autofluorescence signals. Furthermore, it allows the simultaneous acquisition and separation of stroma SHG and autofluorescence photons.

The multiphoton tomograph *MPTflex* is a compact certified clinical system with an articulated mirror arm and a scan head (Fig. 16.3 (c)). An 80 MHz NIR Ti:sapphire 100 fs laser tunable between 710 nm and 920 nm is used for sample excitation. Dichroic mirrors are used to separate the generated autofluorescence and SHG signals which are then simultaneously detected with two individual PMTs. A time-correlated single-photon counting SPC-830 module (Becker & Hickl GmbH, Berlin, Germany) is coupled with the autofluorescence detector for lifetime imaging. Using this system, NAD(P)H and flavins can be selectively excited by changing the laser excitation wavelength. For corneal cells, metabolic information based on both metabolic cofactors' autofluorescence must be acquired in two consecutive images.

16.4 Two-photon imaging of the human cornea

Human corneas, provided by the Lion's Cornea Bank Saar-Lor-Lux, Trier/West Pfalz at the Department of Ophthalmology, Saarland University Medical Center, Homburg/Saar, Germany, were imaged by two-photon microscopy. The samples were unsuitable for corneal transplantation but otherwise normal, presenting negative results in serology and microbiology tests. Corneal hydration was maintained during image acquisition using live cell imaging solution (#A14291DJ, Life Technologies, USA). All tenets of the World Medical Association Declaration of Helsinki for research using human corneas were followed.

All layers of the human cornea were discriminated by two-photon excitation microscopy using the autofluorescence and SHG signals of the endogenous fluorophores present in this tissue. In the human cornea epithelium (Fig. 16.4 (a)–(d)), autofluorescence signals from the metabolic cofactors were detected. Using the morphological characteristics of individual cells, they were characterized as superficial (Fig. 16.4 (a) and (b)), wing (Fig. 16.4 (c)), and basal cells (Fig. 16.4 (d)). Due to the

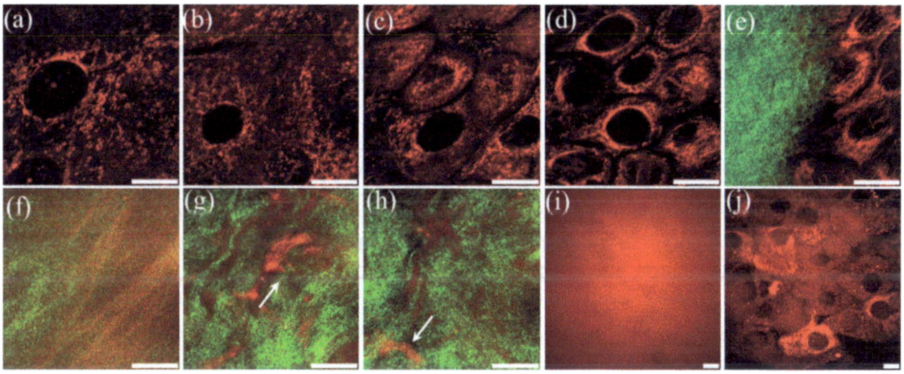

Fig. 16.4: Autofluorescence intensity (red) and SHG (green) images of the human cornea: (a) and (b) epithelial superficial cells; (c) wing cells; (d) basal cells; (e) epithelium and Bowman's layer interface; (f) Bowman's layer; (g) and (h) stroma; (i) Descemet's membrane; and (j) endothelial cells. The signal intensity in each pixel represents the number of detected photons. Images were acquired with the multiphoton tomograph *MPTflex* using an excitation wavelength of 760 nm. Scale bar: 10 μm.

subcellular resolution of this imaging modality a characterization of mitochondria density was also possible. In the basal cells, a high number of mitochondria mostly localized around the cell nuclei were observed (Fig. 16.4 (d)). Toward the corneal epithelial surface, the concentration of mitochondria decreases and their distribution becomes more disperse (Fig. 16.4 (a)–(c)). This is in accordance with the loss of organelles occurring during the basal cell differentiation into wing and superficial cells [2]. The epithelial superficial cells are known to have a reduced number of cellular organelles [1, 2].

At the interface between the corneal epithelium and Bowman's layer, both autofluorescence (red) generated by the basal epithelial cells and SHG (green) generated by the collagen fibers of Bowman's layer were detected (Fig. 16.4 (e)). In the Bowman's layer (Fig. 16.4 (f)) and stroma (Fig. 16.4 (g) and (h)) both autofluorescence and SHG signals from the collagen fibers were detected. The higher degree of collagen fiber organization in the stroma is visible in the SHG images (green). Besides collagen, keratocyte autofluorescence (indicated by the arrows) was also visible in the corneal stroma (Fig. 16.4 (g) and (h)). In the Descemet's membrane, a strong autofluorescence signal was observed (Fig. 16.4 (i)). This corneal layer is mostly composed of collagen [2]. However, its lack of organization prevents SHG generation [62, 63]. The corneal endothelial cells are shown in Fig. 16.4 (j). The regular arrangement of these cells, as well as the high mitochondrial density can be seen by two-photon microscopy.

Along with morphological information, two-photon microscopy enables cells' metabolic assessment based on the autofluorescence lifetime of the metabolic cofactors. The mean autofluorescence lifetimes and free to protein-bound ratios of the nonpathological human cornea epithelium (average of all cell types), stroma, and

Tab. 16.1: Human cornea layers' mean autofluorescence lifetimes and their free to protein-bound ratio. Data was obtained with the 5D multiphoton laser scanning microscope. Values are represented as mean ± standard deviation.

Corneal layer	Epithelium NAD(P)H	Stroma	Endothelium NAD(P)H
τ_m (ns)	1.14 ± 0.14	0.97 ± 0.18	1.02 ± 0.13
Ratio (free : protein-bound)	1.15 ± 0.12	–	1.27 ± 0.22

endothelium are shown in Tab. 16.1. Data acquisition was performed using the 5D multiphoton laser scanning microscope.

An important feature of two-photon microscopy is the possible assessment of the stroma architecture. The regular organization of the stroma collagen fibers is essential to maintain corneal transparency, and information on the collagen fiber organization can be retrieved by SHG. In Fig. 16.5 (a), en face SHG images of the human corneal stroma at 0 μm, 40 μm, 80 μm, and 120 μm are shown. The morphological characteristics of the collagen fibers, in particular their small diameter, prevent individual fibers from being resolved. Thus, backward detected SHG images appear as homogenous and speckle-like intensity images (Fig. 16.5 (a)) and the determination of collagen fiber orientation is a difficult task. However, since in vivo SHG images must be acquired in backward direction, efforts to extract information on the collagen fiber organization from backward detected SHG images have been carried out [64, 65]. The visualization of corneal collagen architecture with depth was possible by creating 3D representations of the SHG signals (Fig. 16.5 (b)).

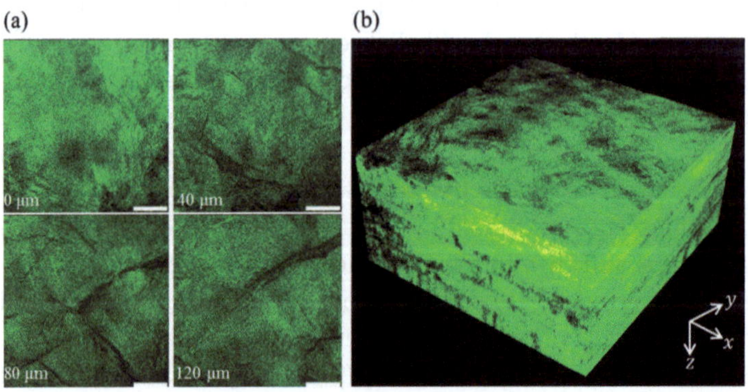

Fig. 16.5: SHG images of the cornea stroma. (a) En face SHG images of the human corneal stroma at 0 μm, 40 μm, 80 μm, and 120 μm. (b) 3D reconstruction of the anterior 120 μm portion of the stroma. Images were acquired with the 5D multiphoton laser scanning microscope. Scale bar: 40 μm.

16.4.1 Viability for corneal transplantation

One of the possible clinical applications of two-photon microscopy is the ex vivo evaluation of human corneas prior to corneal transplantation. Corneal endothelium properties are usually the focus of this evaluation. These cells do not regenerate and their loss is compensated by cell migration and expansion. However, a high density of endothelial cells is required for proper tissue function [2, 3]. Therefore, we focus on the suitability of two-photon microscopy to analyze the corneal endothelium status.

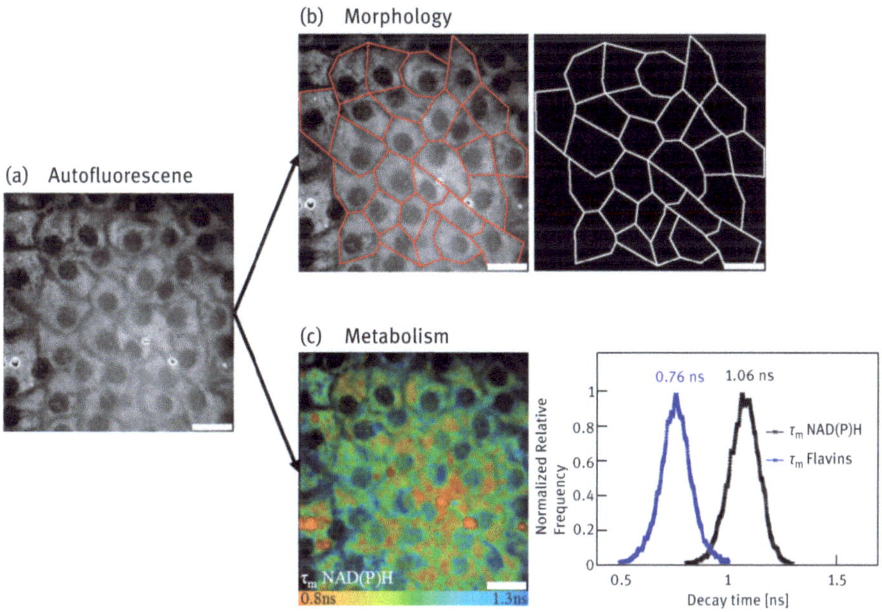

Fig. 16.6: Two-photon microscopy for the analysis of the endothelium status prior to corneal transplantation. Endothelium (a) autofluorescence images can be used (b) to assess the cell's morphology, through the segmentation of individual endothelial cells, and (c) to assess the cell's metabolism using the autofluorescence lifetime of both metabolic cofactors, NAD(P)H and flavins. Images were acquired with the commercial tomograph *MPTflex*. Scale bar: 20 µm.

Parameters such as endothelial cell density, average shape, and size are used to assess the cornea status in clinical practice. Using two-photon microscopy, this information can also be retrieved. The lack of autofluorescence signals from the cell's membrane enables single-cell visualization because the cell boundaries appear dark (Fig. 16.6 (a)). Therefore, the human corneal endothelium can be segmented by highlighting the cell membranes. The overlay of the corneal endothelium autofluorescence with a manual segmentation (shown in red) as well as the binary image of the endothelial cell segmentation are shown in Fig. 16.6 (b). Endothelial cell widths of approximately (27.0 ± 3.8) µm were measured. Based on the number of cells in the field of view, an endothelial cell density of (2444 ± 88) cells/mm^2 was estimated for this human corneal sample.

Along with the cells' morphological characteristics, two-photon microscopy also provides information on the metabolic state of the endothelial cells. Fig. 16.6 (c) shows the autofluorescence lifetime image of the endothelium color-coded for NAD(P)H mean autofluorescence lifetime as well as NAD(P)H and flavins mean autofluorescence lifetime distributions. Since metabolic alterations are known to occur prior to morphological alterations, the analysis of the endothelial cells' metabolism provides an additional level of information for the assessment of the corneal endothelium. Thus, the combined analysis of endothelial cell morphology and metabolism may provide a better assessment of corneal status. Therefore, two-photon microscopy may provide a finer evaluation of the corneas viable for transplant.

16.4.2 Corneal pathologies evaluation

The differential diagnosis of corneal pathologies is one of the possible future clinical applications of two-photon microscopy. Here, we demonstrate its feasibility to characterize human cornea samples with keratoconus, a corneal ectatic disorder characterized by the progressive stepping and thinning of the cornea [66], and *Acanthamoeba* keratitis, a pathology caused by a parasitic infection that induces thinning and necrosis of the corneal epithelium [66]. The samples were obtained within 24 h following corneal keratoplasty from the Department of Ophthalmology, University of Saarland, Medical Center, Homburg/Saar, Germany. Up to 3 different areas were imaged for each sample.

The morphological alterations induced by both corneal pathologies to the epithelial cells were observed using the autofluorescence intensity of the metabolic cofactors (Fig. 16.7). In the nonpathological human cornea, the basal cells have uniform sizes (Fig. 16.7 (a)), while in corneas affected by keratoconus (Fig. 16.7 (b)) and *Acanthamoeba* keratitis (Fig. 16.7 (c)) cell dimensions are irregular. Differences in the mitochondria density and localization were visible for both pathologies (Fig. 16.7 (b) and (c)). Furthermore, an increase in the cell membrane autofluorescence appears to occur in *Acanthamoeba* keratitis (Fig. 16.7 (c)).

The metabolic alterations arising from keratoconus and *Acanthamoeba* keratitis were analyzed using the autofluorescence lifetime of the metabolic cofactor NAD(P)H (Fig. 16.8). A significant decrease in NAD(P)H mean autofluorescence lifetime was observed in both pathological samples. The ratios of free to protein-bound NAD(P)H were increased in both pathologies. Keratoconus and *Acanthamoeba* keratitis corneas had free to protein-bound ratios of 1.63 ± 0.46 and 1.62 ± 0.20, respectively, while the nonpathological corneas had a free to protein-bound ratio of 1.15 ± 0.15 (Tab. 16.1). The

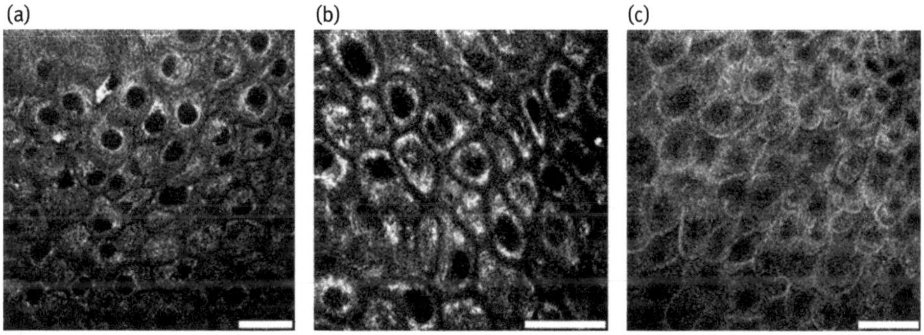

Fig. 16.7: Human corneal epithelium autofluorescence intensity images of (a) nonpathological samples and samples with (b) keratoconus, and (c) *Acanthamoeba* keratitis. Using two-photon microscopy the morphological alterations due to pathology can be assessed. Image acquisition was performed with the 5D multiphoton laser scanning microscope. Scale bar: 30 μm.

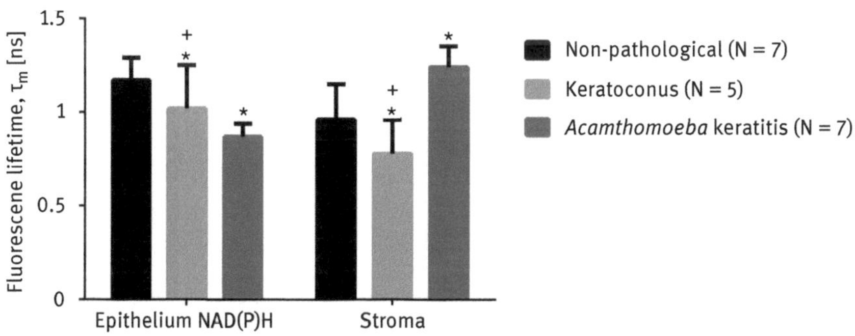

Fig. 16.8: Human corneal epithelium NAD(P)H and flavins mean autofluorescence lifetimes and stroma autofluorescence lifetime of nonpathological and pathological samples. Data is presented as mean ± SD. * $p < 0.05$, comparing with the nonpathological samples; + $p < 0.05$, comparing with *Acanthamoeba* keratitis cornea.

observed alterations indicate that the studied pathologies may induce alterations to the metabolism of epithelial cells.

The initial results indicate that besides allowing the discrimination between non-pathological and pathological samples, NAD(P)H autofluorescence lifetime may also enable the differentiation between different pathologies (Fig. 16.8). However, the number of corneal samples must be increased.

The corneal stroma is the largest layer of the cornea. Therefore, alterations due to pathologies in this layer were also assessed. As aforementioned, both autofluorescence and SHG photons can be detected from the stroma. Here, both were analyzed. A significant increase was observed in the stroma mean autofluorescence lifetime of corneas diagnosed with *Acanthamoeba* keratitis, while for keratoconus samples it was significantly decreased (Fig. 16.8). The stroma structural organization was assessed by SHG. Fig. 16.9 shows en face backward detected SHG images at stromal depths between 0 µm and 80 µm with a 20 µm interval, for nonpathological (Fig. 16.9 (a)), keratoconus (Fig. 16.9 (b)), and *Acanthamoeba* keratitis (Fig. 16.9 (c)) corneas. As visible in the figure, the collagen fiber organization is altered at all stromal depths in both pathologies. This demonstrates that both the autofluorescence lifetime and SHG of the corneal stroma provide additional information that may help improve corneal pathology detection and discrimination.

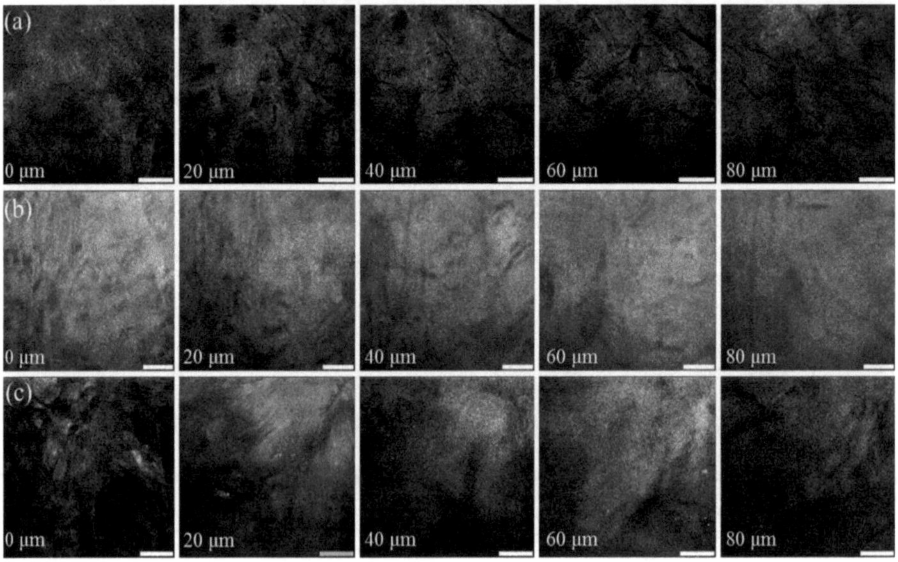

Fig. 16.9: Human corneal stroma SHG images of (a) nonpathological, (b) keratoconus, and (c) *Acanthamoeba* keratitis corneas. The stroma structural alterations due to pathology can be assessed using SHG. Images were acquired with the 5D multiphoton laser scanning microscope. Scale bar: 30 µm.

16.5 Conclusions and outlook

Pathologies of the human cornea are the second major cause of blindness worldwide [67]: their early detection is highly important. Two of the first signs of pathology are alterations of cell metabolism and stroma structural organization. However, the current corneal diagnostic methods fail to provide this information. This indicates that a new diagnostic tool is required.

Two-photon microscopy and fluorescence lifetime imaging of the cornea may be used to improve diagnosis. Resembling the current corneal diagnostic methods, these imaging modalities can be used to evaluate tissue morphology. However, it also allows the evaluation of individual corneal layer metabolism, based on the metabolic cofactors NAD(P)H and flavins autofluorescence lifetimes, and the assessment of stroma collagen fiber organization using the SHG generated by these structures.

The characterization of human cornea cell morphology, metabolism, and stroma architecture is possible using two-photon microscopy and it may be used for sample evaluation before transplantation or pathology diagnosis.

References

[1] Nishida T, Saika S. Cornea and sclera: anatomy and physiology. In: Krachmer JH, Mannis MJ, Holland EJ, editors. Cornea – fundam diagnosis manag (volume 1). 3rd edition. Mosby/Elsevier; 2011.

[2] Remington LA. Clinical anatomy of the visual system. 2nd edition. St. Louis, Missouri: Elsevier; 2005.

[3] Lang GK, Amann J. Ophthalmology: a short textbook. Thieme; 2000.

[4] Bye L, Modi N, Stanford M. Basic sciences for ophthalmology. Oxford, United Kingdom: Oxford University Press; 2013.

[5] Agarwal S, Agarwal A, Buratto L, et al. Textbook of Ophthalmology. Lippincott Williams & Wilkins; 2002.

[6] Minsky M. Memoir on inventing the confocal scanning microscope. Scanning. 1988;10(4): 128–138.

[7] Tavakoli M, Hossain P, Malik RA. Clinical applications of corneal confocal microscopy. Clin Ophthalmol. 2008;2(2):435–445.

[8] Masters BR, Bohnke M. Confocal microscopy of the human cornea in vivo. Int Ophthalmol. 2001;23(4-6):199–206.

[9] Petran M, Hadravsky M, Benes J, Boyde A. In vivo microscopy using the tandem scanning microscope. Ann N Y Acad Sci. 1986;483:440–7.

[10] Messmer EM. Confocal microscopy: when is it helpful to diagnose corneal and conjunctival disease? Expert Rev Ophthalmol. 2008;3(2):177–192.

[11] Huang D, Swanson EA, Lin CP, et al. Optical coherence tomography. Science. 1991;254:1178–1181.

[12] Fujimoto JG, Drexler W. Introduction to optical coherence tomography. In: Drexler W, Fujimoto JG, editors. Opt Coherence Tomogr. Technol Appl. 1st edition. Berlin: Springer Science & Business Media; 2008. p. 1346.

[13] Swanson EA, Izatt JA, Hee MR, et al. In vivo retinal imaging by optical coherence tomography. Opt Lett. 1993; 18(21):1864.

[14] Izatt JA. Micrometer-scale resolution imaging of the anterior eye in vivo with optical coherence tomography. Arch Ophthalmol. 1994; 112(12):1584.

[15] Radhakrishnan S. Real-time optical coherence tomography of the anterior segment at 1310 nm. Arch Ophthalmol. 2001; 119(8):1179.

[16] Huang D, Izatt JA. Physics and fundamentals of anterior segment optical coherence tomography. In: Steinert RF, Huang D, editors. Anterior Segm Opt Coherence Tomogr. SLACK Incorporated; 2008. p. 1–10.

[17] Denoyer A, Labbé A, Baudouin C. Anterior segment OCT imaging. In: Bernardes R, Cunha-Vaz J, editors. Opt Coherence Tomogr A Clin Tech Updat. Springer; 2012. p. 255.

[18] Steinert RF, Huang D. Anterior segment optical coherence tomography. Thorofare (NJ): SLACK Incorporated; 2008.

[19] Kim J, Brown W, Maher JR, et al. Functional optical coherence tomography: principles and progress. Phys Med Biol. 2015;60(10):R211–R237.
[20] Periasamy A, Clegg RM. FLIM microscopy in biology and medicine. Boca Raton, FL: CRC Press; 2009.
[21] Georgakoudi I, Quinn KP. Optical imaging using endogenous contrast to assess metabolic state. Annu Rev Biomed Eng. 2012;14:351–367.
[22] Mason WT. Fluorescent and luminescent probes for biological activity: a practical guide to technology for quantitative real-time analysis. 2nd edition. London: Elsevier Science; 1999.
[23] Karp G. Aerobic respiration and mitochondrion. In: Karp G, editor. Cell Mol Biol Concepts Exp. 6th edition. Hoboken, New Jersey: John Wiley & Sons; 2009. p. 173–205.
[24] Schweitzer D, Kolb A, Hammer M, Anders R. Time-resolved measurement autofluorescence. A method to detect metabolic changes in the fundus. Ophthalmologe. 2002;99(10):774–779.
[25] Schweitzer D, Quick S, Schenke S, et al. Comparison of parameters of time-resolved autofluorescence between healthy subjects and patients suffering from early AMD. Ophthalmologe. 2009;106(8):714–722.
[26] Schweitzer D, Deutsch L, Klemm M, et al. Fluorescence lifetime imaging ophthalmoscopy in type 2 diabetic patients who have no signs of diabetic retinopathy. J Biomed Opt. 2015;20(6):61106.
[27] König K, Wollina U, Riemann I, et al. Optical tomography of human skin with subcellular spatial and picosecond time resolution using intense near infrared femtosecond laser pulses. Proc SPIE 4620; 2002. p. 191–201.
[28] König K. Hybrid multiphoton multimodal tomography of in vivo human skin. IntraVital. 2012; 1(1):11–26.
[29] Ulrich M, Klemp M, Darvin ME, et al. In vivo detection of basal cell carcinoma: comparison of a reflectance confocal microscope and a multiphoton tomograph. J Biomed Opt. 2013;18(6):61229.
[30] Weinigel M, Breunig HG, Uchugonova A, König K. Multipurpose nonlinear optical imaging system for in vivo and ex vivo multimodal histology. J Med Imaging. 2015;2(1):016003–016003.
[31] Lakowicz JR. Principles of fluorescence spectroscopy. 3rd edition. New York: Springer; 2007.
[32] Piston DW, Masters BR, Webb WW. Three-dimensionally resolved NAD(P)H cellular metabolic redox imaging of the in situ cornea with two-photon excitation laser scanning microscopy. J Microsc. 1995;178(Pt1):20–27.
[33] Masihzadeh O, Lei TC, Ammar DA, et al. A multiphoton microscope platform for imaging the mouse eye. Mol Vis. 2012;18:1840–1848.
[34] König K. High-resolution multiphoton imaging and nanosurgery of the cornea using femtosecond laser pulses. In: Fankhauser F, Kwasniewska S, editors. Lasers Ophthalmology. Basic, diagnostic and surgical aspects. Monroe: Kugler Publications; 2003.
[35] König K, Krauss O, Riemann I. Intratissue surgery with 80 MHz nanojoule femtosecond laser pulses in the near infrared. Opt Express. 2002;10(3):171–176.
[36] Teng SW, Tan HY, Peng JL, et al. Multiphoton autofluorescence and second-harmonic generation imaging of the ex vivo porcine eye. Invest Ophthalmol Vis Sci. 2006;47(3):1216–1224.
[37] Aptel F, Olivier N, Deniset-Besseau A, et al. Multimodal nonlinear imaging of the human cornea. Invest Ophthalmol Vis Sci. 2010;51(5):2459–2465.
[38] Wang B, Halbhuber KJ, Riemann I, König K. In-vivo corneal nonlinear optical tomography based on second harmonic and multiphoton autofluorescence imaging induced by near-infrared femtosecond lasers with rabbits. Proc SPIE 5964; 2005:596400–11.
[39] Tan H-Y, Sun Y, Lo W, et al. Multiphoton fluorescence and second harmonic generation microscopy for imaging infectious keratitis. J Biomed Opt. 2007;12(2):024013.

[40] Wang TJ, Lo W, Hsueh CM, et al. Ex vivo multiphoton analysis of rabbit corneal wound healing following conductive keratoplasty. J Biomed Opt. 2008;13(3):34019.
[41] Teng SW, Tan HY, Sun Y, et al. Multiphoton fluorescence and second-harmonic-generation microscopy for imaging structural alterations in corneal scar tissue in penetrating full-thickness wound. Arch Ophthalmol. 2007;125(7):977–978.
[42] Lee JH, Lee S, Gho YS, et al. Comparison of confocal microscopy and two-photon microscopy in mouse cornea in vivo. Exp Eye Res. 2015;132:101–108.
[43] Steven P, Bock F, Hüttmann G, Cursiefen C. Intravital two-photon microscopy of immune cell dynamics in corneal lymphatic vessels. PLoS One. 2011;6(10):e26253.
[44] Lakowicz JR, Szmacinski H, Nowaczyk K, Johnson ML. Fluorescence lifetime imaging of free and protein-bound NADH. Proc Natl Acad Sci USA. 1992;89(4):1271–1275.
[45] Palero JA, Bader AN, de Bruijn HS, et al. In vivo monitoring of protein-bound and free NADH during ischemia by nonlinear spectral imaging microscopy. Biomed Opt Express. 2011;2(5): 1030–1039.
[46] Batista A, Breunig HG, Uchugonova A, et al. Label-free SHG imaging and spectral FLIM of corneas using a sub-15 fs laser microscope. Proc SPIE 8930; 2014 p. 89300V–9.
[47] Batista A, Breunig HG, Uchugonova A, et al. Characterization of porcine eyes based on autofluorescence lifetime imaging. Proc SPIE 9329; 2015. p. 93290E–8.
[48] König K, Raphael AP, Lin L, et al. Applications of multiphoton tomographs and femtosecond laser nanoprocessing microscopes in drug delivery research. Adv Drug Deliv Rev. 2011;63(4–5):388–404.
[49] Batista A, Breunig HG, Uchugonova A, et al. Two-Photon Spectral Fluorescence Lifetime and Second-Harmonic Generation Imaging of the Porcine Cornea with a 12 Femtosecond Laser Microscope. J Biomed Opt. 2016;21(3):11.
[50] Batista A, Breunig HG, Uchugonova A, et al. Two-photon autofluorescence lifetime and shg imaging of healthy and diseased human corneas. Proc SPIE 9307; 2015. p. 93071Q–8.
[51] Steven P, Muller M, Koop N, et al. Comparison of Cornea Module and DermaInspect for noninvasive imaging of ocular surface pathologies. J Biomed Opt. 2009;14(6):64040.
[52] Gehlsen U, Oetke A, Szaszak M, et al. Two-photon fluorescence lifetime imaging monitors metabolic changes during wound healing of corneal epithelial cells in vitro. Graefes Arch Clin Exp Ophthalmol. 2012;250(9):1293–1302.
[53] Steven P, Hovakimyan M, Guthoff RF, et al. Imaging corneal crosslinking by autofluorescence 2-photon microscopy, second harmonic generation, and fluorescence lifetime measurements. J Cataract Refract Surg. 2010;36(12):2150–2159.
[54] Fine S, Hansen WP. Optical second harmonic generation in biological systems. Appl Opt. 1971; 10(10):2350–2353.
[55] Tian L, Qu J, Guo Z, et al. Microscopic second-harmonic generation emission direction in fibrillous collagen type I by quasi-phase-matching theory. J Appl Phys. 2010;108(5):54701–54709.
[56] Bianchini P, Diaspro A. Three-dimensional (3D) backward and forward second harmonic generation (SHG) microscopy of biological tissues. J Biophotonics. 2008;1(6):443–450.
[57] Morishige N, Wahlert AJ, Kenney MC, et al. Second-harmonic imaging microscopy of normal human and keratoconus cornea. Invest Ophthalmol Vis Sci. 2007;48(3):1087–1094.
[58] Bueno JM, Gualda EJ, Giakoumaki A, et al. Multiphoton microscopy of ex vivo corneas after collagen cross-linking. Invest Ophthalmol Vis Sci. 2011;52(8):5325–5331.
[59] McQuaid R, Li J, Cummings A, et al. Second-harmonic reflection imaging of normal and accelerated corneal crosslinking using porcine corneas and the role of intraocular pressure. Cornea. 2014;33(2):125–130.

[60] Latour G, Kowalczuk L, Savoldelli M, et al. Hyperglycemia-induced abnormalities in rat and human corneas: the potential of second harmonic generation microscopy. PLoS One. 2012; 7(11):e48388.
[61] Földes-Papp Z, König K, Studier H, et al. Trafficking of mature miRNA-122 into the nucleus of live liver cells. Curr Pharm Biotechnol. 2009;10(6):569–78.
[62] Theodossiou TA, Thrasivoulou C, Ekwobi C, Becker DL. Second harmonic generation confocal microscopy of collagen type I from rat tendon cryosections. Biophys J. 2006;91(12):4665–4677.
[63] Yeh AT, Choi B, Nelson JS, Tromberg BJ. Reversible dissociation of collagen in tissues. J Invest Dermatol. 2003;121(6):1332–1335.
[64] Latour G, Gusachenko I, Kowalczuk L, et al. In vivo structural imaging of the cornea by polarization-resolved second harmonic microscopy. Biomed Opt Express. 2012;3(1):1–15.
[65] Breunig HG, Batista A, Uchugonova A, König K. Motionless polarization-resolved second harmonic generation imaging of corneal collagen. Proc SPIE 9329; 2015. p. 93292P–6.
[66] Yanoff M, Duker JS, Augsburger JJ. Ophthalmology. Mosby Elsevier; 2009.
[67] Whitcher JP, Srinivasan M, Upadhyay MP. Corneal blindness: a global perspective. Bull WHO. 2001;79(3):214–221.

Anna Letizia Allegra Mascaro, Ludovico Silvestri,
Leonardo Sacconi, and Francesco S. Pavone

17 Multiscale correlative imaging of the brain

Abstract: Recent advancements in optical microscopy are challenging our understanding of the brain. In this chapter, we show the potential of optical microscopy techniques to tackle different aspects of brain structure and function, from whole-brain neuroanatomy to neural network plasticity and functionality. We will first address novel implementations of light microscopy for cellular resolution imaging of neuronal anatomy spanning the whole brain. Afterwards, we will illustrate real-time brain rewiring of single synaptic contacts visualized through two-photon microscopy in vivo. Then, the functionality of microcircuits is investigated with nonlinear microscopy combined with fluorescent indicators of neuronal activity. Nevertheless, a single technique is not enough for targeting the articulate organization of the brain; a wider view is more efficiently gained by combining complementary approaches. In the last section of this chapter, we show examples of this multiscale approach by discussing correlative imaging obtained by combining different microscopy techniques. At the end, we discuss the perspective of a wider methodological framework fusing multiple levels of brain investigation possibly leading to an omni-comprehensive view of brain machinery.

17.1 Introduction

Modern neurobiology is striving to draw a molecular picture of the processes that determine neuronal plasticity. Since the works of Golgi and Ramon y Cajal, neuroscientists took advantage of light microscopy to observe the brain at many levels. The technologies employed in biophotonics have been constantly evolving over the past fifteen years offering always-new experimental capabilities to inspect the nervous system across multiple scales, from whole brain anatomical reconstructions to ultrastructural imaging of synaptic components, and allowing a constant improvement of sensitivity and resolution.

In this chapter, we will show different light microscopy techniques tackling different aspects of brain anatomy and functionality. A "connectomic" perspective is first described, where neuronal projections are visualized through the entire brain with light sheet microscopy. Advancements in nonlinear imaging led to time-lapse microscopy studies in different brain areas of resting and behaving mice, addressing both structural and functional rewiring. This chapter surveys recent innovations in optical imaging techniques of particular interest to neuroscientists.

https://doi.org/10.1515/9783110429985-019

17.2 Brain anatomy

Mapping neuronal connections across the whole brain (the so-called 'connectome' [1]) is a mandatory step to afford a clearer view of the brain itself. However, the structure of the brain challenges imaging techniques to reconstruct macroscopic volumes with microscopic resolution. Indeed, very thin neuronal processes (dendrites and axons, about 1 μm or less in diameter) usually extend for much larger distances, even the whole encephalon in the case of axons [2].

In the last years, a number of optical methods have been implemented to reconstruct large specimens with micron-scale resolution. Combined with different (sparse) labeling methods, such as Golgi staining [3], transgenic expression of fluorescent proteins [4], dye injection [5], viral transfection [6], and whole-mount immunohistochemistry [7], high-throughput optical microscopy has been exploited to reconstruct fine brain anatomy from the mesoscale to the microscale (Fig. 17.1). In the following, we will briefly review three approaches: serial two-photon sectioning (STP), micro-optical sectioning tomography (MOST), and light sheet microscopy (LSM).

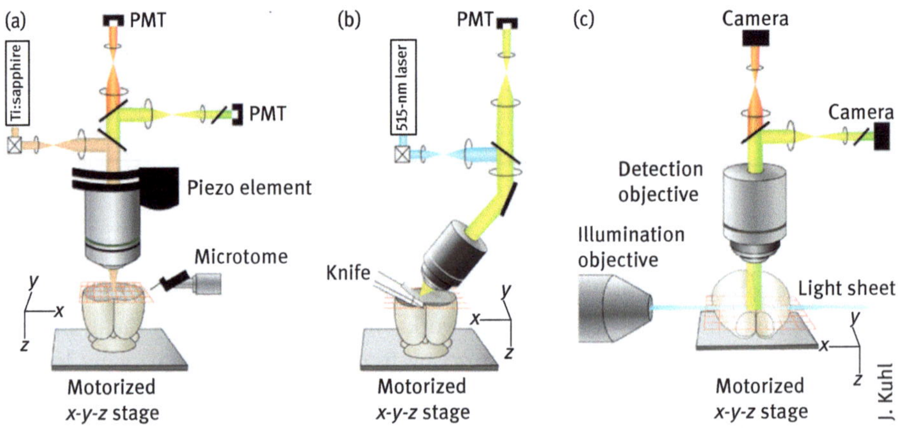

Fig. 17.1: Schematic representation of the three main optical techniques used to reconstruct neuroanatomy on a brain-wide scale: serial two-photon sectioning (a), micro-optical sectioning tomography (b), and light sheet microscopy (c). Reproduced with permission from [9].

17.2.1 Serial two-photon sectioning

STP systems are standard laser-scanning two-photon fluorescence microscopes coupled with automated translation stages and a vibrating blade microtome, which cuts away thin layers of tissue once they have been imaged (Fig. 17.1 (a)) [8]. Two-photon excitation allows imaging below the surface, thus minimizing cutting artefacts. The penetration of two-photon imaging inside aldehyde-fixed mouse brain tissue is limited in practice to about 100 μm; as a consequence, only a single optical section is acquired by STP between consecutive slicing steps (typically at 50 μm depth [8]). This results in poor axial sampling (about 1 μm every 50, discarding about 95 % of the volume), since physical slices cannot be thinner than several tens of microns. Although such sparseness prevents full volumetric imaging, on the plus side it allows imaging an entire mouse brain in a day (a quite short time for a point-scanning microscope) and keeps the size of acquired datasets within reasonable limits (around 100 gigabytes per mouse brain [9]). This method is also very reproducible and scalable, and to date is the only one which has been used for large-scale studies like the mesoscale connectivity atlas realized by the Allen Institute for Brain Sciences, where 469 mouse brains have been scanned [10].

17.2.2 Micro-optical sectioning tomography

MOST exploits a serial sectioning approach similar to STP, but with significant differences in sample preparation. Indeed, the tissue is embedded in a hard resin, allowing cutting very thin slices (about 1 μm) without significant slicing artefacts [11]. The sparse sampling of STP can therefore be overcome in this case. In typical applications, a single line on the sliced section is imaged continuously, directly on the knife edge (Fig. 17.1 (b)) – that's why this technique is also known as knife-edge scanning microscopy (KESM [12]). Image can be acquired using a linear charge-coupled device (CCD) camera [11] or a photomultiplier tube (PMT) in combination with fast acousto-optic scanners [13, 14]. MOST can produce high-resolution and high-contrast images of the whole mouse brain, on which it is possible to trace single axons throughout the entire encephalon (Fig. 17.2 (a)). The main current limitation of this method is the low rates of image acquisition, as about 10 days are required to image a whole mouse brain.

17.2.3 Light sheet microscopy

The fastest optical imaging method for whole brain imaging is probably light sheet microscopy, which provides micron-scale reconstruction of the entire murine encephalon in hours to few (<3) days [15, 16]. Indeed, planar illumination (the 'light sheet') confines excitation to the focal plane of detection optics, allowing optical sectioning in a wide-field detection scheme (Fig. 17.1 (c)) [17]. This technique is also called selective (or single) plane illumination microscopy (SPIM) [18], or ultramicroscopy [15].

To be applied in macroscopic specimens – such as whole mouse brains – LSM needs to be coupled to chemical clearing of the tissue [19]. Classic methods to clear the tissue are based on the substitution of water with high-refractive index (≈ 1.56) or-

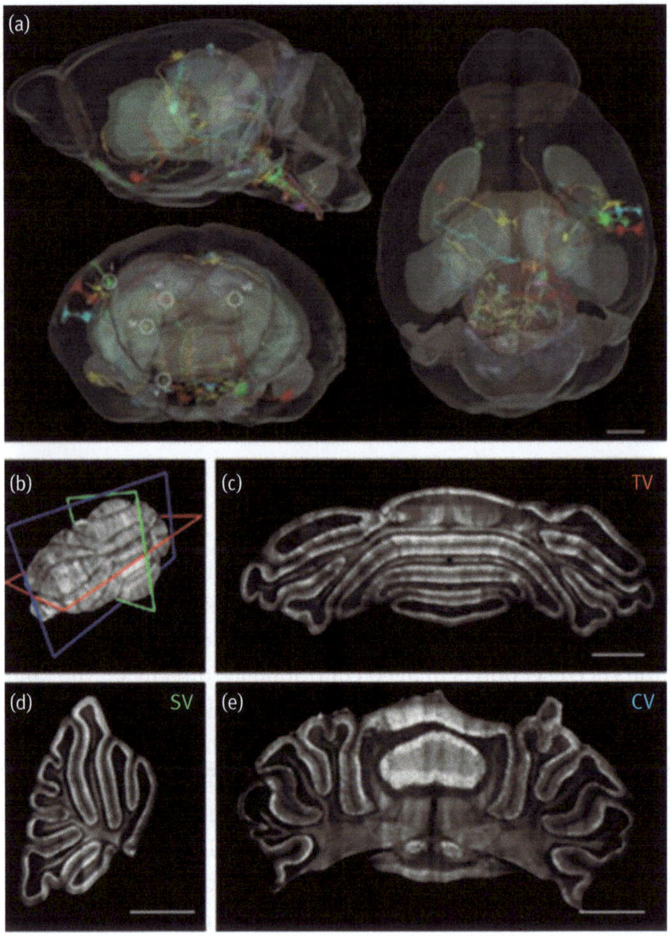

Fig. 17.2: Mapping brain-wide anatomy. (a) Tracing of long-distance projecting neurons in the brain of a Thy1-eYFP-H mouse. Traced neurons (depicted with different colors) are registered on an MRI reference mouse brain. (b)–(e) Full reconstruction of Purkinje cells neuroanatomy in the cerebellum of an L7-GFP mouse. (b) 3D volume rendering of the whole cerebellum. The superimposed planes refer to transverse (red), sagittal (green) and coronal (blue) digital sections shown in (c)–(e) respectively. Each section is a maximum intensity projection of 40 μm thick slabs. Scale bars: 1 mm. Reproduced with permission from [13] (a), and [16] (b)–(e).

ganic solvents and render the brain very transparent, but no objective lens suitable for immersion in these clearing agents is currently available. Air objectives are commonly used in this case, but the refractive index mismatch introduces strong spherical aberrations [20].

Objective lenses are commercially available for immersion in moderate refractive indices (≈ 1.45), which is the one obtained in CLARITY [7], a clearing protocol which results in highly transparent brains and good fluorescence preservation. In this method, the sample is hybridized into a hydrogel, allowing lipids removal while leaving proteins and nucleic acids in place. This approach greatly increases sample porosity, allowing exogenous labeling over macroscopic portions of tissue, paving the way to anatomical mapping of human tissue. However good the transparency of the tissue, some residual scattering is always present, leading to out-of-focus contributions and thus image blur. To recover image quality, different strategies can be used, like structured illumination [21, 22], and confocal line detection [16, 23, 24] (Fig. 17.2 (b)).

17.3 Structural plasticity of cortical neurons: from an historical perspective to recent advances in fluorescence imaging in vivo

Large-scale neuroanatomical data are limited to one time point, and thus need to be complemented with studies addressing the temporal evolution of plastic events in live brain. Fluorescence microscopy brought to light neuronal plasticity and its dynamic timeline. The term "structural plasticity" refers to alteration in anatomical connectivity through modifications of synaptic patterns or densities and reshaping of axons and dendrites [25]. Structural plasticity plays a major role during development, when new connections are continuously established and refined [26–30], and occurs consistently during adulthood, e.g. as a result of experience and learning [31–33]. Two-photon fluorescence microscopy (TPFM), by tracking complex dynamics inside live brain, allows the investigation of structural plasticity at the cellular level in vivo [33–38]. In combination with transgenic mice expressing fluorescent proteins in selected neuronal populations, in vivo TPFM showed events of structural rewiring at synaptic level in the adult brain. Long-term in vivo optical imaging revealed temporal details of dynamic remodeling in the adult brain cortex [26, 31, 39], olfactory bulbs [40–42], and cerebellum [43, 44]. Time-lapse TPFM of cortical dendrites proved the direct correlation between spine turnover and experience-dependent plasticity [45] in different experimental paradigms, like enriched environments [46, 47], long-term stimulation [48, 49] and deprivation [30]. These studies showed how spine plasticity contributes to experience-dependent rewiring of cortical circuits. Fig. 17.3 shows two examples of experience-dependent spine plasticity in the adult neocortex investigated by two-photon microscopy in vivo. Fig. 17.3 (a)–(e) shows how chessboard whisker trimming modifies whisker representational maps in the barrel cortex at the synaptic level. Fig. 17.3 (a) shows examples of spine stability (yellow arrowhead), and turnover (white arrowhead). The authors showed that new persistent spines (orange arrowhead) are more likely to grow after whisker trimming, whilst previously persistent spines (green arrowhead) are more likely to disappear. In a different experimental paradigm, i.e., focal lesion in the retina (Fig. 17.3 (f)–(j)), dendritic spines near or in the lesion projection zone (LPZ) in the neocortex display high turnover rates; in the center of the LPZ almost the entire spine population turns over. Although some spines are present over more than two months of imaging (for example, yellow arrowhead), most spines are lost (for example, green arrowhead) and replaced by new persistent spines (orange arrowhead).

New imaging configurations, by exploiting implanted endoscopes [50] or microprisms [51], three-photon excitation [52] or by means of regenerative amplifiers [53], increased the depth of the imaging area to deep circuits like the hippocampus. Miniaturized portable microscopes [54], fiber optics [50, 55] or head-restrained configurations [56] are currently used for imaging in awake and behaving animals.

TPFM recently provided useful insights into the morphological rearrangements of neurons and glial cells involved in recovery from neuronal injury [57–59]. Technical issues so far limited the investigation of the biological mechanisms of the reactive plasticity after injury in the adult CNS [60]. The main obstacle concerns the targeted manipulation of CNS neurons. The spatial localization of multiphoton absorption scales down lesions dimensions to the sub-micrometer scale. Indeed, the intrinsic confinement of nonlinear excitation to the focal volume can be exploited to perform localized targeted photochemistry and photodamage. Laser nanosurgery has been applied in various experimental paradigms, from the investigation of the biological function of subcellular compartments in living cells [61–63], to the study of neuronal plasticity of adult mice in vivo [64–67]. A recent TPFM study addressed the re-growing potential of adult cerebellar axons [66]; they showed that the laser-induced ablation of a single axonal branch may elicit axonal sprouting and synaptic remodeling in the surviving portion of the axon (Fig. 17.4 (a)). By combining light and electron microscopy, this study disclosed the formation of new synaptic contacts on the sprouted branch. An analogous study showed the regenerative capabilities of laser axotomized pyramidal neurons in the somatosensory cortex [64] (Fig. 17.4 (b)). Laser-induced lesions have been used to study blood brain barrier disruption [68] and microglial migration towards the lesion site, thus confirming their role in the recovery process [69, 70]. Pulsed laser irradiation on single blood vessels [71] provides a method to create targeted highly confined strokes mimicking cerebral microvascular diseases [72].

▸ **Fig. 17.3:** (a) Time-lapse TPF images of dendritic spines in the barrel cortex before and after chessboard whisker trimming. (b) The experimental paradigm: chessboard whisker trimming causes changes in the whisker representational map in the barrel cortex. (c) Spine density in the barrel cortex does not change after whisker trimming. (d) The fraction of surviving spines is slightly decreased after whisker trimming due to an increased loss of persistent spines. (e) The fraction of new persistent spines increases ≈ 2.5 fold after whisker trimming. (f) Time lapse of dendritic spines in the visual cortex after a unilateral focal lesion in the retina. (g) The experimental paradigm: immediately after a focal lesion in the retina, the LPZ is unresponsive to visual stimulation, as measured by intrinsic signal optical imaging (bottom, day 0). The size of the unresponsive region decreases with time. Scale bar, 700 μm. (h) Spine density in the visual cortex remains unchanged after a small retinal lesion. (i) Dendritic spines near or in the LPZ display high turnover rates; in the center of the LPZ almost the entire spine population turns over. (j) In the LPZs of small lesions, the fraction of new persistent spines is increased several fold. Reproduced with permission from [25].

TPF imaging and laser nanodissection demonstrated to be a fine tool to dissect at high resolution the dynamics of many neurodegenerative pathologies.

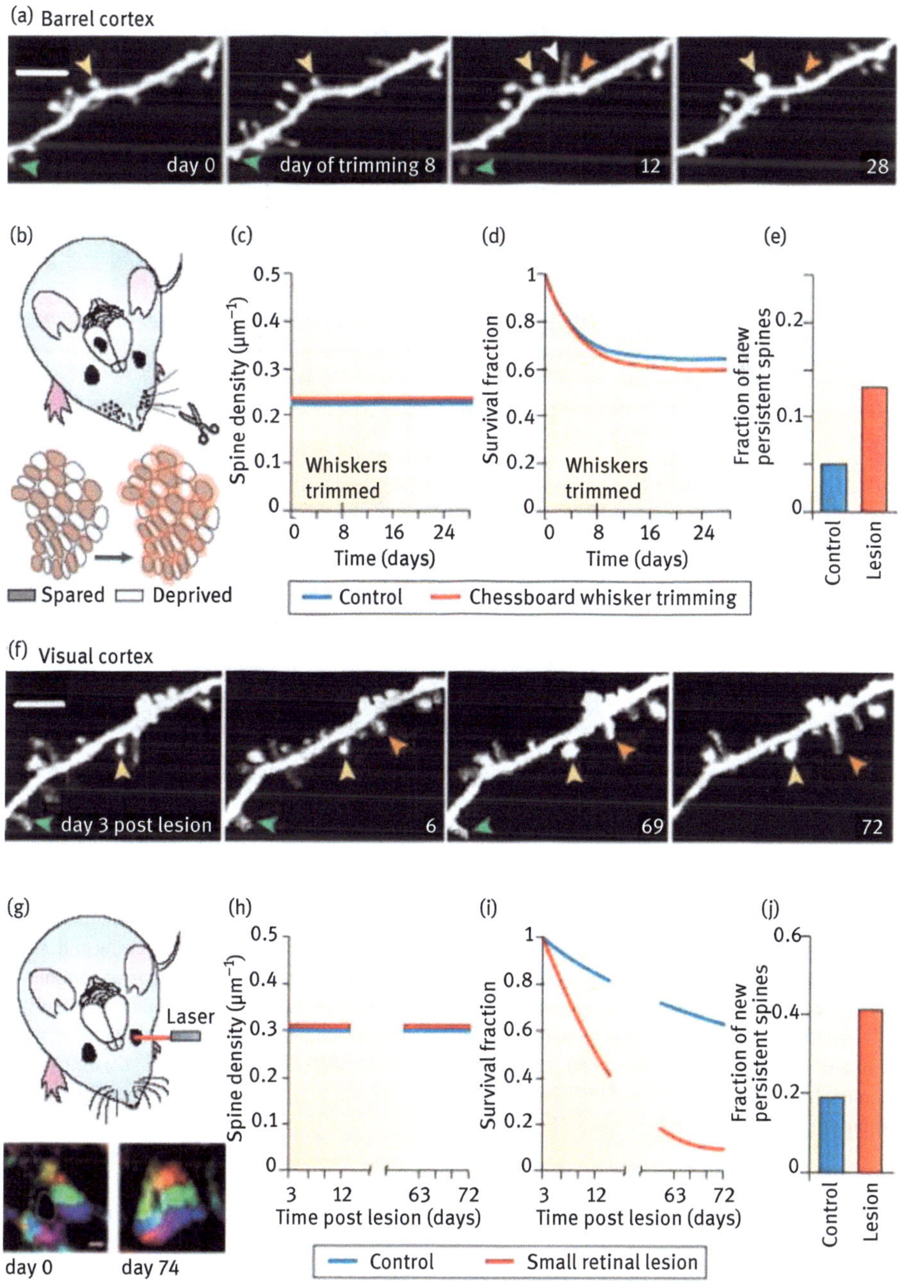

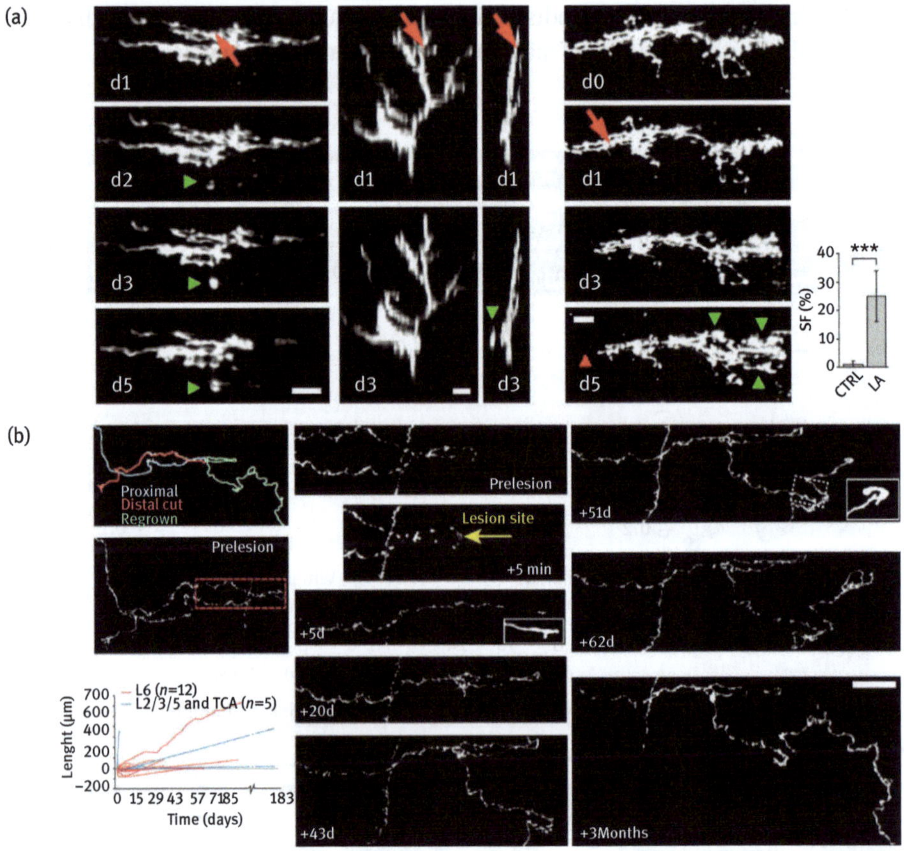

Fig. 17.4: Morphological rewiring of axonal connectivity after laser transection in different brain regions. (a) Time course (from day 0 to day 5) of a cerebellar axon before and after laser dissection. The first image (d 0) was acquired one day before laser irradiation. The laser beam was focused on the axon where the red arrow points on d1. The red and green arrowheads at d5 highlight the degeneration of distal portion and the protrusion of new branches, respectively. Scale bar, 15 μm. The histogram compares the sprouting frequency (SF) in control and laser axotomized climbing fibers. Reproduced with permission from [66]. (b) Following laser dissection (yellow arrow in the +5 min panel) and subsequent degeneration of the axonal ending (red line), few axons of pyramidal layer 6 (L6) neuron demonstrated sustained regrowth over a 3 month time period (green line). The red dotted line box indicates the area followed over time in the time-stamped panels. Insets at +5 d and +51 d show the growth cone. The graph on the lower left shows that 55 % of L6 axons (red) demonstrated regrowth over several weeks compared with only ≈ 20 % of all other axons (blue). Reproduced with permission from [64].

17.4 Functional imaging and stimulation of neural circuits

Advantages in optical microscopy in combination with calcium indicator and voltage-sensitive dyes (VSDs) allow registrations of neuron activity across a population of cells, opening promising prospectives in exploring brain functionality [50, 73, 74]. Further, neurotransmitters uncaging and neuroengineering tools based on microbial opsins provide optical control over the activity of specific populations of neurons. If combined, light can be used to probe and manipulate the electrical activity of neurons singularly, with high spatial-temporal resolution. Here, we will report the state of the art in detecting and manipulating the activity of microcircuits of neurons.

17.4.1 Optical recording of neuronal activity

The first challenge to address for imaging neural activity is the loading of calcium indicators or VSDs into neurons. Fig. 17.5 (a) illustrates the two approaches for dye loading of neurons based on organic and genetically encoded functional probes. The upper panels depict the "acute" network loading: many neurons are labeled simultaneously by pressure injection loading (left panel), by loading with dextran-conjugated dye (middle panel), and by bulk electroporation (right panel). Due to their electrostatic charge, calcium indicators are unable to cross lipid membranes and the use of physical or chemical methods is needed [75]. For example, the esterification of the carboxylic groups with an acetoxymethyl (AM) group makes the dye neutral, thus capable of crossing the cell membrane. Besides AM calcium dyes, dextran-conjugated calcium indicators can also be employed for network labeling [76]. Electroporation can be alternatively used to allow the membrane passage of charged calcium probes [77]. Using similar approaches, bulk loading of VSDs can be achieved by incubating the tissue or using pressure injection [78, 79]. Further, in recent years, genetically encoded calcium probes and voltage sensitive proteins have become a widely used tool in neuroscience [80–84]. There are different possibilities of expressing genetically encoded probes in neurons. The lower row of Fig. 17.5 (a) shows the expression of genetically encoded indicators (GE) by viral transduction (left panel), in utero electroporation (middle panel), and generation of transgenic mouse lines (right panel). Viral constructs can be used to label specific brain areas using stereotaxic injection [85]. In utero electroporation of DNA plasmids encoding for the genetically encoded indicator produces relatively sparse labeling. Finally, generating transgenic mice expressing genetically encoded fluorescent proteins enables long-lasting monitoring of neuronal activities at the cellular, circuit, and system level to address questions of nervous system development and maintenance, learning, and memory.

These labeling methods can be exploited to probe cell activity across a population of neurons in an intact circuit and even in vivo. For example, by incubating an acute brain slice with a novel fluorinated VSDs, single spontaneous action potentials

▸ **Fig. 17.5:** Functional imaging of neuronal circuits with cellular resolution. (A) Dye-loading and genetic-encoded indicators approaches. Reproduced with permission from [75]. (B) Real-time multicellular action potential recording by random access two-photon microscopy. Two-photon fluorescence image of a parasagittal acute cerebellar slice stained with voltage-sensitive dye (VSD) by pressure injection. The multiunit optical recording was carried out from the lines drawn (red) on the five Purkinje cells (PCs). The electrical activity (cell-attached recording) of PC1 was also monitored. Reproduced with permission from [86]. (C) Calcium imaging of place cells in the CA1 hippocampal region of mice which are placed on a spherical treadmill. Calcium traces are shown in black. Red traces indicate significant calcium transients. In parallel, the position of the mouse along the virtual linear track is recorded. Reward times are shown at the bottom. Reproduced with permission from [88].

were recorded from multiple neurons using a random access multiphoton microscope [86, 87] (Fig. 17.5 (b)). Advanced optical methods can also be applied for imaging the activity of neurons with cellular resolution in behaving mice [56, 88, 89] during virtual navigation. Fig. 17.5 (c) shows an example of two-photon images of neuron cell bodies in *stratum pyramidale* of CA1 labeled with the genetically encoded calcium indicator, performed while the mouse is running along a virtual linear track. The experimental setup the authors used, consisting of a spherical treadmill, a virtual reality apparatus and a custom-made two-photon microscope, is displayed in the left panel.

Besides monitoring neuronal activity in behaving animals, an elegant alternative to dissect neuronal circuitry implies controlling network firing.

17.4.2 Optical stimulation of neurons

Caged compounds are widely used in neuroscience enabling the optical manipulation of neuronal circuits [90, 91] (Fig. 17.6 (a)). Glutamate can be uncaged by focusing light at a specific position in the brain slice, while in the meantime the electrical responses from a selected neuron can be monitored. For example, by scanning the uncaging beam in different position, one can map the territories that generate excitatory or inhibitory responses in the recorded cell.

The capability of stimulating neurons with light has been recently improved by new neuroengineering tools, channelrhodopsin-2 (ChR2) and halorhodopsin (NpHR) [92, 93]. ChR2 is a monovalent cation channel that allows cations to flow into the cell following illumination to blue light, while NpHR is a chloride pump that activates upon exposure with yellow light (Fig. 17.6). The use of ChR2 in neurons has been applied to a number of investigations on circuit mapping [94], memory storage [95], and Parkinson's disease [96]. A fiber optic-based system has also been developed for modulating the activity of ChR2- and NpHR-expressing neurons of freely moving animals [97]. If combined with nonlinear excitation, optogenetic tools benefit from increased control of neuronal activation within the intact tissue, even at single-cell level. However, standard ChR2 has low channel conductance and displays fast kinetics; there-

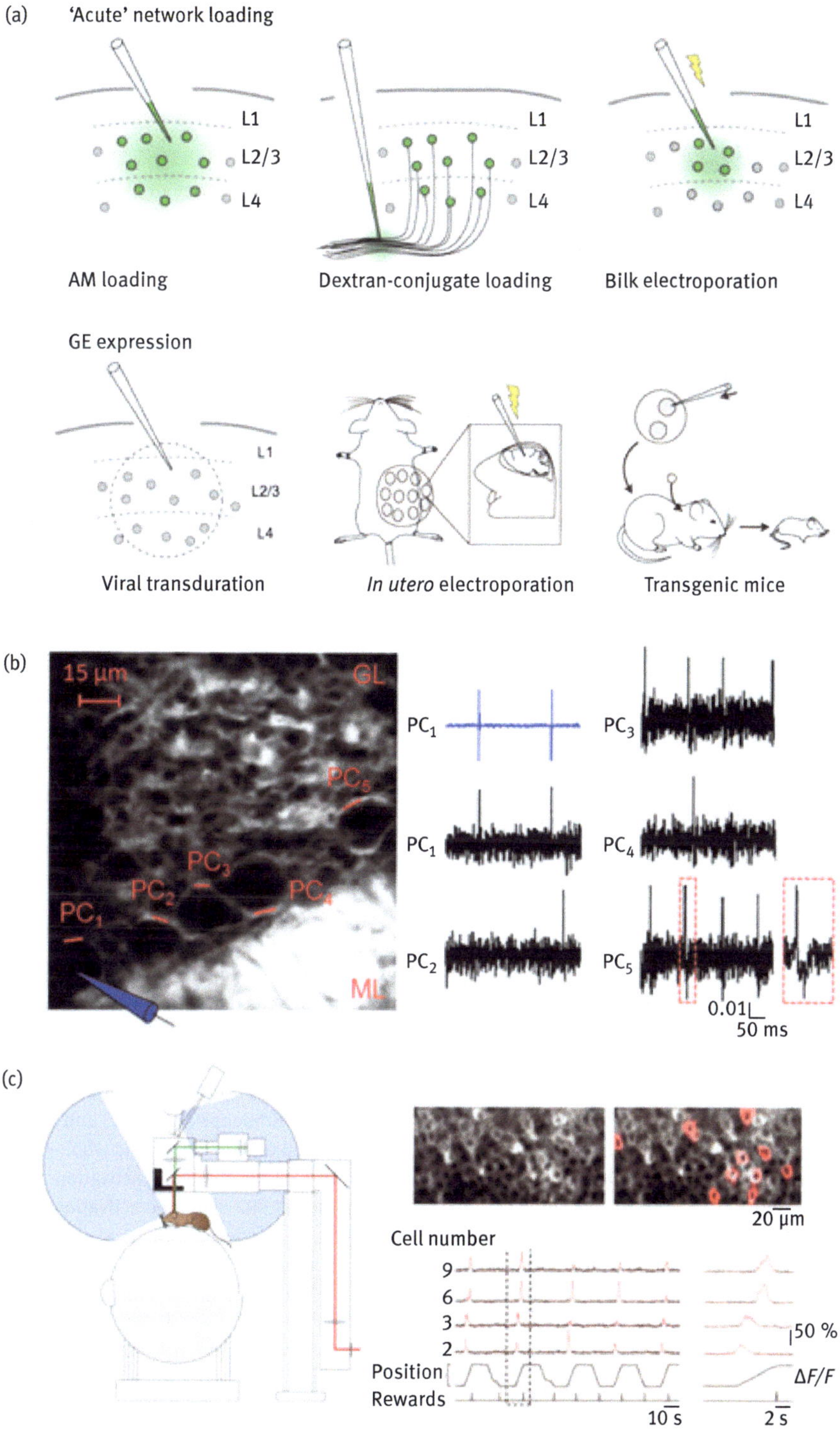

(a)
'Acute' network loading
L1
L2/3
L4
AM loading
Dextran-conjugate loading
Bilk electroporation
GE expression
Viral transduration
In utero electroporation
Transgenic mice
(b)
15 µm
GL
ML
PC1
PC2
PC3
PC4
PC5
0.01
50 ms
(c)
20 µm
Cell number
9
6
3
2
Position
Rewards
10 s
2 s
50 %
ΔF/F

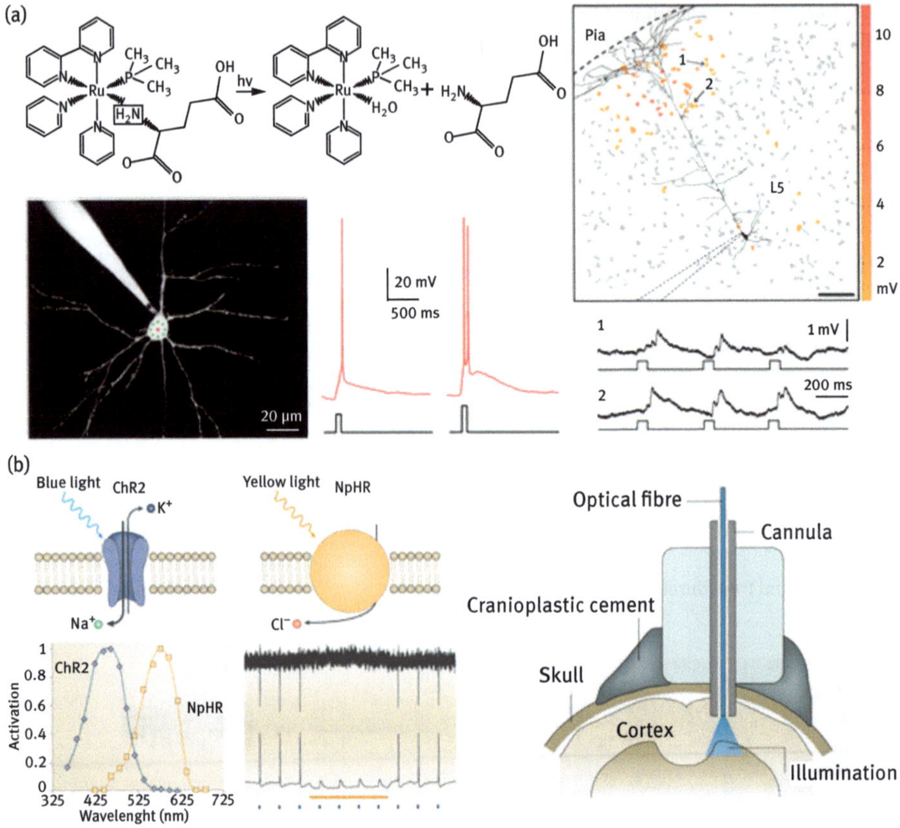

Fig. 17.6: Optical interrogation of neuronal circuit. (a) Neuron excitation by glutamate uncaging. Structure of Ruthenium-bipyridine-trimethylphosphine-Glutamate (RuBi-Glutamate) and glutamate photorelease reaction. Layer 2/3 pyramidal cell loaded with Alexa-594 and position of the multiplexed uncaging laser targets (eight subtargets) on the soma of the cell. Action potentials triggered by uncaging of RuBi-Glutamate in a layer 2/3 pyramidal neuron. Input map for a layer 5 pyramidal neuron and superimposed morphological reconstruction of its dendritic tree. Gray areas outline stimulated neurons. Colored areas are outlines of all positive cells, color-coded according to peak EPSP amplitude. Dotted outline marks the patch pipette location. Scale bar, 100 μm. Voltage recordings from the patched cell during uncaging in locations labeled by arrows. Reproduced with permission from [90, 91]. (b) Optogenetic tools. Schematic of channelrhodopsin-2 (ChR2) and the halorhodopsin (NpHR) pump. Action spectra for ChR2 and NpHR. Cell-attached (top) and whole-cell current-clamp (bottom) traces from hippocampal neurons showing all-optical neural activation and inhibition. Blue pulses represent the blue light flashes used to drive ChR2-mediated activation and the yellow bar denotes NpHR-mediated inactivation. A cannula is implanted into the head of the experimental animal to guide an optical fiber to the targeted brain region. The optical fiber is coupled to a strong light source to bring blue or yellow light into the brain. Reproduced with permission from [93].

fore, two-photon excitation of a neuron requires complex stimulation strategies [98]. The optimization of a new red-shifted chimeric opsin that displays slower off-kinetics and larger photocurrents has been performed and applied for single neuron stimulation in in vivo preparation [99]. Recently, simultaneous two-photon optogenetic activation and calcium imaging by coexpression of a red-shifted opsin and a genetically encoded calcium indicator have been performed enabling high-throughput and long-term optical interrogation of specific neural circuits with single-cell and single-spike resolution in vivo [100].

17.5 Correlative microscopy

17.5.1 Understanding brain machinery requires multilevel investigation

The structure of the brain encompasses several orders of magnitude, ranging from synapses to neurons, from cortical columns to the whole encephalon. Consequently, neuronal dynamics occurs across different timescales, from the milliseconds of action potential propagation to the continuous remodeling of the brain, which lasts the entire lifespan. Structural and functional data can be collected across all these scales by using a large repertoire of experimental tools (Fig. 17.7).

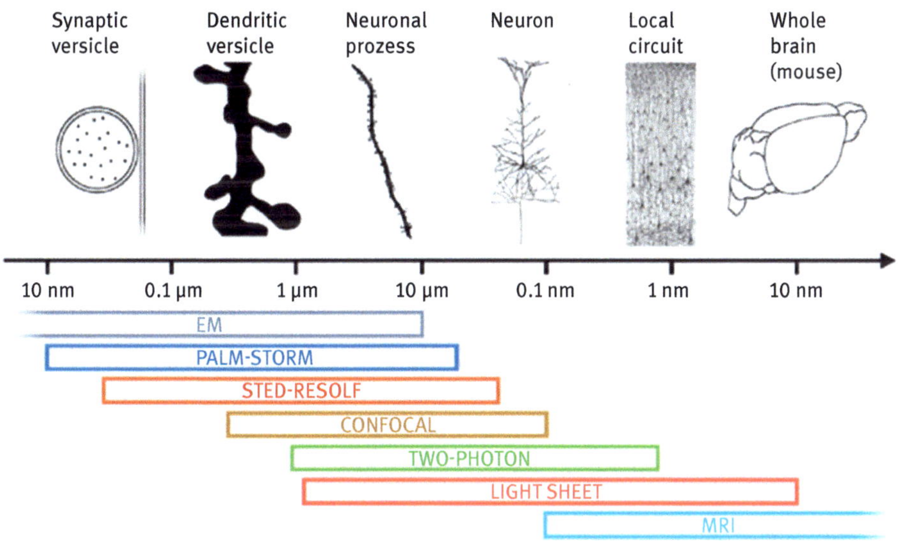

Fig. 17.7: Multiple spatial scales in the brain. On the top, different structures are depicted in proximity to their typical size, showing how relevant spatial scales in the brain span several orders of magnitude. On the bottom, typical working ranges of state-of-the-art imaging technologies.

Minimally invasive neuroimaging methods can provide whole-brain data also in living humans, albeit at low resolution (no more than 100 μm). For instance, functional Magnetic Resonance Imaging (fMRI) allows inferring neuronal activity from local blood oxygenation levels, based on the different magnetic responses of oxygenated and deoxygenated hemoglobin [101–103]. On the other hand, diffusion Magnetic Resonance Imaging (dMRI) can be used to reconstruct axonal fiber tracts throughout the encephalon, exploiting the anisotropic diffusion of water in the presence of lipid-rich myelin sheaths [101, 104, 105].

Finer structural details on smaller – though still macroscopic – samples can be explored with high-throughput light microscopy (see Section 17.2). However, given the high density of neuronal processes (which can be separated by just few nm), standard optical methods have to be coupled with sparse fluorescent labeling; spectral multiplexing in Brainbow methods might overcome this limitation [106]. Dense neuronal circuits can be fully disentangled using electron microscopy (EM) [107]. However, the EM staining is usually nonspecific, and data acquisition rates are very slow, preventing (up to now) its use on the whole mouse brain. EM automated volumetric imaging with resolution of tens of nm can be afforded with approaches based on tissue slicing, like Serial Block-face Scanning Electron Microscopy (SBSEM) [108] – where a scanning electron microscope is coupled with automatic slicing and removal of the most superficial layer after imaging – or Automatic Tape-Collecting Lathe UltraMicrotome (ATLUM) [109] – where ultrathin sample slices are automatically collected on a tape for subsequent high-throughput imaging with transmission electron microscopy. Higher resolution (near-isotropic voxels of ≈ 4 nm), albeit at slower speed, can be achieved with Focused-Ion Beam Scanning Electron Microscopy [110] (FIBSEM). The nm realm can also be explored with optical super-resolution techniques like Photo-Activated Localization Microscopy (PALM) [111] and Stimulated Emission Depletion (STED) [112].

Optical methods such as two-photon fluorescence microscopy can also provide valuable functional data with cellular and subcellular resolution over spatial scales of few hundreds of μm (see Section 17.4). The same approaches can also be used to investigate neuronal reshaping in vivo over weeks to months (see Section 17.3).

The vast collection of imaging techniques developed in the last decades allows researchers to investigate many different aspects of brain architecture and activity with unprecedented detail. Nonetheless, we still lack a comprehensive description of brain machinery across all its organizational levels. A promising approach in this sense is to combine multiple imaging techniques, each one providing different information on different scales. This correlative framework can overcome intrinsic limitations of single methods.

17.5.2 Correlative imaging overcomes the limitation of single techniques

The challenge of bridging the gap between post mortem microscopic and in vivo macroscopic worlds can be tackled by novel approaches that, by combining the spatio-temporal resolution of complementary techniques, cross several orders of magnitudes. The following section shows examples of correlative measures that linked different temporal/spatial scales by exploiting the resolution of different techniques, from whole brain fMRI to sub-synaptic imaging with FIBSEM.

Different neuroimaging techniques provide functional connectivity data over different spatial and temporal scales. fMRI, by estimating brain activation based on the alterations in blood oxygenation levels (BOLD contrast), enables non-invasive monitoring of activity in healthy and diseased brains with sub-millimeter spatial resolution in humans and animals [101, 113, 114]. The causal link between BOLD signals and the neural activity has not been well described yet [103]. Helmchen et al. demonstrated the possibility of combining fluorescence measures of brain activity with fMRI. By using an optical fiber compatible with the fMRI scanner, the authors could perform BOLD fMRI and simultaneous recording of calcium activity (Fig. 17.8 (a)). This correlative approach allowed defining features of the complex fMRI BOLD signals in terms of neuronal and glial activation. This hybrid method could further disclose crucial features of neurovascular coupling.

TPFM longitudinal studies on the structural plasticity of selected neuronal populations in vivo proved otherwise inaccessible data on micro-connectivity reshaping in different conditions [25, 32, 73, 102, 104]. These dynamic measures on long timescales can only access a limited portion of the brain (hundredths of microns). By combining TPFM with ex vivo imaging techniques like light sheet microscopy that can explore large volumes it is possible to reframe the in vivo visualized neurons within long-range connectivity settings. In detail, neuroanatomical tracing of the neuron previously observed in vivo allows associating the dynamic fingerprint of a neuron to its connectivity partners. Silvestri et al. used the blood vessel pattern as internal reference to retrieve a neuron whose dendritic arbor was previously imaged in a live mouse using TPFM, and traced the entire neuron from confocal light sheet images [115] (Fig. 17.8 (b)). This correlative two-photon and light sheet microscopy approach is a useful tool to contextualize in vivo long-term imaging within a wider neuroanatomical framework.

Linking data on brain functionality with structural connectivity details at high resolution (nm) gives answers to otherwise impenetrable neurobiological questions. Briggman et al. combined in vivo calcium imaging in the intact retina and SBSEM-based reconstruction of the circuitry in the same piece of tissue to explain the behavior of retinal ganglion cells (DSGCs) [116] (Fig. 17.8 (c)). The authors directly correlated neuronal wiring asymmetry with the computation of direction selectivity. High-resolution 3D maps of neuronal connectivity will soon be a "*conditio sine qua non*" for the correlation of functionality and structure.

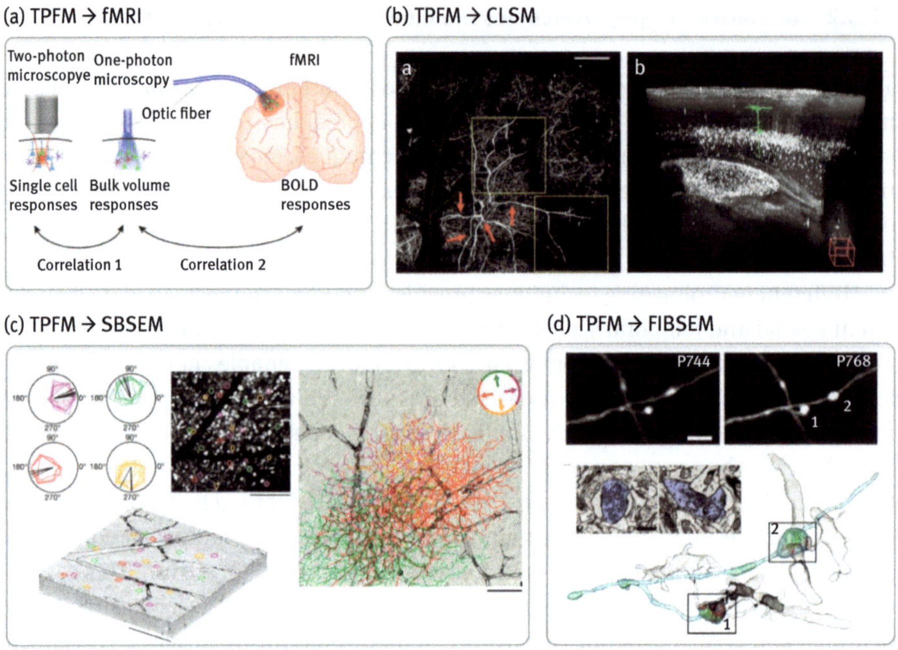

Fig. 17.8: Correlative microscopy in neuroscience. (a) Experimental approach for examining the relationship between single-cell activity and fMRI signals. Single-cell Ca^{2+} measured by two-photon microscopy was correlated with bulk responses in the same region measured by one-photon fiber-optic microscopy. Then the bulk one-photon responses were correlated with the BOLD fMRI signals. Figure modified with permission from [126]. (b) On the left, maximum intensity projections of several stacks TPF stitched together in a single image. The red dashed line highlights blood vessel shadows. Red arrows highlight characteristic features of a dendritic arbor, to help finding it back in the CLSM images. On the right, CLSM imaging of the same neuron observed with TPF. Starting from the apical portion of the dendritic tree, the neuron has been segmented and is shown inside a maximum-projection 3D rendering. The scale of the figure can be inferred from the red cube down on the right, which has 100 µm sides. Figure modified with permission from [115]. (c) Functional characterization of direction-selective retinal ganglion cells (DSGCs) and their localization within SBSEM volume. On the left, Polar tuning curves for 25 DSGCs sorted and color-coded by preferred direction. The corresponding soma locations superimposed onto a two-photon image from the recorded region of the ganglion cell layer and the acquired SBSEM volume (scale bars: 100 nm). On the right, skeleton reconstructions of DSGCs. DSGCs, color-coded by preferred direction (inset), normal to the plane of the retina (scale bars: 50 µm). Figure modified with permission from [116]. (d) Correlative in vivo TPF and focused ion beam scanning electron microscopy of cortical neurons. In vivo TPF imaging of an axon showing two stable boutons (scale bar: 5 µm). Both boutons make multiple synaptic contacts, as visible in a single plane of the correspondent EM images, with multiple dendritic spines (scale bar: 500 nm). 3D rendering of the same axon imaged in TFP microscopy. The cytoplasm of the axon is represented in light blue, mitochondria in green, synaptic vesicles in yellow and synapses in red. The postsynaptic spiny neurons are shown in gray. Figure modified with permission from [119].

The correlation of in vivo imaging of neurons and their synaptic connectivity with electron microscopy allows combining dynamic and ultrastructural information [37, 45]. The appearance of FIBSEM significantly increased the level of automation, reliability and acquisition speed of EM [110]. Recent studies provided evidence of the benefits of correlating in vivo TPFM and FIBSEM for targeting synapses at the micro- and ultrastructural level [117–119]. Recently, Allegra Mascaro et al. [66] and Canty et al. [64] combined two-photon imaging in vivo with FIBSEM to study the possible formation of synaptic contacts by axonal branches regenerated after injury (Fig. 17.8 (d)). EM images of the distribution of synaptic vesicles and post-synaptic densities allowed deducing the structural and functional relationship between the regenerated axon and post-synaptic dendrites.

17.5.3 Fusing multiple levels of investigation might boost our understanding of the brain

The correlative methods presented here showed fundamental insight into different spatio-temporal scales of brain functioning. This approach can be promisingly expanded towards a unified method covering most perspectives. In vivo imaging by non-invasive fMRI could be the starting point of a long pipeline that investigates the long-term plasticity of small populations of neurons through in vivo TPF imaging. Once functional and structural data have been obtained in vivo at the small-circuitry level, the same sample shall be processed with LSM for ex vivo long-range anatomical analysis. This contextualization into a wider framework is refined up to the synaptic scale when imaged through super-resolution techniques or electron microscopy. This multilevel approach could be useful in a variety of investigations, including neurological disease studies like stroke. Stroke alters and triggers changes in intra- and inter- hemispheric connectivity; this rewiring aims at compensating for the loss of function [120]. fMRI can easily tell the progression of the pathology, showing the plastic remapping over the whole brain [121]. Simultaneously, TPF imaging of fluorescent labeled neurons can highlight the structural and functional rewiring underlying fMRI signals in the newly activated cortical area of the same mouse with cellular detail [58]. Nonlinear imaging on stroke animal models is capable of providing fundamental insight into axonal rewiring and spine plasticity [58, 122, 123] while accurately depicting the functional remapping of the damaged cortical area [124, 125]. Then, alterations in long-range projections underlying inter-hemispheric plasticity can be explored ex vivo on the same cleared brain using LSM. Stroke-induced expression of several molecules and proteins, like growth-associated factors and inflammatory chemokines, can be addressed additionally with multi-round immunohistochemistry. At this point, as discussed above, fine details like the presence of synaptic contacts on regenerated axons are available by electron microscopy on targeted regions of the same sample. This multidimensional hybrid strategy can be extremely useful in the investigation of complex brain diseases, and would speed up the translation of neurobiology studies to clinical settings.

References

[1] Sporns O, Tononi G, Kotter R. The human connectome: A structural description of the human brain. PLoS Comput Biol. 2005;1(4):e42.

[2] Lichtman JW, Denk W. The big and the small: challenges of imaging the brain's circuits. Science. 2011;334(6056):618–623.

[3] Golgi C. Sulla fina anatomia degli organi centrali del sistema nervoso. Reggio Emilia: Calderini; 1885.

[4] Feng G, Mellor RH, Bernstein M, Keller-Peck C, Nguyen QT, Wallace M, et al. Imaging neuronal subsets in transgenic mice expressing multiple spectral variants of GFP. Neuron. 2000;28(1):41–51.

[5] Wang Q, Sporns O, Burkhalter A. Network analysis of corticocortical connections reveals ventral and dorsal processing streams in mouse visual cortex. The Journal of Neuroscience: the official journal of the Society for Neuroscience. 2012;32(13):4386–4399.

[6] Harris JA, Oh SW, Zeng H. Adeno-associated viral vectors for anterograde axonal tracing with fluorescent proteins in nontransgenic and cre driver mice. Current protocols in neuroscience/ editorial board, Jacqueline N Crawley et al. 2012; Chapter 1 (Unit 1 20):1–18.

[7] Chung K, Wallace J, Kim SY, Kalyanasundaram S, Andalman AS, Davidson TJ, et al. Structural and molecular interrogation of intact biological systems. Nature. 2013;497(7449):332–337.

[8] Ragan T, Kadiri LR, Venkataraju KU, Bahlmann K, Sutin J, Taranda J, et al. Serial two-photon tomography for automated ex vivo mouse brain imaging. Nat Methods. 2012;9(3):255–258.

[9] Osten P, Margrie TW. Mapping brain circuitry with a light microscope. Nat Methods. 2013;10(6):515–523.

[10] Oh SW, Harris JA, Ng L, Winslow B, Cain N, Mihalas S, et al. A mesoscale connectome of the mouse brain. Nature. 2014;508(7495):207–214.

[11] Li A, Gong H, Zhang B, Wang Q, Yan C, Wu J, et al. Micro-optical sectioning tomography to obtain a high-resolution atlas of the mouse brain. Science. 2010;330(6009):1404–1408.

[12] Mayerich D, Abbott L, McCormick B. Knife-edge scanning microscopy for imaging and reconstruction of three-dimensional anatomical structures of the mouse brain. J Microsc. 2008;231(Pt1):134–143.

[13] Gong H, Zeng S, Yan C, Lv X, Yang Z, Xu T, et al. Continuously tracing brain-wide long-distance axonal projections in mice at a one-micron voxel resolution. Neuroimage. 2013;74:87–98.

[14] Zheng T, Yang Z, Li A, Lv X, Zhou Z, Wang X, et al. Visualization of brain circuits using two-photon fluorescence micro-optical sectioning tomography. Optics express. 2013;21(8):9839–9850.

[15] Dodt HU, Leischner U, Schierloh A, Jahrling N, Mauch CP, Deininger K, et al. Ultramicroscopy: three-dimensional visualization of neuronal networks in the whole mouse brain. Nat Methods. 2007;4(4):331–336.

[16] Silvestri L, Bria A, Sacconi L, Iannello G, Pavone FS. Confocal light sheet microscopy: micron-scale neuroanatomy of the entire mouse brain. Optics express. 2012;20(18):20582–20598.

[17] Keller PJ, Dodt HU. Light sheet microscopy of living or cleared specimens. Curr Opin Neurobiol. 2012;22(1):138–143.

[18] Huisken J, Swoger J, Del Bene F, Wittbrodt J, Stelzer EH. Optical sectioning deep inside live embryos by selective plane illumination microscopy. Science. 2004;305(5686):1007–1009.

[19] Richardson DS, Lichtman JW. Clarifying Tissue Clearing. Cell. 2015;162(2):246–257.

[20] Silvestri L, Sacconi L, Pavone FS. Correcting spherical aberrations in confocal light sheet microscopy: A theoretical study. Microscopy Research and Technique. 2014;77(7):483–491.

[21] Kalchmair S, Jahrling N, Becker K, Dodt HU. Image contrast enhancement in confocal ultramicroscopy. Optics Letters. 2010;35(1):79–81.

[22] Mertz J, Kim J. Scanning light-sheet microscopy in the whole mouse brain with HiLo background rejection. Journal of Biomedical Optics. 2010;15(1):016027.

[23] Fahrbach FO, Rohrbach A. Propagation stability of self-reconstructing Bessel beams enables contrast-enhanced imaging in thick media. Nature Communications. 2012;3:632.

[24] Baumgart E, Kubitscheck U. Scanned light sheet microscopy with confocal slit detection. Optics Express. 2012;20(19):21805–21814.

[25] Holtmaat A, Svoboda K. Experience-dependent structural synaptic plasticity in the mammalian brain. Nat Rev Neurosci. 2009;10(9):647–658.

[26] Majewska AK, Newton JR, Sur M. Remodeling of synaptic structure in sensory cortical areas in vivo. The Journal of Neuroscience: the official journal of the Society for Neuroscience. 2006;26(11):3021–3029.

[27] Antonini A, Stryker MP. Rapid remodeling of axonal arbors in the visual cortex. Science. 1993;260(5115):1819–1821.

[28] Portera-Cailliau C, Weimer RM, De Paola V, Caroni P, Svoboda K. Diverse modes of axon elaboration in the developing neocortex. PLoS Biol. 2005;3(8):e272.

[29] Ruthazer ES, Akerman CJ, Cline HT. Control of axon branch dynamics by correlated activity in vivo. Science. 2003;301(5629):66–70.

[30] Zuo Y, Yang G, Kwon E, Gan WB. Long-term sensory deprivation prevents dendritic spine loss in primary somatosensory cortex. Nature. 2005;436(7048):261–265.

[31] Trachtenberg JT, Chen BE, Knott GW, Feng G, Sanes JR, Welker E, et al. Long-term in vivo imaging of experience-dependent synaptic plasticity in adult cortex. Nature. 2002;420(6917):788–794.

[32] De Paola V, Holtmaat A, Knott G, Song S, Wilbrecht L, Caroni P, et al. Cell type-specific structural plasticity of axonal branches and boutons in the adult neocortex. Neuron. 2006;49(6):861–875.

[33] Stettler DD, Yamahachi H, Li W, Denk W, Gilbert CD. Axons and synaptic boutons are highly dynamic in adult visual cortex. Neuron. 2006;49(6):877–887.

[34] De Paola V, Arber S, Caroni P. AMPA receptors regulate dynamic equilibrium of presynaptic terminals in mature hippocampal networks. Nature Neuroscience. 2003;6(5):491–500.

[35] Denk W, Strickler JH, Webb WW. Two-photon laser scanning fluorescence microscopy. Science. 1990;248(4951):73–76.

[36] Grinvald A, Hildesheim R. VSDI: a new era in functional imaging of cortical dynamics. Nat Rev Neurosci. 2004;5(11):874–885.

[37] Holtmaat A, Wilbrecht L, Knott GW, Welker E, Svoboda K. Experience-dependent and cell-type-specific spine growth in the neocortex. Nature. 2006;441(7096):979–983.

[38] Gogolla N, Galimberti I, Caroni P. Structural plasticity of axon terminals in the adult. Curr Opin Neurobiol. 2007;17(5):516–524.

[39] Grutzendler J, Kasthuri N, Gan WB. Long-term dendritic spine stability in the adult cortex. Nature. 2002;420(6917):812–816.

[40] Kopel H, Schechtman E, Groysman M, Mizrahi A. Enhanced synaptic integration of adult-born neurons in the olfactory bulb of lactating mothers. The Journal of Neuroscience: the official journal of the Society for Neuroscience. 2012;32(22):7519–7527.

[41] Neuner J, Ovsepian SV, Dorostkar M, Filser S, Gupta A, Michalakis S, et al. Pathological alpha-synuclein impairs adult-born granule cell development and functional integration in the olfactory bulb. Nature Communications. 2014;5:3915.
[42] Mizrahi A, Katz LC. Dendritic stability in the adult olfactory bulb. Nature Neuroscience. 2003;6(11):1201–1207.
[43] Nishiyama H, Fukaya M, Watanabe M, Linden DJ. Axonal motility and its modulation by activity are branch-type specific in the intact adult cerebellum. Neuron. 2007;56(3):472–487.
[44] Nishiyama N, Colonna J, Shen E, Carrillo J, Nishiyama H. Long-term in vivo time-lapse imaging of synapse development and plasticity in the cerebellum. Journal of Neurophysiology. 2014;111(1):208–216.
[45] Holtmaat A, Bonhoeffer T, Chow DK, Chuckowree J, De Paola V, Hofer SB, et al. Long-term, high-resolution imaging in the mouse neocortex through a chronic cranial window. Nature Protocols. 2009;4(8):1128–1144.
[46] Aigner L, Arber S, Kapfhammer JP, Laux T, Schneider C, Botteri F, et al. Overexpression of the neural growth-associated protein GAP-43 induces nerve sprouting in the adult nervous system of transgenic mice. Cell. 1995;83(2):269–278.
[47] Moser MB, Trommald M, Andersen P. An increase in dendritic spine density on hippocampal CA1 pyramidal cells following spatial learning in adult rats suggests the formation of new synapses. Proceedings of the National Academy of Sciences of the United States of America. 1994;91(26):12673–12675.
[48] Knott GW, Quairiaux C, Genoud C, Welker E. Formation of dendritic spines with GABAergic synapses induced by whisker stimulation in adult mice. Neuron. 2002;34(2):265–273.
[49] Yang G, Lai CS, Cichon J, Ma L, Li W, Gan WB. Sleep promotes branch-specific formation of dendritic spines after learning. Science. 2014;344(6188):1173–1178.
[50] Wilt BA, Burns LD, Wei Ho ET, Ghosh KK, Mukamel EA, Schnitzer MJ. Advances in light microscopy for neuroscience. Annual Review of Neuroscience. 2009;32:435–506.
[51] Andermann ML, Gilfoy NB, Goldey GJ, Sachdev RN, Wolfel M, McCormick DA, et al. Chronic cellular imaging of entire cortical columns in awake mice using microprisms. Neuron. 2013;80(4):900–913.
[52] Horton NG, Wang K, Kobat D, Clark CG, Wise FW, Schaffer CB, et al. In vivo three-photon microscopy of subcortical structures within an intact mouse brain. Nature Photonics. 2013;7(3).
[53] Mittmann W, Wallace DJ, Czubayko U, Herb JT, Schaefer AT, Looger LL, et al. Two-photon calcium imaging of evoked activity from L5 somatosensory neurons in vivo. Nature Neuroscience. 2011;14(8):1089–1093.
[54] Helmchen F, Fee MS, Tank DW, Denk W. A miniature head-mounted two-photon microscope. High-resolution brain imaging in freely moving animals. Neuron. 2001;31(6):903–912.
[55] Engelbrecht CJ, Johnston RS, Seibel EJ, Helmchen F. Ultra-compact fiber-optic two-photon microscope for functional fluorescence imaging in vivo. Optics Express. 2008;16(8):5556–5564.
[56] Dombeck DA, Khabbaz AN, Collman F, Adelman TL, Tank DW. Imaging large-scale neural activity with cellular resolution in awake, mobile mice. Neuron. 2007;56(1):43–57.
[57] Schallert T, Leasure JL, Kolb B. Experience-associated structural events, subependymal cellular proliferative activity, and functional recovery after injury to the central nervous system. J Cereb Blood Flow Metab. 2000;20(11):1513–1528.
[58] Brown CE, Li P, Boyd JD, Delaney KR, Murphy TH. Extensive turnover of dendritic spines and vascular remodeling in cortical tissues recovering from stroke. The Journal of Neuroscience: the official journal of the Society for Neuroscience. 2007;27(15):4101–4109.
[59] Dancause N, Barbay S, Frost SB, Plautz EJ, Chen D, Zoubina EV, et al. Extensive cortical rewiring after brain injury. The Journal of Neuroscience: the official journal of the Society for Neuroscience. 2005;25(44):10167–10179.

[60] Holtmaat A, Randall J, Cane M. Optical imaging of structural and functional synaptic plasticity in vivo. European Journal of Pharmacology. 2013;719(1–3):128–136.
[61] Sacconi L, Tolic-Norrelykke IM, Antolini R, Pavone FS. Combined intracellular three-dimensional imaging and selective nanosurgery by a nonlinear microscope. Journal of Biomedical Optics. 2005;10(1):14002.
[62] Tolic-Norrelykke IM, Sacconi L, Thon G, Pavone FS. Positioning and elongation of the fission yeast spindle by microtubule-based pushing. Current Biology: CB. 2004;14(13):1181–1186.
[63] Kumar S, Maxwell IZ, Heisterkamp A, Polte TR, Lele TP, Salanga M, et al. Viscoelastic retraction of single living stress fibers and its impact on cell shape, cytoskeletal organization, and extracellular matrix mechanics. Biophysical Journal. 2006;90(10):3762–3773.
[64] Canty AJ, Huang L, Jackson JS, Little GE, Knott G, Maco B, et al. In-vivo single neuron axotomy triggers axon regeneration to restore synaptic density in specific cortical circuits. Nature Communications. 2013;4:2038.
[65] Sacconi L, O'Connor RP, Jasaitis A, Masi A, Buffelli M, Pavone FS. In vivo multiphoton nanosurgery on cortical neurons. Journal of Biomedical Optics. 2007;12(5):050502.
[66] Allegra Mascaro AL, Cesare P, Sacconi L, Grasselli G, Mandolesi G, Maco B, et al. In vivo single branch axotomy induces GAP-43-dependent sprouting and synaptic remodeling in cerebellar cortex. Proceedings of the National Academy of Sciences of the United States of America. 2013;110(26):10824–10829.
[67] Allegra Mascaro AL, Sacconi L, Pavone FS. Multi-photon nanosurgery in live brain. Frontiers in Neuroenergetics. 2010;2.
[68] Tsai PS, Blinder P, Migliori BJ, Neev J, Jin Y, Squier JA, et al. Plasma-mediated ablation: an optical tool for submicrometer surgery on neuronal and vascular systems. Curr Opin Biotechnol. 2009;20(1):90–99.
[69] Davalos D, Grutzendler J, Yang G, Kim JV, Zuo Y, Jung S, et al. ATP mediates rapid microglial response to local brain injury in vivo. Nature Neuroscience. 2005;8(6):752–758.
[70] Nimmerjahn A, Kirchhoff F, Helmchen F. Resting microglial cells are highly dynamic surveillants of brain parenchyma in vivo. Science. 2005;308(5726):1314–1318.
[71] Nishimura N, Schaffer CB, Friedman B, Tsai PS, Lyden PD, Kleinfeld D. Targeted insult to subsurface cortical blood vessels using ultrashort laser pulses: three models of stroke. Nat Methods. 2006;3(2), 99–108.
[72] Shih AY, Blinder P, Tsai PS, Friedman B, Stanley G, Lyden PD, et al. The smallest stroke: occlusion of one penetrating vessel leads to infarction and a cognitive deficit. Nature Neuroscience. 2013;16(1):55–63.
[73] Svoboda K, Yasuda R. Principles of two-photon excitation microscopy and its applications to neuroscience. Neuron. 2006;50(6):823–839.
[74] Peterka DS, Takahashi H, Yuste R. Imaging voltage in neurons. Neuron. 2011;69(1):9–21.
[75] Grienberger C, Konnerth A. Imaging calcium in neurons. Neuron. 2012;73(5):862–885.
[76] Gelperin A, Flores J. Vital staining from dye-coated microprobes identifies new olfactory interneurons for optical and electrical recording. Journal of Neuroscience Methods. 1997;72(1):97–108.
[77] Nagayama S, Zeng S, Xiong W, Fletcher ML, Masurkar AV, Davis DJ, et al. In vivo simultaneous tracing and Ca(2+) imaging of local neuronal circuits. Neuron. 2007;53(6):789–803.
[78] Dombeck DA, Sacconi L, Blanchard-Desce M, Webb WW. Optical recording of fast neuronal membrane potential transients in acute mammalian brain slices by second-harmonic generation microscopy. Journal of Neurophysiology. 2005;94(5):3628–3636.
[79] Sacconi L, Mapelli J, Gandolfi D, Lotti J, O'Connor RP, D'Angelo E, et al. Optical recording of electrical activity in intact neuronal networks with random access second-harmonic generation microscopy. Optics Express. 2008;16(19):14910–14921.

[80] Hires SA, Tian L, Looger LL. Reporting neural activity with genetically encoded calcium indicators. Brain Cell Biology. 2008;36(1–4):69–86.
[81] Looger LL, Griesbeck O. Genetically encoded neural activity indicators. Curr Opin Neurobiol. 2012;22(1):18–23.
[82] Tian L, Akerboom J, Schreiter ER, Looger LL. Neural activity imaging with genetically encoded calcium indicators. Progress in Brain Research. 2012;196:79–94.
[83] Akemann W, Mutoh H, Perron A, Park YK, Iwamoto Y, Knopfel T. Imaging neural circuit dynamics with a voltage-sensitive fluorescent protein. Journal of Neurophysiology. 2012;108(8):2323–2337.
[84] Knopfel T. Genetically encoded optical indicators for the analysis of neuronal circuits. Nat Rev Neurosci. 2012;13(10):687–700.
[85] Cetin A, Komai S, Eliava M, Seeburg PH, Osten P. Stereotaxic gene delivery in the rodent brain. Nature Protocols. 2006;1(6):3166–3173.
[86] Yan P, Acker CD, Zhou WL, Lee P, Bollensdorff C, Negrean A, et al. Palette of fluorinated voltage-sensitive hemicyanine dyes. Proceedings of the National Academy of Sciences of the United States of America. 2012;109(50):20443–20448.
[87] Sacconi L, Ferrantini C, Lotti J, Coppini R, Yan P, Loew LM, et al. Action potential propagation in transverse-axial tubular system is impaired in heart failure. Proceedings of the National Academy of Sciences of the United States of America. 2012;109(15):5815–5819.
[88] Dombeck DA, Harvey CD, Tian L, Looger LL, Tank DW. Functional imaging of hippocampal place cells at cellular resolution during virtual navigation. Nature Neuroscience. 2010;13(11):1433–1440.
[89] Ozden I, Dombeck DA, Hoogland TM, Tank DW, Wang SS. Widespread state-dependent shifts in cerebellar activity in locomoting mice. PloS one. 2012;7(8):e42650.
[90] Fino E, Araya R, Peterka DS, Salierno M, Etchenique R, Yuste R. RuBi-Glutamate: Two-photon and visible-light photoactivation of neurons and dendritic spines. Frontiers in Neural Circuits. 2009;3:2.
[91] Nikolenko V, Poskanzer KE, Yuste R. Two-photon photostimulation and imaging of neural circuits. Nat Methods. 2007;4(11):943–950.
[92] Fenno L, Yizhar O, Deisseroth K. The development and application of optogenetics. Annual Review of Neuroscience. 2011;34:389–412.
[93] Zhang F, Aravanis AM, Adamantidis A, de Lecea L, Deisseroth K. Circuit-breakers: optical technologies for probing neural signals and systems. Nat Rev Neurosci. 2007;8(8):577–581.
[94] Petreanu L, Huber D, Sobczyk A, Svoboda K. Channelrhodopsin-2-assisted circuit mapping of long-range callosal projections. Nature Neuroscience. 2007;10(5):663–668.
[95] Liu X, Ramirez S, Pang PT, Puryear CB, Govindarajan A, Deisseroth K, et al. Optogenetic stimulation of a hippocampal engram activates fear memory recall. Nature. 2012;484(7394):381–385.
[96] Kravitz AV, Freeze BS, Parker PR, Kay K, Thwin MT, Deisseroth K, et al. Regulation of parkinsonian motor behaviours by optogenetic control of basal ganglia circuitry. Nature. 2010;466(7306):622–626.
[97] Aravanis AM, Wang LP, Zhang F, Meltzer LA, Mogri MZ, Schneider MB, et al. An optical neural interface: in vivo control of rodent motor cortex with integrated fiberoptic and optogenetic technology. Journal of Neural Engineering. 2007;4(3):S143–S156.
[98] Papagiakoumou E, Anselmi F, Begue A, de Sars V, Gluckstad J, Isacoff EY, et al. Scanless two-photon excitation of channelrhodopsin-2. Nat Methods. 2010;7(10):848–854.
[99] Prakash R, Yizhar O, Grewe B, Ramakrishnan C, Wang N, Goshen I, et al. Two-photon optogenetic toolbox for fast inhibition, excitation and bistable modulation. Nat Methods. 2012;9(12):1171–1179.

[100] Emiliani V, Cohen AE, Deisseroth K, Hausser M. All-Optical Interrogation of Neural Circuits. The Journal of Neuroscience: the official journal of the Society for Neuroscience. 2015;35(41): 13917–13926.
[101] Stephan KE, Weiskopf N, Drysdale PM, Robinson PA, Friston KJ. Comparing hemodynamic models with DCM. Neuroimage. 2007;38(3):387–401.
[102] Margolis DJ, Lutcke H, Schulz K, Haiss F, Weber B, Kugler S, et al. Reorganization of cortical population activity imaged throughout long-term sensory deprivation. Nature Neuroscience. 2012;15(11):1539–1546.
[103] Schulz K, Sydekum E, Krueppel R, Engelbrecht CJ, Schlegel F, Schroter A, et al. Simultaneous BOLD fMRI and fiber-optic calcium recording in rat neocortex. Nat Methods. 2012;9(6):597–602.
[104] Stettler D, Yamahachi H, Li W, Denk W, Gilbert C. Axons and synaptic boutons are highly dynamic in adult visual cortex. Neuron. 2006;49(6):877–887.
[105] Barlow HB, Hill RM, Levick WR. Retinal Ganglion Cells Responding Selectively to Direction and Speed of Image Motion in the Rabbit. The Journal of Physiology. 1964;173:377–407.
[106] Lichtman JW, Livet J, Sanes JR. A technicolour approach to the connectome. Nat Rev Neurosci. 2008;9(6):417–422.
[107] Kasthuri N, Hayworth KJ, Berger DR, Schalek RL, Conchello JA, Knowles-Barley S, et al. Saturated reconstruction of a volume of neocortex. Cell. 2015;162(3):648–661.
[108] Denk W, Horstmann H. Serial block-face scanning electron microscopy to reconstruct three-dimensional tissue nanostructure. PLoS Biol. 2004;2(11):e329.
[109] Schalek R, Kasthuri N, Hayworth K, Berger D, Tapia J, Morgan J, et al. Development of high-throughput, high-resolution 3D reconstruction of large-volume biological tissue using automated tape collection ultramicrotomy and scanning electron microscopy. Microscopy and Microanalysis. 2011;17(Suppl2):966–967.
[110] Knott G, Marchman H, Wall D, Lich B. Serial section scanning electron microscopy of adult brain tissue using focused ion beam milling. The Journal of Neuroscience: the official journal of the Society for Neuroscience. 2008;28(12):2959–2964.
[111] Betzig E, Patterson GH, Sougrat R, Lindwasser OW, Olenych S, Bonifacino JS, et al. Imaging intracellular fluorescent proteins at nanometer resolution. Science. 2006;313(5793): 1642–1645.
[112] Hell SW, Wichmann J. Breaking the diffraction resolution limit by stimulated emission: stimulated-emission-depletion fluorescence microscopy. Optics Letters. 1994;19(11): 780–782.
[113] Logothetis NK. What we can do and what we cannot do with fMRI. Nature. 2008;453(7197): 869–878.
[114] Heeger DJ, Ress D. What does fMRI tell us about neuronal activity? Nat Rev Neurosci. 2002; 3(2):142–151.
[115] Silvestri L, Allegra Mascaro AL, Costantini I, Sacconi L, Pavone FS. Correlative two-photon and light sheet microscopy. Methods. 2014;66(2):268–272.
[116] Briggman KL, Helmstaedter M, Denk W. Wiring specificity in the direction-selectivity circuit of the retina. Nature. 2011;471(7337):183–188.
[117] Cane M, Maco B, Knott G, Holtmaat A. The relationship between PSD-95 clustering and spine stability in vivo. The Journal of Neuroscience. 2014;34(6):2075–2086.
[118] Mostany R, Anstey JE, Crump KL, Maco B, Knott G, Portera-Cailliau C. Altered Synaptic dynamics during normal brain aging. The Journal of Neuroscience. 2013;33(9):4094–4104.
[119] Grillo FW, Song S, Teles-Grilo Ruivo LM, Huang L, Gao G, Knott GW, et al. Increased axonal bouton dynamics in the aging mouse cortex. Proceedings of the National Academy of Sciences. 2013;110(16):E1514–E1523.

[120] Murphy TH, Corbett D. Plasticity during stroke recovery: from synapse to behaviour. Nat Rev Neurosci. 2009;10(12):861–872.
[121] Ward NS, Brown MM, Thompson AJ, Frackowiak RS. Neural correlates of motor recovery after stroke: a longitudinal fMRI study. Brain. 2003;126(Pt11):2476–2496.
[122] Mostany R, Chowdhury TG, Johnston DG, Portonovo SA, Carmichael ST, Portera-Cailliau C. Local hemodynamics dictate long-term dendritic plasticity in peri-infarct cortex. The Journal of Neuroscience: the official journal of the Society for Neuroscience. 2010;30(42): 14116–14126.
[123] Brown CE, Murphy TH. Livin' on the edge: imaging dendritic spine turnover in the peri-infarct zone during ischemic stroke and recovery. The Neuroscientist. 2008;14(2):139–146.
[124] Sigler A, Mohajerani MH, Murphy TH. Imaging rapid redistribution of sensory-evoked depolarization through existing cortical pathways after targeted stroke in mice. Proceedings of the National Academy of Sciences of the United States of America. 2009;106(28):11759–11764.
[125] Harrison TC, Silasi G, Boyd JD, Murphy TH. Displacement of sensory maps and disorganization of motor cortex after targeted stroke in mice. Stroke. 2013;44(8):2300–2306.
[126] Charpak S, Stefanovic B. Shedding light on the BOLD fMRI response. Nat Methods. 2012;9(6): 547–549.

Amy Holmes, Camilla Thorling, Xin Liu, Xiaowen Liang, Haolu Wang, Hans G. Breunig, Karsten König, Hauke Studier, and Michael S. Roberts

18 Revealing interaction of dyes and nanomaterials by multiphoton imaging

Abstract: A key challenge in pharmacology and toxicology is understanding the exposure, fate and effects of drugs and nanomaterials on the body. Nanoparticles are widely used in topically applied consumer products. For example, titanium dioxide and zinc oxide are used in commercial sunscreen formulations to afford the skin protection from harmful ultraviolet radiation. There are still ongoing safety concerns regarding the long-term toxicity and fate of nanomaterials within complex biological systems with a key question being whether topically applied nanoparticles can penetrate human skin and reach the viable epidermis to cause localized and potentially systemic toxicity under 'in-use' conditions. Recent advances in imaging technology have provided us with the tools to map the distribution of native, unlabeled, nanomaterials after application to biological tissue. Another key organ for *in situ* monitoring is the liver as it is the major site for drug metabolism and excretion in the body. Clearance of drugs from the body can be greatly altered in liver disease, resulting in toxicity and altered effects of the drugs. It is therefore crucial to understand liver functional changes in diseased livers. Recent advances in imaging technology have provided us with insight into mapping the distribution of native, unlabeled, nanomaterials after application to a range of organs within the body. This chapter deals with the use of multiphoton fluorescence microscopy in combination with fluorescence lifetime imaging techniques to assess the safety concerns regarding nanomaterials discussing the key challenges faced.

18.1 Introduction

Nanomaterials defined as particles with dimensions in the range of 1–100 nm exhibit high reactivity due to their high volume-to-surface ratio that bestows on them unique characteristics, such as the endless potential for tailoring of their surface chemistry. While there are a vast number of potential benefits and applications for the use of nanomaterials compared to conventional macromaterials, there have also been numerous reports of nanoparticle toxicity with still little known about their long-term toxicity in biological systems and the environment [1–3]. At present, nanoparticles are contained in a variety of personal care products, e.g., cosmetics, sunscreens, antiseptics and wound dressings and the number of nanocontaining products on the market is rising with global markets expected to exceed $1 trillion in investment within the next

https://doi.org/10.1515/9783110429985-020

couple of years [4–6]. The use of these products present a potential risk of systemic toxicity due to dermal exposure as an occupational hazard with regard to exposure during the manufacturing process but also many of the nanomaterial-containing products are designed as topically applied formulations. Thus, the skin can be exposed to various nanomaterials by either their intentional liberal application to the skin or through unintentional contact in the environment or manufacturing process. Fortunately, the outer layer of the human skin, the stratum corneum, is a very effective barrier against the ingress of nanoparticles even though it is typically only 10–25 µm in thickness [7].

One example of the intentional application of nanoparticles to human skin is the use of sunscreen in which zinc oxide nanoparticles (ZnO NPs) are incorporated to mitigate sunburn and reduce the occurrence of skin cancer by blocking harmful UV radiation from reaching the viable tissue. The potential for the percutaneous absorption of ZnO NPs after liberal and repeated application to human skin may constitute a health hazard, in particular, if the nanoparticles are able to pass through the outer layer of skin, the stratum corneum due to impairment of this barrier or under 'in-use' conditions [8–12]. Thus, with the increasing rate of production and use of nanoparticle-containing products, biosafety is an increasingly important issue requiring investigation. It is of paramount importance to map the distribution of these native nanomaterials within tissues *in situ* and to quantitatively assess any penetration through the skin that may have occurred. In earlier work, inductively-coupled-plasma mass spectrometry (ICP-MS) was employed to determine the concentration of zinc from nanoparticles that penetrated human skin [4]. The limitation of this type of study is that this technique measures the total metal penetration (i.e., zinc ions as well as ZnO NPs) in tissue rather than the actual nanoparticle penetration. Another technique traditionally used for this type of mapping is conventional electron microscopy, however the techniques associated with this form of microscopy are laborious, qualitative, sample destructive and invasive, thus not appropriate for *in situ* tissue imaging. Recently, a different imaging approach was found to be very useful for nanoparticle penetration studies: multiphoton microscopy (MPM) in combination with fluorescence lifetime imaging (FLIM) [13, 14]. This combined technique provides unique non-invasive access to quantitative investigations of nanoparticle penetration and deposition within human skin. With the launch of the DermaInspect (JenLab GmbH, Jena, Germany) *in vivo* studies observing nanoparticle penetration and deposition on human skin became possible [15–17]. This device (or its latest version, the MPT*flex* [18]) can be additionally equipped with a time-correlated single-photon counting (TCSPC) FLIM module. FLIM can be used to differentiate endogenous fluorophores (e.g., nicotinamide adenine dinucleotide (phosphate) (NAD(P)H), collagen, and vitamin A) in tissues based on their unique lifetimes. This can then be exploited to investigate whether applied nanomaterials affect the metabolic state of the tissue (discussed further within this chapter). FLIM can be used to reveal the different binding states of fluorophores, reflect changes in their microenvironment and differentiate the lifetime of exogenously applied nanomaterials from endogenous fluorophores.

As well as being applied to skin, the technique MPM-FLIM with its resultant advantages has also been applied to other organs. Liver is the major site for drug metabolism and excretion in the body. Liver diseases, particularly viral hepatitis, cirrhosis and hepatocellular carcinoma, are frequently encountered in clinical practice with high morbidity and mortality worldwide [19]. Goetz *et al.* have recently reported the application of confocal laser endomicroscopy for the real-time histological examination of liver disease in both animal models and in human volunteers [20]. This technology allows immediate *in vivo* subsurface microscopic imaging and has the potential to dynamically monitor pathologic events of the liver with high resolution. Compared to confocal microscopy, MPM has the advantages of decreased photobleaching and photodamage, and has considerably enhanced imaging penetration depth due to a reduction in scattered multiphoton excitation in samples. MPM-FLIM can be used to detect changes in liver microenvironment, such as pH, protein binding and injury induced fibrosis, which are usually associated with disease but that cannot be revealed by microscopy using fluorescence intensity alone [21]. This chapter explores the use of MPM-FLIM and various other complementary imaging techniques to delineate the biodistribution of nanomaterials possessing unique optical phenomena and various exogenous fluorophores within tissue.

18.2 Multiphoton imaging of nanomaterials within tissue

Multiphoton microscopy has a wide range of applications in biological and biomedical research and is suitable for *in vivo* applications [17, 22]. The technique provides high-resolution 3D-imaging based on the spatially resolved excitation and detection of fluorescence and harmonic signals [23]. In particular the detection of fluorophores with one-photon cross sections in the UV/blue-visible spectral range is possible because these fluorophores can be excited by two-photon absorption of NIR light. The use of NIR light provides numerous advantages including low one-photon absorption and scattering rates inside tissue which allows deeper imaging within tissue ("optical window of tissue in the NIR") and the availability of easy-to-use turnkey-laser systems. Two-photon absorption occurs with sufficient probability only at high enough intensities and such intensities can be exclusively realized in tiny focal volumes by highly focusing the ultrashort pulses. Outside the focal region, the intensity is then too low for two-photon processes to occur. This nonlinear intensity dependence leads to intrinsic 3D-sectioning capability of multiphoton imaging even without a pinhole, as used for confocal imaging. Multiphoton imaging is typically realized by spatially resolved excitation and signal detection in a common laser-scanning setup. For our imaging we currently employ the CE-certified multiphoton tomograph DermaInspect (JenLab GmbH), a mobile imaging system which incorporates a tunable fs laser (Mai Tai, Spectra Physics; tuning range 710–920 nm, pulse duration 80 fs, repetition rate 80 MHz) and a laser-scanning system for the simultaneous pixelwise detection of signals in two (or more) spectral channels. The system generates horizontal en-face images, parallel to the surface with lateral and axial resolutions of about 0.36 µm and 1.7 µm, respectively [18]. The laser pulses are focused onto the skin by a 1.3 NA focusing optics. The imaging procedure requires the use of a metal ring and an adhesive tape to make a (temporal) "rigid" contact between the focusing optics and the skin such that motion artifacts are minimized during the *in vivo* measurements [24]. The elimination of artifacts whilst conducting *in vivo* imaging experiments using live systems presents a significant challenge and can never be totally avoided due to respiration. However, to keep these artifacts to a minimum the measurements should be recorded quickly and within a reduced timeframe. For FLIM this poses a further problem where ideally you would want to have as many photons as possible to attain good photon statistics allowing for ideal analysis (retrospective decay curve fitting procedure), a short image recording time will be a drawback due to reduced photon counts. In this scenario it is important to find an ideal compromise, e.g., laser scanning speed (number of frames per time), and overall experimental time.

A strength of multiphoton imaging is that skin and other organ imaging can be performed without the use of external fluorescent markers, instead relying completely on endogenous fluorophores such as NAD(P)H, flavins, melanin, elastin and several other biomolecules [17, 24]. In addition, a second-harmonic generation (SHG) signal can arise from the interaction between light and matter with a highly-polarizable non-centrosymmetric structure. In tissue, collagen fibers that form the dermal collagen network typically generate an SHG signal. The skin is not a homogenous membrane, instead being strata of various distinct histological layers including the epidermis, dermis, and the subcutaneous tissue. The outer layer of the epidermis, the stratum corneum, is composed of flattened and dead corneocytes that act as the principle barrier to the ingress of exogenous materials. The innermost layer of the epidermis consists of a single layer of cuboidal cells, basal keratinocytes, that undergo a process of terminal differentiation where they differentiate and migrate toward the skin surface forming (from the outer to the innermost layer) the stratum corneum, the stratum granulosum, the stratum spinosum and the stratum basale. Only the dermis below the epidermis is composed of collagen and elastin fiber networks, hence the upper layers of the skin do not give rise to an SHG signal.

In general, it is difficult to resolve the distribution of nanoparticles in skin by (one-photon) fluorescence microscopy because the photoluminescence of nanoparticles often spectrally overlaps with the autofluorescence of the metabolically active epidermal skin layers and also because of the high turbidity of both the nanoparticles and the skin. However, some nanoparticles, e.g., ZnO NPs incorporated into commercial sunscreen formulations, show strong SHG signals when excited by multiphoton. Thus using MPM, by exploiting SHG signals from appropriate nanoparticles (e.g. ZnO, Ag, Au) a spectral separation of SHG signals and fluorescence from endogenous fluorophores is possible by collecting the distinct signals in two separate spectral ranges comprised of SHG signal and autofluorescence in each channel. The SHG channel should be very narrow banded and have maximum transmission at exactly half the near-infrared (NIR) excitation wavelength. For the autofluorescence channel, typically a broader emission range is selected that is red shifted to the SHG band and covers the autofluorescence of the desired skin or other organ biomolecules.

Multiphoton imaging is often employed as a multimodal technique which relies, in addition to fluorescence and SHG signals, on the fluorescence lifetime [25, 26], spectrally-resolved signal detection [27], and Raman imaging techniques [28, 29]. Raman techniques can detect nonfluorescent and non-SHG generating material. Time-correlated single-photon counting (TCSPC) FLIM provides information on the localization of a specific fluorophore or variations in the microenvironment. FLIM can also be used to enhance the contrast of the SHG signal of a nanoparticle to the autofluorescence signal of the tissue in addition to the spectral discrimination. This can be done by selecting the 'early' time channels (e.g. the first 100 ps) that correspond to the SHG signal and separate the photons that were recorded in 'later' time channels (> 100 ps) that correspond to autofluorescence. This method is particularly advantageous when imaging nanoparticles that exhibit SHG and a signal based on localized surface plasmon resonance (LSPR). This can be observed on many noble metal nanoparticles, e.g., Ag and Au. The LSPR signal can be very broad band and also very strong so that it is overlaid with the autofluorescence signal of the surrounding tissue. In this case, TCSPC-FLIM alone enables the signal separation without spectral discrimination [30].

Section 18.3 will focus on using these outlined imaging techniques to assess the penetration of nanoparticles into human skin in more detail, and the sections thereafter will discuss the advantages of using exogenous dyes and nanoparticles within the liver.

18.3 Monitoring zinc oxide nanoparticle (ZnO NPs) penetration into human skin

ZnO nanoparticles (ZnO NPs) and micronized ZnO microparticles (ZnO MPs) are used in a wide range of personal care products including cosmetics, diaper rash ointments and commercial sunscreen formulations to afford the skin protection from harmful ultraviolet (UV) radiation. A key question put to regulatory bodies and scientists [5, 31] by nongovernmental organizations such as Friends of the Earth is whether ZnO NPs/MPs can penetrate human skin, particularly after 'in-use' conditions [32]. To investigate the potential penetration and deposition of ZnO NPs/MPs after topical application of the native and unlabeled particles the optical phenomena of ZnO NPs were exploited and developed.

Initially our group used *in vitro* skin and various techniques to assess nanoparticle penetration, including transmission electron microscopy (TEM). Our first study concerned interaction of ZnO particles with human epidermis where it was found that

ZnO NPs were associated with the superficial layers of the stratum corneum [4]. In 2008, our group employed scanning electron microscopy (SEM) and energy dispersive X-ray (EDX) in conjunction with multiphoton microscopy that mapped the deposition of ZnO NPs within the furrows after application to human skin both *in vitro* and *in vivo* [14]. This study spectrally resolved using a narrow band pass filter (380 ± 20 nm) to obtain the principle luminescence signal of ZnO NP at 385 nm after multiphoton excitation of the tissue at 740 nm. It was found that the narrow band pass filter used to detect the ZnO NP photoluminescence spectrally discriminated against the skin autofluorescence such as NAD(P)H, flavin adenine dinucleotide (FAD), and porphyrins with emission maxima of 450, 520 and 625 nm respectively that exist as endogenous fluorophores within human skin [33]. Luminescence characteristics of numerous endogenous and exogenous components within or applied to the skin are displayed in Tab. 18.1. ZnO is a wide bandgap (≈ 3.3 eV) semiconductor that displays luminescent properties in the near UV range that are well known and also in visible regions which is less well understood [34, 35]. The nature of this UV emission is known to be a result of excitonic emission due to the recombination of electron hole pairs [36], the visible emissions between 500–600 nm are potentially due to numerous interstitial defects as discussed in detail elsewhere [37, 38]. Interestingly, the ZnO NP emission in the visible range has been shown to be dependent on the shape and surface chemistry of the ZnO NPs [39] though the emission in the visible range is of limited value in biomedical imaging due to the presence of fluorescence from endogenous fluorophores within the green-orange spectral bands.

Tab. 18.1: Endogenous and exogenous components imaged in skin.

Component	Ex (nm)	Em (nm)	Lifetime (ns)
NAD(P)H free	720–760	460	0.3
NAD(P)H-protein	720–760	440	2.0–2.3
FAD	860–920	530	5.2
Keratin	720–800	420–580	0.4/2.3
Elastin	> 860–880	420–460	0.2/2.5
Collagen	> 860–880	420–460 (λ/2, SHG)	0.3/2.0, (0)
Melanin	720–760 (< 920)	440, 520, 575	0.1/1.9/8
ZnO NP	710–840	370–600 (λ/2, SHG)	> 21 (0)
Ag NP	760–840	450–690 (LSPR), (λ/2, SHG/HRS)	< 0.001 (0)

Ex: two-photon excitation wavelength (range of high absorption); Em: emission wavelength (center)

Further, in 2008 our group also used the photoluminescence lifetime of ZnO NPs to detect the NPs after topical application *in vivo* [40]. As well as spectrally resolving the ZnO NP luminescence away from endogenous fluorophores it was determined using TCSPC-FLIM that the different lifetimes of the ZnO NPs and the autofluorescence within the 350–450 nm spectral band can be used to further resolve the ZnO NP signal. This important combination of selective filters and short characteristic lifetimes of ZnO NPs compared to the autofluorescence of the tissue was used to illustrate no penetration of ZnO NPs after topical application to a large number of volunteers ($n = 60$) [40]. In 2009, Kuo *et al.* illustrated using mouse skin that the second-harmonic generation (SHG) signal could be used to distinguish ZnO NPs from the autofluorescence of skin and the authors showed an increase in deposition within murine skin after co-application of ZnO NPs with chemical penetration enhancers [41]. The fluorescent lifetime of ZnO NPs as a result of SHG signal is known to be instantaneous and the emission wavelength is exactly half of the two-photon excitation wavelength [42, 43]. The SHG signal that arises from nanocrystalline ZnO is discussed in detail in a recent review by Lariprete and Centini [44].

In an attempt to quantify the ZnO NPs within the superficial layers of the stratum corneum and deep within the furrows of the skin, a further study was conducted by our group using nonlinear optical microscopy which was calibrated using increasing concentrations of immobilized ZnO NPs within a 90 nm poly-vinyl-alcohol layer [45]. After topical application of ZnO NPs in caprylic capric triglycerides (CCT) the luminescence (spectral emission 390 ± 40 nm) of the ZnO NPs after excitation at 770 nm was determined and the concentration was elucidated from the calibration curve. CCT is a refined coconut oil that is commonly used as a base within topically applied consumer products including sunscreen formulations. It was confirmed that the ZnO NP deposition was not homogenous across the skin surface but that the NPs aggregate within the skin furrows, these aggregates were estimated to contain ≈ 800 ZnO particles per μm^3 but when averaged across the length of the human skin cross section there were between 3–7 ZnO particles per $\mu m^3/mm$ depending on the depth [45]. Also in 2011, our group published on the simultaneous detection of ZnO NPs with metabolic state imaging on human skin *in vivo* using TCSPC-FLIM [46]. Within this study the luminescence of ZnO NPs after excitation at 740 nm and emission detected between 350–450 nm was found to possess a much shorter fluorescent lifetime value than NAD(P)H or keratin and thus the fluorescence lifetime amplitude $\alpha_1\%$ of ZnO NP luminescence was distinct from that of autofluorescence. The typical $\alpha_1\%$ of endogenous keratin/NAD(P)H within human skin was found to be between 45–85 % whereas $\alpha_1\%$ above 90 % was indicative of ZnO NP luminescence. The ZnO NPs in CCT were topically applied *in vivo* to both intact and impaired barrier skin (tape stripped and psoriasis/atopic dermatitis lesions) and the ZnO NP distribution was quantified within the skin using an *in vitro* standard calibration curve. It was found that though the ZnO NP signal increased within the stratum corneum, particularly in the furrows for barrier-impaired skin compared to intact skin, no ZnO NP signal was detected within the viable epidermis *in vivo*. In terms of free NAD(P)H signal, no change was observed when comparing barrier-impaired lesional skin even with an increase in ZnO NP signal within the stratum corneum but a significant change was observed within the viable epidermis after 4 h application of ZnO NPs to tape stripped skin compared to its vehicle control [46].

The benefit of non-invasive imaging *in vivo* detecting the presence of both ZnO NPs and the metabolic state of the skin was then further exploited to incorporate important variables such as application of ZnO NPs in different formulations [47]. Within this study it was found that coated or uncoated ZnO NPs had no effect on hydration of the human skin nor had any effect on the redox state of the viable epidermis. It was elucidated that when coated ZnO NPs were applied to the skin in a water-in-oil emulsion, the ZnO NPs had penetrated deeper into the skin strata [47]. Further, other 'in-use' conditions have been investigated more recently *in vivo* including the effect of occlusion, i.e., sunscreen application under clothing [48] and the effects of massage and flexing of sunscreen application [49] on the penetration of coated and uncoated ZnO NPs. The use of non-invasive MPM-FLIM has allowed us to delineate the deposition and penetration of a commercially used NP on human skin *in vivo* where we have shown repeatedly that the NPs do not penetrate to the viable epidermis nor cause toxicity within the viable epidermis. A study in 2012 by Darvin *et al.* also showed no penetration of ZnO NPs within the epidermis after the application of three sunscreen formulations, containing three different concentrations of ZnO NPs, to the forearms of volunteers. They reported a limit of detection of 0.08 $fg/\mu m^3$ using the SHG signal and hyper-Rayleigh scattering signal, and the authors also showed that the ZnO NPs were retained within the furrows and hair follicle orifices of the skin [42]. In another recent study in 2014, a commercial sunscreen, labeled as an "anti-aging sun-care product" with a sun protection factor (SPF) of 50, containing ZnO (but no TiO_2), was applied to *in vivo* human skin in order to investigate penetration of the ZnO NPs [50]. Unfortunately, the manufacturer did not provide the exact size and portion of the ZnO NPs, but common grain sizes used in cosmetic sunscreen products are typically between 12 and 80 nm which may form aggregations of sizes of $\approx$ 200 nm [12, 45, 51]. Fig. 18.1 shows a pseudo-colored composite multiphoton image of the sunscreen formulation *in vitro*. Regions showing nanoparticles of different sizes (NPs, NP aggregates and MPs) as well as other fluorescent components of the sunscreen can be seen. The signal arising from the ZnO NPs can be observed (red signal in Fig. 18.1) and most likely corresponds to agglomerated and aggregated particles. To obtain the image shown in Fig. 18.1, two detection channels were employed with ranges between 370 and 410 nm ("channel 1") and 410 and 565 nm ("channel 2"), respectively. The center wavelength of the fs pulses was set to 780 nm. The images were recorded in 7 s with 512×512 pixels and a mean laser power of $\approx$ 14 mW applied to the skin. The spectra were recorded with a fiber-coupled thermoelectrically cooled CCD-array spectrometer (BTC112E, B & W TEK) which provided a wavelength-dependent resolution of a few nm.

Fig. 18.2 shows emission spectra of a sunscreen sample (top) for different laser wavelengths between 750 nm and 815 nm. The ranges of the spectral channels used for SHG (channel 1) and fluorescence detection (channel 2), respectively, are indicated. The spectra of the sunscreen are dominated by sharp peaks at half the laser wavelength, i.e., between 375.0 nm and 407.5 nm, and can be attributed to SHG light generated by the laser-light interaction with the ZnO NPs within the sunscreen. Fluores-

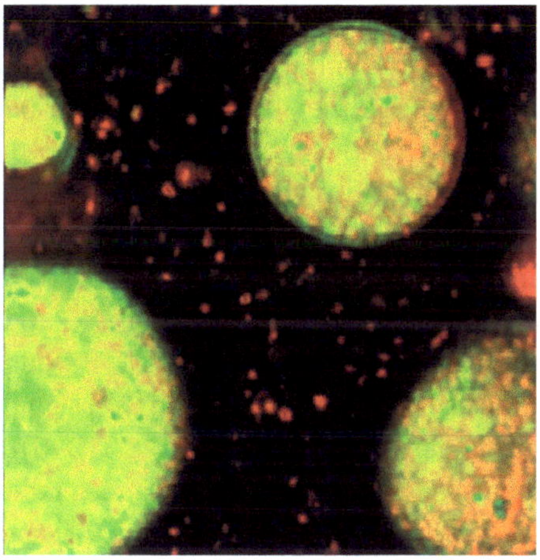

Fig. 18.1: Composite multiphoton image of a commercial sunscreen containing ZnO nanoparticles with two detection channels for differentiating between the emitted second-harmonic generation signal and the fluorescence. The SHG light from the sub-resolution nanoparticles which form agglomerates is pseudo-color coded in red. Other components of the sunscreen which fluoresce appear in yellow. The imaged area has a size of 41 μm × 41 μm.

cence arising from the sunscreen formulation is present in the spectral range of channel 2 (420–650 nm) and the intensity that increases with decreasing laser wavelength does not disturb the detection of ZnO NPs. The spectra of the skin differs depending on the skin strata analyzed, the stratum corneum emits significant fluorescence > 400 nm whereas the collagen-rich epidermal-dermal junction produces a strong SHG signal < 400 nm (Fig. 18.2, bottom). The spectra illustrate that, with the exception of the deeper epidermal-dermal junction, no SHG is present within the viable epidermis, thus any detected SHG signal stems exclusively from the ZnO NPs.

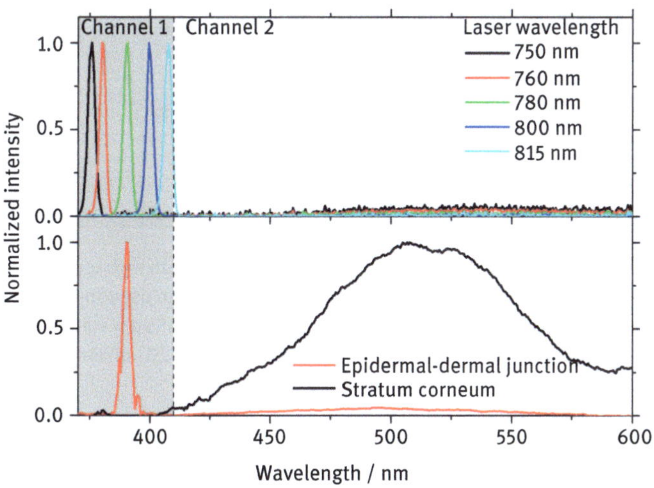

Fig. 18.2: Signal spectra of sunscreen *in vitro* (top) and human skin *in vivo* (bottom) under fs laser light illumination. The laser wavelength applied to the skin was 790 nm. The labels "channel 1" and "channel 2" indicate the spectral detection ranges which were used to record SHG and autofluorescence, respectively. From [50].

Fig. 18.3 shows *in vivo* multiphoton-photon composite images (channel 1: red, channel 2: green) from different epidermal layers, starting from the outermost epidermal layer (stratum corneum, Fig. 18.3 (a) and (b)) to the stratum spinosum (Fig. 18.3 (e) and (f)) and stratum basale (Fig. 18.3 (h) and (i)), down to the dermis (Fig. 18.3 (j)–(l)). The images are horizontal en-face visualizations of the distributions of SHG-generating and autofluorescent biomolecules. The SHG signal from the ZnO NPs within the sunscreen is clearly visible and forms a thin film across the skin surface observed as yellow/red (Fig. 18.3 (a)–(c); the yellow color results from the overlay of SHG and autofluorescence signals which are red and green color-coded, respectively). Beyond the stratum corneum, i.e., below 15–20 μm of the skin (Fig. 18.4 (e)–(h)) no SHG signal arising from ZnO NPs is visible. At the epidermal-dermal boundary (Fig. 18.3 (j)–(l)), SHG signal from collagen structures start to become visible. This image series indicates no penetration of ZnO nanoparticles across the stratum corneum which forms an effective nanoparticle barrier. The detection limit for pure ZnO NPs *in vitro* was determined with a similar multiphoton imaging system to be as low as $\approx 0.08\,\mathrm{fg/\mu m^3}$ (1 mg/ml) by SHG intensity measurements and $\approx 1\,\mathrm{fg/\mu m^3}$ (1 mg/ml) in fluorescence lifetime imaging, respectively [46, 52]. Even within the skin furrows with a thinner stratum corneum (reduced path length) no penetration into the viable epidermis was observed. The study reiterates the usefulness of multiphoton imaging for investigating the penetration of xenobiotics such as ZnO NPs into skin *in vivo*.

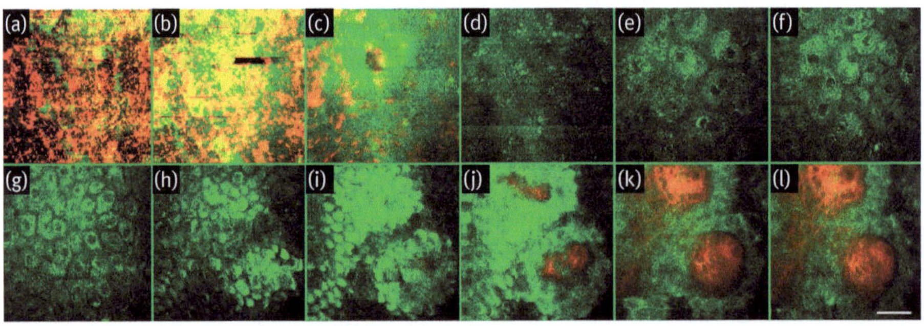

Fig. 18.3: *In vivo* two-photon composite images of human skin from different depth with topically applied sunscreen. SHG (red) and autofluorescence signals (green) are overlaid in pseudo-colors. The images represent horizontal en-face images with depth increasing in steps of 5 μm starting from the stratum corneum (a) up to the epidermal/dermal boundary (i)–(l). Scale bar: 20 μm. From [50].

More recently our groups focus has been on the potential penetration of zinc ions from ZnO NPs that accumulate within the furrows of the skin [53]. The study combined multiphoton microscopy with selective filters for the second-harmonic generation (SHG) signal of ZnO NPs (two-photon excitation at 800 nm, emission at exactly half the excitation wavelength 400 nm) and confocal microscopy of a labile zinc selective fluorophore, ZinPyr-1 (one-photon excitation at 488 nm and emission detected at

520–560 nm). To reduce the interference of autofluorescence, *ex vivo* heat-separated human skin was used as a membrane to study the penetration of ZnO NPs and ionic zinc species simultaneously. Fig. 18.4 is an example of the images obtained within the study that illustrate the accumulation of ZnO NPs within the furrow whilst also mapping the distribution of labile zinc. The study highlighted that the pH of the formulation used to apply ZnO NPs (uncoated) to the skin will affect the degree of penetration of ionic zinc species within the viable epidermis. The study revealed that application of ZnO NPs in a lower pH formulation will increase the penetration of zinc species when compared to a higher pH or a nonaqueous CCT formulation where the zinc species have a lower solubility and dissolution rate to that of the ZnO NPs [53]. Current investigations within our laboratory are focusing on the uptake of zinc species in human keratinocytes *in vitro* to compare findings with *ex vivo* human skin using multimodal imaging techniques. This includes exploitation of the intense SHG signal of ZnO NPs and the less intense but more specific phosphorescence signal of ZnO NPs using FLIM and phosphorescence lifetime imaging (PLIM) to enhance both the sensitivity and specificity of mapping ZnO NPs within complex biological matrices. For PLIM the pulsed laser (e.g. Ti:Sa) is modulated on/off while scanning each pixel and this procedure enables the user to simultaneously record short and long luminescence lifetimes (cf. Tab. 18.1, penultimate row).

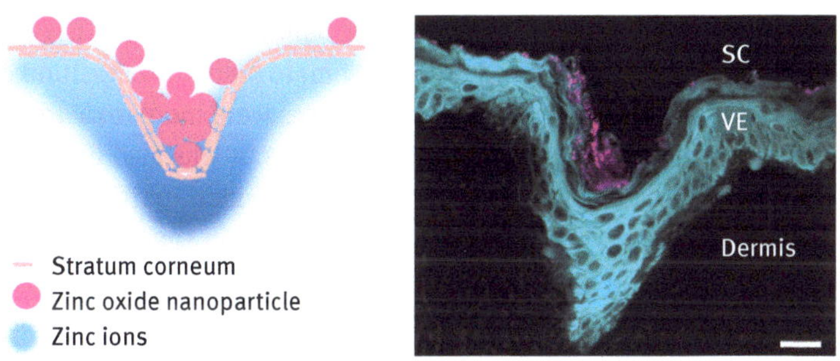

Fig. 18.4: Diagram of the simultaneous detection of ZnO NPs and labile zinc species after application of uncoated commercial ZnO NPs (10 wt%) in CCT to human skin for 48 h. Blue corresponds to ZinPyr-1, a zinc specific dye, (10 μM, single-photon Ex: 488 nm, Em: 520–560 nm) and pink corresponds to SHG of ZnO NPs (two-photon Ex: 800 nm, Em: 370–420 nm). Scale bar represents 20 μm [53].

18.4 Monitoring silver nanoparticles (Ag NPs) penetration into human skin

Imaging silver nanoparticles (Ag NPs) after topical application on human skin is of interest due to their widespread use in wound dressings and personal care products. Interestingly, despite Ag NPs of various shapes and sizes being commercially available, there seems to be a gap in knowledge regarding the shape dependency on human skin penetration and toxicity. Our group has recently conducted a safety assessment of various shaped Ag NPs using human keratinocytes *in vitro* and after topical application to *ex vivo* burned and intact human skin [30]. Noble metal NPs show interesting optical properties that allow us to detect the NPs within biological tissue with the advanced imaging method of high performance TCSPC-FLIM together with MPM. These properties can be divided into two categories based on their different physical nature of elastic and inelastic phenomena. The latter effect originates from the so-called localized surface plasmon resonance (LSPR) [54]. The LSPR signal can be tailored by modifying the size or the shape of the NPs. This signal can be very broad band (inelastic phenomenon) and strong, but is also characterized by a very rapid optical response in the femtosecond range [55]. The TCSPC-FLIM technique can thus be employed with its unique capability of enabling us to distinguish between 'early arriving' photons (short lifetime e.g. LSPR) and 'late arriving' photons (longer lifetime, e.g. autofluorescence of endogenous biomolecules or dyes within tissue) [56]. Ag NPs show pronounced LSPR signals within the endogenous fluorescence spectral band [30, 57]. Ag NPs also show SHG and hyper-Rayleigh scattering (HRS). SHG and HRS are also nonlinear optical processes, but are based on elastic scattering; therefore, the optical response is firstly quasi-instantaneous, and secondly at the double frequency of the incident excitation light of the femtosecond laser that might be narrow banded, e.g. Mai Tai laser.

Our group exploited the use of LSPR (Ex: 800 nm, Em band: 520–560 nm) in combination with SHG/HRS (Ex: 800 nm, Em band: 370–420 nm) to map the deposition of various shaped Ag NPs. It was found that the truncated plates were more toxic to human keratinocytes *in vitro* when compared to cubes or spheres. It was also observed that the shape of the Ag NP influenced the substantivity on the skin, particularly in regards to truncated plate-shaped Ag NPs that accumulated within the undulations and furrows of the skin to a higher degree than Ag NP spheres or to a lesser degree Ag NP cubes. Ag NPs were found within the wound bed of burned human skin, thus they were in direct contact with viable keratinocytes, though no penetration of any shaped Ag NPs was found in the viable epidermis after application to intact human skin for 24 h [30].

This was supported by a previous study which also showed that when Ag NPs were applied to intact or tape stripped skin no statistically significant difference in metabolic state was shown in the viable epidermis, *in vivo*, as measured by the inverse ratio of free to bound NAD(P)H (α_1/α_2) [58]. Further, no difference in penetration was observed due to incomplete SC removal, though reflectance confocal microscopy showed Ag NPs were retained as aggregates for up to ten days [58] supporting similar findings by Lademann *et al.* [59]. It was noted that spherical Ag NPs preferentially accumulated in the superficial layers of the stratum corneum particularly around and within the hair follicle orifice as seen in Fig. 18.5. Our data was also supported by Zhu *et al.* who also found no penetration of Ag NPs into the viable epidermis after application to intact porcine ear skin hair follicles using various imaging techniques including

MPM-FLIM, confocal Raman microscopy and surface enhanced Raman microscopy [60]. The potential for Ag NPs to penetrate human skin is still a controversial point of discussion as George *et al.* using scanning electron microscopy with energy dispersive X-ray microanalysis showed penetration of nanocrystalline silver from a commercially available wound dressing into the dermis after application *in vivo* to intact human skin [61]. In a recent paper, Tak *et al.* showed that the Ag^+ blood concentration after topical application of various shaped Ag NPs to hairless mice for 5 days was highly shape dependent. The authors observed that rod-shaped Ag NPs were found to penetrate the stratum corneum and accumulate within the dermis resulting in higher Ag^+ concentrations compared to truncated triangular and spherical Ag NPs [62]. Further imaging studies are required to elucidate the fate of topically applied nanoparticles.

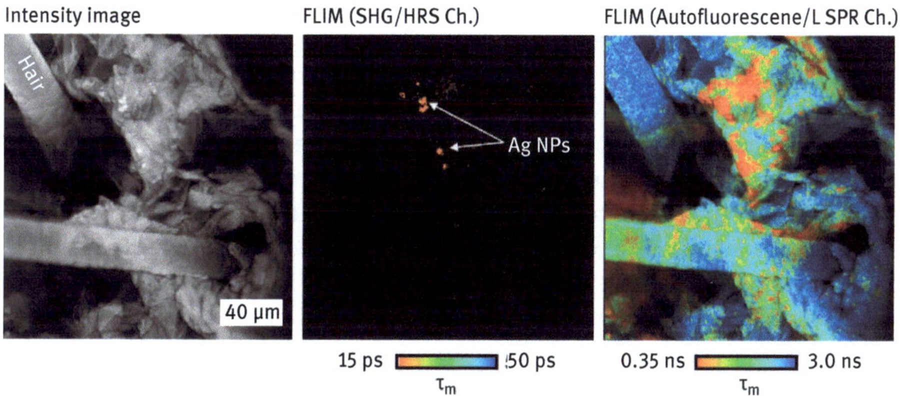

Fig. 18.5: Fluorescent lifetime image of (a) acriflavine (10 µM, aq) stained human skin (Ex: 800 nm, Em: 520–560 nm), blue illustrating shorter lifetime of the spherical Ag NPs applied to the skin and green illustrating the longer lifetime of bound acriflavine and (b) second-harmonic generation signal and hyper-Rayleigh scattering of Ag NPs (Ex: 800 nm, Em: 370–420 nm).

18.5 Detection of dyes within the liver of rodent models

Clearance of drugs from the body can be grossly changed as a result of liver disease, resulting in toxicity or impaired function of the drug. It is therefore crucial to understand liver functional changes in diseased livers [63–65]. Intravital imaging offers an opportunity to study progression of liver disease in a larger sampling area than biopsies and blood enzymes tests [66]. Utilizing the cell's autofluorescent properties only reveals limited information and more specific imaging of liver function requires injection of various dyes. The most commonly used dyes in order to study morphology, microcirculation, apoptosis, oxidative stress, metabolic function and tumor angiogenesis of the liver are shown in Tab. 18.2. Sodium fluorescein can stain plasma [67], sinusoidal perfusion [68–71] and diameter [72] in microvessels of the liver. Fluorescein isothiocyanate (FITC)-labeled dextran can be used to visualize blood perfusion [73–76] and red blood cells (RBC) [77]. In addition, FITC-labeled dextran is also able to quantitate the hepatic microvascular morphology in cirrhotic livers. Rhodamine 6G has proved to be useful in staining leukocytes in sinusoids and postsinusoidal venules [67, 68, 73, 74, 76] and platelets [75]. In addition, rhodamine 6G can also be used to assess leukocyte-endothelial cell interaction [71]. Rhodamine B isothiocyanate-labeled dextran can monitor tumor neocapillary in HepG2 cells with hepatocellular carcinoma [78, 79]. Another way of assessing microvascular perfusion is to inject bisbenzamide H33342, which is useful in detecting perfusion failure [80]. Oxygen distribution in the liver can be studied using an oxygen-sensitive dye, Tris(1,10-phenanthroline)ruthenium(II) chloride hydrate = Ru(phen)2+3. It can be measured in conjunction with NAD(P)H to provide additional information regarding the metabolic state of the cells [81]. Intravital visualization of apoptosis has been limited by the lack of appropriate method and imaging systems in animal models and patients [82]. Recently, miniaturized confocal microscopy has allowed *in vivo* imaging in animal models [83] and humans [84, 85] with subcellular resolution [86]. A novel method imaging apoptosis in hepatocytes is based on translocation of Bax to mitochondria and the release of cytochrome c using biomolecular fluorescence complementation [87]. Another common method in studying apoptosis is with propidium iodide, which labels nuclei of nonviable cells [88]. Rhodamine 123 measures mitochondrial depolarization [89, 90] and calcein measures mitochondrial permeability transition [88, 89]. Dichlorofluorescein diacetate, a hydroperoxide-sensitive fluorogenic probe, is used to study oxidative stress, because it becomes fluorescent when in contact with hydroperoxide [91].

Tab. 18.2: The most commonly used fluorescent probes in liver imaging.

Fluorescent probe	MW	Ex (nm)	Em (nm)	Application
Sodium fluorescein	376	495	517	Plasma, sinusoidal perfusion, sinusoidal diameter
FITC labeled dextran		492	518	Sinusoidal perfusion, red blood cells
bisBenzamide H 33342	562	346	460	Sinusoidal perfusion
Rhodamine 6G	479	528	551	Leukocytes
Rhodamine B isothio-cyanate-labeled dextran		540	573	Tumor neocapillary
$Ru(phen)_3^{2+}$	713	480	580	Oxygen distribution
Propidium iodide	668	482, 540	608	Apoptosis
Rhodamine 123	381	505	534	Mitochondrial depolarization
Calcein	623	470	509	Mitochondrial permeability transition
Dichlorofluorescein diacetate	485	504	524	Oxidative stress

MW: molecular weight; Ex: excitation wavelength (center); Em: emission wavelength (center)

Not only are fluorescent dyes important in studying liver function, but they can also be used to study drug transport in order to determine dosage regimen in different diseases of the liver. Fluorescence microscopy can be useful in pharmacokinetic studies of the liver. Intracellular uptake of benzo[a]pyrene (BaP) [92] and resveratrol [93] was studied in clone 9 and HepG2 cell lines, respectively. Biliary excretion of various fluorescence molecules, such as fluorescein, asculin, thioflavine S, acriflavine, proflavine [94] and carboxyfluorescein [95] was studied using fluorescence microscopy. In addition, fluorescence microscopy in combination with digitized image analysis was used to study the transport of fluorescein in intact perfused rat liver [96]. The liver acinus is the structural and functional unit in the liver [97]. It is divided into three zones, where zone 1 is closest to the portal vein, zone 3 is closest to the central vein and zone 2 is located in between [97, 98]. Fluorescence microscopy can study transport of fluorescent molecules in the different zones of the liver acinus [81, 99–103]. The fluorescent molecules used as model compounds in these studies were Rhodamine B [99, 100], fluorescein diacetate [100], acridine orange, FITC [100], sodium fluorescein, sodium glycholate [102] and Ru(phen)2+3 [81]. Results show that lipid-soluble (fluorescein diacetate) and positively charged water soluble molecules (Rhodamine B and acridine orange) were located in zone 1, however, increased concentration of the latter made them more visible in the other zones. FITC is a water soluble molecule with negative charge and is distributed mostly in zone 3. Zone 1 is exposed to the highest concentration of solutes, meaning that solutes taken up by diffusion or by transporters expressed similarly across the acinus would predominantly be taken up in zone 1 [100]. The reason for this is a decreased concentration of the solute as it moves down the acinus [100]. Further proof of this is that Ru(phen)2+3 is distributed mostly in zone 1, where the highest concentration of oxygen resides [81]. These results suggest that zonal distribution cannot solely be explained by sinusoidal blood flow and simple diffusion but rather a complex transport system within the liver acinus [100].

Another way of studying drug transport is to measure the function of the transporters involved in moving drugs in and out of the liver. Hepatic transporters facilitate transport of endogenous substrates including drugs that resemble their endogenous substrate [104]. The transporters are located on the sinusoidal and canalicular membranes and can facilitate uptake or excretion of their substrates. Fig. 18.6 describes the liver function in drug transport and transporter expression and location in hepatocytes. Uptake transporters are located on the sinusoidal membrane and include organic anion transporting polypeptides (Oatp), sodium-dependent taurocholate cotransporting protein (Ntcp), organic anion transporters (Oat) and organic cation transporters (Oct) [104]. The excretion transporters are located on the canalicular membrane, transporting their substrates to the bile, or on the lateral membranes, excreting their substrates to the sinusoids [104]. The transporters located on the canalicular membrane are the bile salt export pump (Bsep), multidrug resistance-associated protein 2 (Mrp2), multidrug resistance protein 3 (Mdr3) and P-glycoprotein (Pgp) [105]. Located on the lateral membrane are the multidrug resistance-associated proteins 1,

3, 4 and 5 (Mrp1, Mrp3, Mrp4, Mrp5) [104]. Although the transporter system has been studied extensively in various diseased conditions, limited reports use imaging as a tool to measure its function. Lysyl fluorescein conjugated bile acid analogues, e.g. cholyl lysyl fluorescein (CLF) and lithocholyl lysyl fluorescein (LLF), have been used to visualize bile acid transport in the liver. Frozen sections of the liver were examined by fluorescence microscopy [106]. MPM is superior to single-photon microscopy in several ways including:

(1) better effective resolution due to fluorescence arising from the focus;
(2) better suited for deep tissue imaging due to less scattering [107];
(3) photodamage in MPM is limited to the focal plane.

The function of multidrug resistance-associated protein 2 (Mrp2) transporter was examined using MPM by monitoring the biliary excretion of the fluorophore carboxyfluorescein in cholestasis. Our group has explored the role of P-glycoprotein (Pgp) on the disposition of Rhodamine 123 administering an inhibitor for Pgp, cyclosporine A, in

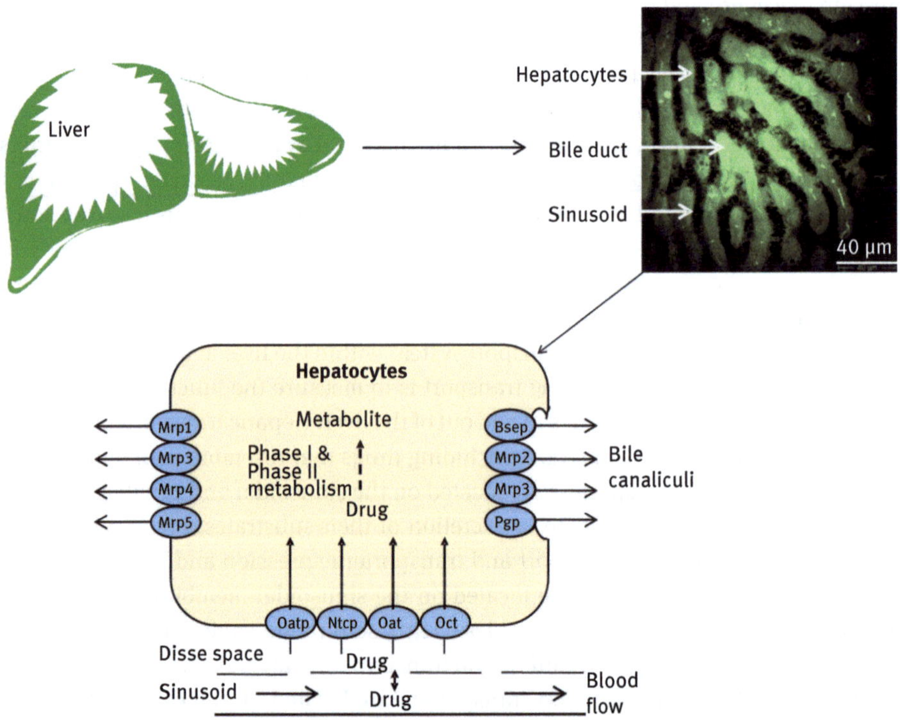

Fig. 18.6: Characteristic function of the liver in drug disposition. Drugs enter the hepatocytes via the sinusoids, which are the small vessels surrounding the cells. The uptake transporters are detailed in the schematic overview of the hepatocyte. Drugs are can be metabolized and excreted back into the sinusoids or into the bile via the export transporters. Adapted from [104, 112].

normal livers [108]. In addition, the function of Pgp was also investigated in hepatic I/R injury using Rhodamine 123 as a probe [109]. Results showed a slower elimination of Rhodamine 123 from the cells in the presence of cyclosporine A and an impaired transport as well as reduced levels of Pgp in long-term I/R injury. In addition, fluorescein is taken up by the organic anion transporter peptide (Oatp) and excreted by multidrug resistance-associated protein 2 (Mrp2) [110]. We examined its distribution in the liver in hepatic I/R injury and found an altered distribution of fluorescein, due to a slower uptake and excretion from the liver.

In conclusion, injection of various dyes, such as fluorescein, rhodamine, propidium iodide, etc., is important in order to study detailed microcirculation and apoptosis [67, 88]. Fluorescein and Rhodamine are currently very useful in studying drug transport in the liver and the function of liver transporters. Drug transport studies provide significant information regarding drug dosage regimen in liver diseases.

18.6 Detection of nanoparticles within the liver of rodent models

Quantum dots (QD) are ultra-small nanoparticles with size range of 2–10 nm in diameter, which have unique light emitting properties such as broad excitation coupled with narrow emission, high quantum yields, good photostability and long fluorescence lifetime. They have unique electronic, optical, and chemical properties and have been used extensively in biomedicine. We recently tracked spatio-temporal disposition of QDs in rat liver at cellular level by intravital imaging using multiphoton microscopy coupled with FLIM [111]. As the fluorescence lifetime of QDs (typically ≥ 10 ns) is much longer than those of tissue autofluorescence (2–3 ns) and most organic dyes (1–5 ns), we could easily differentiate QDs from their environments in biological tissues.

Fig. 18.7 shows representative images of rat liver before and after intravenous injection of QDs via the tail vein. The images were pseudo-colored based on the mean fluorescence lifetimes (τ_m) in individual pixels, which is the weighted average lifetime calculated from short (τ_1) and long (τ_2) lifetime and their relative amplitudes. With an excitation wavelength of 740 nm, we captured signals from the emission channels of 350–450 nm (channel 1) (Fig. 18.7 (a)) and 515–620 nm (channel 2) (Fig. 18.7 (c)). Channel 1 captures the autofluorescence signal of hepatocytes that arises from the presence of NAD(P)H. The color and thus the fluorescence mean lifetime (τ_m) of the whole liver image did not change before or after QD injection (Fig. 18.7 (b)) since no fluorescence from the QDs was detected in this channel. In channel 2, the color in the liver sinusoid (Fig. 18.7 (c)) was found to change to orange and τ_m significantly increased after QD injection (Fig. 18.7 (d)), while the lifetime in the region of hepatocytes did not change before and after QD injection. Channel 2 collected both the fluorescence signal of QDs (maximum emission wavelength of 530 nm) and the autofluorescence of hepatocytes (from flavin adenine dinucleotide (FAD) and of hepatic stellate cells (from Vitamin A). Further examination reveals that the short lifetime component τ_1 (Fig. 18.7 (e), left) did not change either in the hepatocyte or in sinusoid area before and after QDs injection, while the long lifetime component τ_2 (Fig. 18.7 (e), right) increased after QD injection but only for the sinusoid region, not for the hepatocyte region. These results further confirm the observation from MPM fluorescence imaging alone that QDs are retained within the liver sinusoids after intravenous injection and are not taken up by hepatocytes.

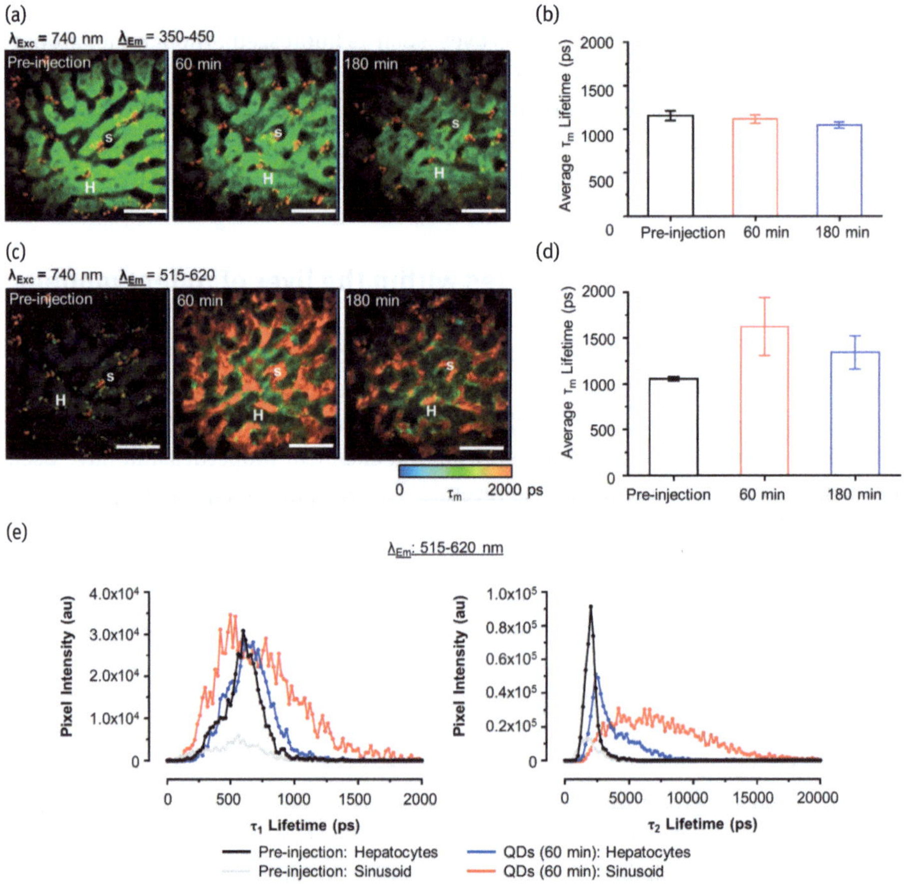

Fig. 18.7: FLIM images of representative rat liver before, 60 min and 180 min after QD bolus injection in emission channel of 350–450 nm (a) and 515–620 nm (c). The pseudo-color is based on the fluorescence mean lifetime τ_m (0–2000 ps; blue–green–red). (The symbol S indicates sinusoid, H indicates hepatocyte, scale bar: 20 µm). The τ_m values at different time points before and after QD injection are displayed in (b) (in channel of 350–450 nm) and (d) (in channel of 515–620 nm). The error bar indicates standard deviation ($n = 3$). The distribution of short lifetime component (τ_1) (e, left panel) and long lifetime component (τ_2) (e, right panel) in emission channel 515–620 nm was analyzed separately in hepatocyte area and sinusoid area.

In summary, we demonstrated that MPM-FLIM is a powerful tool for investigating *in vivo* liver disposition of nanoparticles at the cellular level as these nanoparticles are luminescent and can be differentiated from autofluorescence by spectral selection and TCSPC-FLIM.

References

[1] Choksi AN, Poonawalla T, Wilkerson MG. Nanoparticles: a closer look at their dermal effects. J Drugs Dermatol. 2010;9(5):475–481.

[2] Samberg ME, Oldenburg SJ, Monteiro-Riviere NA. Evaluation of silver nanoparticle toxicity in skin in vivo and keratinocytes in vitro. Environ Health Perspect. 2010;118(3):407–413.

[3] Lewinski N, Colvin V, Drezek R. Cytotoxicity of nanoparticles. Small. 2008;4(1):26–49.

[4] Cross SE, Innes B, Roberts MS, Tsuzuki T, Robertson TA, McCormick P. Human skin penetration of sunscreen nanoparticles: in-vitro assessment of a novel micronized zinc oxide formulation. Skin Pharmacol Phys. 2007;20(3):148–154.

[5] Nohynek GJ, Lademann J, Ribaud C, Roberts MS. Grey goo on the skin? Nanotechnology, cosmetic and sunscreen safety. Crit Rev Toxicol. 2007;37(3):251–277.

[6] Robertson TA, Sanchez WY, Roberts MS. Are commercially available nanoparticles safe when applied to the skin? J Biomed Nanotechnol. 2010;6(5):452–468.

[7] Baroli B, Ennas MG, Loffredo F, Isola M, Pinna R, Lopez-Quintela MA. Penetration of metallic nanoparticles in human full-thickness skin. J Invest Dermatol. 2007;127(7):1701–1712.

[8] Nohynek GJ, Dufour EK. Nano-sized cosmetic formulations or solid nanoparticles in sunscreens: a risk to human health? Arch Toxicol. 2012;86(7):1063–1075.

[9] Baroli B. Penetration of nanoparticles and nanomaterials in the skin: fiction or reality? Journal of Pharmaceutical Sciences. 2010;99(1):21–50.

[10] Serpone N, Dondi D, Albini A. Inorganic and organic UV filters: Their role and efficacy in sunscreens and suncare product. Inorg Chim Acta. 2007;360(3):794–802.

[11] Sharma V, Shukla RK, Saxena N, Parmar D, Das M, Dhawan A. DNA damaging potential of zinc oxide nanoparticles in human epidermal cells. Toxicol Lett. 2009;185(3):211–218.

[12] Smijs TG, Pavel S. Titanium dioxide and zinc oxide nanoparticles in sunscreens: focus on their safety and effectiveness. Nanotechnol Sci Appl. 2011;4:95–112.

[13] Roberts MS, Roberts MJ, Robertson TA, et al. In vitro and in vivo imaging of xenobiotic transport in human skin and in the rat liver. J Biophotonics. 2008;1(6):478–93.

[14] Zvyagin AV, Zhao X, Gierden A, Sanchez W, Ross JA, Roberts MS. Imaging of zinc oxide nanoparticle penetration in human skin in vitro and in vivo. Journal of Biomedical Optics. 2008;13(6):064031.

[15] König K, Riemann I. High-resolution multiphoton tomography of human skin with subcellular spatial resolution and picosecond time resolution. J Biomed Opt. 2003;8(3):432–439.

[16] König K, Ehlers A, Stracke F, Riemann I. In vivo drug screening in human skin using femtosecond laser multiphoton tomography. Skin Pharmacol Physiol. 2006;19(2):78–88.

[17] König K. Clinical multiphoton tomography. J Biophotonics. 2008;1(1):13–23.

[18] Weinigel M, Breunig HG, Kellner-Höfer M, et al. In vivo histology: optical biopsies with chemical contrast using clinical multiphoton/coherent anti-Stokes Raman scattering tomography. Laser Phys Lett. 2014;11(5).

[19] Wang HL, Thorling CA, Liang XW, et al. Diagnostic imaging and therapeutic application of nanoparticles targeting the liver. J Mater Chem B. 2015;3(6):939–958.

[20] Goetz M, Kiesslich R, Dienes HP, et al. In vivo confocal laser endomicroscopy of the human liver: a novel method for assessing liver microarchitecture in real time. Endoscopy. 2008; 40(7):554–562.
[21] Wang HL, Liang XW, Mohammed YH, et al. Real-time histology in liver disease using multiphoton microscopy with fluorescence lifetime imaging. Biomed Opt Express. 2015;6(3):780–792.
[22] König K, Raphael AP, Lin L, et al. Applications of multiphoton tomographs and femtosecond laser nanoprocessing microscopes in drug delivery research. Adv Drug Deliv Rev. 2011; 63(4–5):388–404.
[23] Denk W, Strickler JH, Webb WW. Two-photon laser scanning fluorescence microscopy. Science. 1990;248(4951):73–76.
[24] Breunig HG, Studier H, König K. Multiphoton excitation characteristics of cellular fluorophores of human skin in vivo. Optics Express. 2010;18(8):7857–7871.
[25] Becker W, Bergmann A, Biscotti G, Rück A. Advanced time-correlated single photon counting technique for spectroscopy and imaging in biomedical systems. Commercial and Biomedical Applications of Ultrafast Lasers IV. 2004;5340:104–112.
[26] Suhling K, Hirvonen LM, Levitt JA, et al. Fluorescence lifetime imaging (FLIM): Basic concepts and some recent developments. Medical Photonics. 2015;27:3–40.
[27] Palero JA, de Bruijn HS, van der Ploeg van den Heuvel A, Sterenborg HJ, Gerritsen HC. Spectrally resolved multiphoton imaging of in vivo and excised mouse skin tissues. Biophys J. 2007;93(3):992–1007.
[28] Breunig HG, Bückle R, Kellner-Höfer M, et al. Combined in vivo multiphoton and CARS imaging of healthy and disease-affected human skin. Microsc Res Techniq. 2012;75(4):492–498.
[29] Weinigel M, Breunig HG, Darvin ME, et al. Impact of refractive index mismatches on coherent anti-Stokes Raman scattering and multiphoton autofluorescence tomography of human skin in vivo. Phys Med Biol. 2015;60(17):6881–6899.
[30] Holmes AM, Lim J, Studier H, Roberts MS. Varying the morphology of silver nanoparticles results in differential toxicity against micro-organisms, human keratinocytes and affects skin deposition. Nanotoxicology. 2016;10(10):1503–1514.
[31] European Commission. Submissions scientific hearing on nanotechnology; 2009.
[32] Friends of the Earth. Nano ingredients in sunscreen, the need to regulation; 2012 Jul.
[33] Na RH, Stender IM, Ma LX, Wulf HC. Autofluorescence spectrum of skin: component bands and body site variations. Skin Research and Technology. 2000;6(3):112–117.
[34] Huang MH, Mao S, Feick H, et al. Room-temperature ultraviolet nanowire nanolasers. Science. 2001;292(5523):1897–1899.
[35] Srikant V, Clarke DR. Optical absorption edge of ZnO thin films: the effect of substrate. Journal of Applied Physics. 1997;81(9):6357–6364.
[36] Brus LE. Electron–electron and electron–hole interactions in small semiconductor crystallites: The size dependence of the lowest excited electronic state. The Journal of Chemical Physics. 1984;80(9):4403–4409.
[37] Cheng HM, Hsu HC, Chen SL, et al. Efficient UV photoluminescence from monodispersed secondary ZnO colloidal spheres synthesized by sol–gel method. Journal of Crystal Growth. 2005;277(1):192–199.
[38] van Dijken A, Meulenkamp EA, Vanmaekelbergh D, Meijerink A. The kinetics of the radiative and nonradiative processes in nanocrystalline ZnO particles upon photoexcitation. The Journal of Physical Chemistry B. 2000;104(8):1715–1723.
[39] Kahn ML, Cardinal T, Bousquet B, Monge M, Jubera V, Chaudret B. Optical properties of zinc oxide nanoparticles and nanorods synthesized using an organometallic method. ChemPhysChem. 2006;7(11):2392–2397.

[40] Roberts MS, Roberts MJ, Robertson TA, et al. In vitro and in vivo imaging of xenobiotic transport in human skin and in the rat liver. Journal of Biophotonics. 2008;1(6):478–493.
[41] Kuo TR, Wu CL, Hsu CT, et al. Chemical enhancer induced changes in the mechanisms of transdermal delivery of zinc oxide nanoparticles. Biomaterials. 2009;30(16):3002–3008.
[42] Darvin ME, Koenig K, Kellner-Hoefer M, et al. Safety assessment by multiphoton fluorescence/second harmonic generation/hyper-Rayleigh scattering tomography of ZnO nanoparticles used in cosmetic products. Skin Pharmacol Phys. 2012;25(4):219–226.
[43] Neumann U, Grunwald R, Griebner U, Steinmeyer G, Seeber W. Second-harmonic efficiency of ZnO nanolayers. Applied Physics Letters. 2004;84(2):170–172.
[44] Larciprete MC, Centini M. Second harmonic generation from ZnO films and nanostructures. Applied Physics Reviews. 2015;2(3):031302.
[45] Song Z, Kelf TA, Sanchez WH, et al. Characterization of optical properties of ZnO nanoparticles for quantitative imaging of transdermal transport. Biomed Opt Express. 2011;2(12):3321–3333.
[46] Lin LL, Grice JE, Butler MK, et al. time-correlated single photon counting for simultaneous monitoring of zinc oxide nanoparticles and NAD(P)H in intact and barrier-disrupted volunteer skin. Pharm Res-Dordr. 2011;28(11):2920–2930.
[47] Leite-Silva VR, Lamer ML, Sanchez WY, et al. The effect of formulation on the penetration of coated and uncoated zinc oxide nanoparticles into the viable epidermis of human skin in vivo. Eur J Pharm Biopharm. 2013;84(2):297–308.
[48] Leite-Silva VR, Sanchez WY, Studier H, et al. Human skin penetration and local effects of topical nano zinc oxide after occlusion and barrier impairment. Eur J Pharm Biopharm. 2016;104: 140–147.
[49] Leite-Silva VR, Liu DC, Sanchez WY, et al. Effect of flexing and massage on in vivo human skin penetration and toxicity of zinc oxide nanoparticles. Nanomedicine (Lond). 2016;11(10): 1193–1205.
[50] Breunig HG, Weinigel M, König K. In vivo imaging of ZnO nanoparticles from sunscreen on human skin with a mobile multiphoton tomograph. BiNanoSci. 2014:1–6.
[51] Lewicka ZA, Yu WW, Oliva BL, Contreras EQ, Colvin VL. Photochemical behavior of nanoscale TiO2 and ZnO sunscreen ingredients. J Photoch Photobio A. 2013;263:24–33.
[52] Darvin ME, König K, Kellner-Höfer M, et al. Safety assessment by multiphoton fluorescence/ second harmonic generation/hyper-Rayleigh scattering tomography of ZnO nanoparticles used in cosmetic products. Skin Pharmacol Phys. 2012;25(4):219–226.
[53] Holmes AM, Song Z, Moghimi HR, Roberts MS. Relative penetration of zinc oxide and zinc ions into human skin after application of different zinc oxide formulations. ACS Nano. 2016; 10(2):1810–1819.
[54] Mock JJ, Barbic M, Smith DR, Schultz DA, Schultz S. Shape effects in plasmon resonance of individual colloidal silver nanoparticles. J Chem Phys. 2002;116(15):6755–6759.
[55] Talbot CB, Patalay R, Munro I, et al. Application of ultrafast gold luminescence to measuring the instrument response function for multispectral multiphoton fluorescence lifetime imaging. Optics Express. 2011;19(15):13848–13861.
[56] Thorling CA, Crawford D, Burczynski FJ, Liu X, Liau I, Roberts MS. Multiphoton microscopy in defining liver function. Journal of Biomedical Optics. 2014;19(9):90901.
[57] Podlipensky A, Lange J, Seifert G, Graener H, Cravetchi I. Second-harmonic generation from ellipsoidal silver nanoparticles embedded in silica glass. Opt Lett. 2003;28(9):716–718.
[58] Prow TW, Grice JE, Lin LL, et al. Nanoparticles and microparticles for skin drug delivery. Adv Drug Deliv Rev. 2011;63(6):470–491.
[59] Lademann J, Richter H, Teichmann A, et al. Nanoparticles–an efficient carrier for drug delivery into the hair follicles. Eur J Pharm Biopharm. 2007;66(2):59–64.

[60] Zhu Y, Choe CS, Ahlberg S, et al. Penetration of silver nanoparticles into porcine skin ex vivo using fluorescence lifetime imaging microscopy, Raman microscopy, and surface-enhanced Raman scattering microscopy. Journal of Biomedical Optics. 2015;20(5):051006.
[61] George R, Merten S, Wang TT, Kennedy P, Maitz P. In vivo analysis of dermal and systemic absorption of silver nanoparticles through healthy human skin. Australasian Journal of Dermatology. 2014;55(3):185–190.
[62] Tak YK, Pal S, Naoghare PK, Rangasamy S, Song JM. Shape-dependent skin penetration of silver nanoparticles: does it really matter? Scientific Reports. 2015;5:16908.
[63] Ban D, Kudo A, Sui S, et al. Decreased Mrp2-dependent bile flow in the post-warm ischemic rat liver. J Surg Res. 2009;153(2):310–316.
[64] Fouassier L, Beaussier M, Schiffer E, et al. Hypoxia-induced changes in the expression of rat hepatobiliary transporter genes. Am J Physiol Gastrointest Liver Physiol. 2007;293(1): G25–G35.
[65] Tanaka Y, Chen C, Maher JM, Klaassen CD. Kupffer cell-mediated downregulation of hepatic transporter expression in rat hepatic ischemia-reperfusion. Transplantation. 2006;82(2): 258–266.
[66] Nativ NI, Maguire TJ, Yarmush G, et al. Liver defatting: an alternative approach to enable steatotic liver transplantation. Am J Transplant. 2012;12(12):3176–3183.
[67] Yamagami K, Hutter J, Yamamoto Y, et al. Synergistic effects of brain death and liver steatosis on the hepatic microcirculation. Transplantation. 2005;80(4):500–505.
[68] Abshagen K, Eipel C, Menger MD, Vollmar B. Comprehensive analysis of the regenerating mouse liver: an in vivo fluorescence microscopic and immunohistological study. J Surg Res. 2006;134(2):354–362.
[69] Le MK, Berger A, Eipel C, et al. Uncoupling protein-2 deficient mice are not protected against warm ischemia/reperfusion injury of the liver. J Surg Res. 2011;171(2):742–748.
[70] Mabuchi A, Mullaney I, Sheard PW, et al. Role of hepatic stellate cell/hepatocyte interaction and activation of hepatic stellate cells in the early phase of liver regeneration in the rat. Journal of Hepatology. 2004;40(6):910–916.
[71] Vollmar B, Pradarutti S, Richter S, Menger MD. In vivo quantification of ageing changes in the rat liver from early juvenile to senescent life. Liver. 2002;22(4):330–341.
[72] Vanheule E, Geerts AM, Reynaert H, et al. Influence of somatostatin and octreotide on liver microcirculation in an experimental mouse model of cirrhosis studied by intravital fluorescence microscopy. Liver Int. 2008;28(1):107–116.
[73] Dold S, Laschke MW, Lavasani S, Menger MD, Thorlacius H. Cholestatic liver damage is mediated by lymphocyte function antigen-1-dependent recruitment of leukocytes. Surgery. 2008; 144(3):385–393.
[74] Dold S, Laschke MW, Zhau Y, et al. P-selectin glycoprotein ligand-1-mediated leukocyte recruitment regulates hepatocellular damage in acute obstructive cholestasis in mice. Inflamm Res. 2010;59(4):291–298.
[75] Khandoga A, Stampfl A, Takenaka S, et al. Ultrafine particles exert prothrombotic but not inflammatory effects on the hepatic microcirculation in healthy mice in vivo. Circulation. 2004; 109(10):1320–1325.
[76] Laschke MW, Dold S, Jeppsson B, Schilling MK, Menger MD, Thorlacius H. Rho-kinase inhibitor attenuates cholestasis-induced CXC chemokine formation, leukocyte recruitment, and hepatocellular damage in the liver. J Surg Res. 2010;159(2):666–673.
[77] Sherman IA, Pappas SC, Fisher MM. Hepatic microvascular changes associated with development of liver fibrosis and cirrhosis. Am J Physiol. 1990;258(2 Pt2):H460–H465.

[78] Patumraj S, Yoysungnoen P, Kachonrattanadet P, Wirachwong P. Tumor neocapillary density in hepatocellular carcinoma cells implanted nude mice model. Clinical Hemorheology and Microcirculation. 2005;33(2):137–144.
[79] Yoysungnoen P, Wirachwong P, Bhattarakosol P, Niimi H, Patumraj S. Antiangiogenic activity of curcumin in hepatocellular carcinoma cells implanted nude mice. Clinical Hemorheology and Microcirculation. 2005;33(2):127–135.
[80] Vollmar B, Rucker M, Menger MD. A new method for the intravital microscopic quantification of hepatic sinusoidal perfusion failure using the dye bisbenzamide H33342. Microvasc Res. 1996;51(2):250–259.
[81] Paxian M, Keller SA, Cross B, Huynh TT, Clemens MG. High-resolution visualization of oxygen distribution in the liver in vivo. Am J Physiol Gastrointest Liver Physiol. 2004;286(1):G37–G44.
[82] Blankenberg FG. In vivo imaging of apoptosis. Cancer Biol Ther. 2008;7(10):1525–1532.
[83] Goetz M, Fottner C, Schirrmacher E, et al. In-vivo confocal real-time mini-microscopy in animal models of human inflammatory and neoplastic diseases. Endoscopy. 2007;39(4):350–356.
[84] Goetz M, Deris I, Vieth M, et al. Near-infrared confocal imaging during mini-laparoscopy: a novel rigid endomicroscope with increased imaging plane depth. Journal of Hepatology. 2010;53(1):84–90.
[85] Kiesslich R, Burg J, Vieth M, et al. Confocal laser endoscopy for diagnosing intraepithelial neoplasias and colorectal cancer in vivo. Gastroenterology. 2004;127(3):706–713.
[86] Goetz M, Ansems JV, Galle PR, Schuchmann M, Kiesslich R. In vivo real-time imaging of the liver with confocal endomicroscopy permits visualization of the temporospatial patterns of hepatocyte apoptosis. Am J Physiol Gastrointest Liver Physiol. 2011;301(5):G764–G772.
[87] Yivgi-Ohana N, Eifer M, Addadi Y, Neeman M, Gross A. Utilizing mitochondrial events as biomarkers for imaging apoptosis. Cell Death Dis. 2011;2:e166.
[88] Zhong Z, Ramshesh VK, Rehman H, et al. Activation of the oxygen-sensing signal cascade prevents mitochondrial injury after mouse liver ischemia-reperfusion. Am J Physiol Gastrointest Liver Physiol. 2008;295(4):G823–G832.
[89] Rehman H, Ramshesh VK, Theruvath TP, et al. NIM811 (N-methyl-4-isoleucine cyclosporine), a mitochondrial permeability transition inhibitor, attenuates cholestatic liver injury but not fibrosis in mice. J Pharmacol Exp Ther. 2008;327(3):699–706.
[90] Theruvath TP, Zhong Z, Pediaditakis P, et al. Minocycline and N-methyl-4-isoleucine cyclosporin (NIM811) mitigate storage/reperfusion injury after rat liver transplantation through suppression of the mitochondrial permeability transition. Hepatology. 2008;7(1):236–246.
[91] Suematsu M, Kato S, Ishii H, et al. Intralobular heterogeneity of carbon tetrachloride-induced oxidative stress in perfused rat liver visualized by digital imaging fluorescence microscopy. Lab Invest. 1991;64(2):167–173.
[92] Barhoumi R, Mouneimne Y, Ramos KS, et al. Analysis of benzo[a]pyrene partitioning and cellular homeostasis in a rat liver cell line. Toxicol Sci. 2000;53(2):264–270.
[93] Lancon A, Delmas D, Osman H, Thenot JP, Jannin B, Latruffe N. Human hepatic cell uptake of resveratrol: involvement of both passive diffusion and carrier-mediated process. Biochem Biophys Res Commun. 2004;316(4):1132–1137.
[94] Grafflin AL, Bagley EH. Studies of hepatic structure and function by fluorescence microscopy. Bull Johns Hopkins Hosp. 1952;90(6):395–437.
[95] Kudo A, Kashiwagi S, Kajimura M, et al. Kupffer cells alter organic anion transport through multidrug resistance protein 2 in the post-cold ischemic rat liver. Hepatology. 2004;39(4):1099–1109.
[96] Schmidt R, Buscher HP. Hepatic uptake of fluorescein, investigated by video fluorescence microscopy and digital image analysis. J Hepatol. 1991;13(2):208–212.

[97] Rappaport A, Wilson W. The structural and functional unit in the human liver (liver acinus). The Anatomical Record. 1958;130(4):673–689.
[98] Lamers WH, Hilberts A, Furt E, et al. Hepatic enzymic zonation: a reevaluation of the concept of the liver acinus. Hepatology. 1989;10(1):72–76.
[99] Braakman I, Groothuis GM, Meijer DK. Acinar redistribution and heterogeneity in transport of the organic cation rhodamine B in rat liver. Hepatology. 1987;7(5):849–855.
[100] Gumucio JJ, Miller DL, Krauss MD, Zanolli CC. Transport of fluorescent compounds into hepatocytes and the resultant zonal labeling of the hepatic acinus in the rat. Gastroenterology. 1981;80(4):639–646.
[101] Nijssen HM, Pijning T, Meijer DK, Groothuis GM. Influence of albumin on the net sinusoidal efflux of the organic anion dibromosulfophthalein from rat liver. Hepatology. 1992; 15(2):302–309.
[102] Sherman IA, Fisher MM. Hepatic transport of fluorescent molecules: in vivo studies using intravital TV microscopy. Hepatology. 1986;6(3):444–449.
[103] Vollmar B, Siegmund S, Menger MD. An intravital fluorescence microscopic study of hepatic microvascular and cellular derangements in developing cirrhosis in rats. Hepatology. 1998; 27(6):1544–1553.
[104] Li P, Wang GJ, Robertson TA, Roberts MS. Liver transporters in hepatic drug disposition: an update. Curr Drug Metab. 2009;10(5):482–498.
[105] Lecureur V, Courtois A, Payen L, Verhnet L, Guillouzo A, Fardel O. Expression and regulation of hepatic drug and bile acid transporters. Toxicology. 2000;153(1–3):203–219.
[106] Milkiewicz P, Mills CO, Hubscher SG, et al. Visualization of the transport of primary and secondary bile acids across liver tissue in rats: in vivo study with fluorescent bile acids. J Hepatol. 2001;34(1):4–10.
[107] Butko MT, Drobizhev M, Makarov NS, Rebane A, Brinkman BC, Gleeson JG. Simultaneous multiple-excitation multiphoton microscopy yields increased imaging sensitivity and specificity. BMC Biotechnol. 2011;11:20.
[108] Liu X, Thorling CA, Lu J, Roberts MS. Intravital multiphoton imaging of rhodamine 123 in the rat liver after intravenous dosing. IntraVital. 2012;1(1):54–59.
[109] Thorling C, Liu X, Burczynski F, Fletcher L, Roberts M, editors. Effect of hepatic ischemia reperfusion injury on the function of p-glycoprotein in transporting rhodamine 123. Hepatology. 2011;54:695A-695A.
[110] De BT, Fattah S, Stieger B, Augustijns P, Annaert P. Sodium fluorescein is a probe substrate for hepatic drug transport mediated by OATP1B1 and OATP1B3. Journal of pharmaceutical sciences. 2011;100(11):5018–5030.
[111] Liang XW, Grice JE, Zhu Y, et al. intravital multiphoton imaging of the selective uptake of water-dispersible quantum dots into sinusoidal liver cells. Small. 2015;11(14):1711–1720.
[112] Roberts MS, Magnusson BM, Burczynski FJ, Weiss M. Enterohepatic circulation: physiological, pharmacokinetic and clinical implications. Clin Pharmacokinet. 2002;41(10):751–790.

Ana-Maria Pena, Etienne Decencière, Sébastien Brizion, Steeve Victorin, Serge Koudoro, Thérèse Baldeweck, and Emmanuelle Tancrède-Bohin

19 Multiphoton FLIM in cosmetic clinical research

Abstract: There is an increasing need in cosmetic clinical research for non-invasive, high content, skin imaging techniques offering the possibility on the one hand, to avoid performing invasive biopsies, and on the other hand, to supply a maximum of information on the skin state throughout a study, especially before, during and after product application. Multiphoton microscopy is one of these techniques compatible with *in vivo* human skin investigations, allowing human skin three-dimensional (3D) structure to be characterized with sub-μm resolution. In association with fluorescence lifetime imaging (FLIM) and specific 3D-image processing, one can extract several quantitative parameters characterizing skin constituents in terms of morphology, density and organization. Various intracellular and extracellular constituents present specific endogenous signals enabling a non-invasive visualization of the 3D structure of epidermal and superficial dermal layers. Multiphoton FLIM applications in the cosmetic field range from knowledge to evaluation studies. Knowledge studies aim to acquire a better knowledge of skin differences appearing with aging, solar exposure or between the different skin phototypes. Evaluation studies deal with the efficacy of cosmetic anti-aging or whitening ingredients. The goal of this chapter is not to give a literature review of multiphoton FLIM applications in cosmetic clinical research, but rather to acquaint the reader with the quantitative 3D information afforded by multiphoton FLIM imaging of human skin and its interest in cosmetic clinical research.

19.1 Multiphoton fluorescence lifetime imaging of *in vivo* human skin

In dermatology, skin imaging has always found a privileged position in the scientific literature, for teaching purposes or for the monitoring of the patient, owing to the direct access to the skin. Techniques such as ultrasound, magnetic resonance imaging, optical coherence tomography or *in vivo* confocal microscopy have offered new opportunities for skin imaging not only at its surface but also in depth. Today, the development of non-invasive skin imaging methods, which can be applied *in vivo* on human volunteers, allowing the replacement of biopsies in a number of situations, avoiding their renewal for treatment monitoring, permitting to guide a surgical procedure or to acquire new knowledge on skin constituents, is still a major objective for dermatologists.

https://doi.org/10.1515/9783110429985-021

Similar needs exist in cosmetic clinical research: we are looking for non-invasive, high content, large field of view, high resolution 3D skin imaging techniques offering the possibility, on the one hand, to avoid performing invasive biopsies and, on the other hand, to supply a maximum of information on the skin state throughout a study: before, during and after product application. The most recent imaging techniques offer the subcellular resolution required for the detailed observation of skin. Some of them may also provide access to a three-dimensional *in vivo* functional imaging in real time [1]. Among these, multiphoton microscopy has gradually emerged as the most suitable technique to image skin constituents at the subcellular level *in vivo* on human volunteers. Indeed, multiphoton microscopy is compatible with *in vivo* human skin investigations, allowing human skin 3D structure to be characterized with sub-μm resolution. It offers the possibility to image the cellular and extracellular matrix components non-invasively, by taking advantage of intrinsic multiphoton signals: second-harmonic generation (SHG) created by fibrillar collagens and two-photon excited fluorescence (2PEF) emitted by keratin, nicotinamide adenine dinucleotide (NADH), flavin adenine dinucleotide (FAD), melanin, or elastin [2–7].

Multiphoton microscopy is a form of laser scanning microscopy that uses localized two-photon excitation to produce signals only at the focal point of the objective [8, 9]. The principle of two-photon absorption phenomenon was theoretically demonstrated by Maria Göppert-Mayer in 1931 [10, 11], experimentally proved 30 years later by Kaiser and Garrett [12] with the advent of laser sources. In the 1970s, C. Sheppard and R. Kompfner proposed the construction of a nonlinear laser microscope, built in the 1990s by Denk *et al.* [13]. The concept is based on the idea that two energy-equal near-infrared (NIR) photons can simultaneously be absorbed by a molecule, and produce an excitation equivalent to the absorption of a single photon in the ultraviolet/visible range. Each NIR photon provides half the energy needed to excite the molecule from the fundamental state to the excited state (see Fig. 19.1). The excitation results in the subsequent emission of a fluorescence photon, typically at a higher energy, smaller wavelength, as compared to the infrared (IR) photons. As two photons are required for excitation, the two-photon excited fluorescence signal intensity is proportional to the square of the excitation intensity. This nonlinear dependence on excitation intensity affords a 3D localization of the excitation and emission, confined at the focal point of the objective lens in a sub-femtoliter volume. Indeed, as the probability of two-photon absorption is extremely low, such phenomenon is only possible if the sample is excited with a high power density (MW/cm^2 to GW/cm^2). This can be achieved by focusing the photons both temporally (using IR femtoseconds–fs pulsed lasers that deliver ultra-short laser pulses with a duration of ≈ 100 fs) and spatially (at the focal point of the objective lens). This intrinsic "optical sectioning" is the key advantage of multiphoton microscopy over confocal microscopy, which employs additional elements such as pinholes to reject out-of-focus fluorescence. The excitation volume is then scanned in two or three dimensions in the sample to acquire 2D or 3D images.

Multiphoton microscopy presents several other advantages: superior imaging depth within biological tissues as IR light penetrates deeper in the skin than UV/VIS light, less photobleaching and phototoxicity as confined to the excitation volume. Furthermore, it provides additional contrast modes with increased specificity, such as the second harmonic generation, a coherent second-order nonlinear process, forbidden in centrosymmetric media and implying no absorption phenomenon. In the skin, the SHG signal is obtained from dense and ordered macromolecular structures such as fibrillar collagens [14]. SHG microscopy is sensitive to the organization of collagen molecules rather than to the collagen type [14] and probes collagen macromolecular structure at the micrometer scale, which makes it a powerful technique to study collagen organization in biological tissues. Both 2PEF and SHG signals can be excited using the same laser source and detected separately based on their spectral difference. Simultaneous recording of 2PEF and SHG signals enable multimodal imaging of epidermis and superficial dermis up to a depth of about 160–200 µm depending on the body region.

Fluorescence lifetime imaging microscopy [15] brings another dimension to multiphoton microscopy: in association with 2PEF imaging it offers an "additional" mode of contrast with increased specificity offering an insight into the type of fluorophores contributing to the detected signal. Fluorophores with similar emission spectra can be differentiated by FLIM provided that they present different fluorescence lifetimes. Fluorescence lifetime (τ) is a measure of the average time that the fluorophore remains in the excited state following excitation. The excited state decays back to the ground state in a statistical manner. This fluorescence decay has an exponential shape for simple molecules. In skin, the autofluorescence signal within the sub-femtoliter excitation volume comes from a variety of fluorophores and the fluorescence decay typically shows a bi-exponential behavior:

$$Photon\ count(t) = a_1 \mathrm{e}^{-t/\tau_1} + a_2 \mathrm{e}^{-t/\tau_2},$$

where τ_1 and τ_2 are respectively the fast and slow lifetimes and a_1 and a_2 their amplitudes.

The skin endogenous fluorophores present lifetimes on the order of hundreds of picoseconds (e.g. melanin, free NADH, bound FAD) to nanoseconds (see [16] and included references). The fluorescence lifetime is independent of fluorophore concentration, but depends on the local microenvironment of the molecule that is, for example, pH, binding status, conformational changes.

As an example, the lifetimes of NADH and FAD, two coenzymes involved in cell metabolic activity, differ substantially depending on whether they are free or bound to proteins. The two-photon excited fluorescence intensities of NADH and FAD and especially the "RedOx ratio" (defined as the ratio of the 2PEF intensities of NADH and FAD, or NADH/(NADH + FAD)) can be used as an indicator of the oxidation-reduction potential of a cell, a non-invasive optical metabolic index allowing to assess the changes in metabolic activity. This type of functional information was initially evidenced in the 1960s by Chance *et al.* in isolated cells [17]. Fluorescence lifetime imaging offers additional functional information in terms of preferred energy production pathways (glycolysis, Krebs cycle, oxidative phosphorylation) by the measuring of the ratio of free and protein-bound NADH and free and protein-bound FAD. A detailed review of the role of NADH and FAD in the metabolic process as well as of the advantages and limitations of two-photon excited fluorescence lifetime imaging to assess the metabolic state is given by Georgakoudi and Quinn in reference [18] and included citations.

Usually NADH and FAD redox ratio imaging and FLIM imaging both require long acquisition times: firstly, due to the fact that an optimal excitation of these two chromophores implies a sequential imaging at ≈ 740 nm for NADH and ≈ 900 nm for FAD, secondly, because NADH and FAD imaging implies different image acquisition times due to their different mitochondrial concentration (orders of magnitude higher for NADH as compared to FAD) and lastly, because FLIM analysis requires an increased number of photons per pixel (hundreds of photons/pixel for a bi-exponential decay). For all these reasons, clinical trials dealing with multiphoton FLIM assessment of the metabolic activity in human volunteers involve mostly a single excitation wavelength, around 760 nm, convenient for NADH imaging.

Fluorescence lifetime imaging affords also a specific detection of melanin, the pigment that gives the color of the skin. Melanin detection can of course be performed by 2PEF imaging in the basal layers of epidermis (highly concentrated melanin regions show 2PEF intensities stronger than that of other endogenous fluorophores). This works satisfactorily in the basal and supra-basal layers as demonstrated with the study of corticosteroids effects [19]. But this type of intensity-based detection is not always satisfactory. First, the signal intensity may be disrupted by other fluorophores also creating intense 2PEF signals (e.g., keratin in the *stratum corneum*), and second, it does not take into account pixels with low melanin concentration of a fluorescence intensity comparable to that of other endogenous fluorophores. A more specific detection can be done by FLIM, as melanin presents an extremely fast decay component (< 0.1 ns). However, melanin detection in the whole epidermis with FLIM is too long to be performed on human volunteers, and in practice is limited to selected 2D slices, obtained at a depth chosen by the operator.

Another approach, combining multiphoton and FLIM, called Pseudo-FLIM, can be used to specifically detect melanin from multiphoton FLIM-like data compatible with 3D *in vivo* acquisitions on human volunteers [20, 21]. The main idea is to bin the fluorescence photons in a reduced number of time channels either at the acquisition or afterwards during processing and to calculate the slope of the decay. By applying a threshold to keep the high slope values (i.e. short lifetimes), one can identify melanin pixels as melanin presents a very short ($\tau_1 < 0.1$ ns) and predominant ($a_1 > 90\,\%$) lifetime component. As this analysis requires only a few photons per pixel (≈ 30), it results in a short image acquisition time (16 μs/pixel) compatible with 3D imaging on humans.

In the literature, many contributions focus on multiphoton/FLIM imaging of normal human skin, characterized either *ex vivo* or *in vivo*. The reader can have an insight on the *in vivo* work that started more than 20 years ago and at the evolution of the imaging instruments by consulting the references [2, 6, 8, 16, 22–44]. The first multiphoton/FLIM data of human skin *in vivo* were obtained in the 1990s by and Masters *et al.* with a homemade microscope on the forearm of one of the authors [2]. Nowadays commercial microscopes clinically/medically approved for their use on humans exist, in either tabletop (DermaInspect™, since 2002) or in a more flexible, compact and "portable" configuration (MPT*flex*™, since 2011) more suitable for clinical use [45]. The clinical applications of *in vivo* multiphoton/FLIM microscopy range from the characterization of age-related or photoaging changes [38, 46–55]], dermatological disorders and melanoma [28, 56–71], up to the assessment of penetration and effects of pharmaceutical/cosmetic products on human skin [19, 21, 55, 72–81].

19.2 Clinical multiphoton FLIM systems

Two types of multiphoton FLIM microscopes classified as class 1M devices, medically approved by certified bodies of the European Union for their use on humans are nowadays commercially available (JenLab GmbH, Germany).

The first system on the market was the DermaInspect™ in 2002 [6], a tabletop microscope that offers high quality simultaneous 2PEF and SHG imaging of epidermis and superficial dermis up to a depth of about 160–200 µm depending on the body region [19]. It also enables fluorescence lifetime imaging that is performed using a time-correlated single-photon counting (TCSPC) unit (SPC830 from Becker & Hickl GmbH, Berlin, Germany). Most of the clinical trials performed with this system were done on the forearm. Within this region, good signal-to-noise ratio images can be acquired up to a depth of 160 µm with an average excitation power increasing from 10 mW at the skin surface up to 50 mW when reaching the dermis [19]. The excitation wavelength can also be varied as this system is equipped with tunable femtosecond pulsed lasers, but an optimal excitation of a maximum of endogenous fluorophores is obtained when working at wavelengths around 760 nm [5]. This excitation wavelength also enables second harmonic generation, as this signal is generated at exactly half the excitation wavelength ($\lambda/2$). An example of a z-stack of 2PEF/SHG images acquired on normal human forearm skin can be found in [19] and images acquired at different depths are given in Fig. 19.1.

In the epidermis, typically, the first 10 µm reveal the organization of the *stratum corneum disjunctum* – SCD (made of corneocytes – polygonal-shaped dead keratinocytes emitting a 2PEF signal that mainly arises from keratin) and the next 10 µm correspond to *stratum corneum compactum* – SCC (this epidermal layer creates less intense 2PEF signals and presents an almost homogenously distributed pattern with no distinguishable cell structures). The presence of cell nuclei indicates the tran-

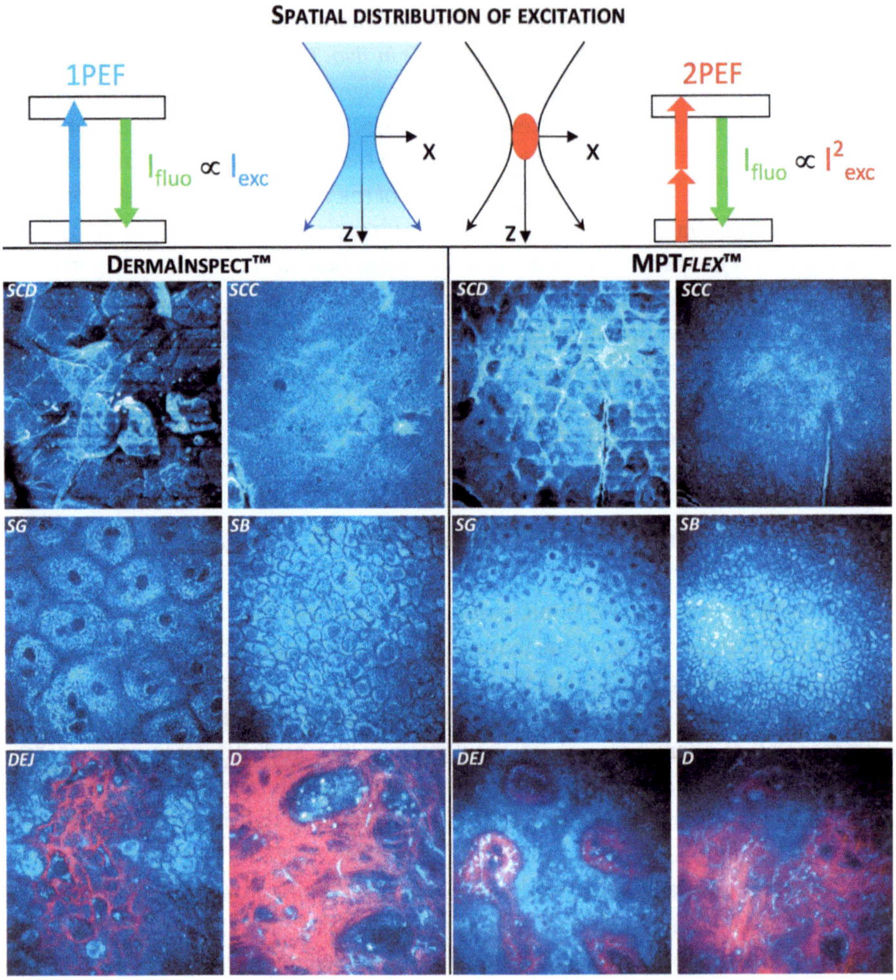

Fig. 19.1: (Top) Spatial distribution of illumination in one-photon and two-photon excitation process. (Bottom) Example of multiphoton 2PEF/SHG images acquired *in vivo* on human skin forearm with the DermaInspect™ and MPT*flex*™ systems at different depths. 2PEF signal (cyan hot color) reveals the endogenous fluorophores distribution inside the epidermis and dermis (elastic fibers, fibroblasts) whereas the SHG signal (red color) reveals the collagen fibers. Acquisition conditions: DermaInspect™ – 511 × 511 pixels, 0.255 μm/pixel; 130 × 130 μm^2, 7.4 s/image; MPT*flex*™ – 511 × 511 pixels, 0.4 μm/pixel; 205 × 205 μm^2, 4.4 s/image.

sition to living epidermis (*stratum granulosum, spinosum* and *basale*). The living cells, mainly keratinocytes, show homogenous fluorescent cytoplasm and nonfluorescent membranes and nuclei. Their diameter is about 30 μm in *stratum granulosum* and reaches 10 μm in *stratum basale* (both cell diameter and the ratio between the cytoplasmic and nuclear volumes decrease with depth). The cytoplasm fluorescence of the epidermal living cells is mainly due to mitochondrial NADH upon ≈ 760 nm excitation.

A 2PEF signal is also detected from melanin, which is found in different concentrations within some epidermal cells and mainly in the basal layers of the epidermis. As previously mentioned, a highly intense 2PEF signal, greater than the one created by the other cellular fluorophores, can be obtained in the basal layers where melanin is highly concentrated. FLIM can help distinguishing the less concentrated melanin regions inside the epidermis presenting comparable fluorescence intensities with the other cytoplasmic fluorophores. But, unfortunately, 2PEF-FLIM imaging does not allow distinguishing keratinocytes from melanocytes or from other epidermal cells such as Langerhans cells. Melanocytes can only be identified *in vivo* by multiphoton imaging in diseases such as melanoma due to the abnormal melanin accumulation in these cells and especially into their dendrites [58].

In the dermis, SHG and 2PEF contrast modes offer an insight into the 3D structure of superficial dermis ($\approx 100\,\mu m$). In young volunteers, this region corresponds to the papillary dermis, but in older volunteers, the superficial part of the reticular dermis can also be imaged (due to epidermal atrophy with age, flattening of the dermal-epidermal junction and thinning of the papillary dermis).

In multiphoton images, the interface between the epidermis and dermis is revealed by the appearance of an SHG signal. This signal is created by the fibrillar collagens and allows visualizing the collagen fiber organization inside the dermal papilla. The other visible constituents are the elastic fibers, the dermal cells (probably fibroblasts), the blood capillaries and blood cells often seen inside these capillaries.

In clinical trials, one needs to acquire several z-stacks of multiphoton images in different regions of an area of interest. In practice, in our clinical trials, we limited the measures to a minimum of 2 acquisitions as 3D imaging is time consuming. Typically, with the DermaInspect™, a multiphoton 3D (x, y, z) image of $130 \times 130 \times 162\,\mu m^3$ volume, corresponding to a z-stack of 70 *en face* images, can be acquired in less than 10 minutes (z-stack of 511×511 pixels images with 0.255 µm/pixel acquired every 2.35 µm; 7.4 s/image). But almost an hour per two areas of interest is starting to become tiring for the volunteers, not to mention the additional time required for image reacquisition, in case the volunteer moved during the acquisition. Hopefully, technology evolves and the new MPT*flex*™ system overcomes this difficulty by employing more sensitive detectors which offer the possibility of reducing the pixel dwell time.

MPT*flex*™, a second generation CE-marked, medically-certified multiphoton FLIM microscope available since 2011, presents several advantages over the DermaInspect™ [45]. First of all, it is a compact and therefore easily transportable system. It benefits from improved ergonomics due to the flexible articulated mechano-optical arm that opens the way to an *in vivo* investigation of human body areas less accessible such as face, scalp and armpits. The detection efficiency has been improved with the use of more sensitive detectors, but also by their implementation in a "miniaturized" scanning head at the end of the articulated arm. When comparing images acquired with similar pixel dwell times, the MPT*flex*™ offers an increased imaging depth even though the excitation power is limited to a maximum of 30 mW. The field of view is

≈ 4 times higher reaching $250 \times 250\,\mu m^2$ (see Fig. 19.1) and wide-field images can also be acquired within a $5 \times 5\,mm^2$ region by mosaic scanning using the motorized scan head. In practice, to keep a reasonable image acquisition time per z-stack, one has to reduce the FOV and/or the pixel dwell time. A z-stack of $205 \times 205 \times 162\,\mu m^3$ (65 *en face* 511×511 pixels images, $dx = 0.4\,\mu m$, $dz = 2.54\,\mu m$; 4.4 s/image, 16.8 µs/pixel) can be acquired in less than 5 minutes.

19.3 Quantitative data afforded by multiphoton imaging of human skin *in vivo*

The routine use of multiphoton microscopy in clinical practice requires specific automated image processing tools that would allow the investigator to extract pertinent quantitative 3D-information on skin layers and constituents. Up to now, very few image processing methods have been developed for characterizing multiphoton images of human skin *in vivo*. In the dermis, the proposed methods are mainly based on the computation of the mean intensity of 2PEF and SHG signals, or of the density of pixels occupied by these signals on conveniently chosen 2D slices within dermis at a fixed depth from the skin surface [46–49]. Parameters linked to the epidermis thickness [32] and keratinocytes morphology estimated on 2D slices within granulosum and spinosum layers [65] have also been proposed. But neither of these parameters was assessed in the 3D epidermal or dermal volume.

In the past years, our team focused on segmenting the multiphoton images in order to automatically separate the epidermis from the dermis [38] and validating 3D quantification parameters of interest for the study of photoaging or whitening phenomena [19–21, 50]. We developed a user-friendly software, the Multiphoton Skin Tools Suite (MPSTS), allowing to choose a set of 3D acquisitions, build the list of measures to compute and run a batch processing to compute all the segmentations as well as the measures for an entire clinical study. The tools that we propose open the possibility of characterizing in 3D the skin layers and sublayers, the cells, their melanin content or the elastic and collagen fibers in terms of quantity, size and organization.

19.3.1 Automatic 3D segmentation of skin layers

The aim of the segmentation is to classify each image voxel in one of the following compartments: coupling medium (CM), epidermis (ED) or dermis (D). Such a task is all but trivial, especially if the dermal-epidermal junction (DEJ) is neither flat nor parallel to the image plane. Moreover, given that a typical multiphoton clinical study may contain several hundred stacks of images, the segmentation method should not be time consuming. From a methodological image processing viewpoint, the main difficulties encountered with multiphoton skin images are:

- The signal tends to be more intense near the optical axis and decreases with the imaging depth, due to attenuation and diffusion of excitation and emission light.
- The SHG signal is specific to the dermis, but not enough to delimit it, as at the DEJ level, dermal regions can be found with elastic fibers only and no collagen fibers, *i.e.*, an absence of SHG signal.
- The sampling steps are not isotropic and the vertical resolution and sampling steps are large with respect to some structures, such as cells at the top of the living epidermis.

To overcome these difficulties, we use a marker-based approach. The general strategy is summed up by the flow chart shown in Fig. 19.2. Markers, sets of pixels considered as specific to a compartment, are afterwards propagated to the entire image volume, in order to have a complete partition.

The propagation is done in two steps:

(i) a fast watershed algorithm produces a low-level segmentation, made of super-voxels (small, homogeneous regions);
(ii) the super-voxels are merged, taking into account the markers, thanks to a graph cut algorithm.

All processing is done in 3D, except when otherwise specified. Moreover, the current version of the segmentation method, which has not undergone any specific computational optimization, takes an average of one minute to process a 3D multiphoton image on a classical personal computer (Intel core i7, 8 GB memory). For details, we refer the reader to [38].

At this stage, an epidermis mask, *i.e.*, a 3D binary image corresponding to the epidermis compartment is available for further processing. Using modern mathematical morphology operators, it is possible to further segment the epidermal layers, namely *stratum corneum* (SC) and *living epidermis* (LED). This automatic segmentation method allows quantifying independently the different compartments inside the epidermis (see Fig. 19.2). Moreover, a more refined segmentation of the epidermis can also be calculated: by defining 12 normalized layers inside the epidermis, one can have access, for example, at the melanin density distribution along the z-axis inside the epidermis.

19.3.2 Pseudo-FLIM specific melanin detection

So far, available melanin detection methods lacked specificity, were limited to 2 dimensions and/or needed invasive biopsies. The classical assessment method is based on Fontana–Masson histological staining associated with 2D white light imaging. But this approach is invasive and inappropriate in clinical studies where follow-up of melanin changes is needed. Confocal microscopy enables a non-invasive melanin visualization based on its higher reflectivity, but cellular membranes and corneocytes exhibit similar reflection signals, making melanin detection and quantification impossible.

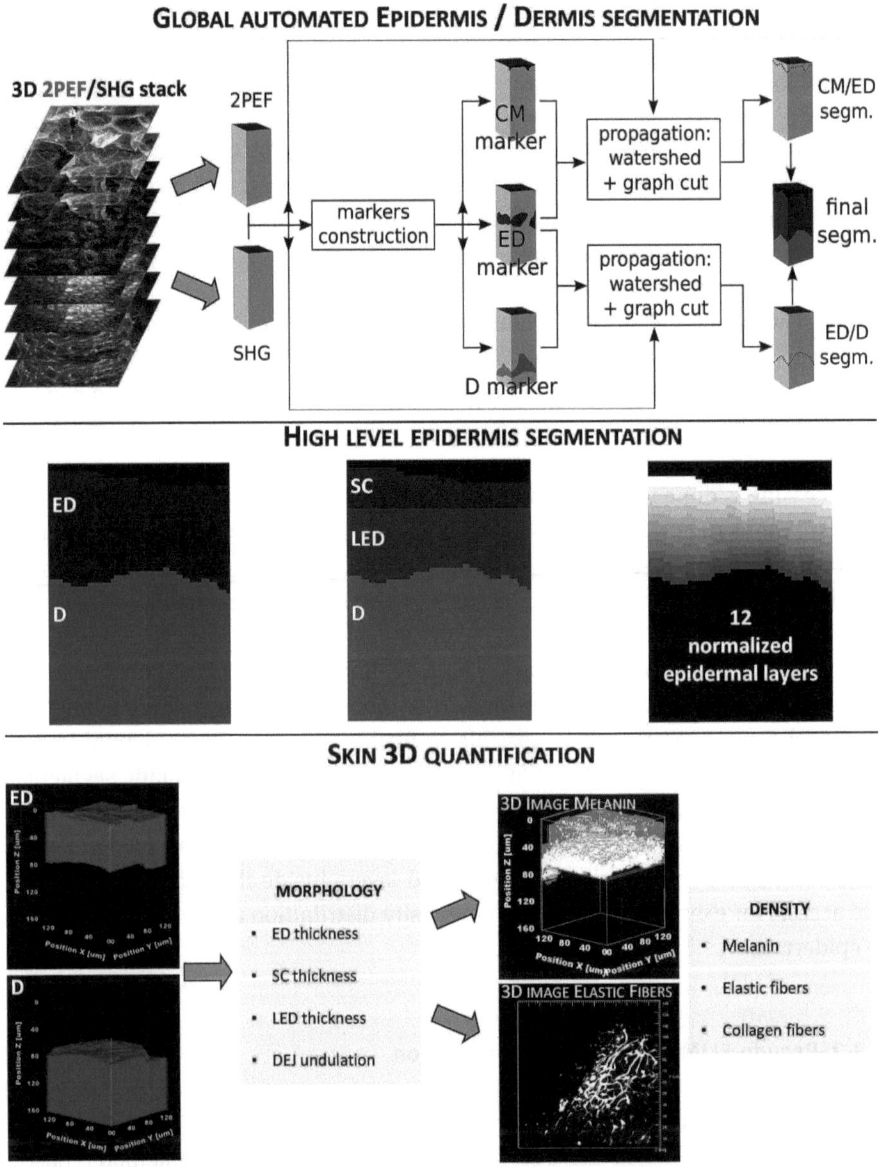

Fig. 19.2: General automated segmentation and quantification strategy for *in vivo* multiphoton FLIM images of human skin implemented in the MPSTS software.

In multiphoton microscopy, melanin quantification based on the 2PEF signal intensity can be performed in the basal layers of epidermis, where melanin is often highly concentrated [19]. But methods based upon fluorescence intensity levels are not always satisfactory: high 2PEF signals can also be detected from other fluorophores (e.g., keratin in the *stratum corneum*), and regions with low melanin concentration (low intensities) will not be taken into account.

FLIM affords specific melanin detection based on its lifetime, but the long acquisition times are incompatible with 3D imaging on human volunteers. In practice it is limited to selected 2D slices at a depth chosen by the user.

We proposed another approach based on multiphoton FLIM, called Pseudo-FLIM, to specifically detect melanin in the whole epidermis [20]. This requires binning the fluorescence photons in a reduced number of time channels (4 instead of 256, 2 ns/time channel). This can be done either at the acquisition or afterwards during processing. In our studies, we use the first option (see Fig. 19.3). The main steps are:
(i) apply a 2D-Gaussian filter on the raw 4-time channels images;
(ii) calculate a log linear regression on the first 3-time channels for each (x, y) pixel;
(iii) calculate a slope parametric image;
(iv) create a melanin mask by applying a threshold that keeps only the high slope values.

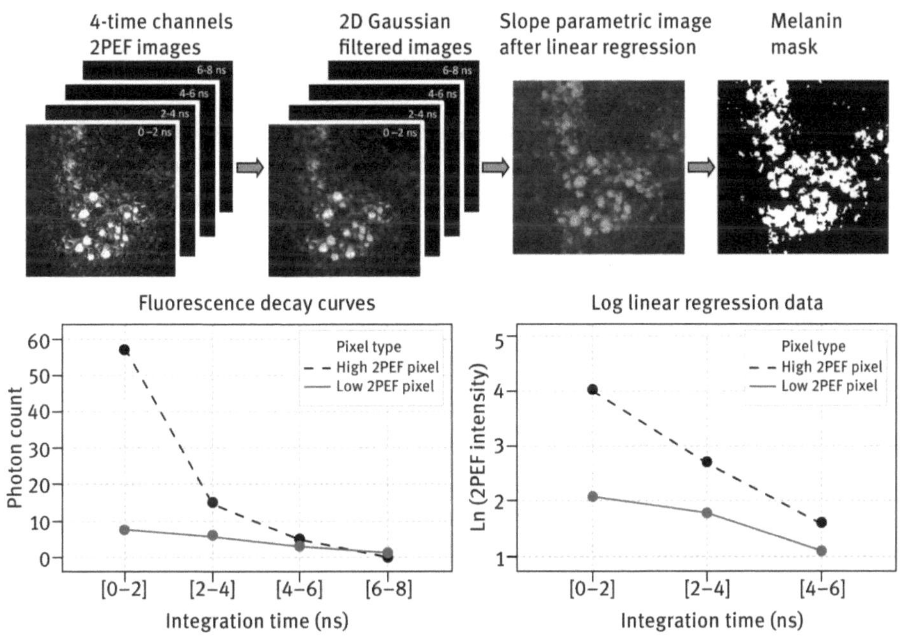

Fig. 19.3: Pseudo-FLIM Melanin quantification process. Specific melanin detection is done using an approach, called Pseudo-FLIM, which allows detecting melanin from multiphoton FLIM-like data compatible with 3D *in vivo* acquisitions on human volunteers as only a few photons per pixels are needed.

The slope of the decay (the higher the slope, the shorter the lifetime) allows to identify melanin pixels, as melanin presents a very short ($\tau_1 < 0.1$ ns) and predominant ($a_1 >$ 90 %) lifetime component (i.e., high slope values). As this analysis requires only a few photons per pixel, it results in a short image acquisition time (≈ 30 photons per pixel; ≈ 16 µs pixel dwell time) compatible with 3D imaging on humans.

19.3.3 Quantitative parameters

The 3D automatic segmentation opens the possibility of automatically computing quantitative measurements, on the segmentation frontiers (skin surface and DEJ), and within each layer (ED, SC, LED, D). Each measure is generically defined as the combination of a modality, a region of interest and an operator. Modalities are, for example, 2PEF, SHG or Pseudo-FLIM. Regions of interest (ROI) can be layers such as the dermis or epidermis, but also frontiers such as the DEJ. Operators compute the final measure using the modality and the ROI. For example, by choosing 2PEF as modality, ED as ROI, and the mean operator, one can compute the mean value of 2PEF intensity within the epidermis. We classify the quantitative parameters extracted from skin multiphoton images into morphological and density measures (see Fig. 19.2) [38].

19.3.3.1 Morphological measures

Morphological measures characterize the regions and frontiers shapes. They are directly derived from the 3D segmentation, without further need of the raw multiphoton signals. One can, for example, estimate the thickness (T) of the living epidermis in µm, T(LED), by calculating it along the z-axis at each (x, y) position and computing its mean value. It is also possible to quantify the DEJ undulation by estimating its normalized area, NA(DEJ), defined as the ratio between the areas of the DEJ surface and its projection on a horizontal plane. A value of 1 implies a totally flat and horizontal DEJ and higher values indicate a more undulated DEJ.

19.3.3.2 Density measures

The density measures can be estimated in 3D in the whole epidermal and dermal layers and sublayers. They aim at characterizing the melanin density in the epidermis or the elastic and collagen fiber density in the superficial dermis. The density parameter requires a binary mask computed by applying a threshold to 2PEF, SHG or Pseudo-FLIM modalities in order to detect the pixels containing melanin, elastic or collagen fibers. It is defined as the ratio of voxels occupied by these constituents to the total number of voxels of the ROI onto which the density is estimated. We also compute in the whole 3D imaged superficial dermis the SHG-to-AF aging index of dermis (3D–SAAID (D)), which has only been calculated in 2D in the literature [46, 82].

19.4 Cosmetic applications

The domain of application is dictated by the type of endogenous signals/constituents that can be evidenced by multiphoton FLIM microscopy. Cosmetic applications range from knowledge to evaluation studies aiming to acquire a better knowledge of skin differences appearing with aging, UV or solar exposure or between the different skin phototypes, as well as to evaluate the efficacy of cosmetic anti-aging or whitening ingredients. The penetration of cosmetic compounds could also be investigated by multiphoton FLIM techniques provided that their 2PEF emission spectrum and/or fluorescence lifetime are different compared to the ones of endogenous skin constituents.

In the following, we will discuss the most straightforward cosmetic clinical applications. The clinical trials described hereafter were performed with the DermaInspect™ device. The experimental protocols were approved by the Saint Louis Hospital ethics committee, complying with the Declaration of Helsinki. All volunteers gave written, informed consent.

19.4.1 Photo-aging

The study of age-induced skin modifications with multiphoton FLIM microscopy is of particular importance if one wants to non-invasively assess the anti-aging effects of a cosmetic compound at the microscopic level and follow these effects over time, before and after product application. To explore the 3D skin differences that can be evidenced by multiphoton microscopy, we performed a clinical trial on 15 young (18–25 a) and 15 aged (70–75 a) Caucasian human female volunteers, focusing on ventral and dorsal forearm sides (sun protected vs. sun-exposed) [38, 50].

As previously mentioned, with multiphoton microscopy epidermis and superficial dermis can be characterized in 3D up to a depth of approximately 160 μm. In young subjects, dermis includes papillary dermis and a limited fraction of reticular dermis, whereas in elderly subjects it mainly corresponds to reticular dermis as defined in histology due to epidermis atrophy and loss of DEJ undulation. In spite of the very limited field of view and imaging depth, as compared to histology, multiphoton imaging and image processing provide quantitative parameters in agreement with what is known for skin aging and photoaging, without the need for skin biopsy.

The 3D quantification data on epidermal thickness and DEJ undulation are shown in Fig. 19.4. We observed no difference in mean thickness of the epidermis between dorsal and volar side of the forearm, but a decrease with age, in agreement with the literature.

Similarly, the normalized area of the dermal-epidermal junction, which is related to the shape of the DEJ, decreases with age without any differences between dorsal and volar side. As expected, with age the ridges of the DEJ flatten out. The melanin content

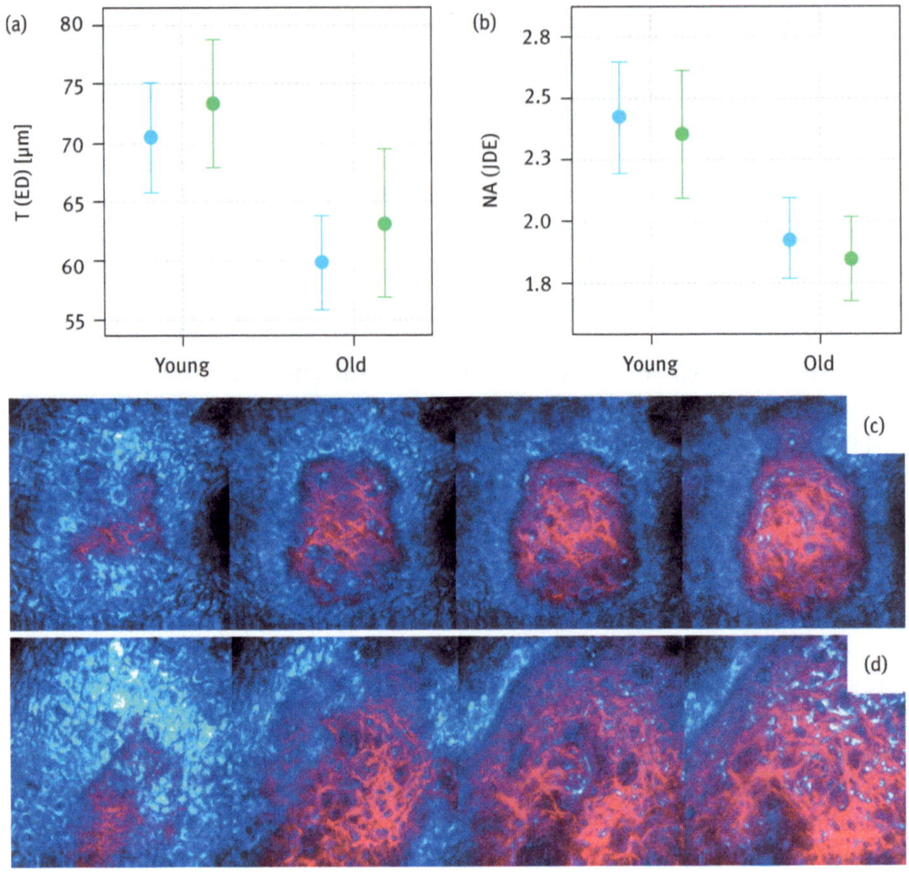

Fig. 19.4: Epidermal changes induced with photo-aging: (a) epidermal thickness; (b) normalized area of DEJ. The 2PEF/SHG images (ventral forearm side) give insights into the shape of the DEJ: round papillae in young volunteers, and flattened ones in old volunteers (d). For each age class, the left blue bars correspond to the ventral measures, and the right green ones to the dorsal measures. Error bars correspond to 95 % confidence intervals of the mean.

(data not shown) is higher on dorsal versus volar side of the forearm, but without any difference between young and old volunteers.

Fig. 19.5 shows the changes occurring with age in the superficial dermis, namely the changes in the elastic and collagen fiber densities that can be evidenced with multiphoton microscopy.

For the mean density of elastic fibers, 3D-2PEF Density (D), an increase is observed with age due to solar elastosis and no differences between dorsal and ventral forearm sides. The mean density of collagen fibers, 3D-SHG Density (D), shows no difference between young and old groups, but is higher on ventral sides compared to dorsal ones.

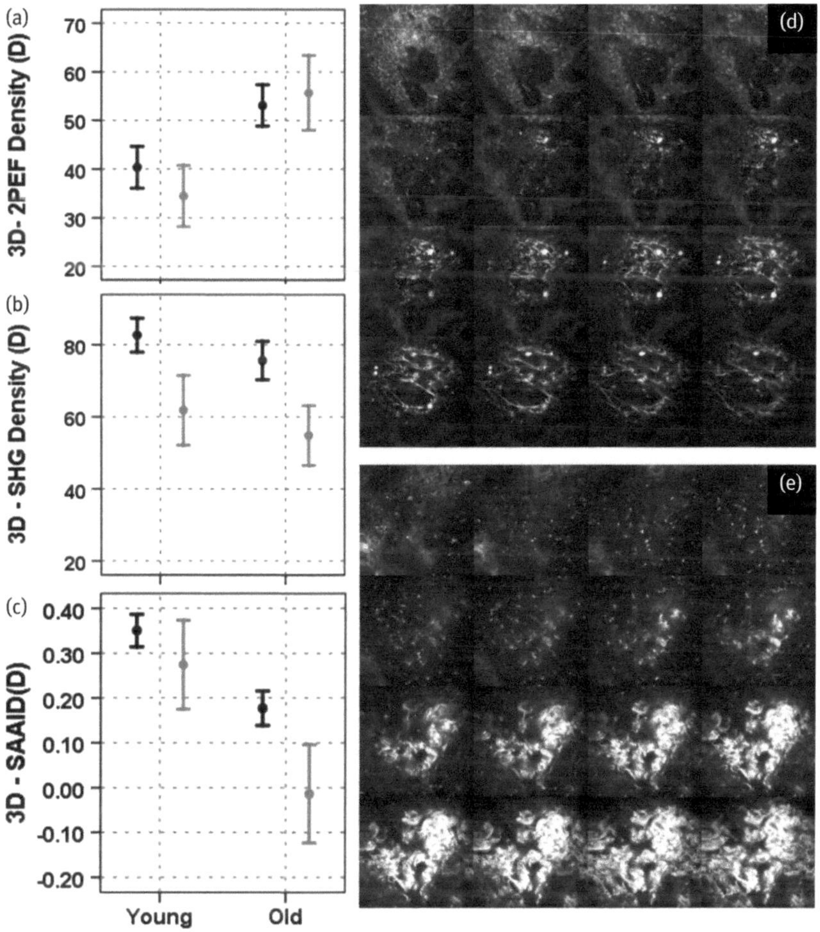

Fig. 19.5: Dermal changes induced with photo-aging in the 3D density of (a) elastic fibers and (b) collagen fibers. The two densities can be used to calculate in 3D the SHG-to-AF aging index of dermis – 3D-SAAID (c). The 2PEF images (dorsal forearm side) show the elastic fibers differences between (d) young and (e) old volunteers. The increase in elastic fiber density with age, due to solar elastosis, is clearly seen (e). For each age class, the left black bars correspond to the ventral measures, and the right gray ones to the dorsal measures. Error bars correspond to 95 % confidence intervals of the mean.

The SHG-to-AF aging index of dermis 3D-SAAID (D), estimated here in the whole imaged volume of superficial dermis, highly decreases with age as also evidenced in 2D at a fixed depth from the skin surface [46, 82]. Moreover, its mean value is also lower on the sun-exposed dorsal forearms compared with the sun-protected ventral side. Negative values mean that there are more significant pixels with 2PEF than with SHG signals, and a value close to zero means that both proportions are similar.

The changes with age in the elastic and collagen fibers can also be addressed in terms of organization, by quantifying the relative position of 2PEF and SHG signals [38].

The estimation of all these density-based measures is more representative of the superficial dermis in 3D than in 2D. The 2D measurements of elastic or collagen fiber density done at a specific fixed imaging depth can be impacted, for example, by the changes in epidermis thickness or DEJ undulation. This is problematic when one wants to assess the effects of cosmetic products or any other effects, especially if the morphology of the epidermis is also modified. In this case, at a fixed depth from the skin surface, the variations in elastic or collagen fiber densities cannot be attributed to the product alone, as the dermal regions investigated before and after product application will not be the same. Not only does the 3D approach avoid the selection of a 2D plane, but the measurements are based on a larger, more representative, volume of tissue. The density measurements are also more robust than the intensity measurements, as these are directly dependent on the excitation power. The excitation power delivered at a certain depth will depend upon epidermal thickness, the presence of melanin or other constituents that can attenuate or diffuse both the laser and the emitted multiphoton signals. More generally, the interpretation of a quantitative parameter should be done by taking into account all the available information in a study.

With this photo-aging study, we showed for the first time that it is possible to process and extract in 3D quantitative measurements revealing expected aging effects.

19.4.2 Study of constitutive pigmentation

Melanin-specific detection by multiphoton FLIM opens the possibility to non-invasively characterize the changes between the different skin phototypes, the pigmentation disorders or to assess compounds able to modulate skin melanin content. When combining the Pseudo-FLIM specific melanin detection with the high level segmentation of the epidermis, one can not only have non-invasive access to the global melanin density, but also to its z-distribution through the epidermis.

In a clinical trial on 45 female volunteers (18–55 a) with different skin color (Individual Typology Angle [83] – ITA values of their ventral forearm ranging from very light – grade I to brown skin – grade V), we evidenced an increase in the global mean melanin density with ITA group, as expected and in agreement with histology (see Fig. 19.6). Moreover, we studied the melanin density normalized z-profile from the dermal-epidermal junction up to the *stratum corneum*. Our results show an increased melanin density at the basal layers for ITA groups II to V, almost no differences at the level of *stratum granulosum* and a slight increase in melanin density at the level of *stratum corneum* for darker skin color.

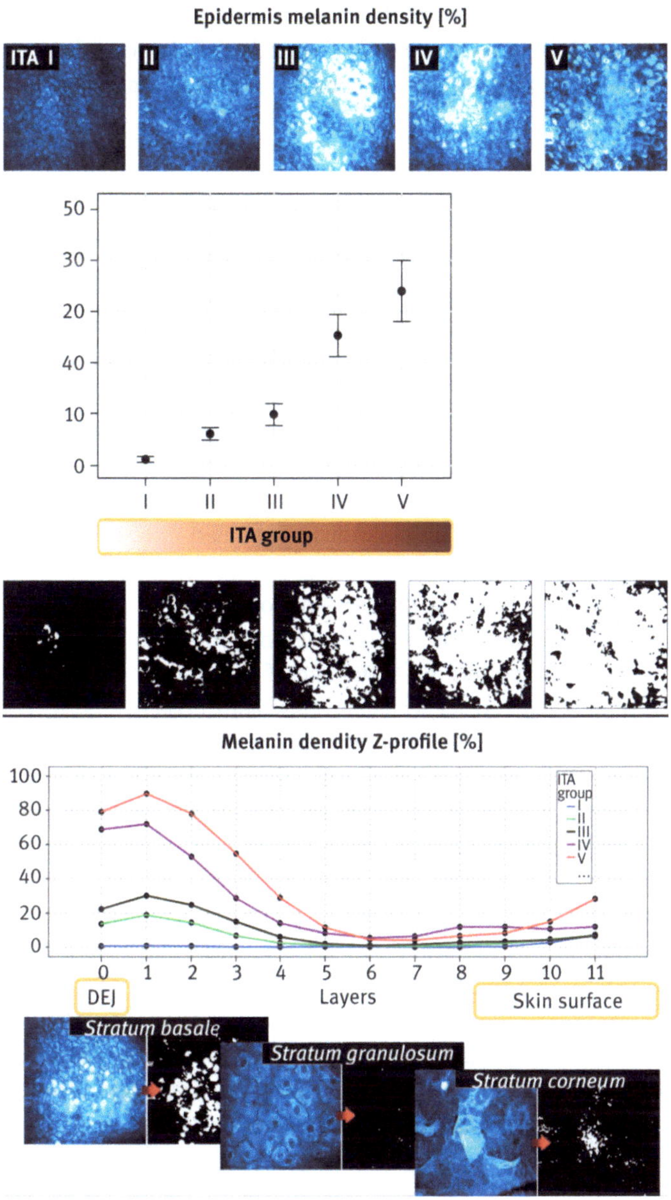

Fig. 19.6: Change with skin color in (top) mean global melanin density of the epidermis and (bottom) in the normalized melanin density *z*-profile (12 normalized layers from 0 – DEJ level to 11 – SC level). The data are expressed in % and error bars correspond to 95 % confidence intervals of the mean. Raw 2PEF images are shown in cyan hot color and melanin masks images obtained by Pseudo-FLIM method in white color.

This innovative approach could have large applications in both the dermatological and cosmetic fields, since easily applied to the characterization of pigmentation disorders or to the assessment of compounds able to modulate skin melanin content.

19.4.3 Efficacy assessment of anti-aging or whitening cosmetic ingredients

Multiphoton FLIM microscopy, as shown above, is able to evidence the age-induced skin changes and the differences in constitutive pigmentation. One of the most straightforward applications in cosmetics is the overtime follow up of the efficacy of anti-aging and whitening ingredients. This can be done in either long-term studies in which the products are applied for weeks or months or in short-term studies during about one month. The ability of multiphoton microscopy to continuously monitor and evaluate epidermal effects of pharmaceutics compounds in short-term study is shown in [19].

In the following, we will give an example of efficacy evaluation of anti-photoaging agents in a short-term screening protocol based on an occlusive patch test. The occlusive patch test was initially developed for assessing topical retinoid activity in human skin and has been extended as a short-term screening protocol for anti-aging agents. In this model, biopsies are performed at the end of the occlusion period for morphological and immunohistochemistry analysis. Our main objective with this study was firstly, to demonstrate that multiphoton microscopy, in association with a patch test, could be used as a short-term screening protocol for anti-aging and whitening agents, and secondly, to validate the relevance of this method for kinetic and quantitative assessment with gold standards of anti-aging that are retinoids.

For that, 20 women, aged 50–65 years, were enrolled. Retinol 0.3 % (RO) and Retinoic acid 0.025 % (RA) were applied to the dorsal photodamaged side of their forearm under occlusive patches for 12 days. A patch alone was applied to a third area as control. Evaluation was performed at day D0, D12 (end of treatment), D18 and D32 using multiphoton microscopy. Epidermal thickness, normalized area of the DEJ and melanin density were estimated using the 3D image processing tools described in Section 19.3. Both RO and RA induced changes in these 3 parameters (see the complete study results in [21]). The anti-aging (epidermal thickening, increase in DEJ undulation) and whitening (decrease in melanin density) effects of RO are shown in Fig. 19.7.

Retinol induced an epidermal thickening due to living epidermis thickening with no effect on SC. This effect was significant at D12 (end of the treatment), D18 and D32 vs. baseline and vs. control.

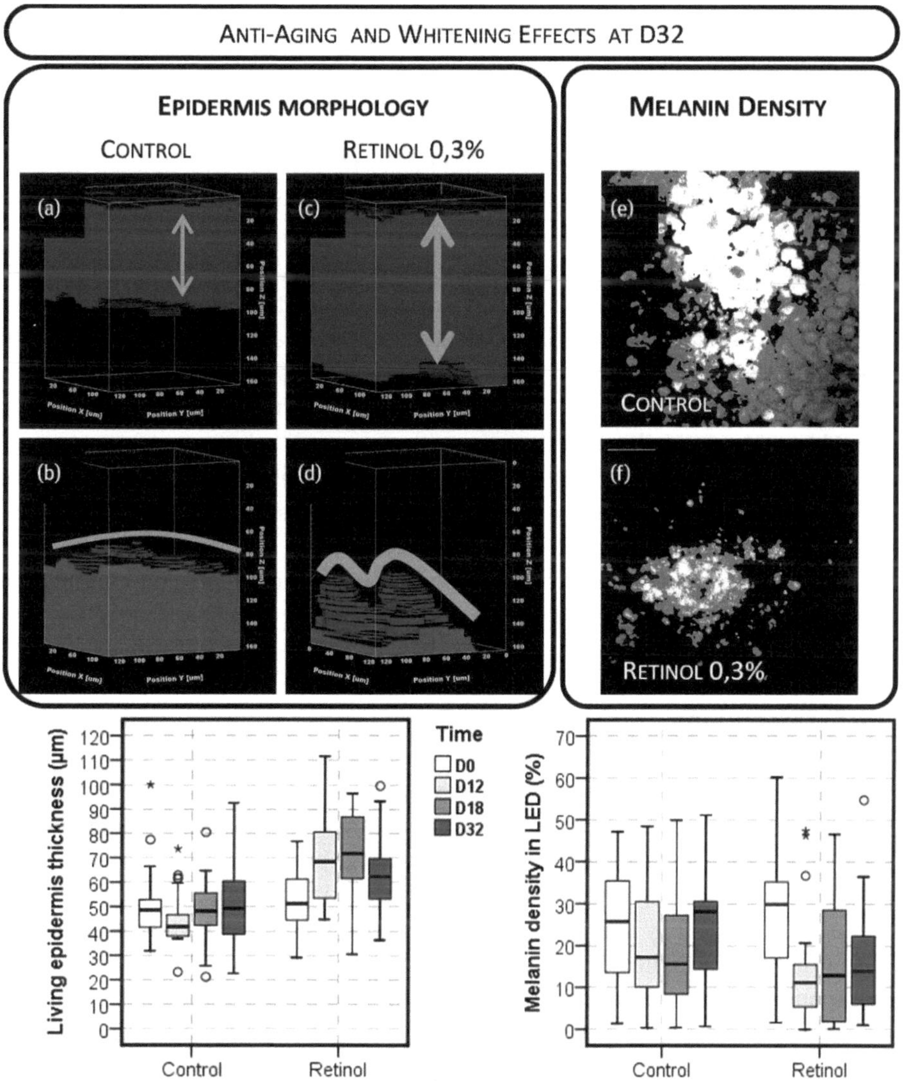

Fig. 19.7: Quantification results of Retinol effects obtained with multiphoton microscopy. The epidermal thickening and the increase in DEJ undulation at D32 for control and Retinol 0.3 % can be visualized on the 3D volume renderings of the segmented epidermal (a, c) and dermal (b, d) compartments. Images (e) and (f) acquired at the basal layers exemplify the melanin density changes. The 3D quantification data are expressed as boxplots with fences. The boxes contain 50 % of the data; the intervals between the lower limit of the box and the lower inner fence contain 25 %, and vice versa for the other 25 %. – indicates the median that divides the population in two groups with equal numbers of data points; ° the outliers and * the extreme data points.

Additionally, multiphoton microscopy evidenced an increase in DEJ undulation with retinoid treatment, the effect of retinol being, to our knowledge, a new finding.

Retinoid-induced decrease in melanin content is also observed for the first time in such a short-term protocol. In some subjects, visible skin whitening was observed on retinoid-treated areas which correlate with the decrease in melanin density. Changes in melanin density are clearly visible on melanin masks obtained using Pseudo-FLIM specific melanin detection method (Fig. 19.7 (e) and (f)).

Dermal changes could not be addressed in this study, in which epidermal thickness increased drastically and exceeded the imaging depth at some time points. We believe that cosmetic product-induced dermal changes that could be addressed with multiphoton microscopy are not likely to be observed in such a short 12 days period of treatment. Possible effects in the superficial dermal structure using this technique should be addressed in long-term studies with open applications.

This study shows that short-term protocol in combination with non-invasive *in vivo* multiphoton microscopy allows epidermal effects induced by retinoids, including melanin content, to be accurately detected and quantified over time. This is the first application of innovative specific 3D quantification tools for multiphoton images to the evaluation of a cutaneous treatment.

19.5 Conclusion

Multiphoton FLIM in cosmetic clinical research is an exciting area of application of this innovative and non-invasive imaging tool, compatible with *in vivo* studies on human volunteers. In association with specific 3D image processing, one can extract several representative quantitative parameters characterizing skin constituents in terms of morphology, density and organization in both epidermis and superficial dermis. In spite of the very limited field of view and imaging depth, as compared to histology, multiphoton FLIM imaging and image processing provide quantitative data of interest in skin photo-aging and pigmentation, without the need for skin biopsy. These innovative approaches opened the way to the kinetic evaluation of the efficacy of cosmetic molecules at the subcellular level. With the arrival of flexible and compact imaging systems, future work in this domain will focus on human face skin, the preferred body region for the application of cosmetic products.

References

[1] Wilhelm KP, Elsner P, Berardesca E, Maibach HI. Bioengineering of the skin: skin imaging and analysis. 2nd edition. New York: Informa Healthcare USA Inc.; 2007.

[2] Masters BR, So PTC, Gratton E. Multiphoton excitation fluorescence microscopy and spectroscopy of in vivo human skin. Biophysical Journal. 1997;72:2405–2412.

[3] Teuchner K, Freyer W, Leupold D, Volkmer A, Birch DJS, Altmeyer P, Stucker M, Hoffmann K. Femtosecond two-photon excited fluorescence of melanin. Photochemistry & Photobiology. 1999;70:146–151.

[4] Campagnola PJ, Millard AC, Terasaki M, Hoppe PE, Malone CJ, Mohler WA. Three-dimensional high-resolution second-harmonic generation imaging of endogenous structural proteins in biological tissues. Biophysical Journal. 2002;82:493–508.

[5] Zipfel WR, Williams RM, Christiet R, Nikitin AY, Hyman BT, Webb WW. Live tissue intrinsic emission microscopy using multiphoton-excited native fluorescence and second harmonic generation. Proceedings of the National Academy of Sciences of the United States of America. 2003;100:7075–7080.

[6] König K, Riemann I. High-resolution multiphoton tomography of human skin with subcellular spatial resolution and picosecond time resolution. Journal of Biomedical Optics. 2003; 8:432–439.

[7] Pena AM, Strupler M, Boulesteix T, Godeau G, Schanne-Klein MC. Spectroscopic analysis of keratin endogenous signal for skin multiphoton microscopy. Opt Express. 2005;13:6268–6274. Erratum 13(17):6667.

[8] König K. Multiphoton microscopy in life sciences. Journal of Microscopy. 2000;200:83–104.

[9] Zipfel WR, Williams RM, Webb WW. Nonlinear magic: multiphoton microscopy in the biosciences. Nature Biotechnology. 2003;21:1369–1377.

[10] Göppert-Mayer M. Über Elementarakte mit zwei Quantensprüngen. Annalen der Physik. 1931; 401:273–294.

[11] Grzybowski A, Pietrzak K. Maria Goeppert-Mayer (1906–1972): Two-photon effect on dermatology. Clinics in Dermatology. 2013;31:221–225.

[12] Kaiser W, Garrett CGB. Two-photon excitation in CaF2: Eu2+. Physical Review Letters. 1961;7: 229–231.

[13] Denk W, Strickler JH, Webb WW. Two-photon laser scanning microscopy. Science. 1990;248: 73–76.

[14] Pena AM, Boulesteix T, Dartigalongue T, Schanne-Klein MC. Chiroptical effects in the second harmonic signal of collagens I and IV. J Am Chem Soc. 2005;127:10314–10322.

[15] Periasamy A, Clegg RM. FLIM microscopy in biology and medicine. 1st edition. Boca Raton, Florida: Chapman and Hall/CRC, Taylor & Francis Group; 2010.

[16] Dancik Y, Favre A, Loy CJ, Zvyagin AV, Roberts MS. Errata: Use of multiphoton tomography and fluorescence lifetime imaging to investigate skin pigmentation in vivo. Journal of Biomedical Optics. 2013;18:029802.

[17] Chance B, Estabrook RW, Ghosh A. Damped sinusoidal oscillations of cytoplasmic reduced pyridine. Proceedings of the National Academy of Sciences of the United States of America. 1964;51:1244–1251.

[18] Georgakoudi I, Quinn KP. Optical imaging using endogenous contrast to assess metabolic state. Annual Review of Biomedical Engineering. 2012;14:351–367.

[19] Ait El Madani H, Tancrède-Bohin E, Bensussan A, Colonna A, Dupuy A, Bagot M, Pena AM. In vivo multiphoton imaging of human skin: assessment of topical corticosteroid-induced epidermis atrophy and depigmentation. Journal of Biomedical Optics. 2012;17:026009.

[20] Pena AM, Baldeweck T, Tancrede E, Decencière E, Koudoro S, inventors; Non-invasive method for specific 3D detection, visualization and/or quantification of an endogenous fluorophore such as melanin in a biological tissue. France patent FR2982369. International publication number WO2013068943; 2011.

[21] Tancrède-Bohin E, Baldeweck T, Decencière E, Brizion S, Victorin S, Parent N, Faugere J, Souverain L, Bagot M, Pena AM. Non-invasive short-term assessment of retinoids effects on human

skin in vivo using multiphoton microscopy. Journal of the European Academy of Dermatology and Venereology. 2015;29:673–681.

[22] Masters BR, So PTC, Gratton E. Optical biopsy of in vivo human skin: Multi-photon excitation microscopy. Lasers in Medical Science. 1998;13:196–203.

[23] Masters BR, So PTC, Gratton ENRI. Multiphoton Excitation Microscopy of In Vivo Human Skin: Functional and Morphological Optical Biopsy Based on Three-Dimensional Imaging, Lifetime Measurements and Fluorescence Spectroscopy. Annals of the New York Academy of Sciences. 1998;838:58–67.

[24] Masters BR, So PTC. Multi-photon excitation microscopy and confocal microscopy imaging of in vivo human skin: A comparison. Microscopy and Microanalysis. 1999;5:282–289.

[25] Masters BR, So PTC. Multiphoton excitation microscopy of human skin in vivo: Early development of an optical biopsy. Proc SPIE 4001; 2000. p. 156–164.

[26] Masters BR, So PTC. Confocal microscopy and multi-photon excitation microscopy of human skin in vivo. Optics Express. 2001;8:2–10.

[27] König K, Ehlers A, Riemann I, Schenkl S, Buckle R, Kaatz M. Clinical two-photon microendoscopy. Microscopy Research and Technique. 2007;70:398–402.

[28] König K. Clinical multiphoton tomography. Journal of Biophotonics. 2008;1:13–23.

[29] Graf BW, Jiang Z, Tu H, Boppart SA. Dual-spectrum laser source based on fiber continuum generation for integrated optical coherence and multiphoton microscopy. Journal of Biomedical Optics. 2009;14.

[30] Breunig HG, Studier H, König K. Multiphoton excitation characteristics of cellular fluorophores of human skin in vivo. Optics Express. 2010;18:7857–7871.

[31] Wang BG, König K, Halbhuber KJ. Two-photon microscopy of deep intravital tissues and its merits in clinical research. Journal of Microscopy. 2010;238:1–20.

[32] Koehler MJ, Vogel T, Elsner P, Konig K, Buckle R, Kaatz M. In vivo measurement of the human epidermal thickness in different localizations by multiphoton laser tomography. Skin Research and Technology. 2010;16:259–264.

[33] König K. High-resolution multimodal clinical multiphoton tomography of skin. Progress in Biomedical Optics and Imaging – Proc SPIE 7883; 2011. p. 78830D-1.

[34] Benati E, Bellini V, Borsari S, Dunsby C, Ferrari C, French P, Guanti M, Guardoli D, Koenig K, Pellacani G, et al. Quantitative evaluation of healthy epidermis by means of multiphoton microscopy and fluorescence lifetime imaging microscopy. Skin Research and Technology. 2011; 17:295–303.

[35] Liang X, Graf BW, Boppart SA. In vivo multiphoton microscopy for investigating biomechanical properties of human skin. Cellular and Molecular Bioengineering. 2011;4:231–8.

[36] Graf BW, Boppart SA. Multimodal in vivo skin imaging with integrated optical coherence and multiphoton microscopy. IEEE Journal on Selected Topics in Quantum Electronics. 2012;18: 1280–1286.

[37] Seidenari S, Arginelli F, Bassoli S, Cautela J, French PMW, Guanti M, Guardoli D, König K, Talbot C, Dunsby C. Multiphoton laser microscopy and fluorescence lifetime imaging for the evaluation of the skin. Dermatology Research and Practice. 2012:810749.

[38] Decencière E, Tancrède-Bohin E, Dokladal P, Koudoro S, Pena AM, Baldeweck T. Automatic 3D segmentation of multiphoton images: A key step for the quantification of human skin. Skin Research and Technology. 2013;19:115–124.

[39] Balu M, Mazhar A, Hayakawa CK, Mittal R, Krasieva TB, König K, Venugopalan V, Tromberg BJ. In vivo multiphoton NADH fluorescence reveals depth-dependent keratinocyte metabolism in human skin. Biophysical Journal. 2013;104:258–267.

[40] Weinigel M, Breunig HG, Fischer P, Kellner-Höfer M, Bückle R, König K. Clinical multiphoton endoscopy with FLIM capability. Progress in Biomedical Optics and Imaging – Proc SPIE 8588; 2013.
[41] Wang H, Lee AMD, Frehlick Z, Lui H, McLean DI, Tang S, Zeng H. Perfectly registered multiphoton and reflectance confocal video rate imaging of in vivo human skin. Journal of Biophotonics. 2013;6:305–309.
[42] Koenig K, Weinigel M, Breunig HG, Uchugonova A. Quantitative multiphoton imaging. Progress in Biomedical Optics and Imaging – Proc SPIE 8948; 2014.
[43] Yew E, Rowlands C, So PTC. Application of multiphoton microscopy in dermatological studies: a mini-review. Journal of Innovative Optical Health Sciences. 2014;7(5):1330010.
[44] Darvin ME, Richter H, Zhu YJ, Meinke MC, Knorr F, Gonchukov SA, König K, Lademann J. Comparison of in vivo and ex vivo laser scanning microscopy and multiphoton tomography application for human and porcine skin imaging. Quantum Electron. 2014;44:646–651.
[45] König K. Hybrid multiphoton multimodal tomography of in vivo human skin. IntraVital. 2012;1: 11–26.
[46] Koehler MJ, Konig K, Elsner P, Buckle R, Kaatz M. In vivo assessment of human skin aging by multiphoton laser scanning tomography. Optics Letters. 2006;31:2879–2881.
[47] Koehler MJ, Hahn S, Preller A, Elsner P, Ziemer M, Bauer A, Konig K, Buckle R, Fluhr JW, Kaatz M. Morphological skin aging criteria by multiphoton laser scanning tomography: Non-invasive in vivo scoring of the dermal fibre network. Experimental Dermatology. 2008;17:519–523.
[48] Koehler MJ, Preller A, Kindler N, Elsner P, König K, Bückle R, Kaatz M. Intrinsic, solar and sunbed-induced skin aging measured in vivo by multiphoton laser tomography and biophysical methods. Skin Research and Technology. 2009;15:357–363.
[49] Kaatz M, Sturm A, Elsner P, König K, Bückle R, Koehler MJ. Depth-resolved measurement of the dermal matrix composition by multiphoton laser tomography. Skin Research and Technology. 2010;16:131–136.
[50] Baldeweck T, Tancrède E, Dokladal P, Koudoro S, Morard V, Meyer F, Decencière E, Pena AM. In vivo multiphoton microscopy associated to 3D image processing for human skin characterization. Progress in Biomedical Optics and Imaging – Proc SPIE 8226; 2012.
[51] Koehler MJ, Preller A, Elsner P, König K, Hipler UC, Kaatz M. Non-invasive evaluation of dermal elastosis by in vivo multiphoton tomography with autofluorescence lifetime measurements. Experimental Dermatology. 2012;21:48–51.
[52] Puschmann S, Rahn CD, Wenck H, Gallinat S, Fischer F. Approach to quantify human dermal skin aging using multiphoton laser scanning microscopy. Journal of Biomedical Optics. 2012; 17:036005.
[53] Sanchez WY, Obispo C, Ryan E, Grice JE, Roberts MS. Erratum: Changes in the redox state and endogenous fluorescence of in vivo human skin due to intrinsic and photo-aging, measured by multiphoton tomography with fluorescence lifetime imaging. Journal of Biomedical Optics. 2013;18(6):069802.
[54] Miyamoto K, Kudoh H. Quantification and visualization of cellular NAD(P)H in young and aged female facial skin with in vivo two-photon tomography. British Journal of Dermatology. 2013; 169:25–31.
[55] Darvin ME, Richter H, Ahlberg S, Haag SF, Meinke MC, Le Quintrec D, Doucet O, Lademann J. Influence of sun exposure on the cutaneous collagen/elastin fibers and carotenoids: Negative effects can be reduced by application of sunscreen. Journal of Biophotonics. 2014;7:735–743.
[56] Leppert J, Krajewski J, Kantelhardt SR, Schlaffer S, Petkus N, Reusche E, Hüttmann G, Giese A. Multiphoton excitation of autofluorescence for microscopy of glioma tissue. Neurosurgery. 2006;58:759–767.

[57] König K, Speicher M, Bückle R, Reckfort J, McKenzie G, Welzel J, Koehler MJ, Elsner P, Kaatz M. Clinical optical coherence tomography combined with multiphoton tomography of patients with skin diseases. Journal of Biophotonics. 2009;2:389–397.
[58] Dimitrow E, Riemann I, Ehlers A, Koehler MJ, Norgauer J, Elsner P, König K, Kaatz M. Spectral fluorescence lifetime detection and selective melanin imaging by multiphoton laser tomography for melanoma diagnosis. Experimental Dermatology. 2009;18:509–515.
[59] Dimitrow E, Ziemer M, Koehler MJ, Norgauer J, König K, Elsner P, Kaatz M. Sensitivity and specificity of multiphoton laser tomography for in vivo and ex vivo diagnosis of malignant melanoma. Journal of Investigative Dermatology. 2009;129:1752–1758.
[60] Paoli J, Smedh M, Ericson MB. Multiphoton laser scanning microscopy – a novel diagnostic method for superficial skin cancers. Seminars in Cutaneous Medicine and Surgery. 2009;28: 190–195.
[61] König K, Speicher M, Kohler MJ, Scharenberg R, Kaatz M. Clinical application of multiphoton tomography in combination with high-frequency ultrasound for evaluation of skin diseases. Journal of Biophotonics. 2010;3:759–773.
[62] Huck V, Gorzelanny C, Thomas K, Niemeyer V, Luger TA, Konig K, Schneider SW. Intravital multiphoton tomography as a novel tool for non-invasive in vivo analysis of human skin affected with atopic dermatitis. Progress in Biomedical Optics and Imaging – Proc SPIE 7548; 2010.
[63] Huck V, Gorzelanny C, Thomas K, Mess C, Dimitrova V, Schwarz M, Riemann I, Niemeyer V, Luger TA, König K, et al. Intravital multiphoton tomography as an appropriate tool for non-invasive in vivo analysis of human skin affected with atopic dermatitis. Progress in Biomedical Optics and Imaging – Proc SPIE 7883; 2011.
[64] Koehler MJ, Speicher M, Lange-Asschenfeldt S, Stockfleth E, Metz S, Elsner P, Kaatz M, Konig K. Clinical application of multiphoton tomography in combination with confocal laser scanning microscopy for in vivo evaluation of skin diseases. Experimental Dermatology. 2011;20: 589–594.
[65] Koehler MJ, Zimmermann S, Springer S, Elsner P, König K, Kaatz M. Keratinocyte morphology of human skin evaluated by in vivo multiphoton laser tomography. Skin Research and Technology. 2011;17:479–486.
[66] Seidenari S, Arginelli F, Dunsby C, French P, König K, Magnoni C, Manfredini M, Talbot C, Ponti G. Multiphoton laser tomography and fluorescence lifetime imaging of basal cell carcinoma: Morphologic features for non-invasive diagnostics. Experimental Dermatology. 2012;21: 831–836.
[67] Seidenari S, Arginelli F, Bassoli S, Cautela J, Cesinaro AM, Guanti M, Guardoli D, Magnoni C, Manfredini M, Ponti G, et al. Diagnosis of BCC by multiphoton laser tomography. Skin Research and Technology. 2013;19:e297–e304.
[68] Patalay R, Talbot C, Alexandrov Y, Lenz MO, Kumar S, Warren S, Munro I, Neil MAA, König K, French PMW, et al. Multiphoton Multispectral Fluorescence Lifetime Tomography for the Evaluation of Basal Cell Carcinomas. PLoS ONE. 2012;7:e43460.
[69] Balu M, Kelly KM, Zachary CB, Harris RM, Krasieva TB, König K, Tromberg BJ. Clinical studies of pigmented lesions in human skin by using a multiphoton tomograph, Progress in Biomedical Optics and Imaging – Proc SPIE 8588; 2013. p. 858812-1–858812-6.
[70] Ulrich M, Klemp M, Darvin ME, König K, Lademann J, Meinke MC. In vivo detection of basal cell carcinoma: Comparison of a reflectance confocal microscope and a multiphoton tomograph. Journal of Biomedical Optics. 2013;18:061229-1–061229-7.
[71] Balu M, Kelly KM, Zachary CB, Harris RM, Krasieva TB, König K, Durkin AJ, Tromberg BJ. Distinguishing between benign and malignant melanocytic nevi by in vivo multiphoton microscopy. Cancer Research. 2014;74:2688–2697.

[72] König K, Ehlers A, Stracke F, Riemann I. In vivo drug screening in human skin using femtosecond laser multiphoton tomography. Skin Pharmacology & Physiology. 2006;19:78–88.
[73] Stracke F, Weiss B, Lehr C-M, König K, Schaefer UF, Schneider M. Multiphoton microscopy for the investigation of dermal penetration of nanoparticle-borne drugs. Journal of Investigative Dermatology. 2006;126:2224–2233.
[74] Schenke-Layland K, Riemann I, Damour O, Stock UA, Konig K. Two-photon microscopes and in vivo multiphoton tomographs – Powerful diagnostic tools for tissue engineering and drug delivery [Review]. Advanced Drug Delivery Reviews. 2006;58:878–896.
[75] Le Harzic R, Colonna A, Bückle R, Ehlers A, Hadjur C, Leroy F, Flament F, Bazin R, Piot B, Riemann I, et al. In vivo multiphoton tomography: a non invasive powerful tool for biochemical investigation of human skin. Proc SPIE 6630; 2007. p. 66300V-7.
[76] Zvyagin AV, Zhao X, Gierden A, Sanchez W, Ross JA, Roberts MS. Imaging of zinc oxide nanoparticle penetration in human skin in vitro and in vivo. Journal of Biomedical Optics. 2008;13:064031.
[77] Bazin R, Flament F, Colonna A, Le Harzic R, Buckle R, Piot B, Laize F, Kaatz M, Konig K, Fluhr JW. Clinical study on the effects of a cosmetic product on dermal extracellular matrix components using a high-resolution multiphoton tomograph. Skin Research and Technology. 2010;16: 305–310.
[78] König K, Raphael AP, Lin L, Grice JE, Soyer HP, Breunig HG, Roberts MS, Prow TW. Applications of multiphoton tomographs and femtosecond laser nanoprocessing microscopes in drug delivery research. Advanced Drug Delivery Reviews. 2011;63:388–404.
[79] Roberts MS, Dancik Y, Prow TW, Thorling CA, Lin LL, Grice JE, Robertson TA, Konig K, Becker W. Non-invasive imaging of skin physiology and percutaneous penetration using fluorescence spectral and lifetime imaging with multiphoton and confocal microscopy. European Journal of Pharmaceutics and Biopharmaceutics. 2011;77:469–488.
[80] Darvin ME, Konig K, Kellner-Hoefer M, Breunig HG, Werncke W, Meinke MC, Patzelt A, Sterry W, Lademann J. Safety assessment by multiphoton fluorescence/second harmonic generation/hyper-Rayleigh scattering tomography of ZnO nanoparticles used in cosmetic products. Skin Pharmacology and Physiology. 2012;25:219–226.
[81] Leite-Silva VR, Lamer ML, Sanchez WY, Liu DC, Sanchez WH, Morrow I, Martin D, Silva HDT, Prow TW, Grice JE, et al. The effect of formulation on the penetration of coated and uncoated zinc oxide nanoparticles into the viable epidermis of human skin in vivo. European Journal of Pharmaceutics and Biopharmaceutics. 2013;84:297–308.
[82] Sugata K, Osanai O, Sano T, Takema Y. Evaluation of photoaging in facial skin by multiphoton laser scanning microscopy. Skin Research and Technology. 2011;17:1–3.
[83] Del Bino S, Sok J, Bessac E, Bernerd F. Relationship between skin re-sponse to ultraviolet exposure and skin color type. Pigment Cell Research. 2006;19:606–614.

Sven R. Kantelhardt

20 Multiphoton microscopy and fluorescence lifetime imaging for resection guidance in malignant glioma surgery

Abstract: At a rate of 9/100 000 newly diagnosed cases per year malignant glioma is one of most feared diseases in the western hemisphere. The WHO grading system classifies gliomas according to their aggressiveness. WHO grade III and IV are called high-grade glioma, grades I and II low-grade glioma. Standard therapy of high-grade gliomas includes surgical resection, radio- and chemotherapy. Recent clinical trials highlighted the significance of the extent of surgical resection, while the surrounding, often still functional brain tissue limits the extent of any surgical procedure. In low-grade gliomas most experts likewise consider surgery the treatment of first choice. Unfortunately it can be difficult to discriminate tumor and adjacent brain tissue intraoperatively. Therefore techniques for intraoperative resection guidance have been developed. 5-ALA, which is metabolized in most high-grade gliomas to the fluorescent metabolite PpIX, is the most frequently used option. While this has shown to improve the rate of complete or rather complete surgical resections, the technique does not stain low-grade gliomas and does not stain the actual tumor cells, so that at high magnifications tumor margins are not accurately delineated. We have therefore been looking for alternative technologies which would allow an in vivo identification of glioma cells. Multiphoton intensity imaging of endogenous fluorophores has proven to provide high definition structural images of anatomical and pathological landmarks. We could show that the technology is able to clearly discriminate different cells and tissues within healthy mouse brain, as well as experimental and bioptic human brain tumors (high- and low-grade glioma, metastasis, meningioma). The application of fluorescence lifetime imaging (FLIM) provides further information and corresponds with the metabolic activity of the examined cells (more aggressive tumors show generally longer fluorescence lifetimes). Finally we could introduce multiphoton intensity and FLIM imaging into the operating room and show that in vivo applications are feasible, safe and provide a similar image quality compared to ex vivo examinations. Multiphoton microscopy carries the potential to provide near real-time, high definition of label-free glioma tissue ex vivo and in vivo.

https://doi.org/10.1515/9783110429985-022

20.1 Some facts about malignant glioma

Within the industrialized nations malignant glioma belongs to the most dreaded diseases. They account for about 77 % of all malignancies of the central nervous system. The number of newly diagnosed cases in the US alone exceeds 18 000 or 9/100 000 inhabitants per year. 13 000 of these patients die every year and the loss of life time caused by malignant glioma is statistically higher than that of the sum of all other tumor entities together [1]. With 45–50 % of all malignant gliomas the most aggressive histopathological subtype, the so-called glioblastoma multiforme (GBM), is also the most frequent one. GBMs are undifferentiated to a degree which does not allow to identify which kind of glial cells they are derived from. Better differentiated (and less malignant) gliomas are named after their suspected progenitor cells. Astrocytic gliomas (anaplastic, differentiated and pilocytic astrocytoma) arise from astrocytic progenitor cells (type-I-astrocytes), oligodendroglial tumors and mixed gliomas (oligoastrocytoma) arise from O-2-A progenitor cells (oligodendrocytes) and ependymomas derive from the ependym of the ventricles [2]. The WHO tumor classification grades gliomas (and other tumors) according to their suspected aggressiveness and their malignant potential. GBM is graded with the highest level (grade IV, Fig. 20.1 (c)), while the only benign glioma, the pilocytic astrocytoma is grade I [3].

20–30 % of gliomas are anaplastic, or WHO grade III, while only about 16–18 % are classified as low-grade and benign gliomas. 10 % are differentiated astrocytomas (Fig. 20.1 (a) and (b)), 4–6 % are differentiated oligodendrogliomas and the mixed glioma, also called oligoastrocytoma, 3–4 % are ependymomas and about 2 % are pilocytic astrocytomas [4].

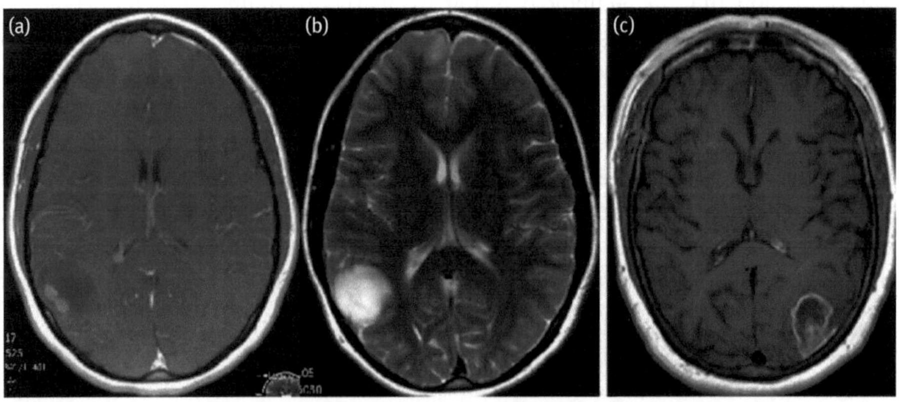

Fig. 20.1: (A) shows an axial gadolinium enhanced T1 weighted MRI scan of a low-grade glioma (WHO grade II) – note the faint contrast enhancement. (B) shows a T2 T1 weighted scan of the same region – now the whole tumor extent can be seen. (C) shows an axial gadolinium enhanced T1 weighted MRI scan of a glioblastoma multiforme (GBM, WHO grade IV) with typical gadolinium enhancement of the tumor margins and central necrosis.

20.2 Standard therapy and prognosis

The therapy of gliomas depends on the WHO grade and several other factors, such as surgical accessibility, molecular factors (MGMT promotor methylation) etc. For high-grade gliomas (WHO grade III and IV) large clinical trials could establish a standard therapy. This includes gross total resection followed by concomitant radiochemotherapy and chemotherapy [5]. Despite this regimen median survival time for GBM is about 15–18 months only [5, 6]. However, before introduction of this regimen (using surgery and radiation only) the survival time was about 12 months only with a 5-year survival rate of 3–4 % [7, 8].

Corresponding to the histopathological diagnosis, the prognosis of WHO grade III tumors is better with a 5-years survival rate of 30–31 % [7, 8].

Recent investigations could show that the extent of the surgical resection is an important predictor of progression-free and overall survival in glioma WHO grade III and IV and that a more complete resection leads to a better outcome [9–11].

Low-grade gliomas, as suggested by their name, show a much better prognosis. Average survival following diagnosis can be expected to be 7–8 years. As with anaplastic tumors, the individual prognosis depends on the histopathological subtype. Astrocytoma generally has a worse prognosis than oligodendrogliomas or ependymomas [2]. In contrast to the situation in high-grade glioma no level I or II evidence exists defining a standard treatment protocol.

Pilocytic astrocytomas (WHO grade I), which are considered benign lesions, can be cured by complete resection [12, 13]. If this cannot be achieved patients have to be monitored with frequent MRI scans and re-resection should be considered if the tumor progresses [14].

Infiltrating low-grade gliomas have a median survival of 3.2–7.7 years, primarily dependent on the histopathological subtype but also on other factors like age at diagnosis, tumor size, infiltration of one or both hemispheres, and neurological status at diagnosis [15]. Recent studies and case series (mainly level V evidence) favor surgery directly after diagnosis and a radical resection if achievable at reasonable risk for neurological function [11, 16, 17].

Radiotherapy is an additional option, but literature on the optimal timing of radiation is controversial [18]. Besides any deficits caused by neurotoxic side effects of radiotherapy are more severely experienced by low-grade patients, because of the rather long survival times [19–21].

As with high-grade glioma, chemotherapy shows some effect, but like for radiotherapy the benefit/risk ratio is affected by the longer survival times and therapy has to be defined on an individual basis [22, 23].

20.3 Limitations for surgical resection

One major problem for the treatment of gliomas is the fact that they arise from normal glial cells which are part of the regular brain histoarchitecture. Depending on the grade of differentiation these cells still share many features with their progenitor and neighboring cells. This makes it hard to target them by chemo- or radiotherapy and even harder to identify them during surgery. It can be extremely difficult for the surgeon to discern tumor and regular brain cells at the tumor margins. Especially rather well differentiated low-grade gliomas or less malignant regions of high-grade gliomas present a challenge to the operating neurosurgeon. Accordingly, resections are often less radical than optimal and parts of an in principle resectable tumor frequently remain.

The other main problem is the fact that most gliomas (apart from the pilocytic astrocytoma WHO grade I) tend to migrate along fiber tracts and blood vessels into the neighboring (often still functional) brain tissue and even into the contralateral hemisphere [24]. These migrating cells beyond the margin of the solid tumor cannot be cured surgically. Even so-called super-radical resections, which were performed in the 1960s (in some cases the complete hemispherectomy of the tumor-bearing hemisphere was performed) did not finally cure gliomas [25]. Nevertheless, even though this limits the value of surgery we have good level I and II evidence that more complete resections of the solid part of a tumor does improve symptoms, progression-free survival and overall survival [9–11]. Further, we know that the majority of recurrent glioma arise from the resection site or its direct neighborhood [25] and might thus have been prevented by more radical resection.

Unfortunately most gliomas are surrounded by functional brain tissue. 53.9 % of newly diagnosed GBM are located within or close to functionally important centers in the brain such as speech areas or motor cortical areas [26]. In this vicinity a resection with a "safety margin", as for example performed in intestinal tumor surgery, is not an option.

Neurosurgeons around the world have therefore focused their research to optimize the intraoperative visualization and identification of gliomas. A first important step was the introduction of microsurgical techniques and application of the operating microscope [27]. Conventional operating microscopes have a magnification of about 5–80× and furthermore offer optimal illumination of the operating field. Using these operating microscopes, experienced neurosurgeons can distinguish characteristic features of malignant gliomas as different shades of color, thrombotic blood vessels or necrotic areas. Besides the haptic appearance of the tissue, as experienced by the surgeon during resection, provides valuable additonal information. Nevertheless the rate of suboptimal resection remains rather high and new tools helping with the intraoperative identification of tumor margins are in demand.

20.4 Fluorescence imaging for resection guidance in glioma surgery

An important step to optimize the extent of resection was the fluorescence labeling of glioma tissue. The most widely applied technique uses a natural intermediate of heme biosynthesis: 5-amino laevolinic acid (5-ALA) [10, 28, 29]. Oral application of 5-ALA leads to accumulation of its fluorescent metabolite protoporphyrin IX (PpIX) in many tumors including the majority of high-grade gliomas. Under UV light excitation PpIX shows a marked red fluorescence. PpIX has a maximum absorption at about 400 nm and emits red light visible to the human eye and to modified operating microscopes [29]. In their above cited prospective multicenter study Stummer and colleagues could not only show that a more radical resection leads to prolonged progression-free survival of high-grade glioma patients. They further demonstrated that the application of 5-ALA-based fluorescence resection guidance improved the rate of complete resections [10]. Drawbacks of the technique are that most low-grade gliomas (WHO grade II) and low-grade components of high-grade gliomas do not show detectable 5-ALA-derived fluorescence. Likewise, tumor necrosis does not show fluorescence due to the lack of viable cells and sufficient synthesis of PpIX.

20.5 Multiphoton intensity imaging of native (murine) brain tissue

These problems gave rise to the idea of using the autofluorescence of brain and glioma cells for the identification of tumor tissue.

We have been investigating upcoming optical technologies like optical coherence tomography [30–33] or two-photon [34, 35] and multiphoton microscopy [36–38] for this purpose.

Of these technologies, multiphoton microscopy presently carries the greatest potential to develop into a real-time imaging tool of label-free, fluorescence guided resection at a cellular resolution. It does not alter the exposed tissues and provides insight into the examined tissue to a certain depth. It is an established diagnostic tool in dermatology [39–43] and has been used to study physiological processes in living intestinal cells [44, 45] and ophthalmology [46, 47].

We used a multiphoton tomograph designed for application in dermatology (DermaInspect MPT, JenLab, Jena, Germany) for the first experiments on brain and brain tumor tissue.

The instrument provided multiphoton intensity images of 300 nm lateral resolution, 1–2 µm axial resolution, and 200 ps temporal resolution (for fluorescence lifetime imaging). The light source is a titanium:sapphire near-infrared (NIR) laser (pulse width of 100 fs, repetition frequency 80 MHz). In situ mean power is 2–50 mW, with a tunable wavelength of 710–920 nm. Image processing is performed using a beam-scanning module with galvanometer scanners and piezodriven optics. Fluorescence of endogenous fluorophores is recorded by a dual channel photomultiplier (PMT) detector with single photon sensitivity. Typical scan range is 350 μm × 350 μm (horizontal); 200 μm (vertical) using a focusing optics [37, 38].

We started with ex situ examination of freshly explanted, native brain tissue from healthy mice. Multiphoton intensity images of different regions within the mouse brain allowed differentiation of several tissues and cells. The tissue could be clearly identified by certain histoarchitectural characteristics closely resembling the appearance in HE stained conventional histologies (Fig. 20.2). The magnification of the multiphoton microscope further allowed identification of subcellular compounds and organelles, such as granula, nuclei, etc. Cellular nuclei appeared as hypointense structures against a rather homogeneous matrix-dominated background. Multiphoton intensity imaging showed that metabolically highly active cells and tissues, like the ependyma, the choroid plexus, or vascular endothelial cells tended to show high signal intensity [37, 38].

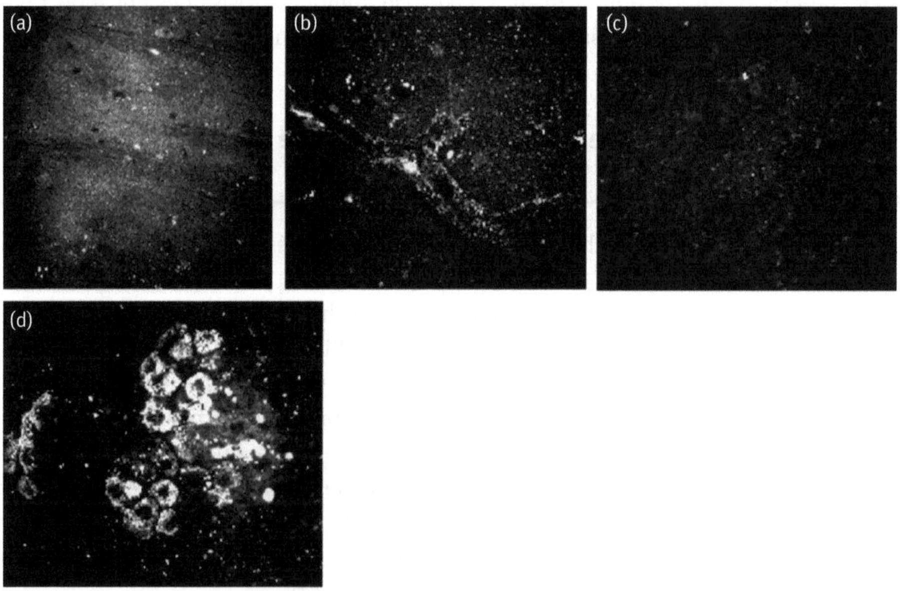

Fig. 20.2: Examples of multiphoton intensity images of prominent structures in the healthy mouse brain: (A) fiber tracts in the white matter, (B) blood vessels in the white matter, (C) white matter of the cerebellum, (D) choroid plexus.

20.6 Multiphoton fluorescence lifetime (FLIM) imaging of native (murine) brain tissue

Fluorescence lifetime (FLIM) imaging provided further means for the differentiation of tissues and cells [37, 38]. FLIM images were recorded by time-correlated single-photon counting starting at the emission of the exciting laser pulse. Spatially resolved autofluorescence decay curves were recorded for 256 × 256 pixels per image field, which typically were 150 × 150 μm^2. Mean fluorescence lifetimes for each pixel were displayed in color-coded images. The lifetime of fluorescence for each region of interest could thus be determined. From specific areas of interest lifetime images were obtained using increasing excitation wavelengths from 720–780 nm at increments of 10 nm (corresponding to the optimal excitation wavelength determined in multiphoton intensity imaging experiments of healthy mouse brain). The distribution of fluorescence lifetime was color-coded, using either a continuous spectrum of red (short-lived) to blue (long-lived, Fig. 20.3) or using discrete colors for manually defined wavelength areas (Fig. 20.6).

Metabolically highly active cells and tissues, which did show high intensity signals in multiphoton intensity imaging, like the ependyma, the choroid plexus, or vascular endothelium also showed the longest lifetimes of endogenous fluorophores (> 1700 ps) excited at 750 nm. In contrast, excitable fluorophores within erythrocytes were the shortest lived (900 ± 72 ps). In gray and white matter the brain parenchyma showed an intermediate fluorescence lifetime (1380 ± 23 ps and 1360 ± 33 ps respectively). Generally, the nuclei of glia showed low fluorescence intensity. The cytoplasm of glia cells frequently contained granules of high fluorescence intensity and relatively short fluorescence lifetime.

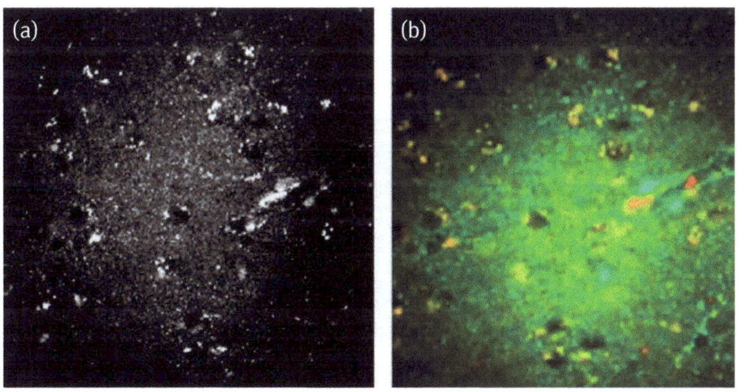

Fig. 20.3: Multiphoton intensity (A) and FLIM image (B) of white matter in the healthy murine brain showing continuous color-coding of fluorescence lifetimes ranging from red (500 ps) to blue (2500 ps).

20.7 Multiphoton microscopy of human cell line derived GBM tissue in an orthotopic mouse glioma model

Following the experiments in healthy mice which demonstrated the ability to examine native brain tissue by multiphoton microscopy (intensity imaging and FLIM), the technique's capability to identify tumor cells and tissues had to be evaluated. After initial experiments on cultured human GBM-derived cell lines we applied an orthotopic glioma mouse model. Cell suspensions from these cell lines were injected into the brain of (immune deficient) nude NMRI mice using a stereotactical frame in a one stage procedure. Tumor-bearing mouse brains were harvested 4–5 weeks post implantation and cut along the coronal plane using a scalpel. Multiphoton microscopy was performed on freshly explanted native samples under conditions resembling the physiological environment.

Corresponding to the results from healthy mice brains, multiphoton intensity imaging provided high definition structural images which allowed the identification of individual tumor cells within the tumor bulk and single invasive tumor cells in surrounding brain tissue [37, 38]. Cytoarchitectural hallmarks typical for GBM, such as hypercellularity and pleomorphism could be demonstrated and multiphoton intensity images again closely resembled conventional H&E stained histopathological sections (Fig. 20.4).

Raster scanning of the tissue and capture of image stacks along the z-axis (into the tissue) allowed reconstruction of 2D and 3D images of fluorescence intensity [36] (Fig. 20.5). Further (automated) segmentation of nuclei and analysis of cellular density was performed on 3D intensity images. The cell count in a given sample volume provided an additional means to differentiate adjacent brain tissue from brain tumor tissue, as the cellular density was found to be much higher in experimental and bioptical high-grade glioma or brain metastasis compared to adjacent or healthy brain tissue [36].

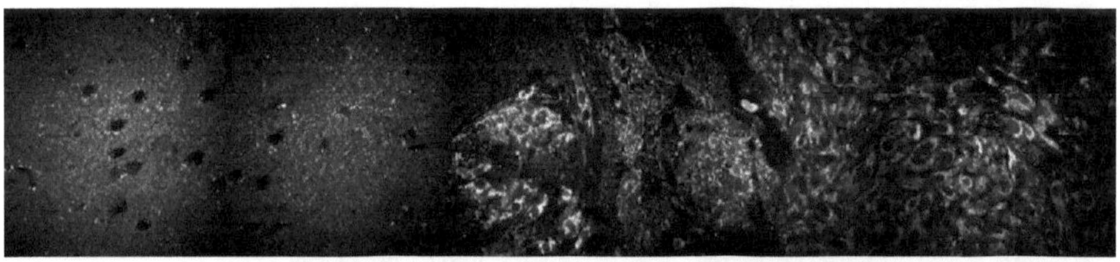

Fig. 20.4: Multiphoton intensity images documenting the brain (left)/tumor (right) interface in an experimental glioma (human glioma-derived cell line U87 in an orthotopic mouse model): autofluorescence of brain and glioma cells 4, concomitant radiochemotherapy 2, fluorescence lifetime imaging 1, gliomas migrate along fiber tracts and blood vessels 3, intraoperative resection guidance 1, intraoperative visualization and identification of gliomas 3, malignant glioma 1, migrating cells beyond the margin of the solid tumor 3, surgical resection 1.

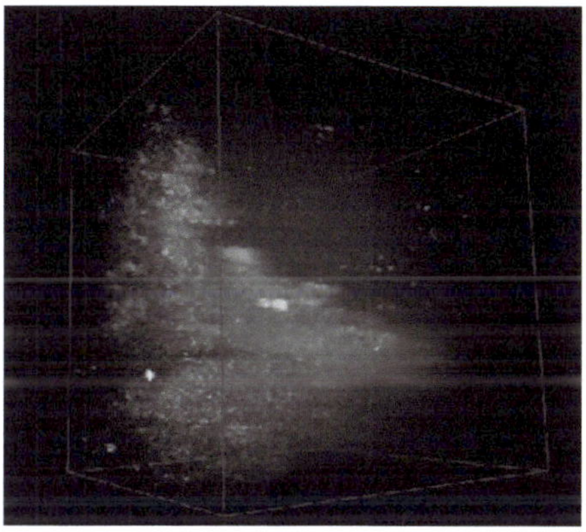

Fig. 20.5: Multiphoton intensity images of tumor/brain interface – 3D image were calculated from stacks of 20 consecutive images at depths intervals of 5 µm.

When examining experimental gliomas of the orthotopic glioma mouse model using FLIM we found that tumor tissue showed longer fluorescence lifetimes in comparison to adjacent brain tissue (≈ 1400 ps for normal brain, ≈ 1600 ps for glioma tissue). This corresponded with the observation on regular brain cells with a high metabolic activity, which showed prolonged fluorescence lifetimes. Tumor cells, which are still faster replicating and thus very active, had even higher fluorescence lifetimes. By selecting distinct colors for different wavelength areas (discrete color-coding) brain tumor and adjacent brain tissue could be clearly delineated [37, 38].

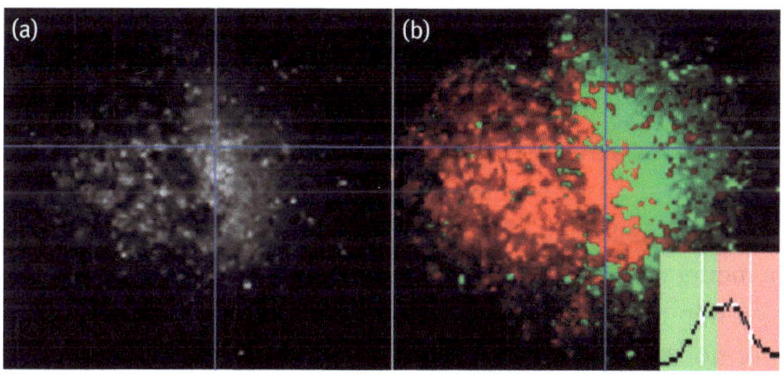

Fig. 20.6: Multiphoton intensity (A) and FLIM image (B) of tumor/brain interface – the FLIM image shows discrete color-coding with fluorescence lifetimes below 1500 ps green (brain tissue) and those above 1500 ps red (tumor).

20.8 Multiphoton microscopy of 5-ALA-stained experimental gliomas

In order to evaluate the accuracy of 5-ALA-staining of glioma tissue we applied the method in an orthotopic glioma mouse model. As expected, the GBM cell line derived tumors showed uptake of 5-ALA and a bright red PpIX fluorescence. At higher resolutions however, it seemed that the fluorescent metabolite leaked into the surrounding brain tissue. PpIX fluorescence showed a marked decrease in intensity towards tumor edges [48]. The metabolite might undergo diffusion processes and not remain in the cell of origin. Besides, the dye shows rapid bleaching under UV-light excitation. This is again problematic in cases of prolonged neurosurgical procedures. Thus we could conclude that while 5-ALA is a very useful and efficient tool under standard conditions of a neurosurgical OR using a conventional operating microscope with maximum magnification of about 80×, it is not suited for resection guidance under higher magnifications and at a cellular level.

When examining 5-ALA-enhanced experimental glioma it could however be noted that the resulting (PpIX) fluorescence was highly reproducible, allowing using the same discrete color-coding modes to differentiate tumor and adjacent brain in different tumors. The individual (case-by-case) selection of corresponding wavelength areas was not required if 5-ALA was applied, thus providing a potential additional means for an automated identification of tumor tissue. Apart from this the application of 5-ALA did however not provide any additional benefit for multiphoton microscopy compared to imaging of autofluorescence alone [48].

20.9 Multiphoton microscopy of human glioma tissue ex vivo

The logical next step after differentiating glioma from adjacent brain tissue in an orthotopic mouse glioma model was the examination of human tumor biopsies.

Mirroring the situation in mice we first investigated whether multiphoton intensity and FLIM images of normal human brain corresponded to those from healthy mice. For this purpose tumor-adjacent brain samples from areas that had to be resected during the neurosurgical approach were examined under physiological conditions (freshly resected, native samples) ex vivo. We found that white and gray matter did generally closely resemble images from healthy mouse brain. Nevertheless some samples showed multiple small vacuoles, which by comparison to conventional H&E stained histopathological sections, could be interpreted as correlates of brain edema. Furthermore, several brightly intense cells were found in tumor-adjacent human brain tissue, which did not appear in samples from the orthotopic glioma mouse model. CD-68 staining of conventional histopathological sections proved that these cells were immune cells which could not be observed in the orthotopic glioma mouse model, as immune deficient mice were used here [36].

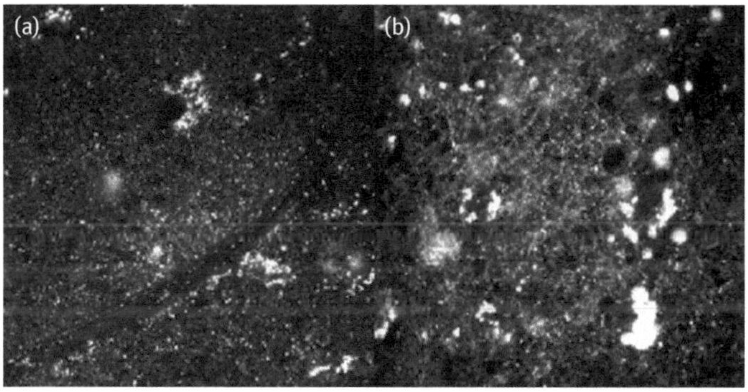

Fig. 20.7: Multiphoton intensity images of human brain tissue: (a) white matter, (b) edema in the white matter.

Following these experiments on native samples from tumor-adjacent brain, tumor biopsies were examined. These included low-grade gliomas (oligodendro-, oligoastro- and astrocytoma WHO grade II), high-grade gliomas (astrocytoma WHO grade III and GBM), as well as other tumor entities like brain metastasis or benign lesions (meningioma WHO grade I) [37, 49].

We found that multiphoton intensity imaging could not only distinguish normal brain and brain tumor, but also different histological (brain) tumor entities. This observation offers the opportunity to develop specific imaging criteria for real-time optical biopsy of brain tumors by multiphoton imaging [49]. Again, as in the experiments using healthy mouse brain and the glioma mouse model, multiphoton images closely resembled conventional histologies. Histopathological hallmarks, such as different degrees of hypercellularity and pleomorphism, necrosis and microvascular proliferation could be clearly identified.

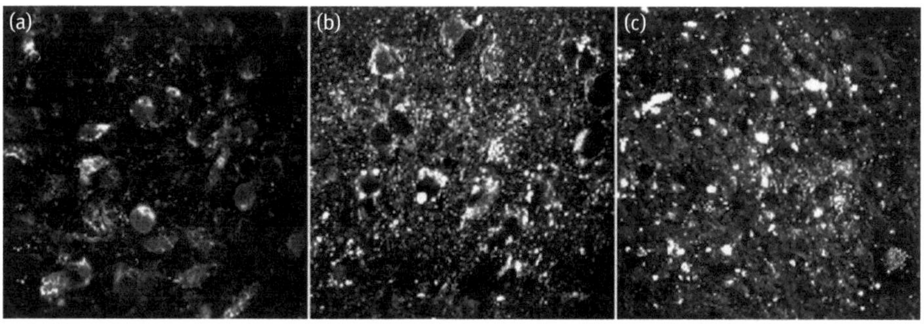

Fig. 20.8: Multiphoton intensity images of human brain tumors: (a) astrocytoma WHO grade II, (b) astrocytoma WHO grade III, (c) GBM. Note the increasing degree of disorganization and variability of cellular components.

In human tumor biopsies FLIM provides further information for the discrimination of glioma and brain metastasis from adjacent brain tissue. As in the glioma mouse model we found a longer fluorescence lifetime in tumor samples compared to normal (or tumor adjacent) brain tissue. As in the murine samples this seemed to be related to a higher metabolic activity. Less aggressive tumors (meningioma WHO grade I or low-grade gliomas) had longer fluorescence lifetimes compared to adjacent brain tissue, but shorter than more aggressive tumors (high-grade glioma and brain metastasis) [2]. This observation was meanwhile also reported by the group of Sun et al. who evaluated a system for endoscopic FLIM and tested it in 9 glioblastoma specimen ex vivo [50].

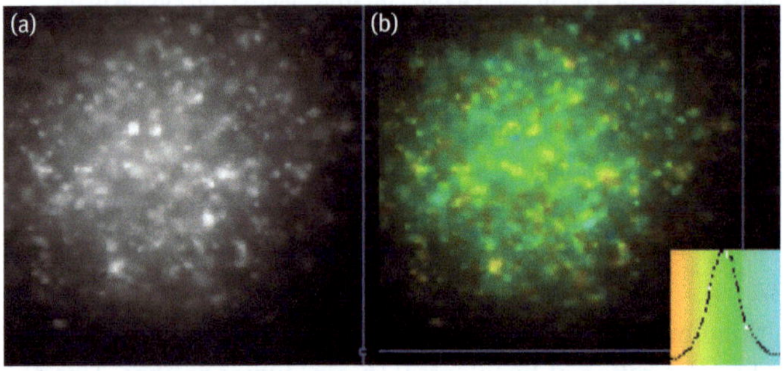

Fig. 20.9: Multiphoton intensity (a) and FLIM image (b) of human GBM tissue – the FLIM image shows continuous color-coding with the red end at about 500 ps and blue end at 2500 ps.

20.10 Multiphoton microscopy of human glioma tissue in vivo and outlook

Recent technical developments gave us the possibility to introduce multiphoton intensity and FLIM imaging in an experimental setting in our OR. The successor model of the DermaInspect®, which we applied in previous studies, the MPTflex™ Multiphoton Laser Tomograph (both JenLab GmbH, Jena, Germany) is equipped with an articulated optomechanical mirror arm with a flexible scan head. This allows positioning of the sterile draped scan head in a resection cavity, enabling high definition imaging of label-free glioma tissue in vivo and in near real time. We found that the instrument could be applied safely and easily (yet because of the dimensions only in cases which required rather large craniotomies for the surgical approach). Image quality resembled that of ex vivo examinations of experimental and bioptic glioma samples [49].

While these results are very encouraging, routine intraoperative application of multiphoton microscopy for fluorescence resection guidance in gliomas is still several steps away. The scan area of recent multiphoton microscopes is still rather small (microscopic fields are about 350 μm square only) and image penetration into the tissue is restricted to about 200 μm, which signifies that tumor compounds isolated from the resection cavity will not be detected. In addition, the system used in our experiments is not yet specifically designed for neurosurgical application and its dimensions restrict the application. Nevertheless, the technique shows great potential as a valuable tool for intraoperative resection guidance in glioma surgery. It can be applied safely, avoiding any damage to the examined tissue and without the need for previous labeling.

References

[1] Schwartzbaum JA, Fisher JL, Aldape KD, Wrensch M. Epidemiology and molecular pathology of glioma. Nat Clin Pract Neurol. 2006;2(9):494–503

[2] Schlegel UWM, Westphal M. Neuroonkologie. Stuttgart: Thieme Verlag; 2003.

[3] Kleihues PCW. World Health Organization classification of tumours: pathology genetics: tumours of the nervous system. Lyon: IARC Press; 2000.

[4] Wrensch M, Minn Y, Chew T, Bondy M, Berger MS. Epidemiology of primary brain tumors: current concepts and review of the literature. Neuro Oncol. 2001;4(4):278–299.

[5] Stupp R, Mason WP, van den Bent MJ, et al. Radiotherapy plus concomitant and adjuvant temozolomide for glioblastoma. N Engl J Med. 2005;352(10):987–996.

[6] Choi, JW, Lee MM, Kim IA, Kim JH, Choe G, Kim CY. The outcomes of concomitant chemoradiotherapy followed by adjuvant chemotherapy with temozolomide for newly diagnosed high grade gliomas : the preliminary results of single center prospective study. J Korean Neurosurg Soc. 2008;44(4):222–227.

[7] Davis FG, McCarthy BJ, Freels S, Kupelian V, Bondy ML. The conditional probability of survival of patients with primary malignant brain tumors: surveillance, epidemiology, and end results (SEER) data. Cancer. 1999;85(2):485–491.

[8] Barnholtz-Sloan JS, Sloan AE, Schwartz AG. Relative survival rates and patterns of diagnosis analyzed by time period for individuals with primary malignant brain tumor, 1973–1997. J Neurosurg. 2003;99(3):458–466.

[9] Lacroix M, Abi-Said D, Fourney DR, et al. A multivariate analysis of 416 patients with glioblastoma multiforme: prognosis, extent of resection, and survival. J Neurosurg. 2001;95(2): 190–198.

[10] Stummer W, Pichlmeier U, Meinel T, Wiestler OD, Zanella F, Reulen HJ. Fluorescence-guided surgery with 5-aminolevulinic acid for resection of malignant glioma: a randomised controlled multicenter phase III trial. Lancet Oncol. 2006;7(5):392–401.

[11] McGirt MJ, Chaichana KL, Gathinji M, et al. Independent association of extent of resection with survival in patients with malignant brain astrocytoma. J Neurosurg. 2009;110(1):156–162.

[12] Fisher BJ, Leighton CC, Vujovic O, Macdonald DR, Stitt L. Results of a policy of surveillance alone after surgical management of pediatric low grade gliomas. Int J Radiat Oncol Biol Phys. 2001;51(3):704–710.

[13] Karim AB, Maat B, Hatlevoll R, et al. A randomized trial on dose-response in radiation therapy of low-grade cerebral glioma: European Organization for Research and Treatment of Cancer (EORTC) Study 22844. Int J Radiat Oncol Biol Phys. 1996;36(3):549–556.

[14] Bowers DC, Krause TP, Aronson LJ, et al. Second surgery for recurrent pilocytic astrocytoma in children. Pediatr Neurosurg. 2001;34(5):229–234.

[15] Pignatti F, van den Bent M, Curran D, et al. Prognostic factors for survival in adult patients with cerebral low-grade glioma. J Clin Oncol. 2002;20(8):2076–2084.
[16] Lang FF, Gilbert MR. Diffusely infiltrative low-grade gliomas in adults. J Clin Oncol. 2006;24(8): 1236–1245.
[17] Schomas DA, Laack NN, Rao RD, et al. Intracranial low-grade gliomas in adults: 30-year experience with long-term follow-up at Mayo Clinic. Neuro Oncol. 2009;11(4):437–445.
[18] van den Bent MJ, Afra D, de Witte O, et al. Long-term efficacy of early versus delayed radiotherapy for low-grade astrocytoma and oligodendroglioma in adults: the EORTC 22845 randomised trial. Lancet. 2005;366(9490):985–990.
[19] Klein M, Heimans JJ, Aaronson NK, et al. Effect of radiotherapy and other treatment-related factors on mid-term to long-term cognitive sequelae in low-grade gliomas: a comparative study. Lancet. 2002;360(9343):1361–1368.
[20] Olson JD, Riedel E, DeAngelis LM. Long-term outcome of low-grade oligodendroglioma and mixed glioma. Neurology. 2000;54(7):1442–1448.
[21] Laack NN, Brown PD, Ivnik RJ, et al. Cognitive function after radiotherapy for supratentorial low-grade glioma: a North Central Cancer Treatment Group prospective study. Int J Radiat Oncol Biol Phys. 2005;63(4):1175–1183.
[22] Baumert BG, Stupp R. Low-grade glioma: a challenge in therapeutic options: the role of radiotherapy. Ann Oncol. 2008;19(Suppl7):217–222.
[23] Tosoni A, Franceschi E, Ermani M, et al. Temozolomide three weeks on and one week off as first line therapy for patients with recurrent or progressive low grade gliomas. J Neurooncol. 2008; 89(2):179–185.
[24] Kelly PJ, Daumas-Duport C, Kispert DB, Kall BA, Scheithauer BW, Illig JJ. Imaging-based stereotaxic serial biopsies in untreated intracranial glial neoplasms. J Neurosurg. 1987;66(6): 865–874.
[25] Giese A, Bjerkvig R, Berens ME, Westphal M. Cost of migration: invasion of malignant gliomas and implications for treatment. J Clin Oncol. 2003;21(8):1624–1636.
[26] Duffau H, Capelle L. Preferential brain locations of low-grade gliomas. Cancer. 2004;100(12): 2622–2626.
[27] Yasargil MG. Personal considerations on the history of microneurosurgery. J Neurosurg. 2010; 112(6):1347.
[28] Uehlinger P, Zellweger M, Wagnieres G, Juillerat-Jeanneret L, van den Bergh H, Lange N. 5-Aminolevulinic acid and its derivatives: physical chemical properties and protoporphyrin IX formation in cultured cells. J Photochem Photobiol B. 2000;54(1):2–80.
[29] Hebeda KM, Saarnak AE, Olivo M, Sterenborg HJ, Wolbers JG. 5-Aminolevulinic acid induced endogenous porphyrin fluorescence in 9L and C6 brain tumours and in the normal rat brain. Acta Neurochir. 1998;140(5):503–512.
[30] Böhringer HJ, Boller D, Leppert J, et al. Time-domain and spectral-domain optical coherence tomography in the analysis of brain tumor tissue. Lasers Surg Med. 2006;38:588–597.
[31] Böhringer HJ, Lankenau E, Rohde V, Huttmann G, Giese A. Optical coherence tomography for experimental neuroendoscopy. Minim Invasive Neurosurg. 2006;49:269–275.
[32] Böhringer HJ, Lankenau E, Stellmacher F, et al. Imaging of human brain tumor tissue by near-infrared laser coherence tomography. Acta Neurochir. 2009;151:507–517.
[33] Giese A, Böhringer H, Leppert J, et al. Non-invasive intraoperative optical coherence tomography of the resection cavity during surgery of intrinsic brain tumors. Proc SPIE 6078; 2006. p. 495–502.
[34] Arndt-Jovin DJ, Kantelhardt SR, Caarls W, et al. Tumor-targeted quantum dots can help surgeons find tumor boundaries. IEEE Transactions on NanoBioscience. 2009;8:65–71.

[35] Kantelhardt SR, Caarls W, de Vries AHB, et al. Specific visualization of glioma cells in living low-grade tumor tissue. Plos One. 2010;5:e11323.
[36] Kantelhardt SR, Leppert J, Kantelhardt JW, et al. Multi-photon excitation fluorescence microscopy of brain-tumour tissue and analysis of cell density. Acta Neurochir. 2009;151: 253–262.
[37] Kantelhardt SR, Leppert J, Krajewski J, et al. Imaging of brain and brain tumor specimens by time-resolved multiphoton excitation microscopy ex vivo. Neuro Oncol. 2007;9:103–112.
[38] Leppert J, Krajewski J, Kantelhardt SR, et al. Multiphoton excitation of autofluorescence for microscopy of glioma tissue. Neurosurgery. 2006;58:759–767.
[39] Ulrich M, Klemp M, Darvin ME, et al. In vivo detection of basal cell carcinoma: comparison of a reflectance confocal microscope and a multiphoton tomograph. J Biomed Opt. 2013;18(6): 61229.
[40] Seidenari S, Arginelli F, Dunsby C, et al. Multiphoton laser tomography and fluorescence lifetime imaging of melanoma: morphologic features and quantitative data for sensitive and specific non-invasive diagnostics. PLoS One. 2013;8(7):e70682.
[41] Arginelli F, Manfredini M, Bassoli S, et al. High resolution diagnosis of common nevi by multiphoton laser tomography and fluorescence lifetime imaging. Skin Res Technol. 2013;19(2): 94–204.
[42] Koehler MJ, Vogel T, Elsner P, et al. In vivo measurement of the human epidermal thickness in different localizations by multiphoton laser tomography. Skin Res Technol. 2010;16(3): 259–264.
[43] Dimitrow E, Ziemer M, Koehler MJ, et al. Sensitivity and specificity of multiphoton laser tomography for in vivo and ex vivo diagnosis of malignant melanoma. J Invest Dermatol. 2009;29(7): 1752–1758.
[44] Yan J, Zhuo S, Chen G, et al. Real-time optical diagnosis for surgical margin in low rectal cancer using multiphoton microscopy. Surg Endosc. 2014;28(1):36–41.
[45] Ying M, Zhuo S, Chen G, et al. Real-time noninvasive optical diagnosis for colorectal cancer using multiphoton microscopy. Scanning. 2012;34(3):181–185.
[46] Gibson EA, Masihzadeh O, Lei TC, Ammar DA, Kahook MY. Multiphoton microscopy for ophthalmic imaging. J Ophthalmol. 2011:870879.
[47] Tang S, Zhou Y, Ju MJ. Multimodal optical imaging with multiphoton microscopy and optical coherence tomography. Biophotonics. 2012;5(5–6):396–403.
[48] Kantelhardt SR, Diddens H, Leppert J, et al. Multiphoton excitation fluorescence microscopy of 5-aminolevulinic acid induced fluorescence in experimental gliomas. Lasers Surg Med. 2008; 40:273–281.
[49] Kantelhardt SR, Kalasauskas D, König K, et al. In vivo multiphoton tomography and fluorescence lifetime imaging of human brain tumor tissue. J Neurooncol. 2016;127(3):473–482.
[50] Sun Y, Hatami N, Yee M, et al. Fluorescence lifetime imaging microscopy for brain tumor image-guided surgery. J Biomed Opt. 2010;15(5):056022.

Aisada Uchugonova and Robert M. Hoffman

21 Non-invasive single-photon and multi-photon imaging of stem cells and cancer cells in mouse models

Abstract: The use of fluorescent protein for in vivo imaging began a revolution in cell biology. In particular, non-invasive imaging of fluorescent-protein expressing cancer cells, stem cells, immune cells and other cell types has enabled longitudinal studies on cancer spread or stem cell differentiation in real time. Fluorescent proteins come in many colors, enabling color-coded imaging of multiple cell types in vivo. Practical applications include evaluation of all types of agents on cancer and metastasis or stem cells in real time. The combination of fluorescent proteins with multiphoton imaging techniques represents a powerful tool for the visualization and tracking of single cancer cells and stem-cell behavior in their natural niche. Multiphoton tomography (MPT) is capable of high-quality recording of fluorescence and second-harmonic generation (SHG) images at submicron resolution, yielding greater structural details than possible with single-photon imaging technology. It also enables optical 3D biopsies without surgery. Most importantly, MPT has been adapted for in vivo imaging. Cell dynamics and subcellular effects of drugs can be investigated in living animals. Autofluorescence/SHG imaging of the extracellular-matrix proteins elastin and collagen is highly complementary to other protein labeling methods, such as green fluorescent proteins (GFP). This chapter discusses aspects of tomographic linear and nonlinear in-vivo optical imaging.

21.1 Fluorescent proteins and non-invasive single-photon imaging in live mice

The Nobel Committee stated in 2008, when they awarded the Nobel Prize in Chemistry,

> The remarkable brightly glowing green fluorescent protein, GFP, was first observed in the beautiful jellyfish, *Aequorea victoria* in 1962. Since then, this protein has become one of the most important tools used in contemporary bioscience. With the aid of GFP, researchers have developed ways to watch processes that were previously invisible, such as the development of nerve cells in the brain or **how cancer cells spread**.

Fluorescent proteins, such as GFP, red fluorescent protein (RFP) and naturally-occurring or engineered variants, have revolutionized biology by enabling what was formerly invisible in an animal to be seen clearly. Fluorescent proteins have allowed imaging, in real time, of important aspects of cancer in living animals, including cancer-cell motility, invasion, metastasis and angiogenesis. These multicolored proteins

https://doi.org/10.1515/9783110429985-023

can be used for the color-coding of cancer cells growing in vivo and enable the distinction of host cells from cancer cells with single-cell or even subcellular resolution. Visualization of many aspects of cancer initiation and progression in vivo are possible with fluorescent proteins [1].

We first reported non-invasive, real-time, imaging of GFP-expressing tumors growing and metastasizing in live mice in 2000 [2]. The simple whole-body optical imaging system was external and non-invasive. It enabled unprecedented continuous visual monitoring of tumor progression within intact animals. Whole-body non-invasive optical images showed metastatic growth in the brain, liver, and bone in real time enabling quantitative measurement of tumor growth in each of these organs. Either a trans-illuminated epifluorescence microscope or a fluorescence lightbox and thermoelectrically cooled color charge-coupled device (CCD) camera was used for imaging. The depth to which metastasis and micrometastasis could be imaged depended on their size [2]. This was a totally unexpected discovery and was reminiscent of Roentgen's first use of X-rays to obtain images, but in the case of fluorescent proteins, only harmless blue light was needed.

We have also described a system for rapidly and non-invasively visualizing transgene expression with GFP in major organs of intact live mice. GFP linked to an adenovirus was expressed in the cells of brain, liver, pancreas, prostate, and bone upon injection, and its fluorescence was visualized non-invasively. For low-magnification images, animals were illuminated in a fluorescence lightbox and directly viewed with a CCD camera. Higher-magnification images are made with the camera focused through an epi-fluorescence dissecting microscope. Within 5–8 h after adenoviral GFP injection, the fluorescence of the expressed GFP in brain and liver became visible non-invasively and whole-body images were recorded at video rates. The GFP fluorescence continued to increase for at least 12 h and remained detectable in liver for up to 4 months. The system's rapidity of image acquisition made it capable of real-time recording. The method requires only that the expressed gene or promoter be fused or operatively linked to GFP [3].

GFP-expressing Lewis lung carcinoma cells were injected into the subcutaneous site of the footpad of nude mice which is relatively transparent, with comparatively few resident blood vessels [4], allowing quantitative non-invasive imaging of tumor angiogenesis in the intact animal. Capillary density increased linearly over a 10-day period as determined by non-invasive imaging. A GFP-expressing human breast cancer MDA-MB-435 was orthotopically transplanted to the mouse fat pad, blood vessel density increased linearly over a 20-week period, shown by non-invasive imaging [4].

In 2003, we described the use of non-invasive imaging of tumors in mice to monitor drug response. RFP-expressing pancreatic cancer tumor fragments were surgically transplanted onto the nude mouse pancreas [5]. The model reliably simulated the aggresive course of human pancreatic cancer. The anti-metastatic efficacy of drugs was followed non-invasively in real time by imaging the RFP fluorescence. Gemcitabine could delay tumor growth and metastasis but not cure the tumor, which matches clinical experience [5].

In 2004, dual-color fluorescent cancer cells with one color in the nucleus and the other in the cytoplasm were developed. RFP was expressed in the cytoplasm of cancer cells, and GFP linked to histone H2B was expressed in the nucleus. These cells enabled real-time nuclear-cytoplasmic dynamics to be visualized in living cells in vivo as well as in vitro [6]. RFP was expressed in the cytoplasm of cancer cells and GFP linked to histone H2B was expressed in the nucleus. Nuclear cytoplasmic ratios as well as simultaneous cell and nuclear shape changes could be imaged. The cell cycle position of individual living cells was determined by the nuclear-cytoplasmic ratio and nuclear morphology. Apoptosis was observed by nuclear size changes and progressive nuclear fragmentation. Dual-color mitotic cells were visualized non-invasively by whole-body imaging after injection in the mouse ear [6].

Non-invasive imaging of RFP-expressing cancer cells growing in the lung, was enabled by spectral separation [7], using the Maestro In-Vivo Imaging System with a liquid tunable filter [7] (Fig. 21.1 and 21.2).

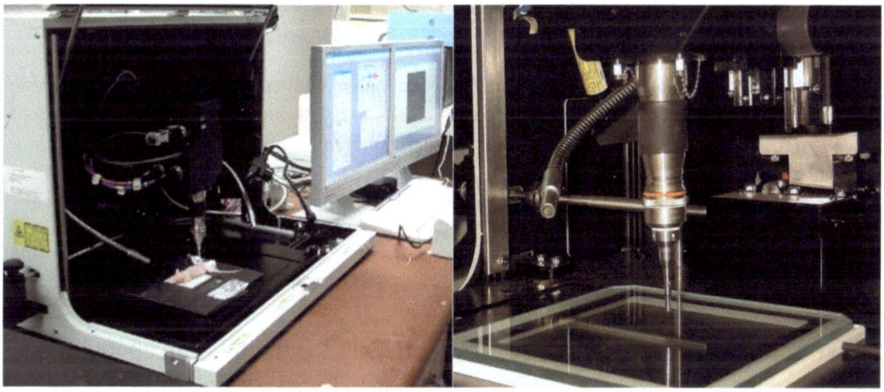

Fig. 21.1: The Olympus IV100 microscope is a scanning laser microscope with a 488 nm argon laser was used and stick objectives (as small as 1.3 mm) which deliver very high resolution images [13].

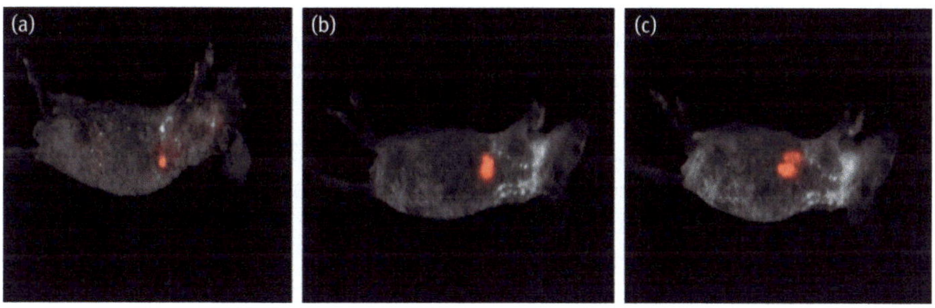

Fig. 21.2: Longitudinal non-invasive imaging of lung tumor growth. Lung tumor growth was monitored non-invasively on days 7 (a), 14 (b) and 21 (c).

A three-color sub-cellular, non-invasive imaging model was developed consisting of GFP-expressing mice transplanted with the dual-color cancer cells labeled with GFP in the nucleus and red fluorescent protein in the cytoplasm [8]. The Olympus IV100 Laser Scanning Microscope, with ultranarrow microscope objectives ("stick objectives"), was used for imaging the two-color cancer cells interacting with the GFP-expressing stromal cells (Fig. 21.3).

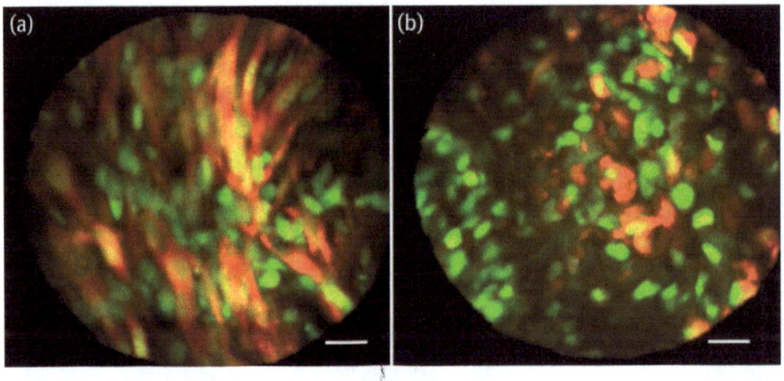

Fig. 21.3: Non-invasive, subcellular imaging of drug response of dual-color mouse mammary cancer cells and GFP stromal cells in the live GFP nude mouse with and without doxorubicin. (a) Dual-color MMT cells were injected in the footpad of GFP transgenic nude mice, non-invasive image of untreated dual-color MMT cells in the footpad of a live GFP mouse. Spindle-shaped dual-color MMT cells are interdispersed among the GFP host cells. (b) Whole-body image of MMT dual-color cancer cells in a live GFP nude mouse 12 h after treatment with doxorubicin (10 mg/kg). The cancer cells lost their spindle shape, and the nuclei appear contracted [14].

Cellular dynamics were non-invasively imaged, including mitotic and apoptotic cancer cells, tumor vasculature, and tumor blood flow. In this model, drug response of both cancer and stromal cells in the intact live animal was also non-invasively imaged in real time [8] (Fig. 21.3).

RFP-expressing human cancer cell lines, including PC-3-RFP prostate cancer, HCT-116-RFP colon cancer, MDA-MB-435-RFP breast cancer, and HT1080-RFP fibrosarcoma were transplanted to transgenic GFP nude mice. Dual-color fluorescence imaging enabled visualization of human tumor-host interaction by non-invasive whole-body imaging [9].

GFP-expressing human cancer cell lines, including HCT-116-GFP colon cancer and MDA-MB-435-GFP breast cancer were orthotopically transplanted to transgenic RFP nude mice. Dual-color non-invasive fluorescence imaging also enabled visualization of human tumor-host interaction [10] (Fig. 21.4).

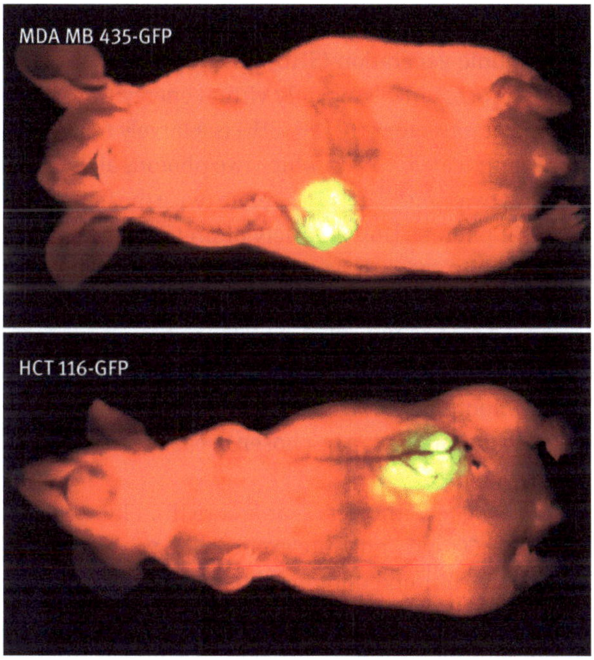

Fig. 21.4: Whole-body images of GFP human cancers growing in RFP nude mice (upper: human MDA-MB-435-GFP breast cancer; lower: human HCT-116-GFP colon cancer).

GFP- or RFP-expressing human pancreatic cancer cells were implanted into the bright blue fluorescent pancreas of the cyan fluorescent protein (CFP) nude mouse. The bright blue fluorescence of the pancreas in the CFP mouse was an ideal background for color-coded imaging of the interaction of implanted cancer cells and the host [11].

Patient-derived orthotopic xenograft (PDOX) models of pancreatic cancer were passaged through transgenic nude mice expressing GFP and RFP [12]. The patient tumors acquired brightly fluorescent stroma from the transgenic host mice, which was stably associated with the tumors through multiple passages. The tumors, with very bright GFP and RFP stroma, were then orthotopically passaged to nontransgenic nude mice. It was possible to image the brightly-fluorescent tumors non-invasively as they progressed [12].

21.2 Advantages of multiphoton imaging of stem cells and cancer cells in live mice

The imaging systems described so far are based on one-photon protein fluorescence excitation. In 2011, we applied multiphoton tomography (MPT) to image live transgenic mice. Our group demonstrated that imaging of specific fluorescence protein expressing cancer and stem cells in live mouse models can be combined with autofluorescence as well as second harmonic generation (SHG) imaging using MPT. Unlike fluorescence, SHG is a coherent process involving only virtual energy transitions and does not induce photobleaching or phototoxicity [15]. SHG is a largely frequency-independent second-order nonlinear optical process in which two photons at the same frequency generate new photons with twice the energy (therefore twice the frequency) and half the wavelength of the initial photons [16]. A variety of proteins, such as extracellular matrix collagen fibrils, tubulin, as well as the muscle-myosin lattices of muscle cells generate SHG [17]. The intrinsic signal permits SHG microscopy of those structures non-invasively with no need for labeling. SHG signals can be collected efficiently and selectively using narrow bandwidth emission filters with minimal contribution from overlapping of any other fluorescence signals [15]. Since SHG is frequency independent, a variety of excitation wavelengths can be chosen to easily separate SHG targets from any other fluorescence signals.

MPT typically employs excitation light in the near infrared (NIR). Fluorophores that have excitation spectra in the 300–600 nm range are excited with wavelengths of 600–1200 nm through absorption of two infrared photons simultaneously. In such a two-photon quantum event, each photon carries approximately half the energy necessary to excite the molecule than that needed for one photon excitation [18] and thus reduces photothermal and photodamage effects. Nevertheless, the simultaneous absorption of two photons requires a high flux of excitation photons which can be effected with femtosecond lasers. Due to the low scattering and absorption of NIR photons, MPT is capable of high-quality acquisition of images in 3D at higher penetration depths than possible with single-photon imaging technology, such as confocal microscopy. Dynamics of specific cells and responses to external agents in the cells' natural environment can be investigated in living small animals with sub-cellular resolution.

21.2.1 Real-time imaging of stem cells and their dynamics with subcellular resolution

Our group first applied high-resolution 3D MPT in order to non-invasively visualize hair follicle-associated pluripotent (HAP) stem cells in their natural environment. HAP stem cells have been found to express nestin, reside in the bulge area and the dermal papilla and have the capacity to develop various tissues of follicular and nonfollicular origin, such as nerves, blood vessels, and smooth muscle [19–22]. MPT was employed to localize these stem cells in unperturbed niches in nestin-driven GFP transgenic athymic nu/nu nude and normal mice [23]. The multiphoton tomograph MPTflex™ (JenLab GmbH, Jena, Germany) equipped with a sealed turn-key tunable 80 MHz titanium:sapphire femtosecond laser (710–920 nm, in situ laser pulse width 250 fs) and with an articulated arm suitable for animal imaging was used (Fig. 21.5 (a) and (b)).

The optical unit consists of an active optical power attenuator to regulate the in situ power of the laser corresponding to tissue depth, an active beam stabilization device, a safety unit and a flexible articulated mirror arm with a compact scan head. The scan head consists of a fast galvoscanning device and a piezodriven z-scanner to generate 3D scans, high NA focusing optics (NA 1.3) and a dual-photon detector unit for the simultaneous recording of a variety of fluorescence and SHG signals. The acquisition time for one optical section is typically 2 seconds. The overall field-of-view of the optical system covers $350 \times 350\ \mu m^2$.

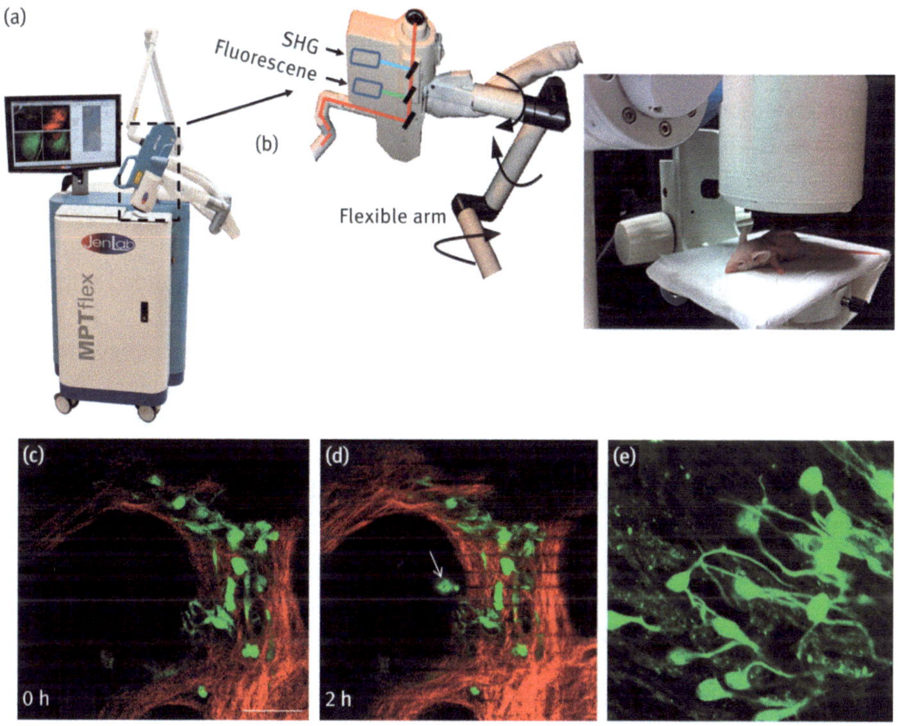

Fig. 21.5: (a) MPTflex with articulated flexible arm for non-invasive high-resolution imaging of living animals. (b) The tunable femtosecond laser (710–920 nm) is transmitted through an articulated mechano-optical arm with a compact scan head. The scan head contains galvoscanners and piezo-driven optics for 3D imaging as wells as single-photon counting detectors for fluorescence and SHG imaging. Filters for fluorescence detection can be varied depending on fluorescent proteins (GFP or RFP). The magnetic interface between animal and scan head allows long-term imaging of live mice. (c) Intratissue hair follicle stem cells detected in live mice. Stem cells express nestin-GFP (green). Extra-cellular matrix (ECM) collagen was detected by SHG (false color coded in red). Typical stem cells in the hair follicles have an oval-shaped body with a typical size of about 7 µm and dendritic-like arms. Scale bar: 50 µm. (d) Long-term tracking of individual intratissue stem cells. Nestin-GFP stem cells moving into the field of view. (e) Nestin-GFP stem cells (bright green) in wounded skin. Autofluorescence of unlabeled cells from the skin tissue are also visible (images from [23, 24]).

Multiphoton imaging provided the possibility of visualization of specific fluorescence protein nestin-GFP expressing stem cells within their intact microenvironment. Surrounding non-stem cells were visualized without exogenous labels based on two-photon excitation of intrinsic fluorophores such as nicotinamide adenine dinucleotide hydrogen (NADH), flavins and melanin in order not to cause any physiological imbalance on stem cells. In addition, SHG images were recorded simultaneously from collagen in the extracellular matrix. These imaging conditions were completely non-invasive and in particular, suitable for in vivo long-term tracking of intratissue stem cells without any significant effect on cell viability [23].

High-resolution multiphoton tomography allowed visualization of nestin-GFP expressing stem cells in their native niche at 930 nm excitation. Autofluorescence and SHG have been recorded simultaneously at 790 nm. 3D optical sectioning without surgical intervention revealed cell morphology, cell size, and stem cell distribution in the tissue (Fig. 21.5 (c)) [23]. Nestin-GFP expressing stem cells often occurred in clusters comprising varying numbers of cells per bulge. Long-term time-lapse imaging allowed detection of stem cell dynamics (Fig. 21.5 (d)) [23–25]. Stem cells migrating from the hair-follicle bulge have been detected inside the skin during optical deep-tissue sectioning. Stem cells migrating along the hair shaft as well as in the wounded skin were also visualized (Fig. 21.5 (e)).

Nestin-expressing stem cells were found to have a different morphology than the main skin cell population in their physiological natural microenvironment. The typical size of stem cells was found to be about 7 μm, whereas the surrounding cells had a typical size of 15 μm [23, 24]. The visualized microenvironment consisted of the extracellular matrix (ECM) components elastin and collagen as well as a variety of nonlabeled cells.

21.2.2 Multiphoton surgery of stem cells

The same near-infrared femtosecond lasers used for multiphoton imaging can also be employed to manipulate cells/cellular organelles with higher precision (sub-100 nm) when applied at higher intensities [27, 28]. Rompolas et al. [29] combined two-photon imaging with laser-induced ablation to investigate stem cell behavior during physiological regeneration of hair follicles. Non-invasive high-resolution imaging allowed long-term observations of the nuclei of stem/progenitor cells in transgenic mice expressing histone H2B-GFP driven by the keratin 14 promoter and revealed oriented cell division and migration of stem cell progeny that accompany hair growth. Additionally, dermal papilla cells in Lef1RFP transgenic mice were targeted to eliminate the mesenchymal cells selectively in order to study hair regeneration. Mesenchyme was ablated with the same NIR femtosecond laser at 900 nm and observed in physiological conditions up to day 7 with multiphoton technique. Mesenchymal dermal papilla cells are required for stem cell activation and hair regeneration [29]. GFP was excited with two-photon absorption at 940 nm, RFP at 1040 nm.

21.2.3 Imaging of GFP-labeled and unlabeled stem cells

MPT was used to compare GFP-labeled and unlabeled stem cell autofluorescence images in mouse whiskers. Autofluorescence multiphoton imaging revealed the same morphology of unlabeled hair follicle stem cells as GFP-labeled hair follicle stem cells [30]. FLIM of stem cells can provide additional information on the differentiation process [31].

21.2.4 High-resolution non-invasive multi-photon tomographic cancer-cell imaging in living animals

MPT was used to visualize dynamics of cancer cells genetically-labeled with specific fluorescent proteins in intact mice as well as stromal cells by autofluorescence within deep heterogenic tumor tissue in live animals [32]. Additionally, migrating rare metastatic cells were identified from nonmetastatic ones during consecutive long-term imaging [33]. Furthermore, the combination of MPT with FLIM can gain insights into cancer cell metabolism due to autofluorescent molecules associated with cellular respiration such as NADH and flavin adenine dinucleotide [34, 35].

21.2.5 Multi-photon imaging of tumor-targeting by *Salmonella typhimurium* A1-R

Our laboratory has developed a new substrain of engineered *Salmonella typhimurium auxotrophs* (*S. typhimurium* A1-R) to selectively eliminate cancer cells in viable as well as necrotic areas of tumor tissue but not in normal tissue [36]. The use of GFP for imaging the bacteria allows real-time visualization of single bacteria in vivo which could lead to the selection of enhanced cancer-cell-targeting variants of *S. typhimurium* [37].

We applied MPT to visualize *S. typhimurium* A1-R targeting Lewis lung cancer cells grown in the skin of nude mice. PMT1924 was used to detect signals from fluorescence and SHG channels. Filter sets LP409 and BP 395/14 were used for GFP and SHG, respectively. To separate Ds-RED fluorescence from GFP- and autofluorescence, filter sets BP593/40 and BP/510/42 were respectively used.

Collagen-extracellular matrix protein was imaged by SHG. Cancer cells expressing Ds-Red *S. typhimurium* A1-R expressing GFP, and stromal cells (autofluorescence) were simultaneously imaged by two-photon excitation (Fig. 21.6). Cancer cells were visualized in their natural microenvironment consisting of stromal cells and extracellular matrix collagen.

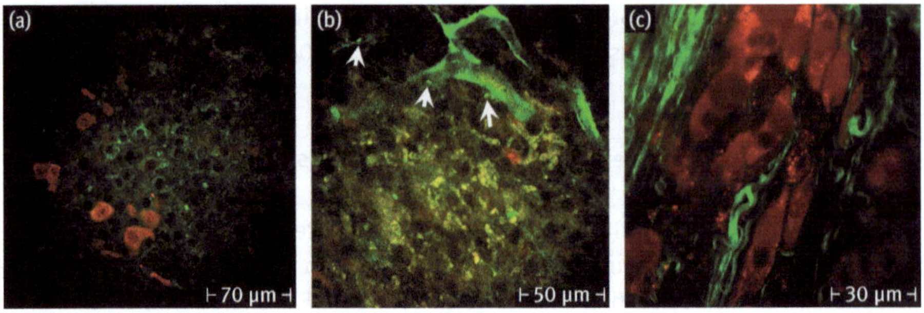

Fig. 21.6: (a) In vivo multiphoton detection of genetically labelled Ds-Red-expressing cancer cells (red color) and autofluorescence of unlabeled stromal cells (green color) within the tumor tissue. In stromal-cells autofluorescence was mainly detected from organelles, e.g., mitochondria in the cytoplasm. Nuclei appear nonfluorescent. (b) Imaging of Ds-Red cancer cells (red), GFP-expressing capillaries (bright green), *S. typhiurium* A1-R-GFP (bright green), and autofluorescent stromal cells (yellow-green). Single bacteria invasion in the vessels of tumor tissue are seen (white arrows) minutes after bacteria injection in the tail vein of live tumor-bearing mice. (c) Bacteria killing of cancer cells monitored in the tumor of live mice. Cancer cells expressing Ds-Red were imaged by two-photon excitation and unlabeled extracellular protein collagen was imaged by SHG. Expanding and bursting cancer cells were detected after some hours of bacteria injection in the tail vein of live tumor-bearing mice (images from [38]).

Spatial distribution of single GFP-expressing bacteria and bacterial colonies inside the cancer cells and in the tumor tissue and tumor vasculature (inside and outside of capillaries) were detected several minutes after injection of bacteria in the tail vein (Fig. 21.6 (b)). Long-term MPT revealed that bacterially-infected cancer cells expanded and burst (Fig. 21.6 (c)). Vascular permeability of tumor-targeting bacteria could also be imaged to gain further mechanistic insights into tumor-targeting bacteria [37, 38].

Previous studies have shown that fibrillary collagen is altered in various types of cancer and invading cells preferentially migrate along collagen fibers [39, 40]. Since MPT visualizes collagen in unperturbed physiological conditions, we applied high-resolution SHG tomography in living mice to study extracellular matrix remodeling during tumorigenesis [41, 42]. The relationship between cancer cells and intratumor collagen, elastin fibrils was visualized in live mice with subcutaneously-grown Colo 26-GFP (GFP expressing colon cancer cells) colon cancer, breast tumor and Lewis lung carcinoma (Fig. 21.7). SHG of collagen fibrils and GFP of cancer cells were acquired simultaneously at an excitation wavelength of 790 nm. High-resolution multiphoton imaging revealed organized extracellular matrix collagen and elastin in tumor tissue of live mice. Well-aligned, parallel, closely packed collagen fibers were detected in Lewis lung carcinoma by SHG. Elastin fibers were seen as parallel structures to collagen structures which are detected by autofluorescence (Fig. 21.7 (a)).

Collagen detected in Colo 26-GFP colon cancer-bearing mice were organized in parallel structures with a wide interfiber spacing. Cancer cells aligned parallel to collagen fibrils. Multiphoton autofluorescence and SHG images of unlabeled breast cancer

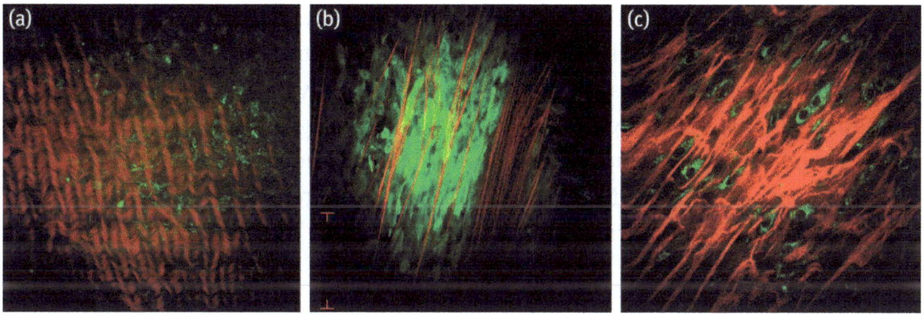

Fig. 21.7: High-resolution imaging of extracellular matrix collagen and elastin in tumor tissue of live mice. (a) Well-aligned, parallel, closely packed collagen fibers were detected in Lewis lung carcinoma by SHG (false color coded in red). Elastin fibers are seen as parallel structures to collagen structures which are detected by autofluorescence. (b) Collagen detected in Colo 26-GFP colon cancer-bearing mice. Cancer cells align parallel to collagen fibrils. Cancer cells express GFP and fluoresce green and collagen fibrils, visualized by SHG, are red (false color coded). The collagen fibrils are organized parallel structures with a wide interfiber spacing. SHG of collagen and GFP were excited simultaneously at an excitation wavelength of 790 nm. (c) Multiphoton autofluorescence and SHG images of unlabeled breast cancer tissue. Breast cancer cells were visualized by autofluorescence and collagen by SHG. Breast cancer cells appeared well-aligned parallel to collagen fibers (images from [41, 42]).

tissue revealed structures of collagen fibers as parallel, straight structures with interfiber spacing and with a different thickness. Breast cancer cell appeared well aligned parallel to collagen fibers. Breast cancer cells were visualized by autofluorescence and collagen by SHG. Our results suggest that collagen fibrils provide the scaffolding for cancer cells to anchor and acquire optimal shape in vivo.

21.2.6 Prospects and limitations of multiphoton tomography

Application of high-resolution MPT allowed the live imaging of stem cells up to depths of approximately 300 µm in nestin-GFP mice. However, localization of stem or cancer cells in deep organs requires future efforts to increase imaging depths. 3D image recordings take a long time due to slow acquisition speed (typically 2–7 seconds per frame). Recording larger image areas is also time consuming. Moreover, animal motion induces artifacts in long-term imaging. Currently, biocompatible fluorescent probes and novel compact laser sources for the second optical window between 1000 nm and 1350 nm and even the third optical window from 1600 nm to 1870 nm are under development. This would allow much deeper tissue imaging than with current Ti:sapphire laser technology (600–1000 nm = first optical window) [44]. Furthermore, there is a lack of information from nonfluorescent tissue components. Therefore a combination of multiphoton imaging tools with other optical label-free imaging technologies such as CARS imaging could provide complementary information on biochemical and structural properties of stem and cancer cells and the intracellular microenvironment [43]. Transition of functional multiphoton imaging systems into the clinic for precise localization of tumor borders, analysis of a variety of individual tumors and stem cells in internal organs, including their metabolism, invasion, differentiation, microenvironment and therapeutic monitoring, would become more realistic with enhanced time and depth resolution. Imaging of hard-to-access areas can be carried out with long-distance adaptive optics and endoscopes [45].

References

[1] Hoffman RM. The multiple uses of fluorescent proteins to visualize cancer in vivo. Nature Reviews Cancer. 2005;5:796–806.

[2] Yang M, Baranov E, Jiang P, Sun FX, Li XM, Li L, Hasegawa S, Bouvet M, Al-Tuwaijri M, Chishima T, Shimada H, Moossa AR., Penman S, Hoffman RM. Whole-body optical imaging of green fluorescent protein-expressing tumors and metastases. Proc Natl Acad Sci USA. 2000;97:1206–1211.

[3] Yang M, Baranov E, Moossa AR, Penman S, Hoffman RM. Visualizing gene expression by whole-body fluorescence imaging. Proc Natl Acad Sci USA. 2000;97:12278–12282.

[4] Yang M, Baranov E, Li XM, Wang JW, Jiang P, Li L, Moossa AR, Penman S, Hoffman RM. Whole-body and intravital optical imaging of angiogenesis in orthotopically implanted tumors. Proc Natl Acad Sci USA. 2001;98:2616–2621.

[5] Katz MH, Takimoto S, Spivack D, Moossa AR, Hoffman RM, Bouvet M. A novel red fluorescent protein orthotopic pancreatic cancer model for the preclinical evaluation of chemotherapeutics. J Surg Res. 2003;113:151–160.

[6] Yamamoto N, Jiang P, Yang M, Xu M, Yamauchi K, Tsuchiya H, Tomita K, Wahl GM, Moossa AR, Hoffman RM. Cellular dynamics visualized in live cells *in vitro and in vivo* by differential dual-color nuclear-cytoplasmic fluorescent-protein expression. Cancer Research. 2004;64:4251–4256.

[7] Zhang Y, Zhang N, Zhao M, Hoffman RM. Real-time noninvasive spectral imaging of orthotopic red fluorescent protein-expressing lung tumor growth in nude mice. Anticancer Res. 2015;35:3755–3759.

[8] Yang M, Jiang P, Hoffman RM. Whole-body subcellular multicolor imaging of tumor-host interaction and drug response in real time. Cancer Res. 2007;67:5195–5199.

[9] Yang M, Reynoso J, Jiang P, Li L, Moossa AR, Hoffman RM. Transgenic nude mouse with ubiquitous green fluorescent protein expression as a host for human tumors. Cancer Research. 2004;64:8651–8656.

[10] Yang M, Reynoso J, Bouvet M, Hoffman RM. A transgenic red fluorescent protein-expressing nude mouse for color-coded imaging of the tumor microenvironment. J Cell Biochem. 2009;106:279–284.

[11] Tran Cao HS, Reynoso J, Yang M, Kimura H, Kaushal S, Snyder CS, Hoffman RM, Bouvet M. Development of the transgenic cyan fluorescent protein (CFP)-expressing nude mouse for "Technicolor" cancer imaging. J Cell Biochem. 2009;107:328–334.

[12] Suetsugu A, Katz M, Fleming J, Truty M, Thomas R, Saji S, Moriwaki H, Bouvet M, Hoffman RM. Noninvasive fluorescent-protein imaging of orthotopic pancreatic-cancer-patient tumorgraft progression in nude mice. Anticancer Research. 2012;32:3063–3068.

[13] Yang M, Jiang P, Hoffman RM. Whole-body subcellular multicolor imaging of tumor-host interaction and drug response in real time. Cancer Res. 2007;67:5195–5200.
[14] Zhang Y, Zhang N, Zhao M, Hoffman RM. Real-time non-invasive spectral imaging of orthotopic red fluorescent protein-expressing lung tumor growth in nude mice. Anticancer Res. 2015;35:3755–3759.
[15] Pantazis P, Maloney J, Wu D, Fraser ES. Second harmonic generating (SHG) nanoprobes for in vivo imaging. Proc Natl Acad Sci. 2010;107(33):14535–14540.
[16] Sutherland RL, McLean DG, Kirkpatrick S. Handbook of nonlinear optics. 2nd edition. New York: Dekker; 2003.
[17] Campagnola PJ, Loew LM. Second-harmonic imaging microscopy for visualizing biomolecular arrays in cells, tissues and organisms. Nat Biotechnol. 2003;21(11):1356–1360.
[18] Benninger RKP, Piston DW. Two-photon excitation microscopy for the study of living cells and tissues. Curr Protoc Cell Biol. 2013;4 (4):1–24.
[19] Amoh Y, Li L, Katsuoka K, Penman S, Hoffman RM. Multipotent nestin-positive, keratin-negative hair-follicle-bulge stem cells can form neurons. Proc Natl Acad Sci USA. 2005;102:5530–5534.
[20] Amoh Y, Li L, Campillo R, Kawahara K, Katsuoka K, Penman S, et al. Implanted hair follicle stem cells form Schwann cells that support repair of severed peripheral nerves. Proc Natl Acad Sci USA. 2005;102:17734–17738.
[21] Liu F, Uchugonova A, Kimura H, Zhang C, Zhao M, Zhang L, et al. The bulge area is the major hair follicle source of nestin-expressing pluripotent stem cells which can repair the spinal cord compared to the dermal papilla. Cell Cycle. 2011;10:830–839.
[22] Hoffman RM. The pluripotency of hair follicle stem cells. Cell Cycle. 2006;5:232–233.
[23] Uchugonova A, Hoffman RM, Weinigel M, Koenig K. Watching stem cells in the skin of living mice noninvasively. Cell Cycle. 2011;10:2017–2020.
[24] Uchugonova A, Hoffman RM, Koenig K. Stem cell imaging in living animals: Stem cells in hairs tracked by multiphoton tomography. Imaging & Microscopy. 2013;3:44–46.
[25] Uchugonova A, Duong J, Zhang N, König K, Hoffman RM. The bulge area is the origin of nestin-expressing pluripotent stem cells of the hair follicle. J Cell Biochem. 2011;112:2046–2050.
[26] Van Thor JJ, Gensch T, Hellingwerf KJ, Johnson LN. Phototransformation of green fluorescent protein with UV and visible light leads to decarboxylation of glutamate 222. Nat Struct Biol. 2002;9(1):37–41.
[27] König K, Riemann I, Fischer P, Halbhuber KJ. Intracellular nanosurgery with near infrared femtosecond laser pulses. Cell Mol Biol. 1999;45(2):195–201.
[28] Uchugonova A, Lessel M, Nietzsche S, Zeitz S, Jacobs K, Lemke C, König K. Nanosurgery of cells and chromosomes using near-infrared 12-femtosecond laser pulses. J Biomed Opt. 2012;17:101502.
[29] Rompalas P, Deschene ER., Zito G, Gonazalez DG, Saotome I, Habermann AM. Live imaging of stem cell and progeny behaviour in physiological hiar-follicle regeneration. Nature. 2012;487:496–500.
[30] Uchugonova A, Cao W, Hoffman RM, Koenig K. Comparison of label-free and GFP two-photon imaging of hair follicle-associated pluripotent (HAP) stem cells in mouse whiskers. Cell Cycle. 2015;14(21):3430–3433.
[31] König K, Uchugonova A, Gorjup E. Multiphoton fluoprescence Lifetime imaging of 3D stem cells spheroids during differentiation. Microscopy Research and Technique. 2011;74:9–17.
[32] Shah AT, Diggins KE, Walsh AJ, Irish JM, Skala MC. In vivo autofluorescence imaging of tumor heterogeneity in response to treatment. Neoplasia. 2015;17(12):862–870.
[33] Tanaka K, Toiyama Y, Okugawa Y, Okigami M, Inoue Y, Uchida K, Araki T, Mohri Y, Mizoguchi A, Kusunoki A. In vivo optical imaging of cancer metastasis using multiphoton microscopy: a short review. Am J Transl Res. 2014;6(3):179–187.

[34] Skala MC, Riching KM, Gendron-Fitzpatrick A, Eickhoff J, Eliceiri KW, White JG, Ramanujam N. In vivo multiphoton microscopy of NADH and FAD redox states, fluorescence lifetimes, and cellular morphology in precancerous epithelia. Proc Natl Acad Sci USA. 2007;104(49):19494–19499.
[35] Walsh AJ, Cook RS, Manning HC, Hicks DJ, Lafontant A, Arteaga CL, Skala MC. Optical metabolic imaging identifies glycolytic levels, subtypes, and early-treatment response in breast cancer. Cancer Research. 2013;73(20):6164–6174.
[36] Hoffman RM. The preclinical discovery of bacterial therapy for the treatment of metastatic cancer with unique advantages. Expert Opin Drug Discov. 2012;7:73–83.
[37] Uchugonova A, Zhang Y, Salz R, Liu F, Suetsugu A, Zhang L, Hoffman RM, Zhang M. Imaging the different mechanism of prostate cancer-cell killing by tumor-targeting Salmonolla typhimurium A1-R. Anticancer Research. 2015;35(10):5225–9.
[38] Uchugonova A, Zhao M, Zhang Y, Weinigel M, Koenig K, Hoffman RM. Cancer cell killing by engineered salmonella imaged by multiphoton tomography in live mice. Anticancer Research. 2012;32(10):4331–7.
[39] Provenzano PP, Eliceiri KW, Campbell JM, Inmar DR, White JG, Keely PJ. Collagen reorganization at the tumor-stromal interface facilitates local invasion. BMC Med. 2006;4:1–16.
[40] Provenzano PP, Inman DR, Eliceiri KW, Knittel JG, Yan L, Rueden CT, White JG, Keely PJ. Collagen density promotes mammary tumor initiation and progression. BMC Med. 2008;6:11.
[41] Uchugonova A, Zhao M, Weinigel M, Zhang Y, Bouvet M, Hoffman R.M, König K. Multiphoton tomography visualizes collagen fibers in the tumor microenvironment maintain cancer cell anchorage and shape. J Cell Biochem. 2013;114(1):99–102.
[42] Uchugonova A, Hoffman RM, König K. Multiphoton tomography with submicron spacila resolution of living tumor-bearing mice. 30th International Congress on High-Speed Imaging & Photonics. South Africa; 2012
[43] Weinigel M, Breunig G, Uchugonova A, König K. Multipurpose nonlinear optical imaging system for in vivo and ex vivo multimodal histology. J Med Imaging. 2015;2(1):016003.
[44] Sordillo LA, et al. Deep optical imaging of tissue using the second and third near-infrared spectral windows. J Biomed. Opt. 2014;19:056004.
[45] Le Harzic R, Riemann I, Weinigel M, König K, Messerschmidt B. Rigid and high-numerical-aperture two-photon fluorescence endoscope. Applied Optics. 2009;48(18):3396–3400.

Christian Mess and Volker Huck

22 Bedside assessment of multiphoton tomography

From skin cell morphology via fluorescence lifetime imaging to clinical pathophysiology

Abstract: The application of multiphoton microscopy in the field of biomedical research and advanced diagnostics promises unique insights into the pathophysiology of skin diseases. Recently, various studies on drug delivery and skin cancer have been successfully completed. With the aim of gaining deeper understanding of the linkage between cellular structure and physiological processes, non-invasive multiphoton-based intravital tomography (MPT) and fluorescence lifetime imaging (FLIM) were combined within the scope of inflammatory skin and chronic wounds in clinical application.

The optical biopsies generated via MPT were morphologically analyzed and aligned with classical skin histology. Because of its subcellular resolution, MPT provided evidence of a redistribution of mitochondria in keratinocytes indicating an altered cellular metabolism. Morphometric algorithms reliably showed a perinuclear accumulation in lesional skin in contrast to an even distribution in healthy skin. MPT-FLIM confirms these observations showing a metabolic shift in lesions. Moreover, detection of the onset and progression of inflammatory processes could be achieved.

In conclusion, the change in the distribution of mitochondria and the verification of an altered cellular metabolism facilitate a better understanding of the processes of wound healing and inflammatory skin diseases envisioning future clinical areas of MPT-FLIM application.

22.1 Multiphoton gleaming – endogenous fluorophores in human skin

In vivo investigations in humans exclude the application of fluorescence labeling of cells or proteins but are open to the use of naturally occurring fluorophores. In human skin, a plethora of these fluorophores has been found: fibrillary structures such as elastin or collagen and, within the epidermis, biomolecules such as melanin, tryptophan, keratin, porphyrin, flavins or NADH [1–4]. Focusing on the latter, a good candidate for both, high-resolution imaging of subcellular morphology [5] and functional characterization of epidermal cells [6, 7], more of which later, has been identified. Multiphoton excitation of this endogenous fluorophore, therefore, enables the non-

https://doi.org/10.1515/9783110429985-024

invasive examination of human skin with all the advantages known for two-photon utilization. With a negligible linear energy transfer at the boundary layer, the excitation almost exclusively takes place within the located target volume of examination without damaging the surrounding tissue [8].

22.2 From morphology to biochemical state – pathophysiological characterization of human skin

The diagnosis and assessment of human skin diseases are mainly based on the skills of the medical specialists: They examine the patient's skin by eye and dermatoscopy or perform the histopathological analysis of biopsies, the current gold standard. Undoubtedly, histological and immunohistochemical processing has a high diagnostic significance but its invasiveness, resulting in the formation of scars, restricts a longitudinal analysis of specific disorders or lesions. Magnifying in vivo techniques, which non-invasively 'penetrate' the surface level, such as ultrasound or confocal laser scanning microscopy [9–11], facilitate a promising closer view into the skin but in terms of their accessible resolution these techniques cannot compete with the validity of histological examination on a cellular level. As MPT and MPT-FLIM allow subcellular in vivo imaging, they open the possibility to validate multiphoton-based tomography as a non-invasive clinical tool for both, the morphological and thus far uninvestigated biochemical characterization of human skin.

In 2014, Yew and co-workers surveyed recent application fields of MPT in clinical research [12]. Many recent studies successfully concentrated on in vivo drug delivery [13, 14] or advanced diagnosis of skin cancer [5, 15]. To bridge the gap between cellular morphology and clinical pathophysiology, in the following we will focus on two highly prevalent disorders, inflammatory skin [16] and chronic wounds [17]. Inflammatory skin diseases have an increasing incidence, mainly in developed countries. Using the example of *Atopic dermatitis*, the first manifestation usually occurs in early childhood, often accompanied by the development of asthma and allergies [18]. According to current data, up to 30 % of newborns are affected by this inflammatory diathesis [19]. The exact underlying pathophysiological mechanisms are under debate; recent data suggest a complex interplay between genetic, epigenetic and environmental factors [20]. The most frequent manifestation of a chronic wound is the *Ulcus cruris*, a deep persistent skin defect with an impaired process of wound healing [21, 22], usually occurring in old age. Subdivided by the main etiologic factor, its pathogenesis is *Too Far Afield* of the pathological persistence in a distinct phase of wound healing. An interplay of diverse components of the coagulation system, an inflammatory response, granulation and a subsequent re-epithelialization is responsible for a smooth healing process or, alternatively, for a wound healing disorder [23, 24].

Common to both diseases is, next to the high prevalence, their chronicity, a fact that necessitates redundant examinations in the longitudinal course. Further knowledge of the pathophysiology is essential to understand the pathogenetic background and to improve the present therapeutic options.

22.3 Hands-on clinical multiphoton tomography

Our first multicenter clinical trial comprised 20 healthy volunteers and 30 patients affected with the inflammatory skin disease *Atopic dermatitis*. For each patient, one lesional and one nonlesional skin area were analyzed during four clinical visits over three months and compared to corresponding skin areas of age-correlated healthy subjects. Thus, we obtained a survey of the diverse states of inflammation in the longitudinal course of different therapeutic regimens. In the second diagnostic study, we focused on ten patients suffering from a chronic wound (chronic leg ulcer) compared to corresponding areas of ten healthy volunteers. The MPT rounds were performed in the same study design as described above to characterize the healing process – or the wound healing disorder – at the wound edge. One must consider that in this case the term 'nonlesional' is advisedly parenthesized: In contrast to the *systemic* disorder of inflammatory skin, in the study group affected with a local chronic wound 'nonlesional' describes the corresponding position at the not-affected leg of the patient. In both studies, medical specialists examined the subjects, gathering clinical data and disease-specific severity scores.

At each skin area during one MPT round, sequences of confocal sections with 10 µm penetration steps and an edge length of 100 µm were recorded. An excitation wavelength of 710 nm and an emission bandpass filter (460/60 nm) were used to match the spectrum of NADH. Starting at the outmost epidermal layer (*Stratum corneum*), defined as a depth of 0 µm, and ending at the papillary dermis at approximately 150 µm, we obtained a multiphoton tomographic insight into a complete skin cube of the investigated area for further analyses.

To cope with the vast amount of images, FLIM data and clinical parameters, a specialized system for structured data storage and retrieval was a prerequisite for further analysis. For this purpose, we used the Open Microscopy Environment (OME). OME consists of OME-XML [25], an open standard format for the description and annotation of microscopy image data, and OMERO (OME Remote Objects) [26, 27], an open source software system for the management and analysis of image data. To enrich the existing image data with histological parameters, all images were first categorized by a medical specialist with expertise in the field of histopathology. For categorization, a region of interest consisting only of cells from one single epidermal layer was defined for each image. Additional examples of parameters were image quality with regard to suitability for automatic analysis and diagnosis from a histological perspective. Categorization parameters, clinical values (e.g., medication and clinical scores) and technical image acquisition data (e.g., excitation wavelength and laser energy), together with the associated images and FLIM data, were then imported to the OMERO server. Various GUI- and Web-clients can be used for image and metadata browsing. Furthermore, the functionality of the OMERO server can be used and extended to implement algorithms for analysis. To obtain a tool for specialized and explorative data analysis and quantification, we have chosen a two-staged approach. In the first step, the user creates a subset of all images by defining various metadata constraints (e.g., image quality: high to very high; epidermal layers: *Stratum granulosum* and *spinosum*, and inflammation severity: 10 to 50) and in a second step, the parameters for quantitative analysis (e.g., inflammation severity vs. τ_m) are chosen. For rapid evaluation of results, different visualization methods, such as box or scatter plots, are available, and the quantitative results can be exported for further analysis.

22.4 The building blocks of intravital multiphoton tomographs

In the present studies, we applied certified CE-marked (five-dimensional) intravital multiphoton tomographic systems (MPT). For image acquisition the *DermaInspect* multiphoton tomograph [1] and the more mobile and flexible successor *MPTflex* [28] were employed (both JenLab GmbH, Jena, Germany). A schematic overview of the principle components of both microscopes and an illustration of the in vivo adapter – the mechanism to fix the objective to the skin – is displayed in Fig. 22.1. Near-infrared laser pulses with a pulse length of 100 fs provided by a titanium:sapphire tunable laser system (Mai Tai, Newport Spectra-Physics, Santa Clara, CA, USA) were used for the excitation of endogenous fluorophores in human skin. The excitation laser beam was attenuated by the use of a Glan calcite polarizer and scanned by two galvanometric mirrors. After passing a beam expander and collimator, the laser pulses are reflected by a dichroic beam splitter into a 40× oil immersion microscope objective with a numerical aperture of 1.3 (Carl Zeiss Jena GmbH, Jena, Germany).

For the *DermaInspect*, the emitted fluorescence light was filtered (F37-490 BrightLine HC 490/, F39-461 BrightLine HC 460/60, F39-390 BrightLine HC 390/40) and cleared by an additional shortwave-pass filter (F75-680 Multiphoton-Emitter HC 680/SP; all Semrock Inc., Rochester, NY, USA). The *DermaInspect* further supports the splitting of the emitted light into three distinct spectral regions by a set of dichroic beam splitters (F43-031 425DCXR, F33-499 495DCXR; Chroma Technology Corp., Bellows Falls, VT, USA). The *MPTflex* configuration used had no support for spectral emission splitting and was equipped with a short- and long-pass filter set (F75-680 Multiphoton-Emitter HC 680/SP and F39-409 BrightLine HC 409/LP; Semrock Inc., Rochester, NY, USA) and a beam splitter (F25-660 660DCXR; Chroma Technology Corp., Bellows Falls, VT, USA) for the autofluorescence pathway.

A photomultiplier (PMT) provided intensity images of the sample, and was readout via time-correlated single-photon counting (TCSPC) to calculate wavelength-range-specific fluorescence lifetime values (MPT-FLIM). The PMT data were processed by a high-resolution TCSPC imaging module (SPC 830, Becker & Hickl GmbH, Berlin,

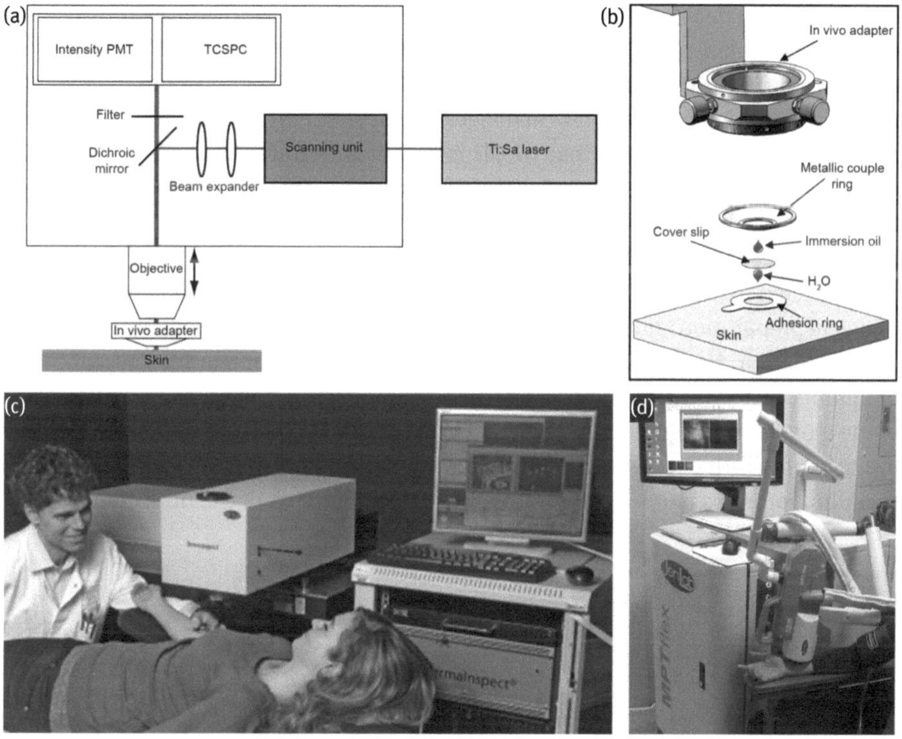

Fig. 22.1: Multiphoton tomographs for in vivo application. (a) Schematic overview of the essential components of the multiphoton tomographic system. (b) Layout of the in vivo skin coupling mechanism. (c) Clinical *DermaInspect* application. (d) *MPTflex* measurement. (a, d) Modified from [29].

Germany). The *DermaInspect* was equipped with three separate PMTs, one for each spectral region and thus providing an extra dimension compared to the *MPTflex*.

In this clinical setup the skin fluorophores were excited by ultrashort laser pulses in the near-infrared spectrum (700–920 nm) leading to two-photon absorption. The two-photon effect means that two photons of approximately half the energy needed for one-photon excitation can excite a fluorophore, if they are absorbed within a very short timeframe [3, 30].

22.5 Alignment of intravital multiphoton tomographic data with classical skin analysis

The naturally first step for establishing multiphoton tomography as a reliable tool for bedside assessment was to prove its imaging and diagnostic validity. Even at first glance, the MPT-derived morphological pattern seemed to be representative for typical keratinocytes within the human epidermis. Focusing exempli gratia on the outermost cellular layer of healthy skin as shown in Fig. 22.2 (MPT) inside the three-dimensional reconstructed skin cube (MPT-3D), characteristic granulated, flat and round-shaped cells within the *Stratum granulosum* appear as presentable high-resolution cross sections, similar to high-class confocal images of biopsies. Previous studies of multiphoton imaging have suggested that NADH is the major natural fluorophore in human cells [5, 31, 32]. NADH, in particular the protein-bound fraction, is mainly located at the inner mitochondrial membrane. As proof, we took punch biopsies immediately after the last clinical MPT round from target skin areas in three healthy subjects and three affected patients. The biopsies were split for standard vertical and horizontal histological preparation and stained with the anti-mitochondrial antibody MTC02 ab3298 (Abcam, Cambridge, MA, USA). After further preparation for immunofluorescence (IF) or immunohistochemical microscopy (IHC), the MPT images and the horizontal layers of the biopsies were aligned. As shown in Fig. 22.2, the bright spots observed in the cytoplasm of keratinocytes surrounding dark nuclei when using MPT are mainly due to multiphoton-excited NADH, ensured by the combination of the utilized excitation

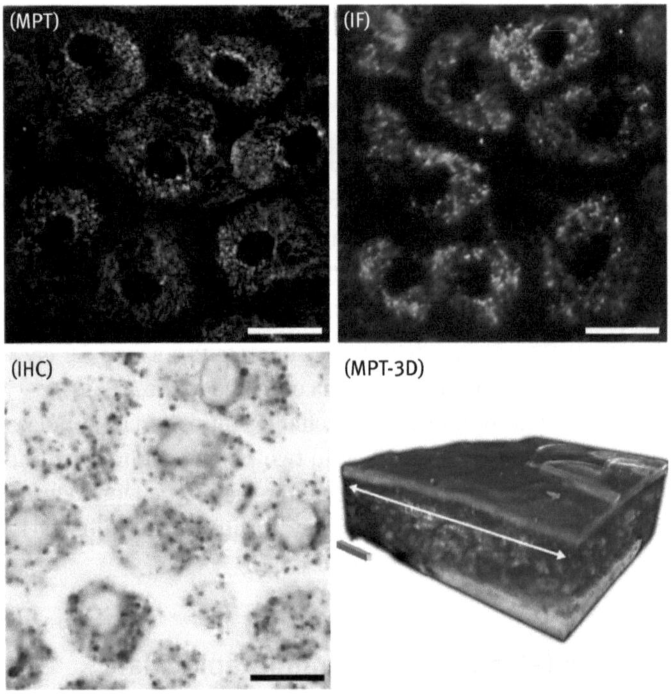

Fig. 22.2: Multiphoton tomographic optical biopsy and mitochondria staining of healthy human skin. A representative multiphoton optical section of the *Stratum granulosum* of healthy skin in vivo (MPT). Alignment with mitochondria-specific immunofluorescence staining (IF) and immunohistochemistry (IHC) of the corresponding skin region performed ex vivo. The white arrow within the multiphoton tomographic three-dimensional reconstructed skin cube (MPT-3D) indicates the examined target section of (MPT). Scale bars correspond to 20 µm. Modified from [7].

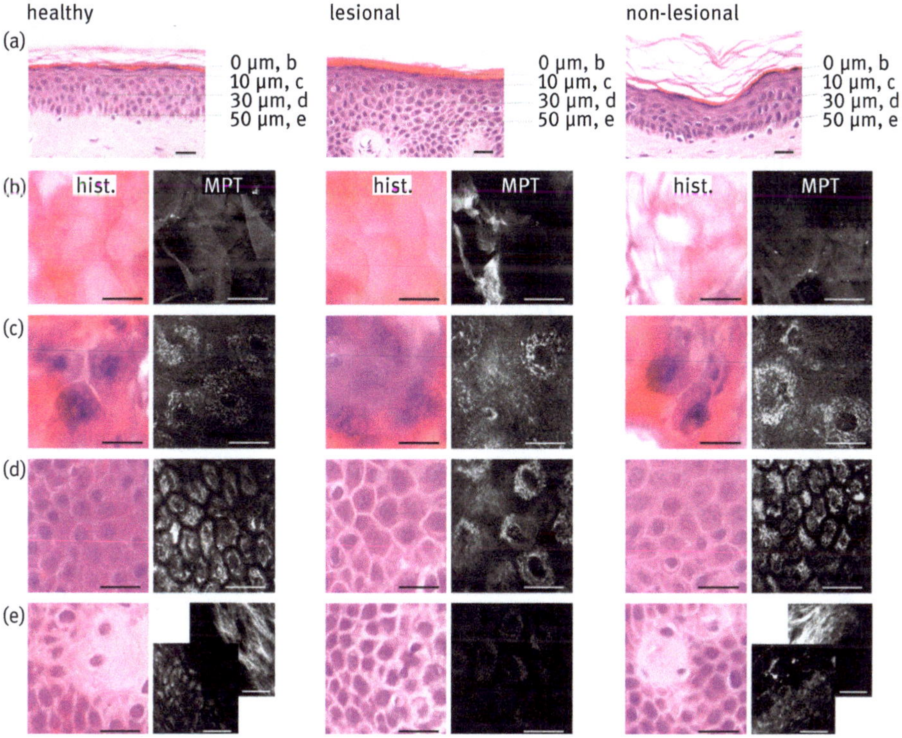

Fig. 22.3: Gold standard alignment of intravital multiphoton tomography. (a) Standard vertical histological sections of the skin from healthy subjects (healthy), and from a patient suffering from inflammatory skin (nonlesional and lesional). (b)–(e) Alignment of horizontal histological sections (hist.-column) with multiphoton tomographic sections (MPT-column) of the same regions of healthy, nonlesional and lesional areas of skin at indicated skin depths. A comparison of hist. with MPT illustrates the identical validity of assessing the morphological skin architecture and detecting characteristic skin morphologies. In the lesional column, the dermis is not accessible at a depth of 50 µm due to typical thickening of the epidermis. The accessibility of the dermis was demonstrated through additional imaging of collagen as shown in the back most image of the composite (e). Scale bars correspond to 20 µm. Modified from [7].

wavelength and spectral emission filtering. As expected, these spots are located within the mitochondria as validated by IF and IHC.

Based on these settings, the same skin region in healthy subjects and patients affected by an inflammatory skin disease were compared via MPT and subsequently via histological analyses out of the described set of punch biopsies. Each subject was clinically examined by a medical specialist and by MPT over a period of three months. During each session, one lesional (inflamed tissue) and one nonlesional (ostensibly healthy) skin region were analyzed in comparison to age-correlated healthy subjects. As shown in Fig. 22.3, the analyses of histological biopsies stained with hematoxilin-eosin and multiphoton-based intravital tomographic images of lesional skin areas were of identical pathognomonic validity. Upon MPT, characteristic morphologies of inflamed skin were detectable, such as intercellular edema and an impaired architecture, accompanied by thickening of the epidermis in lesional skin. Without any artifact-inducing embedding, dehydration and staining procedures, the MPT technique was thereby shown to be suitable for an optical biopsy, meaning a non-invasive in vivo characterization of human skin at patients' bedside.

22.6 Morphological analysis of the cellular mitochondrial distribution

Focusing on the MPT autofluorescence intensity, lesional keratinocytes showed a strikingly altered tomographic pattern compared to healthy cells. In lesional cells, a ring of high intensities in the vicinity of the nucleus was found (Fig. 22.4 (g)), whereas within the cytoplasm of healthy keratinocytes, a relatively homogenous intensity distribution was present (Fig. 22.4 (a)). As described, this signal originates mainly from the NADH within mitochondria. To quantify this change in the mitochondria configuration, we developed a new computational method – the *Radial Profile Analysis* (RPA) – that gives a measure of the radial intensity distribution within arbitrary cell-like objects [7].

In a first step, histopathologists manually segmented cells. In detail, the nucleus and the outer cell membrane of each cell is traced by two nonintersecting polygons P_{nucl} and P_{cell} (see Fig. 22.4 (a) and (g)). The difference $P_{\text{cell}} - P_{\text{nucl}}$ defines the cell cytoplasm which is the region of interest (ROI) for the mitochondrial distribution analysis. Most images suffered from an inhomogeneous illumination with lower intensities near the boundaries. To correct this vignetting effect, we used a previously reported Gaussian filtering-based method [33].

The cell outline can be very irregular (see for example Fig. 22.4 (a)) and often the image plane cuts the cell in a skewed way which can result in very eccentric nuclei (cf. Fig. 22.5 (e)). The naive approach (described as *Annuli Filling Analysis* in [7]) to divide the ROI into concentric segments and calculate the relative mitochondrial count within each segment does not really produce satisfying results for these distorted cells. These issues are circumvented with the RPA method.

The general idea of RPA, which is based on a previously published algorithm [34], is to sample the intensities within the ROI on radial profile lines and a subsequent analysis of the resulting profile plot (Fig. 22.4). As a starting point, l equiangular radial line segments are generated within the ROI from 0 to 2π radians with an angular step size of $2\pi/l$. On each line segment, the subjacent intensity values are sampled at s equidistant points using bilinear interpolation resulting in l radial profiles p. All profiles are then stacked together to form a $s \times l$ matrix which represents the intensity information of the unrolled length-normalized cytoplasm (Fig. 22.4 (e) and (f)). The intensity distribution of each single profile is quantified by summing the sampled values

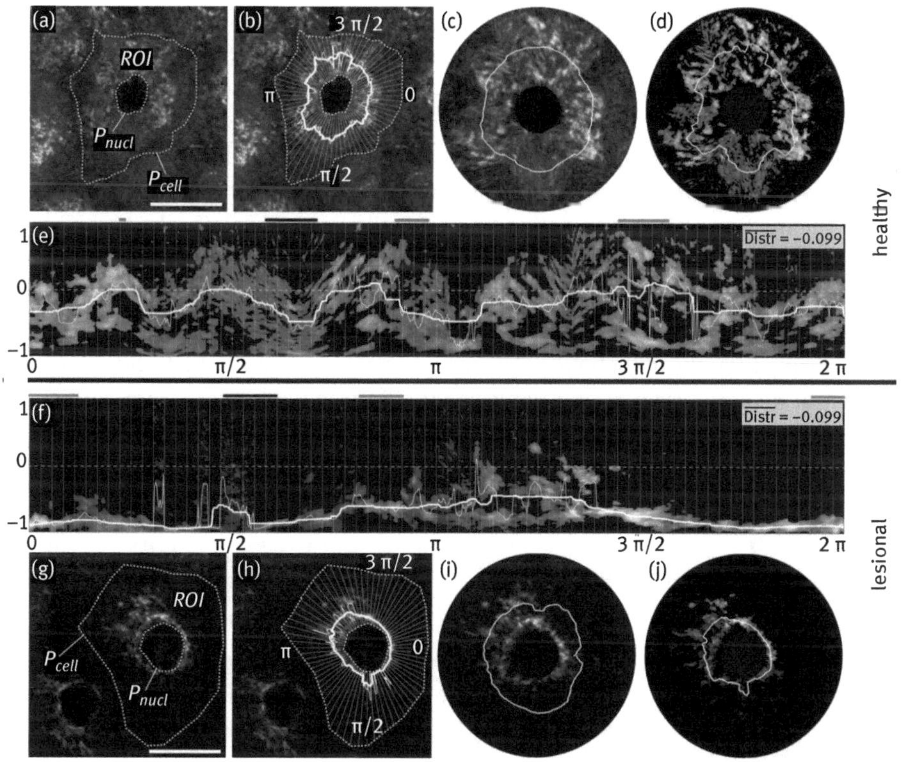

Fig. 22.4: Radial profiles analysis (RPA) of cellular mitochondrial distribution. Archetypical keratinocytes of the *Stratum granulosum* in healthy (a)–(e) and lesional (f)–(j) skin areas have been selected. (a), (g) Representative MPT cell images. The cytoplasm bounded by P_{nucl} and P_{cell} is used as ROI for RPA. (b), (e), (f), (h) Radial profile lines and plots of mitochondrial distribution values (thin: raw; bold: smoothed) are shown. Virtual cell reconstructions based on radial profiles with smoothed distribution values are depicted before (c), (i) and after thresholding (d), (j). Consequently, the unrolled thresholded radial profiles – the distribution value (Distr $\in [-1, 1]$) plotted against the radian measure – for healthy (e) and lesional (f) cells are displayed with the calculated mean values ($\overline{\text{Distr}}$). Outliers are marked with horizontal bars above the plot (gray: short, black: long). Scale bars correspond to 20 μm. Modified from [7].

from inner to outer. If the sum exceeds the half of the cumulative sample value of p, the index of this specific point is denoted as the distribution value Distr of p. To further improve this calculation, the effect of background noise and non-mitochondrial low intensity pixels is reduced by thresholding each column profile with Otsu's method in a preceding step. This algorithm assumes that the intensity histogram of an image consists of two classes and finds the optimal threshold by minimizing the intra-class variance [35]. The benefit of this approach is shown in Fig. 22.4 (d) and (j). The Distr plot (white line) of the thresholded virtual cells reflects the shape of the real intensity distribution in a much better way. The normalized mean $\overline{\text{Distr}} \in [-1, 1]$ of all profiles is the quantifier for the cell's mitochondrial distribution. A value near 0 stands for a homogeneous distribution (Fig. 22.4 (e)), which is typical for healthy cells, whereas a strong negative $\overline{\text{Distr}}$ is found in lesional cells with a centralized mitochondria configuration around the nucleus (Fig. 22.4 (f)). A tilted optical sectioning indicated by distinct eccentric nuclei might result in a single-sided impaired information density. Therefore, extraordinarily long and short radial profile line outliers are filtered out using the median absolute deviation [36]. RPA allows to trade off calculation speed against accuracy by freely selectable parameters s and l. It is even possible to incorporate subpixel information by increasing s and l so that the distance of two adjacent sampling points falls below the pixel distance.

Utilizing RPA, the mitochondrial distribution of keratinocytes in the *Stratum granulosum* of healthy, lesional and nonlesional skin was analyzed in both studies. In Fig. 22.5, the results of representative cells are visualized, and a clear mitochondria centralization resulting in strongly negative $\overline{\text{Distr}}$ values can be observed in lesional

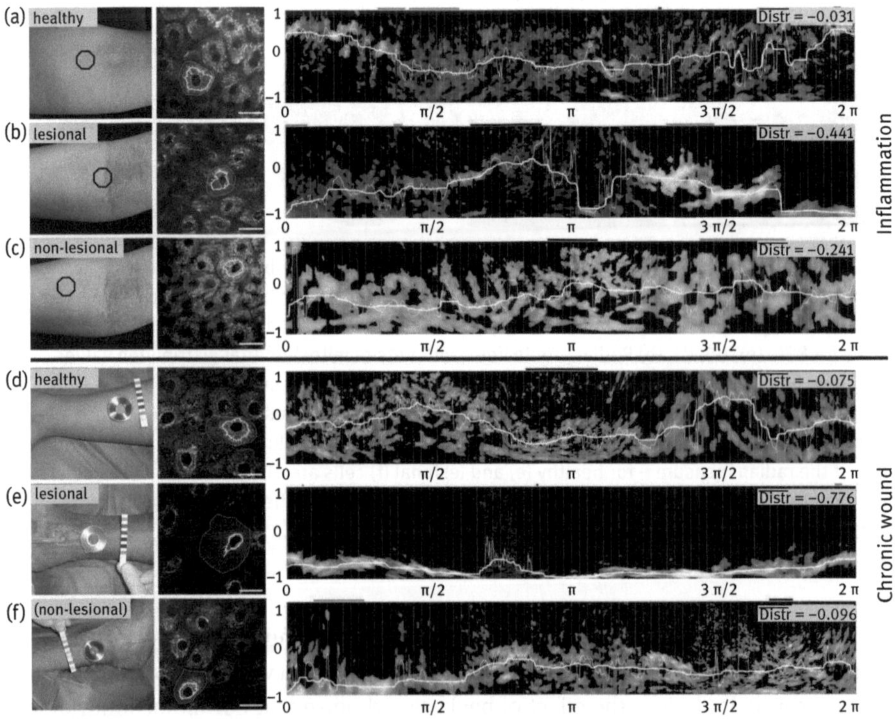

Fig. 22.5: Mitochondrial distribution of patients' epidermal cells in inflammation and chronic wounds. From left to right, photographs of the examination areas, representative MPT intensity images and radial profiles of *Stratum granulosum* cells from patients affected by inflammation (a)–(c) or chronic wounds (d)–(f) are presented. Radial profiles for healthy (e), (d), lesional (b), (e) and non-lesional (c), (f) cells are displayed with the calculated mean distribution values ($\overline{\text{Distr}}$). Lesional cells exhibit an intense mitochondrial centralization in both study groups. Scale bars correspond to 20 μm.

cells within both study groups. On the contrary, the healthy and nonlesional cells show a relatively even distribution reflected by a $\overline{\text{Distr}}$ value near 0.

The statistical analysis of all segmented cells (Fig. 22.6) reveals a highly significant difference comparing the lesional to the healthy and nonlesional group in both studies. Nevertheless, it is not possible to differentiate between healthy and nonlesional cells using RPA.

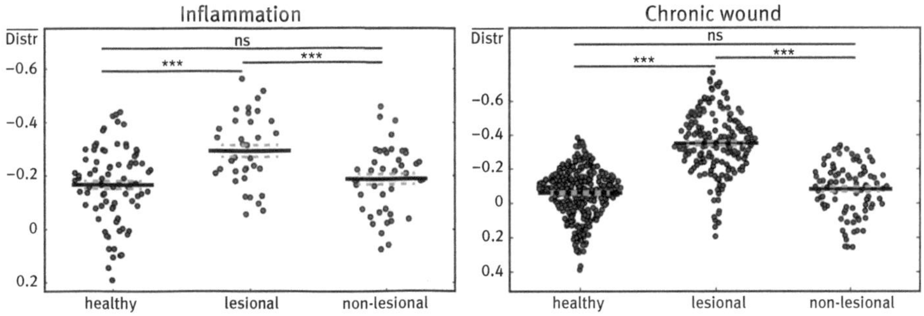

Fig. 22.6: Statistical synopsis of clinical mitochondrial distribution. For both study groups, RPA $\overline{\text{Distr}}$ plots of cells from healthy, lesional and nonlesional patients are shown, demonstrating mitochondrial centralization in lesions. Each point represents one segmented cell.

22.7 Fluorescence lifetime imaging – foundation, calculation, clinical application

The mean fluorescence lifetime (τ_m) of free and protein-bound NADH can be used as an indicator for the metabolic status of the cell. Fluorescence lifetime is an intrinsic quality of fluorophores and does not depend on factors like excitation wavelength, exposure time or photobleaching [37, 38], which makes lifetime analysis a valuable companion method for simple intensity measurements.

For the acquisition of fluorescence lifetime data, frequency-domain and time-domain methods have been developed. These approaches are technically distinct, but the results of both methods are mathematically equivalent as their data is interconvertible by Fourier transform [39]. In the presented studies on clinical pathophysiology, a time-domain method based on time-correlated single-photon counting (TCSPC) was used for data acquisition [40].

The multi-exponential decay function in an idealized experiment is defined as

$$F(t) = I_0 \sum_{i=1}^{n} a_i \, e^{-t/\tau_i},$$

where τ_i are the lifetimes of the various fluorophores and a_i are the corresponding fractions with $\sum a_i = 1$. The presented experimental setup focuses on the analysis of

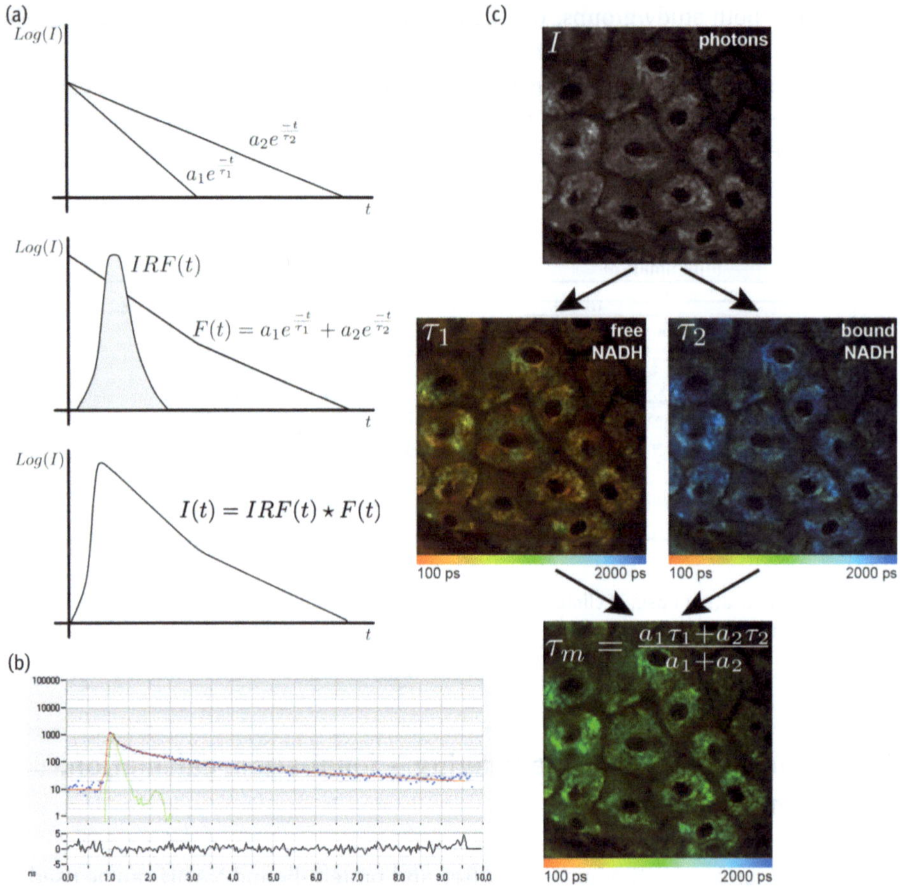

Fig. 22.7: Fluorescence lifetime imaging of cellular NADH. (a) Graphs of the combination of two decay functions and subsequent convolution with the IRF. (b) TCSPC decay trace (blue dots), the experimentally recorded IRF (green line) and the fitting result (red line) are shown (red graph). (c) Illustration of τ_m imaging of free and protein-bound NADH.

two fluorophores – free NADH with a short τ and protein-bound NADH with a long τ – and therefore a 2-component decay model is used:

$$F(t) = I_0(a_1 e^{-t/\tau_1} + a_2 e^{-t/\tau_2})$$

In the following, τ_1 is defined as the lifetime of free NADH and τ_2 of protein-bound NADH (Fig. 22.7).

Another important factor in the calculation of fluorescence lifetimes from experimental data is the incorporation of the optical and electrical properties of the measurement system. These properties influence the measured signal and are reflected by the instrumental response function (IRF) within calculation [41]. The IRF of the system can be obtained by the excitation and subsequent measurement of a nonfluorescent scattering media. The IRF is then combined with the model function F by mathematical convolution leading to

$$I(t) = \mathrm{IRF}(t) * F(t).$$

The measured photon count response is affected by a number of various other factors like light scattering within the tissue, a small rest of ambient light, second harmonic generation or dark noise of the detector, which should be also incorporated as factors to improve the model function.

As mentioned before, the technical basis for the calculation of fluorescence lifetimes is TCSPC, which allows to measure the number of emitted photons at an excited point within very small timeframes. In our setup the focused epidermal section is scanned in a raster $M = 128 \times 128$. At each single point $p \in M$, the photons are counted with a very high resolution in the picosecond range leading to a series of data points representing the fluorescence decay trace at that point (blue dots in Fig. 22.7 (b)). Given this information, the parameters of the model function are fitted to the decay trace by least squares curve fitting using the Levenberg–Marquardt algorithm [42] resulting in an estimate of τ_1, τ_2 and their ratios a_1 and a_2. With this the mean fluorescence lifetime τ_m is defined as

$$\tau_m = \frac{a_1\tau_1 + a_2\tau_2}{a_1 + a_2}$$

and is interpreted as the ratio of free-to-bound NADH at the measured point. The required photon count for the reliability of the fitting process depends on the number of free model parameters. Because of this, the τ_m of the whole image is defined as the intensity weighted mean so that darker image regions with lower information have less influence.

22.8 MPT-FLIM provides evidence of disease-related alteration of cellular metabolism

To investigate whether the altered subcellular localization of mitochondria is associated with an altered cellular metabolism in vivo, MPT-FLIM was applied in the setting of the clinical trials. We calculated the mean τ_m, reflecting the ratio of free-to-bound NADH, for all images of the *Stratum granulosum* obtained for the entire population of both studies. This analysis is illustrated in Fig. 22.8 for typical *Stratum granulosum* cells of all study groups.

The evaluation of all calculated τ_m values, shown in Fig. 22.9, revealed a significant decrease of τ_m in inflamed skin compared with healthy skin, indicating a shift of the cellular metabolic state (Fig. 22.9, left). In contrast to the histological and MPT morphological analyses, the MPT-FLIM analysis even allowed the distinction between the nonlesional skin of patients suffering from *systemic* inflammatory skin and healthy subjects. In the study group affected by a local chronic wound, a similar decrease of

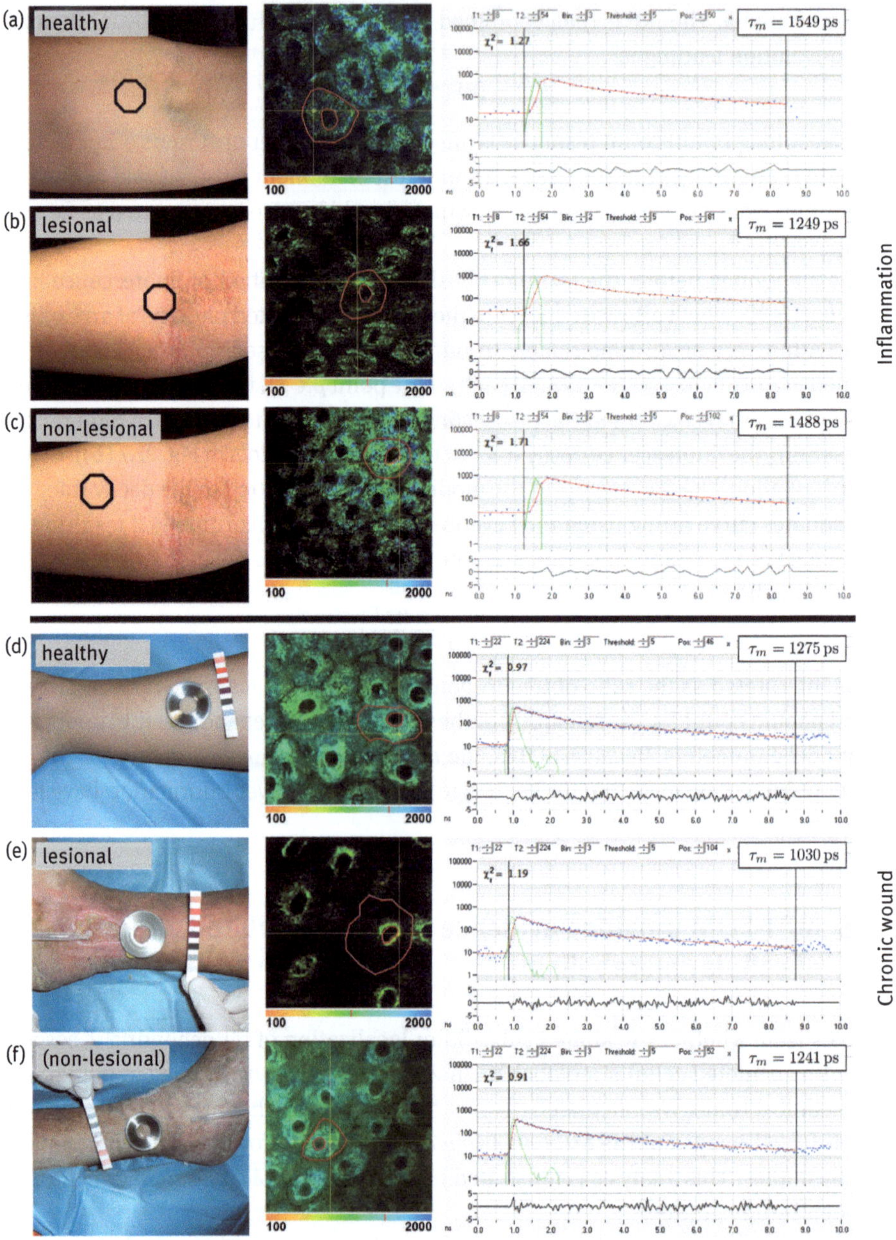

Fig. 22.8: Fluorescence lifetime imaging of patients' epidermal cells in inflammation and chronic wounds. From left to right, photographs of the examination areas, representative τ_m color-coded MPT-FLIM images and corresponding decay graphs of the same *Stratum granulosum* cells as used for RPA (cf. Fig. 22.5) are presented. Patients are affected by inflammation (a)–(c) or chronic wounds (d)–(f). Lesions exhibit a clear τ_m decrease in both study groups. Calculated τ_m values are also denoted within the color bars.

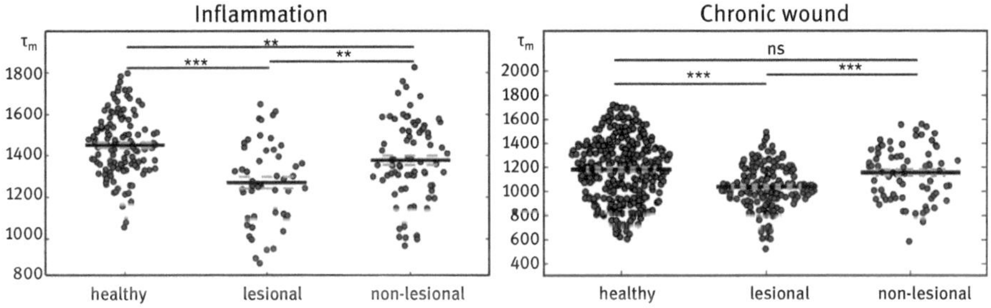

Fig. 22.9: Statistical synopsis of clinical mean fluorescence lifetimes. For both study groups, the distribution of τ_m in healthy, lesional and nonlesional cell layers of patients is displayed. Lesions exhibit a clear τ_m decrease representing a shift of the free-to-bound NADH ratio. In contrast to chronic wounds, the τ_m of nonlesional areas of patients suffering from *systemic* inflammatory skin is significantly decreased compared to healthy areas. Each point represents the mean τ_m of one image.

τ_m of lesional cells could be observed (Fig. 22.9, right). However, as expected, the nonlesional area at the not-affected leg did not show any difference in τ_m compared to healthy subjects.

While the mean inflammatory state of patients' skin slightly decreased during the course of the study (as indicated by a decreasing *Severity Score of Atopic Dermatitis*, inflammation severity), we could detect an increase of τ_m in lesional skin areas from (1166 ± 40) ps to (1378 ± 45) ps. Moreover, we observed a continuous significant difference in the measurements throughout the study compared with τ_m of healthy skin (Fig. 22.10 (a)). Plotting the τ_m values of lesional skin against the corresponding clinical score of chronic inflammation severity, we found a significant linear correlation, in which τ_m decreased with an increasing inflammatory state (PCC 0.65, $p < 0.0001$, Fig. 22.10 (b)).

22.9 Outlook

The present data demonstrate the translation of a medical assessment to in vivo morphometric analyses and MPT-FLIM in the context of clinical application. In line with classical histological analyses, the verifiability of typical pathognomonic characteristics of inflamed skin (Fig. 22.3) confirmed the validity of MPT imaging for novel dermatological purposes. Moreover, in the objective computational analysis of the morphological dataset, MPT provided information on the reorganization of mitochondria (Fig. 22.4). The RPA parameter $\overline{\text{Distr}}$, established in our clinical trials, reliably showed a perinuclear accumulation of mitochondria in human skin lesions in contrast to an even distribution in healthy skin (Fig. 22.5 and 22.6) [7].

This observation suggested a direct metabolic impact. In line with the morphological findings, MPT-FLIM utilization confirmed a shift of τ_m in both, lesional inflamma-

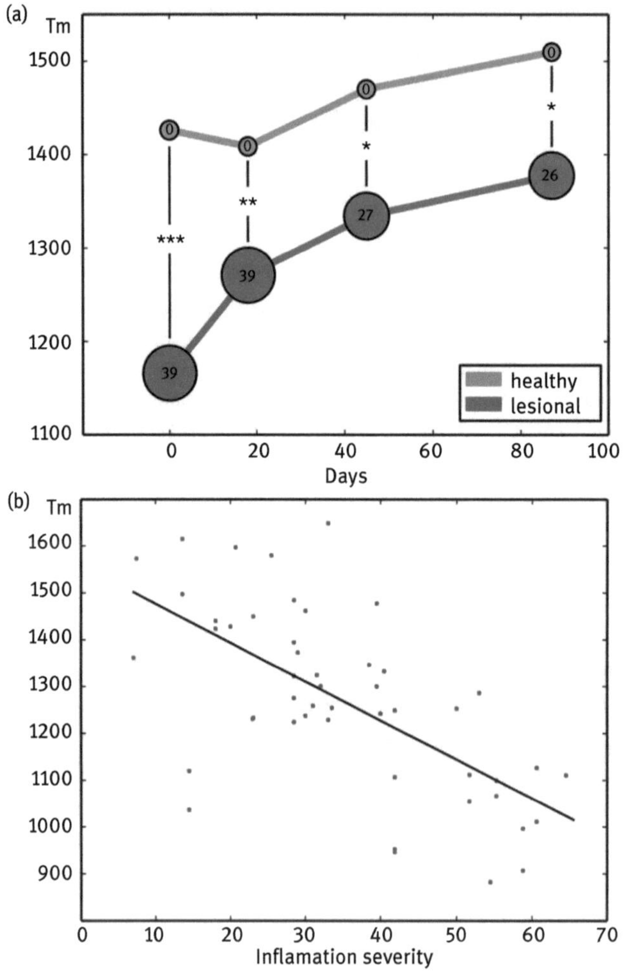

Fig. 22.10: Fluorescence lifetimes in the course of the inflammation. (a) Mean τ_m values of lesional areas in all patients suffering from inflammatory skin continuously show a significant difference to healthy skin over the entire term. The bubble size represents the severity of inflammation ranging from 0 to 103. (b) Plotting the mean τ_m of lesional skin against the corresponding inflammation severity, a significant linear correlation is revealed: τ_m decreases with increasing inflammation. Modified from [7].

tory and wound edge areas (Fig. 22.8 and 22.9). The decrease of τ_m indicated an accumulation of free NADH in the mitochondria, suggesting an overload of the respiratory electron transport. This overload could be related to either a lack of oxygen or hyperproduction of NADH during glycolysis. The underlying mechanism could be a malfunction of the mitochondrial electron transport chain, which is known as a marker of cellular stress accompanied by the production of reactive oxygen species [43]. For the future, the inclusion of the course of the total NADH intensity per cell as a second parameter next to the free-to-bound NADH ratio would allow a more precise attribution to the current biochemical state of the cellular metabolism [7]. Additionally, τ_m showed a clear negative correlation with the individual severity of inflammation (Fig. 22.10 (b)), pointing towards increased stress in keratinocytes within the affected skin regions [44].

Interestingly, lesional areas of the two examined fundamentally diverging disorders – inflammatory skin and chronic wounds – exposed similar reaction patterns on a cellular level. As expected, the so called 'nonlesional' areas in patients suffering from a chronic leg ulcer, representing the corresponding area at the not-affected leg, far from the local manifestation of the chronic wound, did not show any metabolic shift. Nevertheless, in patients affected by *systemic* inflammatory skin, MPT-FLIM – in contrast to the exclusive morphological analysis – was able to demonstrate that even the apparently healthy nonlesional areas exhibited pathological features resembling those of eczematous skin lesions. Therefore, these data support previous genetic, histological and molecular biological findings from nonlesional skin regions [25, 45, 46]. As a unique feature, MPT-FLIM combined with standardized computational analysis generates new in vivo parameters that can be used to detect and monitor pathophysiological alterations in the earliest state of inflammatory processes in human skin, prior to clinical manifestation.

Bridging the gap between the physiological basis of cellular metabolism and its potential impact on a patient's pathogenesis, the described studies validate MPT-FLIM as a useful diagnostic tool in the field of clinical skin pathophysiology and dermatologic therapy control. Thus, the in vivo application of MPT-FLIM offers the possibility of longitudinal examinations of patients' skin in addition to consecutive observation of individual etiopathology in a non-invasive manner.

References

[1] König K, Riemann I. High-resolution multiphoton tomography of human skin with subcellular spatial resolution and picosecond time resolution. J Biomed Opt. 2003;8(3):432–439.

[2] Williams RM, Zipfel WR, Webb WW. Multiphoton microscopy in biological research. Curr Opin Chem Biol. 2001;5(5):603–608.

[3] Lakowicz JR, Szmacinski H, Nowaczyk K, Johnson ML. Fluorescence lifetime imaging of free and protein-bound NADH. Proc Natl Acad Sci U S A. 1992;89(4):1271–1275.

[4] Zieger M, Springer S, Köhler MJ, Kaatz M. Multiphoton tomography. Der Hautarzt. 2015;66(7): 511–521.

[5] Seidenari S, Arginelli F, Bassoli S, Cautela J, French PM, Guanti M, et al. Multiphoton laser microscopy and fluorescence lifetime imaging for the evaluation of the skin. Dermatol Res Pract. 2012;2012:810749.

[6] Balu M, Mazhar A, Hayakawa CK, Mittal R, Krasieva TB, König K, et al. In vivo multiphoton NADH fluorescence reveals depth-dependent keratinocyte metabolism in human skin. Biophys J. 2013;104(1):258–267.

[7] Huck V, Gorzelanny C, Thomas K, Getova V, Niemeyer V, Zens K, et al. From morphology to biochemical state – intravital multiphoton fluorescence lifetime imaging of inflamed human skin. Sci Rep. 2016;6:22789.
[8] Fischer F, Volkmer B, Puschmann S, Greinert R, Breitbart E, Kiefer J, et al. Assessing the risk of skin damage due to femtosecond laser irradiation. J Biophotonics. 2008;1(6):470–477.
[9] Köhler MJ, Speicher M, Lange-Asschenfeldt S, Stockfleth E, Metz S, Elsner P, et al. Clinical application of multiphoton tomography in combination with confocal laser scanning microscopy for in vivo evaluation of skin diseases. Exp Dermatol. 2011;20(7):589–594.
[10] Ulrich M, Lange-Asschenfeldt S. In vivo confocal microscopy in dermatology: from research to clinical application. J Biomed Opt. 2013;18(6):061212.
[11] Rajadhyaksha M, Grossman M, Esterowitz D, Webb RH, Anderson RR. In vivo confocal scanning laser microscopy of human skin: melanin provides strong contrast. J Invest Dermatol. 1995; 104(6):946–952.
[12] Yew E, Rowlands C, So PT. Application of multiphoton microscopy in dermatological studies: a mini-review. J Innov Opt Health Sci. 2014;7(5):1330010.
[13] Schenke-Layland K, Riemann I, Damour O, Stock UA, König K. Two-photon microscopes and in vivo multiphoton tomographs—powerful diagnostic tools for tissue engineering and drug delivery. Adv Drug Deliv Rev. 2006;58(7):878–896.
[14] König K, Raphael AP, Lin L, Grice JE, Soyer HP, Breunig HG, et al. Applications of multiphoton tomographs and femtosecond laser nanoprocessing microscopes in drug delivery research. Adv Drug Deliv Rev. 2011;63(4–5):388–404.
[15] Patalay R, Talbot C, Alexandrov Y, Lenz MO, Kumar S, Warren S, et al. Multiphoton multispectral fluorescence lifetime tomography for the evaluation of basal cell carcinomas. PLoS One. 2012; 7(9):e43460.
[16] Huck V, Gorzelanny C, Thomas K, Mess C, Dimitrova V, Schwarz M, et al. Intravital multiphoton tomography as an appropriate tool for non-invasive in vivo analysis of human skin affected with atopic dermatitis. Proc SPIE 7883; 2011. p. 78830R.
[17] König K, Weinigel M, Bückle R, Kaatz M, Hipler C, Zens K, et al. Monitoring wound healing by multiphoton tomography/endoscopy. Proc SPIE 9303; 2015. p. 93030F.
[18] Pols DH, Wartna JB, Moed H, van Alphen EI, Bohnen AM, Bindels PJ. Atopic dermatitis, asthma and allergic rhinitis in general practice and the open population: a systematic review. Scand J Prim Health Care. 2016:1–8.
[19] Bieber T. Atopic dermatitis. N Engl J Med. 2008;358(14):1483–1494.
[20] Cork MJ, Robinson DA, Vasilopoulos Y, Ferguson A, Moustafa M, MacGowan A, et al. New perspectives on epidermal barrier dysfunction in atopic dermatitis: gene-environment interactions. J Allergy Clin Immunol. 2006;118(1):3–21; quiz 2–3.
[21] Finnie A. The SIGN guideline on the care of chronic leg ulcers: an aid to improving practice. J Wound Care. 2000;9(8):365–367.
[22] Markova A, Mostow EN. US skin disease assessment: ulcer and wound care. Dermatol Clin. 2012;30(1):107–111, ix.
[23] Broughton G, 2nd, Janis JE, Attinger CE. The basic science of wound healing. Plast Reconstr Surg. 2006;117(7 Suppl):12S–34S.
[24] Karrer S. Topical therapy of ulcers. Der Hautarzt. 2005;56(12):1165–1179.
[25] Linkert M, Rueden CT, Allan C, Burel JM, Moore W, Patterson A, et al. Metadata matters: access to image data in the real world. J Cell Biol. 2010;189(5):777–782.
[26] Swedlow JR, Eliceiri KW. Open source bioimage informatics for cell biology. Trends Cell Biol. 2009;19(11):656–660.
[27] Allan C, Burel JM, Moore J, Blackburn C, Linkert M, Loynton S, et al. OMERO: flexible, model-driven data management for experimental biology. Nat Methods. 2012;9(3):245–253.

[28] Miyamoto K, Kudoh H. Quantification and visualization of cellular NAD(P)H in young and aged female facial skin with in vivo two-photon tomography. Br J Dermatol. 2013;169 Suppl 2:25–31.
[29] Kaatz M, König K. [Multiphoton microscopy and in vivo tomography in dermatologic imaging]. Hautarzt. 2010;61(5):397–409.
[30] Denk W, Strickler J, Webb W. Two-photon laser scanning fluorescence microscopy. Science. 1990;248(4951):73–76.
[31] König K. Robert Feulgen Prize Lecture. Laser tweezers and multiphoton microscopes in life sciences. Histochem Cell Biol. 2000;114(2):79–92.
[32] Heikal AA. Intracellular coenzymes as natural biomarkers for metabolic activities and mitochondrial anomalies. Biomark Med. 2010;4(2):241–263.
[33] Leong FJ, Brady M, McGee JO. Correction of uneven illumination (vignetting) in digital microscopy images. J Clin Pathol. 2003;56(8):619–621.
[34] Sanches JM, Figueiredo J, Fonseca M, Duraes C, Melo S, Esmenio S, et al. Quantification of mutant E-cadherin using bioimaging analysis of in situ fluorescence microscopy. A new approach to CDH1 missense variants. Eur J Hum Genet. 2015;23(8):1072–1079.
[35] Otsu N. Threshold selection method from gray-level histograms. IEEE T Syst Man Cyb. 1979; 9(1):62–66.
[36] Leys C, Ley C, Klein O, Bernard P, Licata L. Detecting outliers: Do not use standard deviation around the mean, use absolute deviation around the median. J Exp Soc Psychol. 2013;49(4): 764–766.
[37] Chen Y, Periasamy A. Characterization of two-photon excitation fluorescence lifetime imaging microscopy for protein localization. Microsc Res Tech. 2004;63(1):72–80.
[38] Lakowicz JR, Szmacinski H, Nowaczyk K, Berndt KW, Johnson M. Fluorescence lifetime imaging. Anal Biochem. 1992;202(2):316–330.
[39] Berezin MY, Achilefu S. Fluorescence lifetime measurements and biological imaging. Chem Rev. 2010;110(5):2641–2684.
[40] Becker W, Bergmann A, Hink MA, König K, Benndorf K, Biskup C. Fluorescence lifetime imaging by time-correlated single-photon counting. Microsc Res Tech. 2004;63(1):58–66.
[41] Kolber ZS, Barkley MD. Comparison of approaches to the instrumental response function in fluorescence decay measurements. Anal Biochem. 1986;152(1):6–21.
[42] Marquardt DW. An Algorithm for Least-Squares Estimation of Nonlinear Parameters. J Soc Ind Appl Math. 1963;11(2):431–441.
[43] Bito T, Nishigori C. Impact of reactive oxygen species on keratinocyte signaling pathways. J Dermatol Sci. 2012;68(1):3–8.
[44] Ji H, Li XK. Oxidative stress in atopic dermatitis. Oxid Med Cell Longev. 2016;2016:2721469.
[45] Gros E, Bussmann C, Bieber T, Forster I, Novak N. Expression of chemokines and chemokine receptors in lesional and nonlesional upper skin of patients with atopic dermatitis. J Allergy Clin Immunol. 2009;124(4):753–760. e1.
[46] Suarez-Farinas M, Tintle SJ, Shemer A, Chiricozzi A, Nograles K, Cardinale I, et al. Nonlesional atopic dermatitis skin is characterized by broad terminal differentiation defects and variable immune abnormalities. J Allergy Clin Immunol. 2011;127(4):954–964. e1–4.

Index

Bei Fragen zur Produktsicherheit wenden Sie sich bitte an:
If you have any questions regarding product safety,
please contact:

Walter de Gruyter GmbH
Genthiner Straße 13
10785 Berlin
productsafety@degruyterbrill.com